Lecture Notes in Artificial Intelligence   8481

Subseries of Lecture Notes in Computer Science

LNAI Series Editors

Randy Goebel
  University of Alberta, Edmonton, Canada
Yuzuru Tanaka
  Hokkaido University, Sapporo, Japan
Wolfgang Wahlster
  DFKI and Saarland University, Saarbrücken, Germany

LNAI Founding Series Editor

Joerg Siekmann
  DFKI and Saarland University, Saarbrücken, Germany

Moonis Ali   Jeng-Shyang Pan
Shyi-Ming Chen   Mong-Fong Horng (Eds.)

Modern Advances in Applied Intelligence

27th International Conference
on Industrial Engineering and Other Applications
of Applied Intelligent Systems, IEA/AIE 2014
Kaohsiung, Taiwan, June 3-6, 2014
Proceedings, Part I

 Springer

Volume Editors

Moonis Ali
Texas State University, San Marcos, TX, USA
E-mail: ma04@txstate.edu

Jeng-Shyang Pan
National Kaohsiung University of Applied Sciences, Taiwan
E-mail: jspan@cc.kuas.edu.tw

Shyi-Ming Chen
National Taiwan University of Science and Technology, Taipei, Taiwan
E-mail: smchen@mail.ntust.edu.tw

Mong-Fong Horng
National Kaohsiung University of Applied Sciences, Taiwan
E-mail: mfhorng@ieee.org

ISSN 0302-9743   e-ISSN 1611-3349
ISBN 978-3-319-07454-2   e-ISBN 978-3-319-07455-9
DOI 10.1007/978-3-319-07455-9
Springer Cham Heidelberg New York Dordrecht London

Library of Congress Control Number: 2014939236

LNCS Sublibrary: SL 7 – Artificial Intelligence

Typesetting: Camera-ready by author, data conversion by Scientific Publishing Services, Chennai, India

Printed on acid-free paper

Springer is part of Springer Science+Business Media (www.springer.com)

Preface

On behalf of the Organizing Committee of the 27th International Conference on Industrial, Engineering and Other Applications of Applied Intelligent Systems (IEA-AIE 2014), it is our pleasure to present the proceedings of the conference. IEA-AIE 2014 was hosted by the National Kaohsiung University of Applied Sciences, Taiwan, Chaoyang University of Technology, Taiwan, and Harbin Institute of Technology University, China. IEA/AIE 2014 continued the tradition of emphasizing applications of applied intelligent systems to solve real-life problems in all areas including engineering, science, industry, automation and robotics, business and finance, medicine and biomedicine, bioinformatics, cyberspace, and human–machine interactions. IEA/AIE 2014 included oral presentations, invited speakers, special sessions, and a computer vision competition.

We receive 294 submissions from 27 countries in Africa, Asia, Australia, Europe, North America, and South America. Due to the page limit, only 106 papers were accepted and are published in the proceedings. Each paper was reviewed by at least two reviewers. The reviewers are mainly researchers from universities and industry. We congratulate and thank all the authors of the accepted papers. In addition to the paper presentations, four plenary talks were kindly delivered by Prof. John Roddick from Flinders University, Australia, Prof. Ce Zhu, University of Electronic Science and Technology of China, Prof. Daniel S. Yeung, South China University of Technology, China, and President Han-Chien Chao, Ilan University, Taiwan. We hope that these high-quality and valuable keynotes are beneficial for your future research.

Finally, we sincerely appreciate the contribution of the authors, reviewers, and keynote speakers. We would also like to express our sincere appreciation to the Program Committee members and local committee members for their valuable effort that contributed to the conference's success.

June 2014

Ali Moonis
Jeng-Shyang Pan
Shyi-Ming Chen
Mong-Fong Horng
Rung-Ching Chen
Ngoc Thanh Nguyen
Ali Selamat

Organization

Honorary Chairs

Cheng-Hong Yang	National Kaohsiung University of Applied Sciences, Taiwan
Jen-Chin Chung	Chaoyang University of Technology, Taiwan

General Chairs

Ali Moonis	Texas State University-San Marcos, USA
Jeng-Shyang Pan	National Kaohsiung University of Applied Sciences, Taiwan
Shyi-Ming Chen	National Taiwan University of Science and Technology, Taiwan

Program Committee Chairs

Mong-Fong Horng	National Kaohsiung University of Applied Sciences, Taiwan
Rung-Ching Chen	Chaoyang University of Technology, Taiwan
Ngoc Thanh Nguyen	Wroclaw University of Technology, Poland
Ali Selamat	Universiti Teknologi Malaysia

Advisory Committee Board

Ajith Abraham	Technical University of Ostrava, Czech Republic
Bin-Yi Liao	National Kaohsiung University of Applied Sciences, Taiwan
Yau-Hwang Kuo	National Cheng kung University, Taiwan
Kuan-Ming Hung	National Kaohsiung University of Applied Sciences, Taiwan

Publication Chairs

Chin-Shiuh Shieh	National Kaohsiung University of Applied Sciences
Muhammad Khurram Khan	King Saud University, Saudi Arabia

Invited Session Chairs

Jeng-Shyang Pan   National Kaohsiung University of Applied
                                Sciences, Taiwan

Publicity Chairs

Tien-Tsorng Shih   National Kaohsiung University of Applied
                                Sciences, Taiwan

Keynote Speeches

John Roddick   Dean of the School of Computer Science,
                                Engineering and Mathematics, Flinders
                                University, Australia
Ce Zhu   Professor, University of Electronic Science
                                and Technology of China, China
Daniel S. Yeung   Chair Professor, South China University
                                of Technology, China
Han-Chieh Chao   President, National Han University, Taiwan
Tzung-Pei Hong   Department of Computer Science and
                                Information Engineering, National
                                University of Kaohsiung, Taiwan

Invited Sessions Organizers

Shyi-Ming Chen   National Taiwan University of Science and
                                Technology, Taiwan
Yuh-Ming Cheng   Shu-Te University, Taiwan
Tsong-Yi Chen   National Kaohsiung University of Applied
                                Sciences, Taiwan
Lijun Yan   Harbin Institute of Technology Shenzhen
                                Graduate School, China
Linlin Tang   Shenzhen Graduate School of Harbin Institute
                                of Technology, China
Shu-Chuan Chu   Flinders University, Australia

International Program Committee

Aart Van Halteren   Vrije University Amsterdam, The Netherlands
Ah-Hwee Tan   Nanyang Technological University, Singapore
Alexander Felfernig   Technische Universität Graz, Austria
Alfons Salden   Almende BV, The Netherlands
Al-Mubaid Hisham   University of Houston Clear Lake, USA
Altincay Hakan   Eastern Mediterranean University, Turkey

Amruth N. Kumar	School of Theoretical and Applied Science, USA
Andrea Orlandini	Institute of Cognitive Sciences and Technologies, National Research Council, Italy
Andres Bustillo	University of Burgos, Spain
Antonio Peregrin	Universidad de Huelva, Spain
Anwar Althari	Universiti Putra Malaysia, Malaysia
Ariel Monteserin	Universidad Nacional del Centro de la Provincia de Buenos Aires, Argentina
Arzucan Özgür	University of Michigan, USA
Ashok Goel	Georgia Institute of Technology, USA
Aydogan Reyhan	Delft University of Technology, The Netherlands
Azizi Ab Aziz	Universiti Utara, Malaysia
Bae Youngchul	Chonnam National University, Korea
Barber Suzanne	University of Michigan, USA
Benferhat Salem	Université d'Artois, France
Bentahar Jamal	Concordia University, Canada
Bipin Indurkhya	International Institute of Information Technology, India
Björn Gambäck	Norwegian University of Science and Technology, Norway
Bora İsmail Kumova	Izmir Institute of Technology, Turkey
César García-Osorio	Universidad de Burgos, Spain
Catherine Havasi	Massachusetts Institute of Technology, USA
Chang-Hwan Lee	Dongguk University, South Korea
Chao-Lieh Chen	National Kaohsiung First University of Science and Technology, Taiwan
Chen Ling-Jyh	Academia Sinica, Taiwan
Chen Ping	University of Houston, USA
Cheng-Seen Ho	National Taiwan University of Science and Technology, Taiwan
Chien-Chung Wang	National Defense University, Taiwan
Chih Cheng Hung	Southern Polytechnic State University, USA
Colin Fyfe	Universidad de Burgos, Spain
Correa da Silva, F.S.	University of São Paulo, Brazil
Dan E. Tamir	Texas State University, USA
Daniel Yeung	South China University of Technology, China
Dariusz Krol	Wrocław University of Technology, Poland
Darryl Charles	University of Ulster, UK
Dianhui Wang	La Trobe University, Australia
Dilip Kumar Pratihar	Indian Institute of Technology
Dirk Heylen	University of Twente, The Netherlands
Djamel F. H. Sadok	Universidade Federal de Pernambuco, Brazil
Don-Lin Yang	Feng Chia University, Taiwan
Dounias Georgios	University of the Aegean, Greece

Jose-Maria Peña	DATSI, Universidad Politecnica de Madrid, Spain
Jun Hakura	Iwate Prefectural University, Japan
Jyh-Horng Chou	National Kaohsiung First University of Science and Technology, Taiwan
Kaoru Hirota	The University of Tokyo, Japan
Kazuhiko Suzuki	Okayama University, Japan
Khosrow Kaikhah	Texas State University, USA
Kishan G. Mehrotra	Syracuse University, USA
Kuan-Rong Lee	Kun Shan University, Taiwan
Kurosh Madani	University of Paris, France
Lars Braubach	University of Hamburg, Germany
Leszek Borzemski	Wrocław University of Technology, Poland
Maciej Grzenda	Warsaw University of Technology, Poland
Manton M. Matthews	University of South Carolina, USA
Manuel Lozano	University of Granada, Spain
Marcilio Carlos Pereira De Souto	Federal University of Rio Grande do Norte, Brazil
Marco Valtorta	University of South Carolina, USA
Mark Last	Ben-Gurion University of the Negev, Israel
Mark Neerincx	Delft University of Technology, Netherlands
Mark Sh. Levin	Institute for Information Transmission Problems Russian Academy of Sciences, Russia
Martijn Warnier	Delft University of Technology, The Netherlands
Michele Folgheraiter	Nazarbayev University, Georgia
Miquel Sànchez i Marrè	Technical University of Catalonia, Spain
Murat Sensoy	University of Aberdeen, UK
Nashat Mansour	Lebanese American University, Lebanon
Natalie van der Wal	Vrije Universiteit Amsterdam, The Netherlands
Ngoc Thanh Nguyen	Wroclaw University of Technology, Poland
Nicolás García-Pedrajas	University of Córdoba, Spain
Niek Wijngaards	Delft Cooperation on Intelligent Systems, The Netherlands
Oscar Cordón	Granada University, Spain
Paolo Rosso	Universidad Politecnica de Valencia, Spain
Patrick Brézillon	Laboratoire d'Informatique de Paris 6, France
Patrick Chan	South China University of Technology, China
Paul Chung	Loughborough University, UK
Prabhat Mahanti	University of New Brunswick, Canada
Qingzhong Liu	New Mexico Institute of Mining and Technology, USA

Richard Dapoigny	Université de Savoie, France
Robbert-Jan Beun	University Utrecht, The Netherlands
Rocío Romero Zaliz	Universidad de Granada, Spain
Safeeullah Soomro	Sindh Madresatul Islam University, Pakistan
Sander van Splunter	Delft University of Technology, The Netherlands
Shaheen Fatima	Loughborough University, UK
Shogo Okada	Tokyo Institute of Technology, Japan
Shyi-Ming Chen	National Taiwan University of Science and Technology, Taiwan
Simone Marinai	University of Florence, Italy
Snejana Yordanova	Technical University of Sofia, Bulgaria
Srini Ramaswamy	University of Arkansas at Little Rock, USA
Sung-Bae Cho	Yonsei University, Korea
Takatoshi Ito	Nagoya Institute of Technology, Japan
Tetsuo Kinoshita	Tohoku University, Japan
Tibor Bosse	Vrije Universiteit Amsterdam, The Netherlands
Tim Hendtlass	Swinburne University of Technology, Australia
Tim Verwaart	Wageningen University and Research Centre, The Netherlands
Valery Tereshko	University of the West of Scotland, UK
Vincent Julian	Universidad Politecnica de Valencia, Spain
Vincent Shin-Mu Tseng	National Cheng Kung University, Taiwan
Vincenzo Loia	University of Salerno, Italy
Walter D. Potter	The University of Georgia
Wei-Tsung Su	Aletheia University, Taiwan
Wen-Juan Hou	National Taiwan Normal University, Taiwan
Xue Wang	University of Leeds, UK
Yo-Ping Huang	Asia University, Taiwan
Yu-Bin Yang	Nanjing University, China
Zhijun Yin	University of Illinois at Urbana-Champaign, USA
Zsolt János Viharos	Research Laboratory on Engineering & Management Intelligence, Hungary

Reviewers

Aart Van Halteren	Al-Mubaid Hisham	Antonio Peregrin
Adriane B. de S. Serapião	Altincay Hakan	Anwar Althari
	Amruth Kumar	Ariel Monteserin
Ah-Hwee Tan	Andrea Orlandini	Arzucan Özgür
Alexander Felfernig	Andres Bustillo	Ashok Goel
Alfons Salden	Andrzej Skowron	Aydogan Reyhan

Azizi Ab Aziz
Bae Youngchul
Barber Suzanne
Belli Fevzi
Benferhat Salem
Bentahar Jamal
Bipin Indurkhya
Björn Gambäck
Bora İsmail Kumova
César García-Osorio
Catherine Havasi
Chang-Hwan Lee
Chao-Lieh Chen
Chen Ling-Jyh
Chen Ping
Cheng-Seen Ho
Cheng-Yi Wang
Chien-Chung Wang
Chih Cheng Hung
Ching-Te Wang
Chun-Ming Tsai
Chunsheng Yang
Ciro Castiello
Colin Fyfe
Correa da Silva, F.S.
Dan E. Tamir
Daniel Yeung
Dariusz Krol
Darryl Charles
Dianhui Wang
Dilip Kumar Pratihar
Dirk Heylen
Djamel F.H. Sadok
Don-Lin Yang
Dounias Georgios
Duco Nunes Ferro
Enrique Herrera Viedma
Esra Erdem
Evert Haasdijk
Fariba Sadri
Fernando Gomide
Fevzi Belli
Flavia Soares Correa da
    Silva
Floriana Esposito

Fran Campa
Francesco Marcelloni
Francisco Fernández de
    Vega
Francisco Herrera
François Jacquenet
Frank Klawonn
Gérard Dreyfus
Greg Lee
Gregorio Ismael
    Sainz Palmero
Hamido Fujita
Hannaneh Najd Ataei
Hans Guesgen
Hasan Selim
He Jiang
Henri Prade
Herrera Viedma Enrique
Hiroshi G. Okuno
Honggang Wang
Huey-Ming Lee
Hui-Yu Huang
Humberto Bustince
Hyuk Cho
Jaziar Radianti
Jean-Charles Lamirel
Jeng-Shyang Pan
Jesús Maudes
Jing Peng
João Paulo Carvalho
João Sousa
Joaquim Melendez
    Frigola
John M. Dolan
José Valente de Oliveira
Jose-Maria Peña
Julian Vicent
Jun Hakura
Jyh-Horng Chou
Kaoru Hirota
Kazuhiko Suzuki
Khosrow Kaikhah
Kishan G. Mehrotra
Kuan-Rong Lee
Kurash Madani

Lars Braubach
Lei Jiao
Leszek Borzemski
Linlin Tang
Li-Wei Lee
Lin-Yu Tseng
Maciej Grzenda
Manabu Gouko
Manton M. Matthews
Manuel Lozano
Marcilio Carlos Pereira
    De Souto
Marco Valtorta
Mark Last
Mark Neerincx
Mark Sh. Levin
Martijn Warnier
Michele Folgheraiter
Miquel Sànchez i Marrè
Mong-Fong Horng
Murat Sensoy
Nashat Mansour
Natalie van der Wal
Ngoc Thanh Nguyen
Nicolás García-Pedrajas
Niek Wijngaards
Oscar Cordón
Paolo Rosso
Patrick Brezillon
Patrick Chan
Paul Chung
Pinar Karagoz
Prabhat Mahanti
Qian Hu
Qingzhong Liu
Reyhan Aydogan
Richard Dapoigny
Robbert-Jan Beun
Rocio Romero Zaliz
Rosso Paolo
Safeeullah Soomro
Sander van Splunter
Shaheen Fatima
Shogo Okada
Shou-Hsiung Cheng

Table of Contents – Part I

Agent Technology

Computational Intelligence and Applications

Data Mining and QA Technology

Optimization in Industrial Applications

Pattern Recognition and Machine Learning

Innovations in Intelligent Systems and Applications

Machine Learning

Intelligent Industrial Applications and Control Systems

Multi-objective Optimzation

Table of Contents – Part II

Multi-objective Optimzation

Knowledge Management and Applications

Samrt Data-Mining in Industrial Applications

Innovations in Intelligent Systems and Applications

Media Processing

Media Processing

Smart Living

Smart Living

Information Retrieval

Intelligence Systems for E-Commerce amd Logistics

Multi-agent Learning for Winner Determination in Combinatorial Auctions

Fu-Shiung Hsieh and Chi-Shiang Liao

Department of Computer Science and Information Engineering,
Chaoyang University of Technology
41349 Taichung, Taiwan
fshsieh@cyut.edu.tw

Abstract. The winner determination problem (WDP) in combinatorial double auctions suffers from computation complexity. In this paper, we attempt to solve the WDP in combinatorial double auctions based on an agent learning approach. Instead of finding the exact solution, we will set up a fictitious market based on multi-agent system architecture and develop a multi-agent learning algorithm to determine the winning bids in the fictitious market to reduce the computational complexity in solving the WDP in combinatorial double auctions. In the fictitious market, each buyer and each seller is represented by an agent. There is a mediator agent that represents the mediator. The issue is to develop learning algorithms for all the agents in the system to collectively solve the winner determination problem for combinatorial double auctions. In this paper, we adopt a Lagrangian relaxation approach to developing efficient multi-agent learning algorithm for solving the WDP in combinatorial double auctions. Numerical results indicate our agent learning approach is more efficient than the centralized approach.

Keywords: Agent learning, combinatorial auction, integer programming, winner determination problem.

1 Introduction

Many problems in the real world are notoriously difficult to solve due to dispersed interacting components that seek to optimize a global collective objective through local decision making, limited communication capabilities, local and dynamic information, faulty components, and an uncertain environment. Typical examples of these complex problems include scheduling in manufacturing systems [27, 28, 29, 30], routing in data networks [31, 32, 33, 34, 35], command and control of networked forces in adversarial environments [2, 9, 12] and determination of winners in combinatorial auctions. For some of these problems, it is not feasible to pass all information to a centralized computer that can process this information and disseminate instructions. For others, the complexity of the overall system makes the problem of finding a centralized optimal solution intractable. Motivated by the complexity and distributed nature of these problems, one approach is to model the overall system as a collection of simpler interacting components or agents [26]. Under

A. Moonis et al. (Eds.): IEA/AIE 2014, Part I, LNAI 8481, pp. 1–10, 2014.
© Springer International Publishing Switzerland 2014

such a multi-agent system architecture, the decision making process for any single component or agent in the system is dictated by an optimization problem that is greatly simplified as compared to the centralized problem, but coupled to the decisions of other interconnected components or agents. An important issue in multi-agent systems is the design, analysis and implementation of agent learning algorithms. In [25], Gordon et al. focus on providing provably near-optimal solutions for domains with large numbers of agents under the assumption that individual agents each have limited influence on the overall solution quality. The study of Gordon et al. shows that each agent can safely plan based only on the average behavior of the other agents. Gordon et al. provided a proof of convergence of their algorithm to a near-optimal solution for a theme park crowd management problem.

In this paper, we will study the computational efficiency and scalability of multi-agent learning algorithms for the winner determination problem (WDP) in combinatorial auctions [4, 19, 21]. In a combinatorial auction, a bidder can place a bid on a bundle of items with a price. An important research subject in combinatorial auctions is the WDP, which aims to determine the winners that maximize the seller's revenue based on the set of bids placed. An excellent survey on combinatorial auctions can be found in [3] and [14]. Combinatorial auctions are notoriously difficult to solve from a computational point of view [16], [22] due to the exponential growth of the number of combinations. The WDP can be modeled as a set packing problem (SPP) [20], [1], [5], [10]. Sandholm *et al.* mentions that WDP for combinatorial auction is NP-complete [17], [18], [19]. Many centralized algorithms have been developed for WDP [1], [7], [8],[11],[13], [23], [24]. In this paper, we will propose a multi-agent learning mechanism for WDP. We assume all the players tell truth. The WDP for combinatorial auction can be modeled as an integer programming problem. Instead of finding the exact solution, we will set up a fictitious market based on multi-agent system architecture and develop a multi-agent learning algorithm to determine the winning bids in the fictitious market to reduce the computational complexity in solving WDP. In the fictitious market, each buyer and each seller is represented by an agent. The issue is to develop learning algorithms for all the agents in the system to collectively solve the WDP for combinatorial auctions. In this paper, we adopt a Lagrangian relaxation approach to developing efficient multi-agent learning algorithm for finding approximate solutions. Based on the proposed algorithms, we demonstrate the effectiveness of our method by numerical examples.

The remainder of this paper is organized as follows. In Section 2, we first we formulate the WDP for combinatorial auctions. We describe the multi-agent learning mechanism in Section 3 and develop multi-agent learning algorithms in Section 4. In Section 5, we present our numerical results. We conclude this paper in Section 6.

2   Problem Formulation

In this paper, we first formulate the combinatorial auction problem as an integer programming problem. We assume that there are a set of buyers and a seller. In a combinatorial auction, buyers submit bids to the seller. The revenue of a combinatorial auction is the winning buyers' total payment. To formulate the problem, let's define the notations in this paper.

Notations:

Symbol	
K	the number of different types of items
k	the index of an item, $k \in \{1,2,3,....,K\}$
i	the index of a seller
q_{ik}	the quantity of the $k-th$ items supplied by the $i-th$ seller, where $i \in \{1,2\}$ and $k \in \{1,2,3,....,K\}$.
N_i	the number of potential buyers that place bids on the combinatorial auction of seller i. Each $n \in \{1,2,3,....,N_i\}$ represents a buyer.
p_{in}	a real positive number that denotes the price of the bid placed by buyer n in the combinatorial auction of seller i, where $i \in \{1,2\}$, $n \in \{1,2,3,....,N_i\}$.
d_{ink}	a nonnegative integer that denotes the quantity of the $k-th$ items in the bid submitted by buyer n in the combinatorial auction of seller i, where $i \in \{1,2\}$, $n \in \{1,2,3,....,N_i\}$ and $k \in \{1,2,3,....,K\}$
y_{inh}	y_{inh} is the variable to indicate the bid placed by buyer n in the combinatorial auction CA_i of seller i is a winning bid ($y_{inh}=1$) or not ($y_{inh}=0$).

The winner determination problem is formulated as an Integer Programming problem as follows.

Winner Determination Problem (WDP) for Combinatorial Auction for Seller i : CA_i

$$\max \sum_{n=1}^{N_i} \sum_{h=1}^{H_i} y_{inh}\, p_{inh}$$

$s.t.$

$$\sum_{n=1}^{N_i} \sum_{h=1}^{H_i} y_{inh}\, d_{inhk} \le q_{ik} \quad \forall k \in \{1,2,..., K\} \tag{2.1}$$

$$y_{inh} \in \{0,1\} \quad n \in \{1,2,..., N_i\}, h \in \{1,2,..., H_i\}$$

For any solution of the WDP, the total amount of type of goods supplied by the seller must be greater than or equal to the demands of the buyers. This imposes supply/demand constraints in (2-1). In WDP problem, we observe that the coupling among different bids is caused by the contention for the goods due to supply/demand constraints (2-1).

3   Multi-agent Learning Mechanism

The integer programming problem formulated in the previous section is notoriously difficult to solve from a computational point of view due to the exponential growth of the number of combinations in the solution space [16]. Instead of solving this problem based on a centralized algorithm, a multi-agent learning mechanism is proposed in this paper. The underlying theories for developing the multi-agent learning mechanism proposed in this paper will be detailed later. In this section, we first introduce the idea and concept.

In our proposed approach, we will set up a fictitious market based on multi-agent system architecture and develop a multi-agent learning algorithm to determine the winning bids in the fictitious market to reduce the computational complexity in solving the WDP in combinatorial auctions. In the fictitious market, each buyer and each seller is represented by an agent. Our multi-agent learning mechanism consists of three stages as are shown in Fig. 1(a) through Fig. 1(c). These three stages are executed iteratively by buyers and the seller, respectively. The issue is to develop learning algorithms for all the agents in the system to collectively solve the winner determination problem for combinatorial auctions.

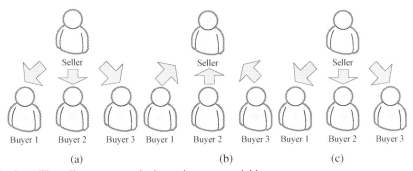

Fig. 1. (a) The seller announce shadow price to potential buyers

(b) Buyers make decisions based on shadow price and respond to the seller

(c) The seller adjusts the shadow price and announces it to buyers

In the aforementioned fictitious market, we need to develop an algorithm for each buyer to make decisions based on shadow price, and an algorithm for the seller to initialize the shadow price and adjust the shadow price iteratively based on the decisions of the buyers and seller. One of the fundamental issues is to determine the shadow price to be used in the fictitious market to guide the buyer agents and seller agents. A good choice of the shadow price will efficiently guide the buyer agents, and seller agent to converge to a good solution. An improper choice of the shadow price will mislead the buyer agents and seller agent and cannot bring them to converge to a good solution.

In this paper, we propose a multi-agent learning algorithm by combining Lagrangian relaxation and a subgradient method to find a solution for the fictitious market. The basic idea of Lagrangian relaxation is to relax some of the constraints of the original problem by moving them to the objective function with a penalty term. That is, infeasible solutions to the original problem are allowed, but they are penalized in the objective function in proportion to the amount of infeasibility. The constraints that are chosen to be relaxed are selected so that the optimization problem over the remaining set of constraints is in some sense easy.

4   Development of Multi-agent Learning Algorithms

The development of our multi-agent learning algorithms is detailed in this section. The proposed multi-agent learning algorithms consist of (i) an algorithm for each buyer agent to make decision according to the fictitious price announced by the seller agent, and (ii) an algorithm for the seller agent to update the fictitious price iteratively based on the decisions of buyer agents. In WDP problem, we observe that the coupling among different operations is caused by the supply/demand constraints (2-1). Let λ_i denote the vector with λ_{ik} representing the Lagrangian multiplier for the $k - th$ items. We define

$$L(\lambda_i) = \max \sum_{n=1}^{N_i} \sum_{h=1}^{H_i} y_{inh} p_{inh} - \sum_{k=1}^{K} \lambda_{ik} (\sum_{n=1}^{N_i} \sum_{h=1}^{H_i} y_{inh} d_{inhk} - q_{ik})$$

$$s.t. \ y_{inh} \in \{0,1\} \quad n \in \{1,2,...,N_i\}, h \in \{1,2,...,H_i\}$$

$$L(\lambda_i) = \max \sum_{n=1}^{N_i} \sum_{h=1}^{H_i} y_{inh} (p_{inh} - \sum_{k=1}^{K} \lambda_{ik} d_{inhk}) + \sum_{k=1}^{K} \lambda_{ik} q_{ik}$$

$$s.t. \ y_{inh} \in \{0,1\} \quad n \in \{1,2,...,N_i\}, h \in \{1,2,...,H_i\}$$

$$= \sum_{n=1}^{N_i} \sum_{h=1}^{H_i} L_{nh}(\lambda) + \sum_{k=1}^{K} \lambda_{ik} q_{ik}, where$$

$$L_{nh}(\lambda_i) = \max \ y_{inh} (p_{inh} - \sum_{k=1}^{K} \lambda_{ik} d_{inhk})$$

$$s.t. \ y_{inh} \in \{0,1\}$$

For a given Lagrangian multiplier λ_i, $L_{nh}(\lambda_i)$ defines a buyer's subproblem (BS). Given λ_i, the optimal solution to BS $L_{nh}(\lambda_i)$ can be solved as follows.

$$y_{inh} = \begin{cases} 1 & if \quad p_{inh} - \sum_{k=1}^{K} \lambda_{ik} d_{inhk} \geq 0 \\ 0 & if \quad p_{inh} - \sum_{k=1}^{K} \lambda_{ik} d_{inhk} < 0 \end{cases}$$

The algorithm for the seller agent to update the fictitious price iteratively is developed based on subgradient method. Although combinatorial auctions can be decomposed into a number of BS that can be solved easily for a given Lagrangian multiplier λ, it must be emphasized that finding the optimal Lagrangian multiplier λ^* is by no means easy. The optimal Lagrange multiplier is determined by solving the dual problem $\min_{\lambda \geq 0} L(\lambda_i)$. Since $L(\lambda_i)$ is not differentiable due to integrality constraints in subproblems [6], we adopt a simple, iterative subgradient method based on the ideas of Polyak [15] to solve the dual problem. Our approach to finding a solution of $\min_{\lambda \geq 0} L(\lambda_i)$ is based on an iterative scheme for adjusting Lagrangian multiplier according to the solutions of BS.

Let l be the iteration index. Let y^l denote the optimal solution to BS, respectively, for given Lagrange multipliers λ^l at iteration l. We define the subgradients of $\min_{\lambda \geq 0} L(\lambda_i)$ with respect to Lagrangian multipliers λ_k as follows.

$$g_1^l(k) = \sum_{n=1}^{N_i} \sum_{h=1}^{H_i} y_{inh}^l d_{inhk} - q_{ik}, where\ k \in \{1,...,k\}$$

$$\lambda_{ik}^{l+1} = \begin{cases} \lambda_{ik}^l + \alpha_1^l g_1^l(k)\ if\ \lambda_{ik}^l + \alpha_1^l g_1^l(k) \geq 0,\ where\ \alpha_1^l = c\ \dfrac{L(\lambda,\pi) - \overline{L}}{\sum_k (g_1^l(k))^2}, 0 \leq c \leq 2 \\ \\ 0\ otherwise \end{cases}$$

5   Numerical Results

Based on the proposed algorithms for combinatorial auctions, we design and implement a software system based on J2EE platform to verify the effectiveness of our multi-agent learning algorithms. We conduct several numerical experiments to illustrate the effectiveness of our method. To demonstrate the effectiveness of the algorithm proposed in this paper, we compare it with an existing centralized solver.

Example: Suppose Seller 1 has a bundle of available items to be sold through combinatorial auctions. The number of available items in each bundle is shown in Table 1 as follows. Suppose Seller 1 holds one combinatorial auction (Combinatorial auction 1) for her bundle available items. Five potential buyers place bids on the combinatorial auction. The bids placed by the potential buyers (Buyer 1~Buyer 5) are shown in Table 2.

Table 1.

Seller \ Item	K#1	K#2	K#3	K#4	K#5
i=1	8	12	17	11	13

Table 2.

Buyer \ Item	K#1	K#2	K#3	K#4	K#5	Price
N #1	5	6	0	0	0	240
N #2	3	6	0	0	0	147
N #3	0	0	15	5	3	31
N #4	0	0	12	11	13	217
N #5	0	0	14	8	9	38

For this combinatorial auction, we have

$q_{11} = 8$, $q_{12} = 12$, $q_{13} = 17$, $q_{14} = 11$, $q_{15} = 13$

$d_{1111} = 5$, $d_{1211} = 6, d_{1311} = 0, d_{1411} = 0, d_{1511} = 0$

$d_{1112} = 3$, $d_{1212} = 6, d_{1312} = 0, d_{1412} = 6, d_{1512} = 10$

$d_{1113} = 0$, $d_{1213} = 0, d_{1313} = 15, d_{1413} = 5, d_{1513} = 3$

$d_{1114} = 0$, $d_{1214} = 0, d_{1314} = 12, d_{1414} = 11, d_{1514} = 13$

$d_{1115} = 0$, $d_{1215} = 0, d_{1315} = 14, d_{1415} = 8, d_{1515} = 9$

$p_{111} = 240$, $p_{121} = 147, p_{131} = 31$, $p_{141} = 217$, $p_{151} = 38$

The solutions found by our learning algorithm for this example are as follows.

The solution for this combinatorial auction is

$$y_{111}^1 = 1, y_{121}^1 = 1, y_{131}^1 = 0, y_{141}^1 = 1, y_{151}^1 = 0.$$

It is an optimal solution for this example.

Although our algorithm does not guarantee generation of optimal solutions, it often leads to optimal or near optimal solutions much more efficiently than the CPLEX integer programming solver (IBM ILOG CPLEX Optimizer). To study the computational efficiency of our proposed algorithm, we conduct the following experiments to compare the computational time of our algorithm with that of the CPLEX integer programming solver with respect to I . Fig. 2 illustrates that our algorithm outperforms the CPLEX integer programming solver in efficiency as I grows. Moreover, the computation time of the CPLEX integer programming solver grows exponentially with I while our algorithm grows approximately linearly with respect to I .

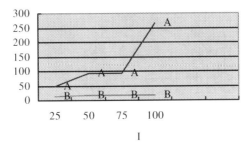

Fig. 1. CPU time (in millisecond) respect to I

A: CPLEX, B: Subgradient algorithm

6   Conclusion

We formulate the WDP for combinatorial auctions as an integer programming problem. The problem is to determine the winners to maximize the total revenue of the seller. Due to computational complexity, it is hard to develop a computationally efficient method to find an exact optimal solution for WDP based on centralized computing architecture. To reduce computational complexity, an alternative way to find a solution for combinatorial auctions is to set up a fictitious market based on multi-agent system architecture and develop multi-agent learning algorithms to determine the winning bids in the fictitious market. In this paper, we propose a multi-agent learning algorithm by combining Lagrangian relaxation and a subgradient method to find a solution for the fictitious market. There are two types of agents in the system, including the seller agent and buyer agents. The proposed multi-agent learning algorithms consist of two parts: (i) an algorithm for the seller agent to initialize the fictitious price and update it iteratively based on the decisions of buyer agents and (ii) an algorithm for each buyer agent to make decision according to the fictitious price announced by the seller agent. In our multi-agent learning algorithms, we consider the fictitious prices associated with the shadow prices corresponding to supply/demand constraints. We conduct experiments to study the performance and computational efficiency of our proposed multi-agent learning algorithm. To study the computational efficiency of our proposed algorithm, we conduct the experiments to compare the computational time of our algorithm with an integer programming solver with respect to the number of buyers. Although our algorithm does not guarantee generation of optimal solutions, the numerical results indicate that our proposed algorithm is significantly more efficient in comparison with the centralized integer programming solver as the size of the problems grows.

Acknowledgement. This paper is supported in part by National Science Council of Taiwan under Grant NSC102-2410-H-324-014-MY3.

References

1. Andersson, A., Tenhunen, M., Ygge, F.: Integer programming for combinatorial auction winner determination. In: Proceedings of the Seventeenth National Conference on Artificial Intelligence, pp. 39–46 (2000)
2. Beard, R.W., McLain, T.W., Goodrich, M.A., Anderson, E.P.: Coordinated target assignment and intercept for unmanned air vehicles. IEEE Transactions on Robotics and Automation 18(6), 911–922 (2002)
3. de Vries, S., Vohra, R.V.: Combinatorial Auctions: A Survey. INFORMS Journal on Computing (3), 284–309 (2003)
4. Perugini, D., Lambert, D., Sterling, L., Pearce, A.: From Single Static to Multiple Dynamic Combinatorial Auctions. In: IEEE/WIC/ACM International Conference on Intelligent Agent Technology, September 19-22, pp. 443–446 (2005)
5. Fujishima, Y., Leyton-Brown, K., Shoham, Y.: Taming the computational complexity of combinatorial auctions: Optimal and approximate approaches. In: Sixteenth International Joint Conference on Artificial Intelligence, pp. 548–553 (1999)
6. Fisher, M.L.: Lagrangian relaxation method for solving integer programming problems. Management Science 27, 1–18 (1981)
7. Gonen, R., Lehmann, D.: Optimal solutions for multi-unit combinatorial auctions: branch and bound heuristics. In: The Proceedings of the Second ACM Conference on Electronic Commerce (EC 2000), pp. 13–20 (2000)
8. Guo, Y., Lim, A., Rodrigues, B., Tang, J.: Using a Lagrangian heuristic for a combinatorial auction problem. In: Proceedings of the 17th IEEE International Conference on Tools with Artificial Intelligence (2005)
9. Murphey, R.A.: Target-based weapon target assignment problems. In: Pardalos, P.M., Pitsoulis, L.S. (eds.) Nonlinear Assignment Problems: Algorithms and Applications, pp. 39–53. Kluwer Academic Publishers, Dordrecht (1999)
10. Hoos, H.H., Boutilier, C.: Solving combinatorial auctions using stochastic local search. In: Proceedings of the Seventeenth National Conference on Artificial Intelligence, pp. 22–29 (2000)
11. Hsieh, F.-S.: Combinatorial reverse auction based on revelation of Lagrangian multipliers. Decision Support Systems 48(2), 323–330 (2010)
12. Ahuja, R.K., Kumar, A., Jha, K., Orlin, J.B.: Exact and heuristic methods for the weapon-target assignment problem. Technical Report #4464-03, MIT, Sloan School of Management Working Papers (2003)
13. Jones, J.L., Koehler, G.J.: Combinatorial auctions using rule-based bids. Decision Support Systems 34, 59–74 (2002)
14. Pekeč, A., Rothkopf, M.H.: Combinatorial auction design. Management Science 49, 1485–1503 (2003)
15. Polyak, B.T.: Minimization of Unsmooth Functionals. USSR Computational Math. and Math. Physics 9, 14–29 (1969)
16. Rothkopf, M., Pekeč, A., Harstad, R.: Computationally manageable combinational auctions. Management Science 44, 1131–1147 (1998)
17. Sandholm, T.: An algorithm for optimal winner determination in combinatorial auctions. In: Proc. IJCAI 1999, Stockholm, pp. 542–547 (1999)
18. Sandholm, T.: Algorithm for optimal winner determination in combinatorial auctions. Artificial Intelligence 135(1-2), 1–54 (2002)
19. Sandholm, T.: Approaches to winner determination in combinatorial auctions. Decision Support Systems 28, 165–176 (2000)

20. Vemuganti, R.R.: Applications of set covering, set packing and set partitioning models: a survey. In: Du, D.-Z. (ed.) Handbook of Combinatorial Optimization, vol. 1, pp. 573–746. Kluwer Academic Publishers, Netherlands (1998)
21. Yanga, S., Segrea, A.M., Codenottib, B.: An optimal multiprocessor combinatorial auction solver. Computers & Operations Research 36, 149–166 (2007)
22. Xia, M., Stallaert, J., Whinston, A.B.: Solving the combinatorial double auction problem. European Journal of Operational Research 164, 239–251 (2005)
23. Hsieh, F.-S., Lin, J.-B.: Assessing the benefits of group-buying based combinatorial reverse auctions. Electronic Commerce Research and Applications 11(4), 407–419 (2012)
24. Hsieh, F.-S., Lin, J.-B.: Virtual enterprises partner selection based on reverse auction. International Journal of Advanced Manufacturing Technology 62, 847–859 (2012)
25. Gordon, G.J., Varakantham, P.R., Yeoh, W., Lau, H.C., Aravamudhan, A.S., Cheng, S.-F.: Lagrangian relaxation for large-scale multi-agent planning. In: Proceedings of the 11th International Conference on Autonomous Agents and Multiagent Systems (AAMAS 2012), vol. 3, pp. 1227–1228 (2012)
26. Geoffrey, J.: Gordon, Agendas for multi-agent learning. Artificial Intelligence 171(7), 392–401 (2007)
27. Akella, R., Kumar, P.R.: Optimal control of production rate in a failure-prone manufacturing systems. IEEE Transactions on Automatic Control 31(2), 116–126 (1986)
28. Gershwin, S.B.: Manufacturing Systems Engineering. Prentice-Hall, Englewood Cliffs (1994)
29. Kimemia, J., Gershwin, S.B.: An algorithm for the computer control of a fexible manufacturing system. IIE Transactions 15(4), 353–362 (1983)
30. Kumar, P.R.: Re-entrant lines. Queueing Systems: Theory and Applications 13, 87–110 (1993)
31. Altman, E., Boulogne, T., El Azouzi, R., Jiménez, T., Wynter, L.: A survey on networking games in telecommunications. Computers and Operations Research 33(2), 286–311 (2005)
32. Altman, E., Shimkin, N.: Individual equilibrium and learning in processor sharing systems. Operations Research 46, 776–784 (1998)
33. La, R., Anantharam, V.: Optimal routing control: Repeated game approach. IEEE Transactions on Automatic Control 47(3), 437–450 (2002)
34. Orda, A., Rom, R., Shimkin, N.: Competitive routing in multi-user communication networks. IEEE/ACM Trans. Networking 1(5), 510–521 (1993)
35. Roughgarden, T.: Selish Routing and the Price of Anarchy. MIT Press, Cambridge (2005)

Agent-Based Modeling of Farming Behavior: A Case Study for Milk Quota Abolishment

Diti Oudendag[1,2], Mark Hoogendoorn[2], and Roel Jongeneel[1]

[1] Wageningen UR, The Hague, The Netherlands
{diti.oudendag,roel.jongeneel}@wur.nl
[2] VU University Amsterdam, Department of Computer Science
De Boelelaan 1081, 1081 HV Amsterdam, The Netherlands
m.hoogendoorn@vu.nl

Abstract. Gaining insight on the effect of policies upon the Agricultural domain is essential to effectively regulate such a sector. A variety of models have therefore been developed that enable a prediction of agricultural development under different policies. Most models do however not make predictions on a fine grained level, making it difficult to see the real effect for individual farmers. This paper presents an agent-based model where each farm is modeled by means of an agent and studies the effect of milk quota abolishment. Simulations are performed for the Netherlands and compared with the predictions of more coarse grained models.

Keywords: Agent Based Modelling, Milk production, Milk Quota System, Agricultural Policies, Dairy Farming Netherlands.

1 Introduction

A lot of agricultural policies in European countries are governed by the European Union. Changes in these policies can have a huge impact on the agricultural development, in particular farm size distribution, as well as on the (production) price developments of agricultural products (milk, meat, etc.) and therefore on farmer's income. In order to decide on policy changes in an adequate manner, policy makers should be able to judge the consequences of these changes in advance. Hence, a need for predictive models exists that enables informed decisions with respect to policy changes.

As a consequence of this need, a variety of predictive models have been developed and published (see e.g. [1];[2];[3]). These models usually take the form of more high-level economic models that tend to strongly focus on market impacts (prices, supply, demand, trade) of the price given particular policy changes but generally ignore impacts on the farm structure. However, adjustments in farm structure, may lead to economies of scale and by that adjust the competitiveness of the farm sector. Therefore these longer term and indirect impacts of policy changes should not be ignored. There are a number of studies which try to analyze the changes in the farm size distribution, for example by modeling farm size evolution by means of a stationary or non-stationary Markov chain process (see e.g. [4]) A problem with these

A. Moonis et al. (Eds.): IEA/AIE 2014, Part I, LNAI 8481, pp. 11–20, 2014.

studies is however that they do not take the inherently spatial nature of agriculture into account. The land market is often not properly taken into account, let alone the heterogeneity and competing claims on land and land use. Furthermore, individual farmers might behave quite different with respect to policy changes, some might be more risk taking because they are still relatively young whereas older farmers might more frequently take a more conservative approach.

To enable a detailed insight into the effect of policy changes, agent-based simulations can be a promising paradigm as they enable simulations on an individual level. Hereby, the individual farmer behavior can be modeled as well as the spatial aspects that were mentioned before.

In this paper, an agent-based model is described which is able to simulate the behavior of individual dairy farmers given certain policy changes. More in specific, the model has been developed to evaluate the impact of the abolishment of the milk quota (for the Netherlands) which is planned for 2015 by the European Union. This paper is organized as follows. In Section 2 related work with respect to predictive models in the agricultural domain is sketched. Thereafter, the agent-based model is presented in Section 3. The experimental setup concerning the milk quota abolishment is presented in Section 4 whereas Section 5 presents the results of simulations and comparison with results of other models. Finally, Section 6 is a discussion.

2   Related Work

A variety of more economic oriented high-level models have been proposed to model the behavior of farmers. The most relevant ones that focus on abolishment of the milk quota are briefly described here. For some, the actual predictions are shown in Section 5 where a comparison is made with the output of the model presented in this paper. There are a number of models that focus on a country level and take the Netherlands into account (which is the country studied in this paper): Helming and Peerlings [2] compare the difference between a scenario with and without a milk quota in 2008 using a demand and supply model. In follow up work, Helming and van Berkum [5] use a similar approach but focus on differences per type of dairy cow. Lips and Reader [3] make predictions for a large selection of European countries (16) and use a so-called global trade analysis model. A European study [6] uses a high-level agricultural model called CAPRI to predict the differences when the milk quota would be abolished. Consortium INRA-Wageningen [7] performed assessments for European alternative dairy policies (including quota abolition) combining two macro- economic models.

Next to more mathematically or sector-oriented approaches, an agent-based model in the domain of agriculture has also been found and is called AgriPoliS (for Agricultural Policy Simulator, see [8]). The main purpose of the approach is to study the effect of policies upon agricultural structure and the model is very extensive. Farms are also represented by means of agents. They use a detailed spatial model for the land of farms to be able to simulate the land market. As a result the model is very extensive and on to this moment only applied in local/regional studies.

3   Agent-Based Model

In this section the agent-based model which has been developed to simulate dairy farm behavior is explained. The model has been developed within the scope of the Dutch dairy sector as ample information is available to develop appropriate models (including a spatial model). Essentially, each dairy farm is represented by an agent which has (1) a land mass at a specific location and a number of dairy cows; (2) certain characteristics related to the owner of the farm (e.g. the age). Based on these characteristics and the land mass the behavior of the agents is expressed. This behavior mainly deals with expansion in terms of buying or selling land from/to other agents and increasing the number of cows. The behavior is based on the economic principle to maximize profit. More details are provided below. Note that this is a simplified specification of the model; the full model is explained in [9].

3.1   Agent Goals, Revenues and Costs

As explained before, the behavior of the agents (i.e. farms) used in the agent-based model is targeted towards profit maximization. Figure 1 shows the theoretical curve for the number of dairy cows owned by a farm versus the cost per unit of milk on the left (see [10]). It is clear that the costs per unit go down as the number of cows increases, however after a certain amount the costs increase again. The declining cost part reflects the economies of scale or size that are known to characterize dairy production ([10]). The increasing part is due to the increasing inefficiencies in management ([10]). The U-shaped

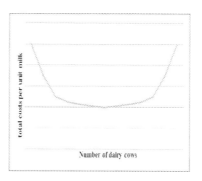

Fig. 1. Cost curve for varying number of cows (see [10])

total average cost curve suggests that a farm has a certain optimal number of cows at which cost per cow or per unit of milk produced is minimized (how the value has been selected for the Netherlands will be shown in the experimental setup). Profit maximization implies that farms will strive to achieve this number of cows. The revenues per farm can be expressed by means of the following equation:

$$revenues = milk\_production \cdot milk\_price + number\_of\_cows \cdot non\_milk\_revenue\_per\_cow + fixed\_premiums \tag{1}$$

Here, the *milk production* of the farm is clearly dependent on the number of cows but also depends on a number of other parameters, the *milk price* results from the supply and demand and is assumed to be fixed during simulations. *Fixed premiums* is an external factor, namely the premiums (e.g. Single Farm Payment) provided by the EU. The *non-milk revenue per cow* represents factors such as the growth and replacement of cattle (which can be sold) as well as other income (e.g. tourism). The cost curve per farm is specified in a way which follows a standard micro-economic framework ([10]) and is expressed as follows:

$$costs = fixed\_costs + costs\_per\_cow \cdot number\_of\_cows + variable\ cost\ per\ farm. \qquad (2)$$

Here, a certain fixed cost is expressed (also depending on whether the farm is fully specialized as a dairy farm or not) whereas a variable cost per cow is specified as well as another variable cost depending on farm characteristics[1].

3.2   Agent Behavior

The behavior of the agents (i.e. farms) is expressed in the form of an activity diagram in Figure 2. Each of the activities is explained in more detail below.

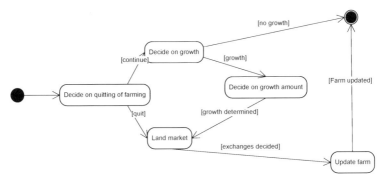

Fig. 2. High level specification of agent behavior

Decide on Quitting of Farming. The first step the agent needs to take is to decide whether or not to quit farming. This decision is made with a certain probability. An agent can decide to quit based on economic reasons as well as non-economic reasons (factors such as disease, disability, divorce). These probabilities are highly dependent on the age of the owner of the farm as well as the size of the farm (see [9]). Finally, the probability clearly depends on whether profit is being made. Overall, different groups are identified based on real data to judge these probabilities (e.g. older farm owners with a less than 20 cows and no profit have a probability of 0.287 to quit farming). Once the owner has decided to quit he tries to sell his land on the land exchange. If the farm does not quit, the probability to grow is assessed.

Decide on Growth. If a farm decides to stay in business the probability to grow is assessed, which is then used to decide whether the farm will actually grow or not. The probability to grow depends on the age of the farm owner (farms with older owners tend to have a lower probability to grow) and also depends on the area in which the farm is located. If the farm is located near or within a nature conservation area the probability of growth will be lower due to the limited opportunity for growth.

$$growth\_prob = age\_growth\_effect \cdot location\_growth \qquad (3)$$

---

[1] Although the cost function is assumed to be quadratic, a quadratic term of dairy cows in the regression is not significant contributing.

When a farm decides to grow, it decides how much it wants to grow. Otherwise, the farm stays as it is and is not considered until the next iteration.

Decide on Growth Amount. As explained before, the farms strive towards a certain optimal amount of dairy cows, if the farm already has more cows than this optimal amount it is assumed that the farmer doesn't have growth potential. In order to reach the optimal amount, a certain amount of land is required. The relationship between the number of cows and the required amount of land is of course crucial. This highly depends upon the intensity of the farm (the number of dairy cows per utilizable agricultural area) as well as the type of farm (specialized dairy farm or not). Once an expansion has been decided upon, a land exchange can take place.

Land Exchange. Once the aforementioned processes have ended, a number of quitting farms and a number of farms that expand will be present. These will try to exchange land on a land market. The market is setup in a relatively simple way: the seller sells only to farms which are sufficiently nearby and selects the one which has the lowest cost-revenue ratio (having the highest shadow value of land).

Update Farms. After all actions have been performed, the farms are updated.

3.3   Parameters

Based on the specification above, the agents have the characteristics as shown in Table 1. In the third column the source of selection of the initial values of the parameters is shown (which will be explained in Section 4).

Table 1. Agent Characteristics

Characteristic	Explanation	Source for simulation
Age	The age of the farm owner.	Annual Census [11]
Location	The location of the farm (X,Y coordinates).	Annual Census [11]
Utilizable agricultural area (UAA)	The amount of land the farm owns that can be used for farming.	Annual Census [11]
Land base	The amount of grass land owned by the farm.	Annual Census [11]
Number of cows	The number of cows the farm owns.	Annual Census [11]
Farm intensity	The intensity of the farm (cows per utilizable agricultural area).	Derived based on autonomous increase (see 3.2). Increases more significant without milk quota system.
Farm type	Dairy farm or non-specialized dairy farm.	Annual Census [11]
Costs	The costs a farm currently has.	Derived (see 3.2)
Revenue	The current revenue of a farm.	Derived (see 3.2)
Nature area	Indicates whether a farm lies in a nature area or not.	Nature 2000 [12]
Premium	The premium the farm receives.	Annual Census [11]
Milk production	The overall milk production on a farm.	Derived (see Table 2 and 3.2)
Fixed costs	The fixed costs for a farm.	Report [13]

Furthermore, the parameters as shown in Table 2 can be distinguished within the model.

Table 2. Parameters of the model

Parameter	Explanation
Optimal number of cows	The number of cows at which the cost per cow is lowest.
Milk price	The (assumed to be) fixed price for a unit of milk.
Milk production per cow	The amount of milk produced per cow given the specific characteristics of a farm.
Non milk revenue per cow	The revenue per cow outside milk (premium, tourism, etc.)
Costs per cow	The costs per cow for a farm.
Costs grass land	The costs of maintaining a certain grassland area.
Probability to quit farming	The probability to quit farming given the age, cost/revenue ratio and number of cows of a farm.
Location growth effect	The probability of farm expansion provided the location of the farm (taking nature areas into account).
Age growth effect	The growth probability of a farm provided the age of the farm owner.
Land expansion required	The (calculated) land expansion required for a farm to grow from x cows to y cows given its intensity and type.
Trade radius	The radius within land trade can take place between farms.

4   Case Study

In order to judge the suitability of the simulation model presented in Section 3 and investigate its predictions, a case study for the dairy farmers in the Netherlands has been conducted to forecast the sector's response to abolishment of the milk quota system. This section describes the setup of the model as well as the manner in which the evaluation of the approach will be performed.

4.1   Model Setup

In this paper, 2006 has been taken as an initial year and 2011 as the year to predict. Beside the years, there a more parameters to be set according to section 3.3.

First of all, initial values for all the agents should be set. The total number of farms that have dairy activities in the Netherlands was 21137 in 2006. For each of those farms an agent has been created, and the values of the characteristics of those agents have been set to values registered for each individual farm according to sources such as the Annual Census [11]. The precise sources for the values are shown in Table 1.

Table 3. Probability of quitting (CRR stands for Cost-revenue ratio)

Age	Number of dairy cows < 20		Number of dairy cows >= 20	
	CRR >= 1	CRR < 1	CRR >= 1	CRR < 1
< 50	0.188	0.163	0.093	0.091
50 – 65	0.269	0.229	0.126	0.139
>= 65	0.287	0.406	0.187	0.200

Second, the global parameters should be set to an appropriate value. In fact these have also been derived based on historic data and are shown in Table 2. Some are however less trivial to specify. The costs for a farm for example was shown to be best described (based on [13]) using the following equation for specialized dairy farms:

$$log(costs) = fixed\_costs + 0.90578 \cdot log(number\_of\_dairy\_cows) - 0.2059 \cdot variable\_costs \qquad (4)$$

Fixed costs depend on specialization of dairy farming (specialized versus non-specialized). Variable costs are farm specific.

The probability of quitting data has been taken from the Census of 2001 and 2006 and is based on categories, see Table 3. Finally, the expansion required to reach a certain number of cows has been made dependent upon the type of farm (specialized dairy farm or not) and the intensity of the farm, this results in a specific equation which specifies the relationship between the number of dairy cows and the required utilizable agricultural area (estimated and based on the Annual Census (see [11]).

Some parameters could not be set with known data. Therefore we tuned these parameters by comparing calculated values for 2011 with milk quota abolishment with known data for 2011(Annual Census [11]). These parameters are the probability on farm expansion in Nature Conservation areas (0.59), the maximum farm expansion (a farm size increase of 20%) and autonomous milk production increase (2.4%).

4.2   Experimental Setup

In order to evaluate the model, three steps are conducted: (1) *Calibrate and compare with known data.* Some parameters are unknown and therefore calibrate the parameters by comparing the prediction of the model for 2011 with the actual data; (2) *Sensitivity Analysis.* A sensitivity analysis can be used to see the influence of changing parameters of the model and investigate whether the resulting change in the model is as expected according to the domain knowledge. And (3) *Prediction with abolishment and comparison with other studies.* Use the model to predict the developments of dairy farmers provided that the milk quota is abolished versus the situation where the system is still in place. Compare these results with other predictive models. For case 3 the milk quota abolishment essentially is reflected by two different settings in the simulation: (1) it is assumed that there is an increased intensity in farming, and (2) it is assumed that there will be a substantial increase in the milk production per cow besides the autonomous development in milk production. The studies used for comparison are the most relevant ones mentioned in Section 2: [2]; [3]; [5]; [6] and [7].

5   Results

The results of the model are described below. The model has been implemented in NetLogo [14]. Furthermore, the numbers presented are averages over 3500 runs given the probabilistic nature of the model. For the sake of brevity a further elaboration on the results of the sensitivity analysis is not possible, see [9] for more details.

5.1   Comparison with Known Data

As stated before, a first comparison addresses the comparison between the prediction of the model for 2011 (after calibration) given initial data from 2006 compared to the actual measurement in 2011. Table 4 presents the results.

Table 4. Predicted value for 2011 with milk quota system present versus the actual values

Attribute	Actual value	Simulation
Number of dairy cows (1000)	1465	1475
UAA (1000 ha)	886	902
Number of farms	18624	18528
UAA per farm	47.6	48.7
Dairy cows per farm	78.8	79.6
Costs per 100 kg milk	53.4	56
Revenues per 100 kg milk	49.5	47.8
Milk production (1000 kg)	11738	11707

Table 4 shows that simulation results for 2011 are quite accurate, suggesting an adequate model calibration. The largest deviation is found for the costs per 100 kg of milk (which in the simulation is 4.8% higher than the value that is actually observed), whereas all other deviations are substantially less than 4%.

5.2   Prediction with Abolishment and Comparison with Other Models

The final set of experiments concern a comparison between the predictions of the model presented in this paper with existing (non-agent based) models in which the development of dairy farming with a milk quota is compared to the developments without having a milk quota. Before going into detail on the comparison with the other models, a graphical illustration of the prediction of the model is presented. Figure 3 presents a representative result of a prediction for the different scenarios. It can be seen that the number of larger farms is certainly predicted to increase from 2006 to 2011.

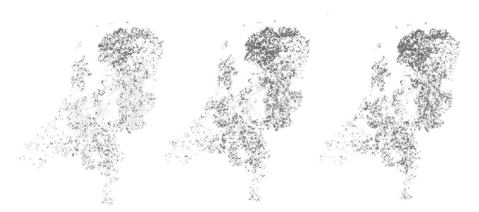

Fig. 3. Comparison of predictions. Light gray indicates an average farm size of < 80 cows for an area of 4 km^2; dark gray ≥ 80 cows. The left figure concerns 2006 (base year); center: 2011 with milk quota and right 2011 without milk quota.

When looking closely, without the milk quota the number of large dairy farms increases a bit more (note that the difference is small and therefore not obvious to see). Table 5 shows the difference in prediction in numbers. It can be seen that the number of dairy cows increases when the milk quota system is not present whereas the number of dairy cows per farm increases as well. The overall milk production increases significantly, and the costs per 100 kg of milk are substantially reduced. This suggests that the milk quota system still constraints Dutch dairy farmers and hampers the sector to find its long run equilibrium with respect to the optimal farm size. This clearly creates an inefficiency (milk is not produced at the lowest possible costs), which in the end also affects consumers.

Table 5. Prediction of model for 2011 with and without milk quota

Attribute	2011 with milk quota	2011 without milk quota
Number of dairy cows (1000)	1475	1525
UAA (1000 ha)	902	899
Number of farms	18528	18527
UAA per farm	48.7	48.5
Dairy cows per farm	79.6	82.3
Dairy cows per UAA	1.6	1.7
Costs per 100 kg milk	56	46
Revenues per 100 kg milk	47.8	47.3
Milk production (1000 kg)	11707	12530

Finally, Table 6 shows the difference between the predicted values in the model introduced in this paper and other works.

Table 6. Comparison between model predictions after milk quota abolishment with other works

Source	Absolute values		Index (our research = 100)	
	Milk production (1000 litre)	Number of dairy cows (in 1000)	Milk production	Number of dairy cows
Helming and Peerlings [2]	16051	1850	128	121
Helming and van Berkum [8]	13229	1430	106	94
Lips and Reader [3]	12573	-	100	
EU [9]	13471	1639	108	108
Inra-Wageningen[10]	13638	1625	109	106
This research	12530	1525	100	100

When looking at the comparison, it can be seen that the predictions of the other models are quite diverse. Note that the research has been conducted in different years and the predictive models are therefore difficult to compare. In addition, they also consider different factors. It can be seen that the model presented in this paper is relatively in line with another model (i.e. the Lips and Reader model [3]). The model presented here does take the spatial aspect into account (including effect of nature areas) which has been ignored in other works. Furthermore, the number generated by the models presented in this paper, are certainly not unrealistic according to domain experts.

6   Discussion

In this paper, an agent-based model of dairy farming has been presented. The approach presented enables a thorough analysis of the effect of policy changes on individual farm(er)s. After calibration and tuning parameters with the actual data in 2011 (milk-quota system present) the forecasting of the effects of the milk quota policy shows that the results are in line with some existing (coarse grained) models whereas for others large differences are seen. The predictions of the model are however not unrealistic according to domain experts.

For future work, the intention is to investigate the influence of other policies as well. Due to the problems with surplus manure production in the Netherlands, a new constraint on the maximum number of dairy cows might be imposed to the Dutch dairy sector even before the quota will be abolished in 2015. The current AB-model is well equipped to analyze the impact of such a policy, also taking into account the different reference levels (referring to farm specific conditions and history) that might be chosen with the implementation of such a policy.

References

1. Bouamra-Mechemache, Z., Jongeneel, R.A., Requilart, V.: EU dairy policy reforms: Luxembourg reform, WTO negotiations and quota regime. EuroChoices 8(1), 13–22 (2009)
2. Helming, J.F.M., van Berkum, S.: Effects of abolition of the EU milk quota system for Dutch Agriculture and environment. Paper presented at the 12th EAAE Congress 'People, Food and Environments: Global Trends and European Strategies', Gent, Belgium
3. Lips, M., Rieder, P.: Abolition of Raw Milk Quota in the European Union: A CGE Analysis at the Member Country Level. Journal of Agricultural Economics 56(1), 1–17 (2005)
4. Tonini, A., Jongeneel, R.: The distribution of dairy farm size in Poland: a Markov approach based on information theory. Applied Economics 41(1), 55–69 (2009)
5. Helming, J.F.M., Peerlings, J.: The impact of Milk Quota Abolishment on Dutch Agriculture and Economy: Applying an Agricultural Sector Model Integrated Into a Mixed Input-Output Model. In: The Xth EAAE Congress, Zaragoza, Spain (2002)
6. European Commission: Economic Impact of the Abolition of the Milk Quota Regime- Regional Analysis of the Milk Production in the EU-. Joint Research Centre for Agriculture and Life Science in the Economy, Seville (2009)
7. Consortium INRA-Wageningen: Study on the impact of future options for the milk quota system and the common market organisation for milk and milk products. Summary report European Commission, DG Agri. 39 p. (2002)
8. Happe, K., Kellerman, K., Balmann, A.: Agent-based Analysis of Agricultural Policies: an Illustration of the Agricultural Policy Simulator AgriPoliS, its Adaption and Behavior. Ecology and Society 11(1), 49 (2006)
9. Oudendag, D.: Effects of Abolition of Milk Quota: an Agent-Based Modelling Approach. Master Thesis. Vrije University, Faculty of Sciences, Amsterdam (2013)
10. Mansfeld, E.: Micro-Economics: Theory and Applications. W.W. Northon and Company, New York (1988)
11. http://www3.lei.wur.nl/ltc/
12. European Commission. Council Directive 92/43/EEC of 21 May 1992 on the conservation of natural habitats and of wild fauna and flora (1992), http://ec.europa.eu/environment/nature/legislation/habitatsdirective/ (retrieved)
13. http://www3.lei.wur.nl/binternet_asp/
14. http://ccl.northwestern.edu/netlogo/

Multi-robot Cooperative Pathfinding: A Decentralized Approach

Changyun Wei, Koen V. Hindriks, and Catholijn M. Jonker

Interactive Intelligence Group, EEMCS, Delft University of Technology,
Mekelweg 4, 2628 CD, Delft, The Netherlands
{c.wei,k.v.hindriks,c.m.jonker}@tudelft.nl

Abstract. When robots perform teamwork in a shared workspace, they might be confronted with the risk of blocking each other's ways, which will result in conflicts or interference among the robots. How to plan collision-free paths for all the robots is the major challenge issue in the multi-robot cooperative pathfinding problem, in which each robot has to navigate from its starting location to the destination while keeping avoiding stationary obstacles as well as its teammates. In this paper, we present a novel fully decentralized approach to this problem. Our approach allows the robots to make real-time responses to the dynamic environment and can resolve a set of benchmark deadlock situations subject to complex spatial constraints in the robots' workspace. When confronted with conflicting situations, robots can employ waiting, dodging, retreating and turning-head strategies to make local adjustments. In addition, experimental results show that our proposed approach provides an efficient and competitive solution to this problem.

Keywords: Cooperative pathfinding, coordination, collision avoidance.

1 Introduction

Autonomous robot teams are now expected to perform complicated tasks consisting of multiple subtasks that need to be completed concurrently or in sequence [1]. In order to accomplish a specific subtask, a robot first needs to navigate to the right location where the subtask can be performed. When multiple robots engage in such teamwork in a shared workspace, there is an inherent risk that the robots may frequently block each other's ways. As a result, the use of multiple robots may result in conflicts or interference, which will decrease overall team performance. In particular, deadlock situations have to be taken into consideration in highly congested settings for distributed or decentralized robots to plan their individual paths. This work is motivated by many practical applications such as unmanned underwater/ground vehicles, autonomous forklifts in warehouses, deep-sea mining robots, etc.

Interference in the Multi-Robot Cooperative Pathfinding (MRCP) (or called Multi-Robot Path Planning) problem stems from conflicting paths, along which multiple robots may intend to occupy the same places at the same time.

A. Moonis et al. (Eds.): IEA/AIE 2014, Part I, LNAI 8481, pp. 21–31, 2014.
© Springer International Publishing Switzerland 2014

The issue of how to plan collision-free paths for multiple robots has been extensively studied in [2,3,4,5,6,7], where the robots are supposed to navigate to distinct destinations from their starting locations. Finding the optimal solution to such a problem, however, is NP-hard and intractable [3]. *Centralized* solutions are usually based on global search and therefore could guarantee completeness, but they do not scale well with large robot teams [4] and cannot be solved in real-time [8]. In many practical applications, robots are expected to be fully autonomous and can make their own decisions, rather than being managed by a central controller [9]. *Decoupled* (or called *distributed*) solutions are fast enough for real-time applications, but they usually cannot guarantee completeness [6], and the robots may easily get stuck in common deadlock situations. Recent decoupled advances [10,2,11] have considered highly congested scenarios. The essential feature of decoupled approaches is that the robots are connected by a common database, e.g., the reservation table in [10] and the conflict avoidance table in [2]. Comparatively, we are interested in fully *decentralized* solutions, where the robots explicitly communicate with one another to keep track of each other's states and intentions, instead of accessing the common database to be aware of the progress of their teamwork in decoupled solutions.

The advantage of decentralized solutions in multi-robot teams is that, as the robots do not use any shared database, they can be fully autonomous and each robot can be a stand-alone system. On the other hand, such a solution may encounter the problem of overload communication in large-scale robot teams. To deal with this problem, a robot in our approach only needs to communicate with the ones locating within its coordinated network. For a team of k robots, the communication load will be reduced to $12k$ times in each time step, rather than $k(k-1)$ times in traditional decentralized systems. We analyze various deadlock situations in a robot's coordinated network, and propose a pairwise coordination algorithm to identify which situation the robot is confronted with and which strategy it should adopt. Specifically, a robot can employ *waiting*, *dodging*, *retreating*, and *turning-head* strategies to make local adjustments to avoid collisions, following the estimated shortest path to its destination.

We begin our work by discussing the state-of-the-art approaches in Section 2. We analyze the models of our decentralized approach in Section 3, and then discuss our proposed algorithms in Section 4. The simulated experiments and results are discussed in Section 5. Finally, Section 6 concludes the paper.

2   Related Work

MRCP problems have also been studied in the context of exploration (e.g., in [12,13]). The robots in [12] do not have long-term destinations to move; instead, they need to continually choose unexplored neighboring vertices to cover. A decision-theoretic approach is presented in [13], where the robots have long-term destinations but do not consider any congested configurations. Our work focuses on general MRCP problems, in which the robots need to navigate to distinct long-term destinations from their starting locations in a highly congested environment.

Earlier work in [14,15,5] presented their decoupled solutions based on prioritized planning, where the robots need to plan respective paths in order of their priorities. The robot with the highest priority first plans its path without considering the other robots' locations and plans, and then this robot is treated as a moving obstacle so that subsequent robots have to avoid it when planning their paths later. In this approach, each robot is expected to find a path without colliding with the paths of higher priority robots, but the overall path planning algorithm will fail if there is a robot unable to find a collision-free path for itself. Particularly, such an approach does not respond well in highly congested scenarios such as intersections. The work in [16] introduced an approach to solving the two-way intersection traffic problem, which will not be considered as a congested setting in our work. Traffic problems usually have two-way roads and thus can be solved by introducing traffic lights, whereas one-way intersections cannot be simply solved using the traffic light theory.

Recent decoupled advances considering highly congested environments include FAR [17], WHCA* [10], MAPP [11], and ID [2]. FAR and WHCA* are fast and can scale well with large robot teams, but they are still incomplete and offer no guarantee with regard to the running time and the solution length. MAPP has an assumption that for every three consecutive vertices in a robot's path to its destination, an alternative path that does not go through the middle vertex should exist, which apparently does not suit a narrow corridor scenario. Only ID claims that it provides a complete and efficient algorithm by breaking a large MRCP problem into smaller problems that can be solved independently, so we only compare our approach with ID in the experimental study.

Our work takes the advantage of *push* and *swap* ideas proposed in [6], which presents a centralized approach that allows one robot to push the other one away from its path, or makes two robots switch their positions at a vertex with degree ≥ 3. However, this approach only allows one robot (or paired robots) to move in each time step. In contrast, we propose a novel decentralized approach that allows a robot to actively make local adjustments to cope with conflicting situations. In order to do so, the robot can employ *waiting, dodging, retreating,* and *turning-head* strategies in its own decision making process. This work focuses on general MRCP problems that are analyzed using graph-based models, so the motion planning problems with regards to low-level control parameters, such as acceleration, deceleration, turning angle, etc., are beyond the scope of this paper.

3   Models of Decentralized Cooperative Pathfinding

3.1   Problem Formulation

A shared workspace can be divided into a set of n discrete cells $\mathcal{V}$ with k robots $\mathcal{R}$, $k < n$, and we use an undirected graph $\mathcal{G}(\mathcal{V}, \mathcal{E})$ to represent it. The edges $\mathcal{E}$ indicate whether two vertices are connected or not in $\mathcal{G}$. In order to model conflicts among the robots, we define a spatial constraint: robots must be present at distinct vertices at any time t:

$$\forall i, j \in [1, k], i \neq j : \mathcal{A}_t[i] \neq \mathcal{A}_t[j], \tag{1}$$

where $\mathcal{A}_t[i] \in \mathcal{V}$ indicates the vertex in which the i-th robot locates at time t. The sets of starting locations and destinations are denoted as $\mathcal{S} \subset \mathcal{V}$ and $\mathcal{T} \subset \mathcal{V}$, respectively. At any time t, each robot can have a next-step plan $\mathcal{P}_t$, either staying at its current vertex $\mathcal{A}_t$ or moving to one of its neighboring vertices $\mathcal{A}_{t+1}$:

$$\forall i \in [1, k], \mathcal{P}_t[i] \Rightarrow \begin{cases} \mathcal{A}_t[i] = \mathcal{A}_{t+1}[i] & \text{(stay at its current vertex,)} \\ (\mathcal{A}_t[i], \mathcal{A}_{t+1}[i]) \in \mathcal{E} & \text{(move to a neighbouring vertex.)} \end{cases} \quad (2)$$

Given the starting and destination vertices, when beginning to carry out the task, a robot (e.g., the i-th robot) needs to plan an individual path $\Pi_0[i] = \{\mathcal{S}[i], \ldots, \mathcal{T}[i]\}$ from its starting location $\mathcal{S}[i]$ to its destination $\mathcal{T}[i]$. At any time t, it may adjust its path to $\Pi_t[i] = \{\mathcal{A}_t[i], \mathcal{A}_{t+1}[i], \ldots, \mathcal{T}[i]\}$. Suppose the robot reaches its destination at time e, then we can know $\Pi_e[i] = \{\mathcal{T}[i]\}$. Therefore, the robot is supposed to minimize e while keeping avoiding collisions with stationary obstacles as well as the other robots.

3.2   Heuristic Estimated Shortest Paths

Fig. 1(a) shows an example of the initial configuration of the environment, where four robots need to go to their destinations. When planning their respective paths to the destinations, if each robot can disregard the presence of the others, they can easily find the shortest paths, as shown in Fig. 1(b). Such a path only considering stationary obstacles provides a *heuristic optimal estimate* with the shortest travel distance that can be calculated by performing a graph search, for example, using Dijkstra's algorithm in our work. Although such estimated paths cannot guarantee that a robot will successfully arrive at its destination without collisions with the other robots, it indeed provides an idealized estimate. We use $\mathcal{H}_t[i]$ to denote the heuristic estimated shortest path of the i-th robot at time t:

$$\mathcal{H}_t[i] = \{\mathcal{U}_1[i], \mathcal{U}_2[i], \ldots, \mathcal{T}[i]\}, \quad (3)$$

where $\mathcal{U}_1[i]$ and $\mathcal{U}_2[i]$ are the first and the second successor vertices in $\mathcal{H}_t[i]$, and thus $(\mathcal{A}_t[i], \mathcal{U}_1[i]) \in \mathcal{E}$, $(\mathcal{U}_1[i], \mathcal{U}_2[i]) \in \mathcal{E}$. When the robot's next step is its destination, then $\mathcal{H}_t[i] = \{\mathcal{T}[i]\}$, $\mathcal{U}_1[i] = \mathcal{T}[i]$ and $\mathcal{U}_2[i] = 0$.

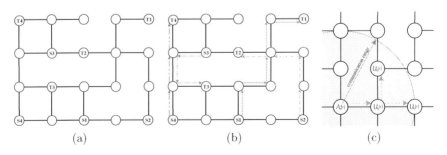

(a)   (b)   (c)

Fig. 1. Heuristic estimated shortest paths & coordinated network

3.3   Coordinated Network

Suppose $\mathcal{G}$ is divided uniformly, a robot takes unit length l to move to its neighboring vertices. For any robot $r \in \mathcal{R}$ at time t, it has the first and second successor vertices, $\mathcal{U}_1[r]$ and $\mathcal{U}_2[r]$, along its estimated shortest path (see Equation 3). At the moment, $\mathcal{U}_1[r]$ might be occupied by one of its teammates, or be free but its teammate(s) is going to occupy it. Thus, we can know that if the next-step plan of robot r is conflicted with the one of its teammates s, s must be locating within the distance of $2l$, as shown in Fig. 1(c). Then each robot can construct a coordinated network with the communication range of $2l$, and it only needs to communicate and coordinate with the ones within its coordinated network. Normally, k decentralized robots may need to communicate $k(k-1)$ times in each time step, whereas in the coordinated network, the communication can be reduced to no more than $12k$ times.

4   Decentralized Multi-robot Coordination

4.1   Main Decision Making Process

As robots make their own decisions in decentralized teams, we will discuss our proposed solution from a single robot's point of view. Algorithm 1 shows the main decision making process as to how robot r makes its next-step plans.

Algorithm 1. Main decision process for a single robot to plan its next step.

1: Given robot r and $\mathcal{T}[r]$ at time t, $\mathcal{A}_t[r] \leftarrow \mathcal{S}[r]$, $\Pi_t[r] \leftarrow \{\mathcal{S}[r]\}$.
2: **while** $\Pi_t[r] \neq \{\mathcal{T}[r]\}$ **do** $\triangleright$ not yet at destination.
3:     **if** TURNING-HEAD $\neq 0$ **then**
4:         $\mathcal{U}_1[r] \leftarrow$ TURNING-HEAD, $\mathcal{U}_2[r] \leftarrow 0$ $\triangleright$ turn head to a new robot.
5:     **else**
6:         Estimate shortest path $\mathcal{H}_t[r] = \{\mathcal{U}_1[r], \mathcal{U}_2[r], \ldots\}$, and $\triangleright$ see Equation 3.
7:         Send $\mathcal{H}_t[r]$ $\triangleright$ broadcast its estimated path.
8:     **end if**
9:     **if** $\exists s \in \mathcal{R}$ such that $\mathcal{U}_1[r] == \mathcal{A}_t[s]$ **then** COORDINATION $\triangleright$ see Algorithm 2
10:    **else if** $\exists s \in \mathcal{R}$ such that $\mathcal{U}_1[r] == \mathcal{P}_t[s]$ **then** $\triangleright$ s is entering the vertex.
11:        $\mathcal{P}_t[r] \leftarrow \mathcal{A}_t[r]$ $\triangleright$ wait at current vertex.
12:    **else** $\mathcal{P}_t[r] \leftarrow \mathcal{U}_1[r]$ $\triangleright$ move to the free successor vertex.
13:    **end if**
14:    Execute $\mathcal{P}_t[r]$ and Send $\mathcal{P}_t[r]$, $t \leftarrow t + 1$
15: **end while**

Robot r will continuously make its next-step plans until it arrives at its destination (line 2). It either can choose the successor vertices along the estimated shortest path (see line 6), or needs to turn its head to a new coordinated robot (line 4) if it took a *turning-head* strategy in the last round of the decision process (see Algorithm 2). If the next successor vertex of robot r is occupied by another robot s, then r needs to coordinate with s (line 9). Sometimes the successor vertex might not be occupied, but one of its teammates is moving to it now (line

10). In this case, the robot needs to wait at its current location for the next round of decision making. Otherwise, it can make the plan to move to the next free successor vertex (line 12). It happens that several robots may make decisions to go to the same vertex synchronously. In our approach, while executing a plan to move to a vertex, if a robot finds that there is the other one moving to the same vertex as well, the robot will immediately stop moving, give up its original plan and inform the other robot about its decisions.

4.2   Pairwise Coordination in Deadlock Situations

When the next successor vertex of a robot is occupied by one of its teammates, pairwise coordination between them is needed to make further progress. The robots may form various environmental configurations, and we list all the instances that may lead to a set of benchmark deadlock situations in Fig. 2.

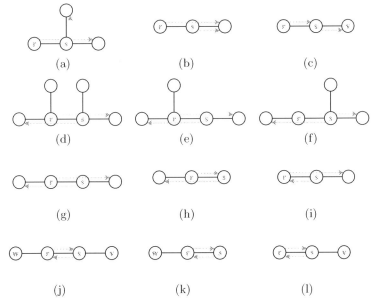

Fig. 2. Various environmental configurations in the pairwise coordination

Trail Following. Fig. 2(a) and 2(b) depict that r is following the trail of s, $(\mathcal{U}_1[r] == \mathcal{A}_t[s]) \bigcap (\mathcal{A}_t[r] \neq \mathcal{U}_1[s])$. After the next successor vertex, r may diverge from s (see Fig. 2(a)), $\mathcal{U}_2[r] \neq \mathcal{U}_1[s]$, or still follow the trail of s (see Fig. 2(b)), $\mathcal{U}_2[r] = \mathcal{U}_1[s]$. In both instances, r will believe that s is trying to move away from the occupied vertex, so it can *wait* for s to succeed. Sometimes they may come to a situation, as shown in Fig. 2(c), where s cannot move away right now because it is blocked by the other robot v and does not have any other free neighboring vertex. In such a case, it is not r who should try to find a solution for s because s must be coordinating with v in its own decision process at the same time.

Algorithm 2. Pairwise coordination for making local adjustments.

1: Given robot r at time t, its current location $\mathcal{A}_t[r]$, and its first and second successor vertices $\mathcal{U}_1[r]$ and $\mathcal{U}_2[r]$, $\exists s \in \mathcal{R}$ such that $\mathcal{U}_1[r] == \mathcal{A}_t[s]$.

2: $\mathcal{F}_t[r] \leftarrow$ set of free neighboring vertices around robot r.

3: $\mathcal{F}_t[s] \leftarrow$ set of free neighboring vertices around robot s.

4: **if** $\mathcal{A}_t[r] \neq \mathcal{U}_1[s]$ **then**   ▷ r follows s.

5:     $\mathcal{P}_t[r] \leftarrow \mathcal{A}_t[r]$, return   ▷ *waiting.*

6: **else if** $\mathcal{A}_t[r] == \mathcal{U}_1[s]$ **then**   ▷ r intends to swap location with s.

7:     **if** $\exists \alpha \in \mathcal{F}_t[r]$ such that $\alpha \neq \mathcal{U}_2[s]$ **then**   ▷ r has the other free α.

8:         $\mathcal{P}_t[r] \leftarrow \alpha$, return   ▷ *dodging.*

9:     **else if** $\exists \beta \in \mathcal{F}_t[s]$ such that $\beta \neq \mathcal{U}_2[r]$ **then**   ▷ s has the other free β.

10:        $\mathcal{P}_t[r] \leftarrow \mathcal{A}_t[r]$, return   ▷ *waiting.*

11:    **end if**

12:    **if** $\exists \alpha \in \mathcal{F}_t[r]$ such that $\alpha == \mathcal{U}_2[s]$ **then**   ▷ r has the only free α.

13:        $\mathcal{P}_t[r] \leftarrow \alpha$, return   ▷ *retreating.*

14:    **else if** $\exists \beta \in \mathcal{F}_t[s]$ such that $\beta == \mathcal{U}_2[r]$ **then**   ▷ s has the only free β.

15:        $\mathcal{P}_t[r] \leftarrow \mathcal{A}_t[r]$, return   ▷ *waiting.*

16:    **end if**

17:    **if** r has the other adjacent robot w **then**

18:        TURNING-HEAD $\leftarrow \mathcal{A}_t[w]$, and   ▷ *turning-head* to new robot.

19:        $\mathcal{P}_t[r] \leftarrow \mathcal{A}_t[r]$, return   ▷ *waiting.*

20:    **else if** s has the other adjacent robot v **then**

21:        $\mathcal{P}_t[r] \leftarrow \mathcal{A}_t[r]$, return   ▷ *waiting.*

22:    **end if**

23: **end if**

Intersections. In the remaining instances (i.e., from Fig. 2(d) to 2(l)), r and s want to swap their positions, $(\mathcal{U}_1[r] == \mathcal{A}_t[s]) \bigcap (\mathcal{A}_t[r] == \mathcal{U}_1[s])$. Fig. 2(d), 2(e) and 2(f) shows T-junction or intersection configurations, where at least one robot has the other free neighboring vertex that is not the second successor vertex of the other one. Such a vertex is available for both r and s in Fig. 2(d), and only available for one of them in Fig. 2(e) and 2(f). As r believes that it has such a free vertex in Fig. 2(d) and 2(e), it can actively *dodge* itself to this vertex. For the instance in Fig. 2(f), r believes that s would apply the same strategy as what it will do in Fig. 2(e), so it just waits for r to succeed in dodging.

Narrow Corridors. Fig. 2(g), 2(h) and 2(i) depict that r and s do not have the other free neighboring vertex, except for each other's second successor vertices. Such a configuration corresponds to usual narrow or dead-end corridors, where one robot has to go backwards in order to make further progress. In Fig. 2(g), each robot can go backwards, whereas only one of them can do so in Fig. 2(h) and 2(i). As r believes that it can go backwards in Fig. 2(g) and 2(h), it will actively *retreat* itself and move backwards. When confronted with Fig. 2(i), r will take it for granted that s would retreat as what it will do in the case of Fig. 2(h), and r just needs to wait for s to make the adjustment.

Clustering Together. Fig. 2(j), 2(k) and 2(l) show that r and s do not have any free neighboring vertex. In Fig. 2(j) and 2(k), r has a neighboring vertex occupied by w, while in Fig. 2(l), apart from the vertex occupied by s, r does not have any other neighboring vertex, but s has the one occupied by v. At the moment, the pairwise coordination is constructed between r and s, and they cannot make any further progress to deal with such a deadlock situation. But in Fig. 2(j) and 2(k), r can *turn* its head and reconstruct the coordination with w, instead of keeping coordinating with s. Similarly, s can reconstruct the coordination with v in its own decision process when confronted with the case of Fig. 2(l).

Algorithm 2 gives the details on how to identify which situation the robot is confronted with and which strategy it should use. A robot can employ *waiting*, *dodging*, *retreating* and *turning-head* strategies to make local adjustments in the pairwise coordination.

5   Experiments and Results

5.1   Experimental Setup

Our experimental study is performed in the Blocks World for Teams (BW4T [1]) simulator. The environment uses grid maps (see Fig. 3), and each cell in the environment only allows one robot to be present at a time. As shown in Fig. 3, the robots start from the bottom of the map, and randomly choose distinct destinations, locating at the top of the map, to navigate towards. The other gray cells represent the stationary obstacles. In the experiment, we test scalable robot teams ranging from 1 to 8, and each simulation has been run for 100 times to reduce variance and filter out random effects. Our experiments run on an Intel i7-3720QM at 2.6GHz with 8 GB of RAM.

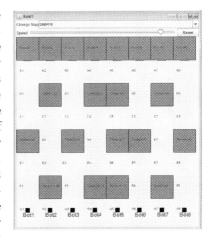

Fig. 3. The BW4T simulator

Robots in BW4T are controlled by agents written in GOAL [19], the agent programming language that we have used for implementing our proposed approach. GOAL also facilitates explicit communication among the agents which make decisions based on its *mental states* consisting of *knowledge*, *beliefs*, and *goals*. Fig. 4 gives an example of how to use GOAL to implement the pairwise coordination module. Of course, the agents also need other modules to work in BW4T, such as modules to obtain environmental percepts, to handle messages, etc. Detailed information about the GOAL agent programming language and the software can be found in [19].

---

[1] BW4T introduced in [18] has been integrated into the agent environments in GOAL [19], which can be downloaded from http://ii.tudelft.nl/trac/goal.

```
1   module pairwiseCoordination( Me, Teammate ) {
2     program {
3       % Me is following the trail of Teammate.
4       if bel( at(Me, MyLocation), at(Teammate, TeammateLocation),
5           firstSuccessor(Teammate, U1), U1 \= MyLocation ) then adopt( waiting ).
6       % Me has the other free neighboring vertex.
7       if bel( freeNeighboringVertex(Me, Free), secondSuccessor(Teammate, U2), U2 \= Free )
8           then adopt( at(Me, Free) ) + send( allother, !at(Free) ).
9       % Teammate has the other free neighboring vertex.
10      if bel( freeNeighboringVertex(Teammate, Free), secondSuccessor(Me, U2), U2 \= Free )
11          then adopt( waiting ).
12      % Me has the only free neighboring vertex.
13      if bel( freeNeighboringVertex(Me, Free), secondSuccessor(Teammate, U2), U2 == Free )
14          then adopt( at(Me, Free) ) + send( allother, !at(Free) ).
15      % Teammate has the only free neighboring vertex.
16      if bel( freeNeighboringVertex(Teammate, Free), secondSuccessor(Me, U2), U2 == Free )
17          then adopt( waiting ).
18      % Me has the other neighboring robot.
19      if bel( neighboringRobot(Me, OtherRobot), at(OtherRobot, Location) )
20          then insert( turninghead(Location) ) + adopt( waiting ).
21      % Otherwise, just wait for Teammate to turn its head.
22      if true then adopt( waiting ). }}
```

Fig. 4. Implementing Algorithm 2 using GOAL agent programming language: bel(Fact) means the agent believes that Fact is true. When the agent adopt(Plan), it will execute an corresponding action to achieve Plan. When the agent insert(Fact), it will insert Fact into its belief base. Using send(allother, !at(Free)), the agent can inform the others within its coordinated network that it wants to move to the Free vertex.

5.2   Results

Fig. 5 shows the results of the experimental study in which we have tested our decentralized approach, called DMRCP, and the ID approach proposed in [2]. The horizontal axis in Fig. 5 indicates the number of robots.

Time-Cost. As the programs of each approach using GOAL agent programming language have different complexities, we get the general impression that for each round of decision processes, DMRCP runs slower than ID. For instance, in the single robot case, DMRCP takes 5.13 seconds, whereas ID only takes 3.92 seconds on average (see Fig. 5(a)). In the cases of 1 to 4 robots, DMRCP always costs more time than ID, but when the number of the robots increases, DMRCP takes less time than ID. Therefore, even though the running time of the DMRCP's program is more expensive than ID, DMRCP can provide a competitive solution. In addition, we can see in Fig. 5(a) that from 1 robot to 8 robots, the time-cost of ID almost linearly goes up from 3.96 to 10.45 seconds. The increasing rate is around $\frac{10.45-3.96}{8-1} \approx 0.92$. Comparatively, the time-cost of DMRCP grows from 5.13 to 9.24 seconds, and its increasing rate is only about 0.59. This result reveals that with the increase of the robots, the time-cost of DMRCP goes up slower than ID. It is because the main strategy of ID is that if two robots have conflicting paths, one of them needs to find an alternative path, the length of which might be much longer than the previous one. It will become much more severe when the team size increases as more robots may need to frequently find

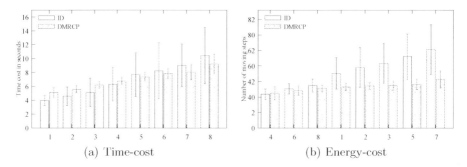

(a) Time-cost   (b) Energy-cost

Fig. 5. Experimental results of DMRCP and ID solutions

alternative paths. Though DMRCP employs the waiting strategy, the robots always take the shortest paths and only need to make local adjustments, and we can see waiting may not increase time-cost in robot teams.

Energy-Cost. Fig. 5(b) shows the fact that the robots take more moving steps along with the increase of the team size. Theoretically, for the single robot case, DMRCP and ID take the same moving steps as there is no other robot interfering it, which can also be seen in Fig. 5(b). With the increase of the number of the robots, the energy-cost of DMRCP goes up slowly, whereas it rises quickly in ID. For instance, in the case of 8 robots, DMRCP needs 15.69 moving steps on average; comparatively, ID takes 25.25 moving steps. The increasing rate of energy-cost in ID is $\frac{25.25-10.94}{8-1} \approx 2.04$, while it is only $\frac{15.69-11.24}{8-1} \approx 0.64$ in DMRCP. This is because DMRCP can benefit greatly from the waiting strategy as well as the local adjustments. Waiting does not increase energy-cost, and local adjustments allow the robots to take as few additional moving steps as possible.

6   Conclusions

In this paper, we analyzed the decentralized multi-robot cooperative pathfinding problem using graph-based models, and then proposed a novel fully decentralized solution to this problem. When confronted with conflicting situations, a robot only needs to coordinate with the ones locating within its coordinated network, which can reduce excessive communication in decentralized teams. Our solution allows the robots to make local adjustments by employing waiting, dodging, retreating and turning-head strategies, and the experimental results have shown that our approach provides a competitive solution to this problem.

References

1. Kaminka, G.A.: Autonomous agents research in robotics: A report from the trenches. In: 2012 AAAI Spring Symposium Series (2012)

2. Standley, T., Korf, R.: Complete algorithms for cooperative pathfinding problems. In: Proceedings of the Twenty-Second International Joint Conference on Artificial Intelligence, pp. 668–673 (2011)

3. Surynek, P.: An optimization variant of multi-robot path planning is intractable. In: AAAI (2010)

4. Desaraju, V.R., How, J.P.: Decentralized path planning for multi-agent teams with complex constraints. Autonomous Robots 32, 385–403 (2012)

5. Van Den Berg, J.P., Overmars, M.H.: Prioritized motion planning for multiple robots. In: IEEE/RSJ International Conference on Intelligent Robots and Systems (IROS), pp. 430–435 (2005)

6. Luna, R., Bekris, K.E.: Efficient and complete centralized multi-robot path planning. In: IEEE/RSJ International Conference on Intelligent Robots and Systems (IROS), pp. 3268–3275. IEEE (2011)

7. de Wilde, B., ter Mors, A.W., Witteveen, C.: Push and rotate: cooperative multi-agent path planning. In: Proceedings of the Twelfth International Conference on Autonomous Agents and Multiagent Systems, pp. 87–94 (2013)

8. Parker, L.E.: Decision making as optimization in multi-robot teams. In: Ramanujam, R., Ramaswamy, S. (eds.) ICDCIT 2012. LNCS, vol. 7154, pp. 35–49. Springer, Heidelberg (2012)

9. Parker, L.E.: Current state of the art in distributed autonomous mobile robotics. In: Distributed Autonomous Robotic Systems 4, pp. 3–12. Springer (2000)

10. Silver, D.: Cooperative pathfinding. In: The First Conference on Artificial Intelligence and Interactive Digital Entertainment (AIIDE), pp. 117–122 (2005)

11. Wang, K.H.C., Botea, A.: Mapp: a scalable multi-agent path planning algorithm with tractability and completeness guarantees. Journal of Artificial Intelligence Research 42, 55–90 (2011)

12. Ryan, M.R.: Exploiting subgraph structure in multi-robot path planning. Journal of Artificial Intelligence Research 31, 497–542 (2008)

13. Burgard, W., Moors, M., Stachniss, C., Schneider, F.E.: Coordinated multi-robot exploration. IEEE Transactions on Robotics 21, 376–386 (2005)

14. Bennewitz, M., Burgard, W., Thrun, S.: Finding and optimizing solvable priority schemes for decoupled path planning techniques for teams of mobile robots. Robotics and Autonomous Systems 41, 89–99 (2002)

15. Zuluaga, M., Vaughan, R.: Reducing spatial interference in robot teams by local-investment aggression. In: IEEE/RSJ International Conference on Intelligent Robots and Systems(IROS), pp. 2798–2805 (2005)

16. Dresner, K., Stone, P.: Multiagent traffic management: a reservation-based intersection control mechanism. In: Proceedings of the Third International Joint Conference on Autonomous Agents and Multiagent Systems, pp. 530–537 (2004)

17. Wang, K.H.C., Botea, A.: Fast and memory-efficient multi-agent pathfinding. In: International Conference on Automated Planning and Scheduling (ICAPS), pp. 380–387 (2008)

18. Johnson, M., Jonker, C., van Riemsdijk, B., Feltovich, P.J., Bradshaw, J.M.: Joint activity testbed: Blocks world for teams (BW4T). In: Aldewereld, H., Dignum, V., Picard, G. (eds.) ESAW 2009. LNCS, vol. 5881, pp. 254–256. Springer, Heidelberg (2009)

19. Hindriks, K.: The goal agent programming language (2013),
http://ii.tudelft.nl/trac/goal

Scheduling Patients in Hospitals Based
on Multi-agent Systems

Fu-Shiung Hsieh and Jim-Bon Lin

Department of Computer Science and Information Engineering,
Chaoyang University of Technology
41349 Taichung, Taiwan
fshsieh@cyut.edu.tw

Abstract. Scheduling patients in a hospital is a challenging issue as it calls for a sustainable architecture and an effective scheduling scheme that can dynamically schedule the available resources. The objectives of this paper are to propose a viable and systematic approach to develop a distributed cooperative problem solver for scheduling patients based on MAS to minimize the patient stay in a hospital. Our solution approach combines contract net protocol (CNP), multi-agent system architecture, process specification model based on time Petri net (TPN) and optimization theories. Our system first automatically formulate patient scheduling problem based on the TPN models. The patient scheduling problem is decomposed into a number of subproblems that are solved by cooperative agents. A collaborative algorithm is invoked by individual agents to optimize the schedules locally based on a problem formulation automatically obtained by the Petri net models. To realize our solution methodology, we proposed a scheme for publication and discovery of agents' services based on FIPA compliant multi-agent system platform. We develop a scheduling system to solve the dynamic patient scheduling problem.

Keywords: Workflow, multi-agent system, scheduling.

1 Introduction

The distributed organizational structure, costly infrastructure, diversified medical procedures and the need for prompt response to emergencies pose challenges in management of hospitals. These challenges call for the development of effective methodologies to reduce the cost and provide efficient and quality services based on a flexible distributed organizational architecture. The issues in hospitals range from allocation of resources, scheduling of patients to delivery of healthcare services. In existing literature, there are several studies on planning and scheduling in hospitals [1], [2], [3], [4], [13], [18]. How to properly handle patients timely under the constraints of resources to reduce risk is an essential issue. Given a set of requests from patients and medical workflows, our objective is to propose a dynamic scheduling scheme to minimize the patient stay in hospital under resource constraints. The problem to determine whether a single patient can be handled by a time

A. Moonis et al. (Eds.): IEA/AIE 2014, Part I, LNAI 8481, pp. 32–42, 2014.
© Springer International Publishing Switzerland 2014

constraint can be stated as a constraint satisfaction problem (CSP) [9], [17] in existing AI literature whereas the problem to determine whether multiple patients can be handled by a time constraint can be described as a scheduling problem. In AI, multi-agent systems (MAS) [6, 7] provide a flexible architecture to dynamically allocate resources in a hospital to handle patients. In this paper, we exploit recent advancements in multi-agent systems (MAS), workflow models, scheduling theory/algorithms and cooperative distributed problem solving (CDPS) [8] to develop a system for scheduling patients.

We first formulate a patient scheduling problem for hospitals based on MAS. To facilitate optimization of patient schedule, we adopt time Petri nets (TPN) [13] as the modeling tool instead of using informal workflow specification languages such as XPDL [10], BPMN [11] and WS-BPEL [12]. Petri net is an effective, formal model for modeling and analysis of workflows [26-28]. The Petri Net Markup Language (PNML) [29] is an XML-based interchange standard for representing Petri nets. Many Petri net tools support PNML format, e.g. WoPeD, Renew, PNK, PEP, VIPtool [30]. Therefore, PNML is used as the format for representing the TPN models of agents in our system. We construct the TPN model for each workflow agent and resource agent. Our system first automatically formulates a patient scheduling problem based on the TPN models. The patient scheduling problem is decomposed into a number of subproblems that are solved by cooperative agents based on minimum cost flow (MCF) algorithm [31]. We propose a scheduling method based on interactions between patient agents, workflow agents and resource agents using contract net protocol (CNP)[14], [20]-[24], a well known protocol for coordination, to efficiently allocate resources, and Java Agent Development Environment (JADE).

The remainder of this paper is organized as follows. In Section 2, we state the patient scheduling problem. In Section 3, we introduce TPN models for agents. In Section 4, we formulate the scheduling problem based on TPN models, present our prototype system implementation in Section 5 and conclude this paper in Section 6.

2   Patient Scheduling Problem

In a health care system, a variety of resources are involved in the medical procedures of patients. Different steps throughout the lifecycle of a medical procedure require distinct resources for processing. Due to the complexity of the tasks performed and the distributed information and resources, medical processes in a hospital require considerable mobility and coordination of resources. A typical medical procedure may consist of a number of operations such as registration, diagnosis, radiology test, blood examination, anesthesia, surgery, intensive and discharge, etc. These operations are performed by different hospital workers such as doctor, staff, specialist and nurse. Multi-agent systems (MAS) [6, 7] provide a flexible architecture to model the interaction of these agents. Therefore, we adopt MAS to dynamically allocate and schedule resources in health care systems to handle patients in this paper. For the entities in a hospital, we define three types of agents, including workflow agents, resource agents (e.g. doctor agents, staff agents, specialist agents and nurse agents), patient agents and scheduling agents. To state the scheduling problem, let ψ denote the time constraint for completing a patient's medical processes, $\boldsymbol{RA}$ denotes the set of resource agents and $\boldsymbol{WA}$ denotes the set of workflow agents. The Patient Scheduling Problem (PSP) is stated as follows. Given a time constraint ψ, a set of resource agents $\boldsymbol{RA}$, a set of workflow agents $\boldsymbol{WA}$ with the processing time for each operation and a set of patient agents, the problem is to schedules each agent in $RA \subseteq \boldsymbol{RA}$ and $WA \subseteq \boldsymbol{WA}$ such that the overdue cost of patients can be minimized subject to the following constraints.

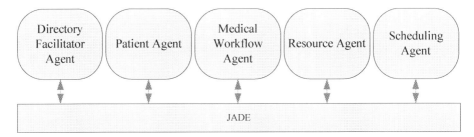

Fig. 1. Architecture to schedule medical workflows in health care systems

(I)The time intervals allocated by each resource agent in RA satisfy the capacity constraints of the resource agent.

(II)The precedence constraints between the workflow agents are satisfied.

(III)All patients can be handled after executing the schedules by the resource agents.

The aforementioned problem will be formulated formally based on transformation of the process models of workflow agents and activity models of resource agents later in this paper. A solution methodology will be proposed. We develop a problem solver based on Java Agent Development Environment (JADE) platform as shown in Fig. 1. To study PSP, a mathematical model for each workflow agent and resource agent is proposed in the next section.

3   Modeling Workflows and Activities in Hospitals

Petri nets have been widely adopted as a tool for modeling processes in industry (Murata, 1989). In this paper, we adopt Petri net [13, 15] to model medical workflows and capture the synchronous/asynchronous activities provided by agents and use PNML format to represent the models. The medical procedures in a hospital usually form complex workflows. A TPN [15] G is a five-tuple $G = (P, T, F, C, m_0)$, where P is a finite set of places, T is a finite set of transitions, $F \subseteq (P \times T) \cup (T \times P)$ is the flow relation, $C : T \to R^+$ is a mapping called static interval, which specifies the time for firing each transition, where R^+ is the set of nonnegative real numbers, and $m_0 : P \to Z^{|P|}$ is the initial marking of the PN with Z as the set of nonnegative integers. A Petri net with initial marking m_0 is denoted by $G(m_0)$. A marking of G is a vector $m \in Z^{|P|}$ that indicates the number of tokens in each place under a state. Firing a transition removes one token from each of its input places and adds one token to each of its output places. A marking m' is reachable from m if there exists a firing sequence s bringing m to m'. To model a medical workflow, we use a place to denote a state and a transition to denote an operation. A medical workflow w_n is defined as follows.

Definition 3.1: A transition is a terminal input transition if it has no input place. A transition is a terminal output transition it has no output place. A connected acyclic time marked graph (CATMG) $w'_n = (P'_n, T'_n, F'_n, C'_n, m'_{n0})$ is a connected marked graph [15] without any cycle and has only one terminal input transition and one terminal output transition.

Definition 3.2: The model of a workflow agent w_n is an acyclic TPN $w_n = (P_n, T_n, F_n, C_n, m_{n0})$ obtained by augmenting a CATMG $w'_n = (P'_n, T'_n, F'_n, C'_n, m'_{n0})$ with an input place ε_n and an output place θ_n and connecting ε_n to the terminal input transition of w'_n and connecting the terminal output transition to θ_n.

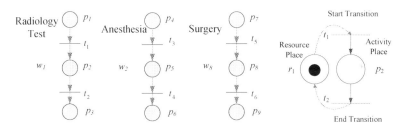

Fig. 2. (a) TPN models of workflow agents $w_1 \sim w_3$ (b) Specification of a Resource Activity based on Petri net

Fig 2(a) shows the Petri net models of three medical workflows (w_1 through w_3). The workflow TPN model for each step has one start state, one or more processing state and one finished state. An activity is a sequence of operations to be performed by a certain type of resources. The Petri net model for the $k - th$ activity of resource r is described by a Petri net H_r^k that starts and ends with the resource idle state place p_r as follows.

Definition 3.3: Petri net $H_r^k(m_r^k) = (P_r^k, T_r^k, F_r^k, C_r^k, m_r^k)$ denotes the $k - th$ activity of resource r. There is no common transition between H_r^k and $H_r^{k'}$ for $k \neq k'$. The initial marking $m_r^k(p_r) = 1$ and $m_r^k(p) = 0$ for each $P_r^k \setminus \{p_r\}$, where p_r is the idle state place of resource r. $C_r^k(t) \cap C_r^k(t') = \Phi \forall t, t' \in T_r^k$ with $t \neq t'$.

Fig. 2(b) illustrates a resource activity associated with the workflow w_1 in Fig. 2(a). To solve the dynamic scheduling problem, a problem formulation based on the proposed TPN models will be detailed in the next section.

4   Algorithm for Scheduling Patients

To formulate the scheduling problem for a scheduling agent, we first obtain the parameters from the corresponding TPN models. Let PA be the set of Patient agents, O be the number of Patient agents and N be the number of workflow agents. That is, $N = |WA|$ and $O = |PA|$. Let K_n be the number of different resource activities involved in W_n. The k-th resource activity in W_n is represented by A_r^k, $k \in \{1, 2, ..., K_n\}$. Suppose the k-th resource activity in W_n is performed by resource agent r_k. Then the processing time π_{nk} of the k-th resource activity A_r^k in W_n is $\pi_{nk} = \mu_r^k(t_s^k) + \mu_r^k(t_e^k)$, where t_s^k and t_e^k denote the starting and ending transitions of the k-th activity of resource agent a_r. Let T be the total number of time periods. Each $t \in \{1, 2, 3, ..., T\}$ represents a time period index. Let C_{rt} be the capacity of resource r at time period t, where $C_{rt} = m_{r0}^k(r)$. Let d_n be the due date of workflow agent w_n. The due date d_n for workflow agent w_n is set by its downstream workflow agents in the negotiation processes. Let v_{okt} be the number of patients of order o that requires resource r_k for processing the k-th resource activity in W_n during time period t, where $v_{okt} \geq 0$ and $v_{okt} \in Z^+$, the set of non-negative integers. Given W_n, A_r^k, π_{nk}, Q_n and d_n, where $n \in \{1, 2, 3, ..., N\}$ and $k \in \{1, 2, 3, ..., K_n\}$, the patient scheduling problem is formulated as follows. Let y_{ot} be the number of patients in patient agent o associated with workflow W_n finished during time period t. Let h_{okt} be the number of patients in patient agent o at the input buffer of the k-th resource activity in W_n at the beginning of period t, where $h_{okt} \geq 0$ and $h_{okt} \in Z^+$, the set of non-negative integers. The scheduling problem for workflow W_n to find an allocation of resource capacities over the scheduling horizon that minimizes the earliness/tardiness cost while satisfying all resource constraints is formulated as

$$\min \sum_{o=1}^{O} \sum_{t=1}^{T} (\omega_{ot} y_{ot})$$

s.t.

$$\sum_{o=1}^{O} \sum_{\tau=t-\pi_{nk}+1}^{t} v_{ok\tau} \leq C_{r_k t} \qquad \forall k \in \{1, 2, ..., K_n\} \qquad (1)$$

$$y_{ot} = v_{oK_n(t-\pi_{nK_n})}, \forall o, \forall t \qquad (2)$$

$$h_{o11} = Q \qquad (3)$$

$$h_{o1(t+1)} = h_{o1t} - v_{o1t}$$

$$h_{ok(t+1)} = h_{okt} - v_{okt} + v_{o(k-1)(t-\pi_{n(k-1)})} \forall k \in \{2, 3, ..., K_n\} \tag{4}$$

$$h_{o(K_n+1)(t+1)} = h_{o(K_n+1)(t+1)} + v_{oK_n(t-\pi_{nk})}, \tag{5}$$

where $\omega_{ot} = \begin{cases} E_n(d_n - t), t \le d_n \\ L_n(t - d_t), t > d_n \end{cases}$

We adopt a divide and conquer approach by applying the Lagrangian relaxation technique to develop a solution algorithm for scheduling patients. The structure of the scheduling problem OP_n faced by each workflow agent can then be represented by a minimum cost flow (MCF) problem [31]. In problem (P), we apply Lagrangian relaxation to relax resource capacity constraints (4.1). Let λ_{okt} be the associated Lagrange multipliers. For resource r at time period t, the dual problem to the scheduling problem can be obtained by Lagrangian relaxation as follows:

$$\max_{\lambda \ge 0} L(\lambda) \equiv \sum_{o=1}^{O} \min_{v_o} MCF_o(v_o, \lambda) - \sum_{k=1}^{K_n} \sum_{t=1}^{T} \lambda_{okt} C_{r_k t},$$

where $MCF_o(v_o, \lambda) \equiv \min[\sum_{t-1}^{T} \{(\omega_{ot} y_{ot}) + \sum_{r=1}^{K_n} \lambda_{ort} \sum_{\tau=t-\pi_{nk}+1}^{t} v_{ok\tau}\}]$

s.t. constraints (4.2) and constraints (4.3), (4.4) and (4.5).

A subgradient algorithm is used in this paper to solve the above problem. Let i be the iteration index. Let v^i denote the optimal solution to MCF subproblems for given Lagrange multipliers λ^i at iteration i. We define the subgradients of $L(\lambda)$ with respect to Lagrangian multipliers λ^i as follows:

$$g_{okt}^i = \sum_{o=1}^{O} \sum_{\tau=t-\pi_{nr}+1}^{t} v_{ok\tau}^i - C_{or_k t}, \quad \forall k = 1, ..., K_n, \forall t = 1, ..., T$$

The subgradient method proposed by Polyak ([25]) is adopted to update λ as follows:

$$\lambda_{okt}^{i+1} = \begin{cases} \lambda_{okt}^i + \alpha^i g_{okt}^i, & \text{if } \lambda_{okt}^i > 0 \text{ or if } \lambda_{okt}^i = 0 \text{ and } g_{okt}^i \ge 0 \\ 0, & \text{if } \lambda_{okt}^i = 0 \text{ and } g_{okt}^i < 0 \end{cases},$$

where $\alpha^i = \dfrac{\beta[\bar{L}(\lambda^*) - L(\lambda^i)]}{\sum\limits_{k,t} (g_{okt}^i)^2}$ and $\bar{L}$ is an estimate of the optimal dual cost and

$0 < \beta < 2$.

Iterative application of the subgradient algorithm will converge to an optimal dual solution. In case the solution is not feasible, we must develop a heuristic algorithm to find a feasible solution. We have implemented a heuristic algorithm that removes the excessive flows from the arcs with capacity violation by setting the arc capacity to zero and move these flows to other part of the network based on MCF algorithm.

5   Implementation of Scheduling System

Our approach to schedule patients relies on interaction between patient agents, workflow agents, resource agents (including doctor agents, nurse agents, staff agents, and specialist agents, etc.) and scheduling agents. Interactions among different types of agents are based on a negotiation mechanism that extends the well-known contract net protocol (CNP) [14]. CNP relies on an infrastructure for individual agents to publish and discover their services and communicate with each other based on the ACL language defined by the FIPA international standard for agent interoperability. We have developed a multi-agent scheduling system based on JADE to realize our methodology. To apply the contract net protocol, a manager must be able to search for the services provided by the other bidders. Each time an agent is added to the system, its services are published through the DF (Directory Facilitator) agent in JADE. Patient agents, workflow agents and resource agents are accompanied with proper graphical user interface (GUI) for users to input the required data. Scheduling agents, which are invoked by workflow agents, have no GUI as they work behind the scenes. Some of the screen shots of our scheduling system are shown in Fig. 3 through Fig. 6.

Fig. 3. (a)GUI for defining a workflow agent (b) GUI for defining a resource agent

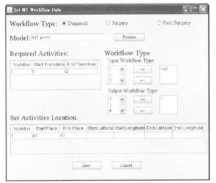

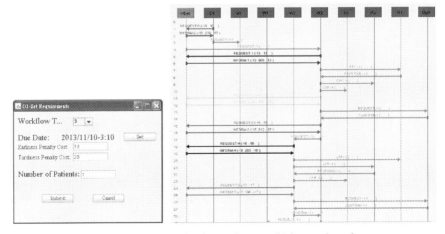

Fig. 4. (a) Setting a patient's requirement (b) Interaction of agents

In the remainder of this section, we use an application scenario to demonstrate the practicality of our scheduling software. Suppose the requests of a patient are received. Handling a patient's request relies on the collaboration of a number of agents. The requirements of a patient agent are specified by a due date, a workflow type, number of patients, and the penalty cost (earliness penalty cost and lateness penalty cost). The three types of workflows can be processed by three workflow agents (W_1 through W_3) and several resource agents (R_1 through R_{10}). A workflow is described by a Petri net model represented in PNML file format. To save space, we will not show all the Petri net models in this paper. The schedule for each resource agent is represented by a Gantt chart in our system. The GUI for defining a medical workflow agent is shown in Fig. 3(a). The GUI for defining a resource agent is depicted in Fig. 3(b). Fig. 4(a) illustrates the GUI for setting the requirement of a patient. Interactions of agents to schedule patients are shown in Fig. 4(b). In Fig. 4(b), patient agent O_1 issues a request and the request is forwarded to W_3. W_3 issues a 'call for proposal' (CFP) message to resource agents and invokes a scheduling agent Opt_1 to allocate resources using our scheduling algorithm. W_3 then requests W_2 to schedule its activities and allocate resources. A similar process occurs to W_1. Finally, patient agent O_1 will receive a proposal, which consist of the schedule, from W_1. Fig. 5 shows the contracts established between agents for handling a patient. The Gantt chart for representing the schedule of a resource agent is shown in Fig. 6.

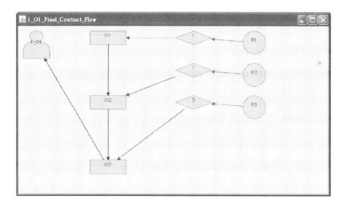

Fig. 5. Contracts established between resource agents and workflow agent for a patient

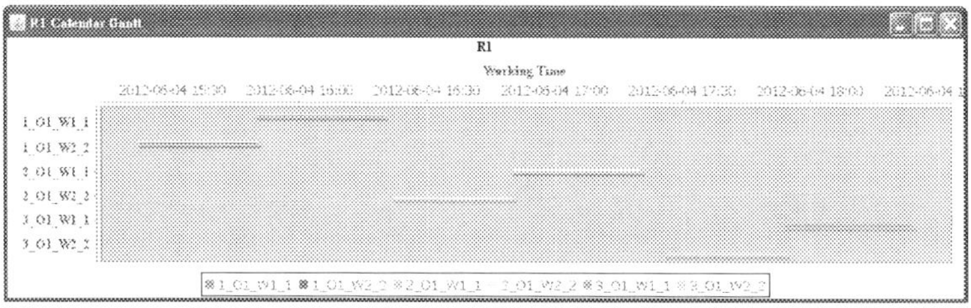

Fig. 6. The Gantt chart for representing the schedule of a resource agent

6   Conclusion

The healthcare sector faces an intense and growing competitive pressure. How to take advantage of the cutting-edge technologies and innovative business models to effectively acquire competitive advantage is critical for healthcare service providers to survive. Scheduling patients in hospitals is an important issue. Recent trends on scheduling operations in hospitals concentrate on the development of dynamic and distributed scheduling techniques to rapidly respond to changing needs and requests from the patients and achieve sustainability. In this paper, we propose a viable and systematic approach to develop a system for scheduling patients in hospitals based on MAS. Resources and workflows are modeled as agents. The key issue is how to make these agents work together coherently to handle multiple patients timely. Our proposed methodology meets the above mentioned requirements by (1) using Petri net as the workflow specification language, (2) adopting the publication/discovery infrastructure provided in the JADE platform and (3) developing algorithms based on distributed computing architecture to solve the scheduling problem. To solve the patient scheduling problem, we combine cooperative distributed problem solving (CDPS) with formal temporal models, optimization theory and MAS. The scheduling problem is divided into several scheduling subproblems that are solved by several cooperative scheduling agents, with one scheduling subproblem being automatically formulated and solved by the optimization agent based on the PNML models for the relevant workflow agent and resource agents. In this way, users not familiar with optimization theories can also apply our tool to schedule patients. The approach to scheduling patients proposed in this paper is different from those reported in [19,22,23,24].

Acknowledgement. This paper is supported in part by National Science Council of Taiwan under Grant NSC102-2410-H-324-014-MY3.

References

1. Decker, K., Li, J.: Coordinated hospital patient scheduling. In: International Conference on Multi Agent Systems, Paris, July 3-7, pp. 104–111 (1998)
2. Kutanoglu, E., Wu, S.D.: On combinatorial auction and Lagrangean relaxation for distributed resource scheduling. IIE Transactions 31, 813–826 (1999)
3. Oddi, A., Cesta, A.: Toward interactive scheduling systems for managing medical resources. Artificial Intelligence in Medicine 20, 113–138 (2000)
4. Daknou, A., Zgaya, H., Hammadi, S., Hubert, H.: A dynamic patient scheduling at the emergency department in hospitals. In: 2010 IEEE Workshop on Health Care Management, February 18-20, pp. 1–6 (2010)
5. Spyropoulos, C.D.: AI planning and scheduling in the medical hospital environment. Artificial Intelligence in Medicine 20, 101–111 (2000)
6. Nilsson, N.J.: Artificial Intelligence: A New Synthesis. Morgan Kaufmann Publishers, Inc., SanFrancisco (1998)
7. Ferber, J.: Multi-Agent Systems, An Introduction to Distributed Artificial Intelligence. Addison-Wesley, Reading (1999)

8. Durfee, E.H., Lesser, V.R., Corkill, D.D.: Trends in cooperative distributed problem solving. IEEE Transactions on Knowledge and Data Engineering 1(1), 63–83 (1989)

9. Russel, S.J., Norvig, P.: Artificial Intelligence—A Modern Approach, 2nd edn. Pearson Education Asia Limited (2006)

10. Workflow Management Coalition, XPDL support and resources (2009), http://www.wfmc.org/xpdl.html

11. Object Management Group, Business process modeling notation (2009), http://www.bpmn.org

12. OASIS, Web services business process execution language version 2.0 (2009), http://docs.oasis-open.org/wsbpel/2.0/OS/wsbpel-v2.0-OS.html

13. Hsieh, F.-S., Lin, J.-B.: Temporal Reasoning in Multi-agent Workflow Systems Based on Formal Models. In: Pan, J.-S., Chen, S.-M., Nguyen, N.T. (eds.) ACIIDS 2012, Part I. LNCS, vol. 7196, pp. 33–42. Springer, Heidelberg (2012)

14. Smith, R.G.: The Contract net protocol: high-level communication and control in a distributed problem solver. IEEE Transactions on Computers 29, 1104–1113 (1980)

15. Murata, T.: Petri Nets: Properties, Analysis and Applications. Proceedings of the IEEE 77(4), 541–580 (1989)

16. Berthomieu, B., Menasche, M.: An Enumerative Approach for Analyzing Time Petri Nets. In: Proc. Ninth Intenational Federation of Information Processing (IFIP) World Computer Congress, vol. 9, pp. 41–46 (September 1983)

17. Conry, S.E., Kuwabara, K., Lesser, V.R., Meyer, R.A.: Multistage negotiation for distributed constraint satisfaction. IEEE Transactions on Systems, Man and Cybernetics 21(6), 1462–1477 (1991)

18. Marinagia, C.C., Spyropoulosa, C.D., Papatheodoroub, C., Kokkotos, S.: Continual planning and scheduling for managing patient tests in hospital laboratories. Artificial Intelligence in Medicine 20, 139–154 (2000)

19. Güray Güler, M.: A hierarchical goal programming model for scheduling the outpatient clinics. Expert Systems with Applications 40(12), 4906–4914 (2013)

20. McFarlane, D.C., Bussmann, S.: Developments in holonic production planning and control. International Journal of Production Planning and Control 11(6), 522–536 (2000)

21. Parunak, H.V.D.: Manufacturing experiences with the contract net. In: Huhns, M. (ed.) Distributed Artificial Intelligence, pp. 285–310. Pitman, London (1987)

22. Saremi, A., Jula, P., ElMekkawy, T., Wang, G.G.: Appointment scheduling of outpatient surgical services in a multistage operating room department. International Journal of Production Economics 141(2), 646–658 (2013)

23. Sauré, A., Patrick, J., Tyldesley, S., Puterman, M.L.: Dynamic multi-appointment patient scheduling for radiation therapy. European Journal of Operational Research 223(2), 573–584 (2012)

24. Ceschia, S., Schaerf, A.: Modeling and solving the dynamic patient admission scheduling problem under uncertainty. Artificial Intelligence in Medicine 56(3), 199–205 (2012)

25. Polyak, B.T.: Minimization of Unsmooth Functionals. USSR Computational Math. and Math. Physics 9, 14–29 (1969)

26. van der Aalst, W.M.P.: The application of Petri nets to workflow management. J. Circuit. Syst. Comput. 8(1), 21–66 (1998)

27. van der Aalst, W.M.P., Kumar, A.: A reference model for team-enabled workflow management systems. Data & Knowledge Engineering 38(3), 3355–3363 (2001)

28. Weske, M., van der Aalst, W.M.P., Verbeek, H.M.W.: Advances in business process management. Data & Knowledge Engineering 50(1), 1–8 (2004)

29. Weber, M., Kindler, E.: The Petri Net Markup Language (2012),
    http://www2.informatik.hu-
    berlin.de/top/pnml/download/about/PNML_LNCS.pdf
30. Billington, J., et al.: The Petri Net Markup Language: Concepts, Technology, and Tools.
    In: van der Aalst, W.M.P., Best, E. (eds.) ICATPN 2003. LNCS, vol. 2679, pp. 483–505.
    Springer, Heidelberg (2003)
31. Edmonds, J., Karp, R.M.: Theoretical improvements in algorithmic efficiency for network
    flow problems. Journal of the ACM 19(2), 248–264 (1972)

Optimal Preference Clustering Technique for Scalable Multiagent Negotiation(Short Paper)

Raiye Hailu and Takayuki Ito

Nagoya Institute of Technology
Nagoya ,Japan
http://www.itolab.nitech.ac.jp

Abstract. We propose protocol for automated negotiations between multiple agents over multiple and interdependent issues. We consider the situation in which the agents have to agree upon one option (contract) among many possible ones (contract space). Interdependency between issues prevents us from applying negotiation protocols that have linear time complexity cost like Hill Climbing implementing mediated text negotiation protocol(HC). As a result most previous works propose methods in which the agents use non linear optimizers like simulated annealing to generate proposals. Then a central mediator can be used to match the proposals in order to find an intersection. But this matching process usually has exponential time cost complexity. We propose multi round HC(MR-HC) for negotiations with multiple and interdependent issues. In each round the normal HC is used to determine a negotiation deal region to be used by the next round. We propose that the agents should cluster their constraints by the cardinality of the constraints in order to get socially optimal contracts before applying MR-HC. To showcase that our proposed clustering technique is an essential one, we evaluate the optimality of our proposed protocol by running simulations at different cluster sizes.

Keywords: Artificial Intelligence, Distributed Artificial Intelligence, Coherence and Coordination, Automated Negotiation.

1 Introduction

We propose protocol for automated negotiations between multiple agents over multiple and interdependent issues. We consider the situation in which the agents have to agree upon one option (contract) among many possible ones (contract space). Each issue of the negotiation has finite number of possible values. A contract is identified by the value it has for each issue of the negotiation.

The issues are interdependent means that for example it is generally not possible for an agent to decide about each issue independently and finally reach at an optimal contract. And more over negotiation protocols that are designed for independent issue negotiations may not result in an optimal deal when the issues are interdependent. We define the optimal deal to be the one that maximizes social welfare that is the total utility.

A. Moonis et al. (Eds.): IEA/AIE 2014, Part I, LNAI 8481, pp. 43–48, 2014.

Interdependency between issues prevents us from applying negotiation protocols that have linear time complexity cost like Hill Climbing implementing mediated text negotiation protocol(HC 2.1). As a result, most previous works propose methods in which the agents use non-linear optimizers like simulated annealing to generate proposals. Then a central mediator can be used to match the proposals in order to find an intersection. But this matching process usually has exponential time cost complexity. Therefore the number of agents that can be supported by such negotiation mechanisms is very limited.

We propose multi round HC for negotiations with multiple and interdependent issues. In each round the normal HC is used to determine a negotiation deal region to be used by the next round. We propose that the agents cluster their constraints by the cardinality of the constraints. . In the first round of multi round HC the cluster which contains the largest constraints should be used by the agents to evaluate the contracts proposed by the mediator, and in the second round the second largest cluster and so on. To showcase that our proposed clustering technique is an essential one, we evaluate the optimality of our proposed protocol by running simulations at different cluster sizes.

In a non linear utility space it is not also possible to locate the optimal contracts of even one agent using HC. Instead non linear optimization technique like simulated annealing(L-SA) is much better(Figure 1). As a result the previous approach for such negotiation was to make agents submit bids (identified by L-SA) to a central mediator which tries to find a match. But such protocols only support a few number of agents due to the computational time complexity of exhaustive matching [1]. Here we want to revisit HC in order to modify it to support non linear negotiations.

We adopt the constraints based utility space model in [1] to represent agents utility spaces. An agent's utility for a contract is the sum of the weights of the constraints the contract satisfies. An optimal contract for an agent is the one that maximizes its utility.

2   Cardinality Based Clustering

Our solution concept is based on the observation that constraints that make up an agent's utility space can be divided into those representing very broad and general criteria , those that represent very specific ones and the rest some where between the two. And we expect that contracts that satisfy the specific constraints also satisfy the general ones. Therefore an agent can iteratively narrow down the search region for optimal the contract(See Algorithm 2.3).

Before running the MR-HC agents cluster their constraints by their cardinality. In cardinality based grouping the criterion for a set of constraints to belong to the same cluster is the similarity of the size of the constraints. Therefore while general constraints that are easily met by many contracts are grouped together, specific constraints that are satisfied by only a few of contracts will be grouped together in another group. Therefore before the beginning of MR-HC each agent is expected to have created $| C |$ number of clusters. Each cluster

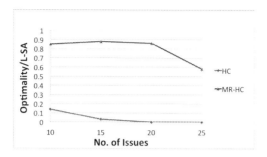

Fig. 1. Optimality for one Agent

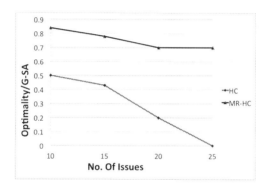

Fig. 2. Optimality for multiple Agents

contains some number of constraints. For each agent the cluster C_1 contains the
largest constraints and $C_{|C|}$ the smallest constraints.

Other researchers ([4])have presented a similar generic idea, but here we have
implemented a concrete algorithm which has the additional advantage of being
efficient. Also [2] have considered iterative narrowing, but while they group constraints based on their positions we cluster them by their cardinality. Author
[5] has considered reducing non linearity of utility spaces by analyzing the issue
interdependencies. We think methods that reduce non linearity can be used as a
pre process for our protocol. Our protocol should be used when there is no way
to avoid non linearity.

The symbols in the algorithms are as follows: I : The issues of the negotiation.
A : A set of agents. V : A set of values for each issue, V_n is for issue n. C : A set
of constraint clusters $C = C_1, C_2...C_c$. Each C_i is characterized by the maximum
cardinality of the constraint it can contain. C_c^a : Set of constraints in Agent a's
cth cluster.DR: DealRegion is a set of set of values for each issues. DR_i:A set
of values for Issue i in DR. Initially DR_i contains all of the V_i, hence the cross
product of the initial DR_is represents the entire contract space.

Algorithm 2.1. $HC(I, V)$

$S \leftarrow initial solution(set randomly)$
for each $i \in I$
$\quad$ **do** $\begin{cases} \text{for each } j \in V_i \\ \quad DO \begin{cases} SS \leftarrow S \text{ with issue } i \text{ value set to } j \\ \text{if } \forall A, U(SS) > U(S) \text{ then } S \leftarrow SS \end{cases} \end{cases}$
return (S)

Algorithm 2.2. $HC(I, V, c, DR, S)$

$\forall A, DR^a \leftarrow DR$
for each $i \in I$
$\quad$ **do** $\begin{cases} \text{for each } j \in DR_i \\ \quad DO \begin{cases} SS \leftarrow S \text{ with issue } i \text{ value set to } j \\ \text{if } \forall A, U(SS, C_c^a, DR^a) > U(S, C_c^a, DR^a) \\ \text{then } S \leftarrow SS \end{cases} \\ //Each \text{ agent updates } DR^a \\ DR^a \leftarrow DR^a \cap satisfied \text{ } cons. \in C_c^a \\ DR \leftarrow DR^{a1} \cap DR^{a2} \cap DR^{a3} .... \\ S \leftarrow random \text{ } contract \in DR \end{cases}$
return (S).

Algorithm 2.3. $MR\text{-}HC(I, V, |C|, DR)$

$S \leftarrow initial solution(set randomly)$
for $c \leftarrow 1$ **to** $|C|$
$\quad$ **do** $\{ S \leftarrow HC(I, V, c, DR, S)$
return (S)

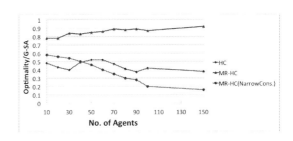

Fig. 3. Optimality at Various Number of Agents(No. of issues 10)

3   Experimentation and Analysis

3.1   Simulation Setup

For a negotiation over I issues we create $4 \times I$ constraints. Each issues has 10 possible values represented by the numbers $1 - 10$. Each agent clusters its constraints into I number of groups. Each group contains 4 constraints that share a common center. This center is randomly chosen contract. Width of each of the constraints in the group is selected from a uniform distribution of the width values 2 to 8. However based on the location of the center point of a group, a constraint's width might be truncated if it exceeds the issue values range. First, each of the constraints in a group are assigned the same of $Ran(1, I) * 100$. Additionally each constraint will have an additional weight of $100/Ran(1, 10)$. It is possible that two or more groups of constraints to overlap.

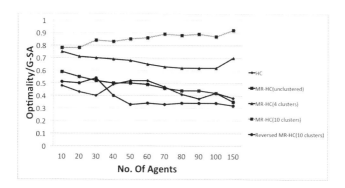

Fig. 4. Optimality at Various Number of Clusters(No. of Issues 10)

3.2   Analysis

Figure 1 shows that MR-HC has better optimality than HC for an agent. The number of issues is varied from 10 to 25 number of issues. This shows that the proposed algorithm MR-HC indeed has better optimality than HC. Moreover from the definition of the algorithm we can see that for fixed number of clusters (C), MR-HC execution time cost increases only linearly with the number of issues.

Figure 2 and Figure 3 show that mediation protocol based on MR-HC has better optimality than HC protocol. G-SA is simulated annealing run on the sum of the utility of the agents. However, MR-HC optimality is affected by the average width of constraints and by the grouping level(See Fig. 3 OP. MR-HC for narrow constraints). That is when most of the constraints have small cardinality and can not be effectively grouped into various size levels, MR-HC can not give optimal results.

Figure 4 shows some experimental result used to prove the essentiality of cardinality based clustering before applying MR-HC. As shown MR-HC with each agent defining ten clusters has the highest optimality followed by MR-HC with 4 clusters per agents. Applying MR-HC without any clustering produces sub optimal results like HC. It is conceptually flawed but we experiment with the reverse MR-HC. That means in the first round the agents use the cluster which contains their smallest constraints, and the second round the cluster containing second smallest constraint sizes and so on. As expected reversed MR-HC has also poor optimality.

4   Conclusions

We note that when MR-HC is run for one agent, it is basically a greedy optimizer. In the first round an agent's task is to select a constraint from cluster 1. It simply greedily chooses the one with the highest weight. This is repeated for each cluster but with the condition that the newly selected constraint should have intersection with the previously chosen constraints.

At these stage we can only say that there is an incentive for both the agents and the mediators to use this protocol. A Mediator that has been using HC has an incentive to start using MR-HC because MR-HC gives more optimal results. Moreover a mediator has incentive to switch from using non linear protocols like bidding based deal identification because it can support more number of agents with MR-HC. The agents as a society have incentive to use MR-HC because they get a chance to collaborate with many other agents while producing optimal social outcomes.

However, we have to investigate at least on average whether agents are better off using MR-HC or not in terms of their final individual utility for the selected deal. Especially we expect that agents to ignore the clustering step in order to manipulate the protocol to end with deal which is most beneficial to them.

References

1. Ito, H.H., Klein: Multi-issue negotiation protocol for agents exploring nonlinear utility spaces. In: International Joint Conference on Artificial Intelligence (2007)
2. Hattori, M.K., Ito: Using Iterative Narrowing to Enable Multi-Party Negotiations with Multiple Interdependent Issues. In: Autonomous Agents and Multi Agents Systems (2007)
3. Ito, H.H., Klein: Multi-issue negotiation protocol for agents exploring nonlinear utility spaces. In: International Joint Conference on Artificial Intelligence (2007)
4. Lopez-Carmona, Marsa-Maestre, I., Hoz, Velasco, J.: A region based multi issue negotiation protocol for non monotonic utility spaces. Studies in Computational Intelligence 27, 166–217 (2011)
5. Fujita, T.I., Klein: An Approach to Scalable Multi-issue Negotiation: Decomposing the Contract Space Based on Issue Interdependencies. In: Intelligent Agent Technology (2010)

Using Common Sense Invariants in Belief Management for Autonomous Agents

Clemens Mühlbacher and Gerald Steinbauer*

Institute for Software Technology
Graz University of Technology
Inffeldgasse 16b, A-8010 Graz, Austria
{cmuehlba,steinbauer}@ist.tugraz.at

Abstract. To design a truly autonomous robot it is necessary that the robot is able to handle unexpected action outcomes. One way to deal with these outcomes is to perform a diagnosis on the history of performed actions. Basically there are two ways to trigger such a diagnosis. One way to trigger it is a direct contradiction between a sensor reading and its expected value. Another way is to use background knowledge if an alternative action outcome does not lead to a direct contradiction. In order to avoid the necessity to hand-code this knowledge we propose to reuse a existing common sense knowledge base. In this paper we present an approach for reusing existing common sense knowledge in a belief management approach for autonomous agents.

1   Introduction

A robot can only be considered to be truly autonomous if it is able to handle unexpected outcomes of actions as well as changes of the environment performed from other agents or humans. The changes which are performed by other agents or human are called exogenous events. To point out more precisely how unreliable actions can hinder a robot to successfully perform its task let's consider a simple example which will be our running example in the remainder of the paper. A robot has to deliver an object O from a room A to a room B. The robot moves to room A and grasps the object. Unfortunately, the object is not fully grasped. Thus the robot does not hold the object. If the robot is not able to detect the fault the object O will not be successful delivered to room B. If the robot performs more complex tasks like setting up a table or cooking a meal a big number of different faults can occur due to imperfect action execution and sensing.

To tackle this problem the robot needs to detect if the world does not evolve as expected. This can be achieved with a check if the belief of the robot is consistent with the sensor readings. The belief of the robot is the model of the world and how this world changes due to the performed actions. Continuing our example from above the unsuccessful pickup of the object causes that the belief of the robot states that the object is held by the robot. If the robot sense the environment to check if it is holding something this discrepancy between belief and the real world can be detected.

---

* The work has been partly funded by the Austrian Science Fund (FWF) by grant P22690.

A. Moonis et al. (Eds.): IEA/AIE 2014, Part I, LNAI 8481, pp. 49–59, 2014.

But often it is not possible to exactly detect a discrepancy with a sensor reading. Let's consider again our running example. If the robot can only sense that it is carrying something it does not know if it carries object O. We need additional background knowledge to describe that carrying object O is the same as still carrying an object.

This background knowledge is the common sense knowledge of the robot about the world. But it is a tedious task to specify all this knowledge for a complex system. To avoid the specification of this background knowledge we propose a mapping technique to reuse common sense knowledge from another knowledge bases.

Beside the detection of an inconsistency it is also important to discover the root cause the inconsistency. Consider again our running example. The robot moves to room B after the unsuccessful pickup action. Afterwards the robot tries to put down the object and senses the environment if the object is in room B as expected. The robot can now detect that the object is not in room B as expected. If we would just use the latest sensor measurement the location of object O will disappear from the agents knowledge. If it is discovers that the pickup action failed it can conclude that the object O is still in A and the agent is able to finish its task.

Gspandl et.al. presented in [1] a method which uses model-based diagnosis to detect what really happened and uses background knowledge to detect inconsistencies between the world and the belief of the agent. We will extend this idea by a mapping which allows to reuse background knowledge from a common sense knowledge basis.

The remainder of the paper is organized as follows. First we discuss the belief management the high-level control as well as how we define the consistency between belief and world. In Section 3 we will discuss the theoretical aspects of the proposed mapping. In the following section we explain how we implemented this mapping using Cyc. In Section 5 we will present preliminary results of an experimental evaluation of the implementation. In Section 6 we discuss briefly some related work. In the last section we will draw some conclusion and point out some future work.

2   Belief Management for High-Level Control

To specify the task of a robot a high-level program can be used. We use IndiGolog [2] to represent this high-level program. To be able to reason about action execution the situation calculus [3] is used. We use Reiter's variation [4] of the situation calculus extend with sensing capabilities as described in [5]. Additionally to model the sensor effects more accurate we use additional helper predicate $RealSense(\alpha, s)$ as it was defined in [6]. We use the following axioms to interpret a high-level program and to evolve the belief of the robot.

1. $\mathcal{D}_{S_0}$: Axioms specifying the initial situation of the world.
2. $\mathcal{D}_{ap}$: An axiom for each precondition of an action α.
3. $\mathcal{D}_{ssa}$: Set of axioms specifying the changes due to the execution of an actions.
4. Σ: Foundation axioms for situations.
5. $\mathcal{D}_{una}$: Unique name axioms for actions
6. $\{Senses\}$: Axioms specifying the sensing. Containing $SF(\alpha, s) \equiv \Psi_\alpha(s)$ which specifies if sensing action α is performed in situation s $\Psi_\alpha(s)$ will hold in this

situation. Additionally the axioms contain $RealSense(\alpha, s)$ specifying the connection between a fluent and the sensing action α. Please note that $SF(\alpha, s) \equiv RealSense(\alpha, s)$ has to hold in a consistent world.

7. $\mathcal{C}$: Axioms for refining programs as terms.

These axioms form the *extended basic action theory* $\mathcal{D}^* = \mathcal{D}_{S_0} \cup \mathcal{D}_{ap} \cup \mathcal{D}_{ssa} \cup \Sigma \cup \mathcal{D}_{una} \cup \mathcal{C} \cup \{Sensed\}$. See [2] for a more detailed discussion on the basics of IndiGolog and [6] for a detailed definition of the sensing axioms $\{Senses\}$.

We will distinguish between actions which yield a sensing result as sensing actions and primitive actions which don't yield a sensor result. To form a consistent theory we use $\Psi_\alpha(s) \equiv \top$ and $RealSense(\alpha, s) \equiv \top$ for primitive actions.

As we motivated above for a robot it is essential to handle unexpected outcomes of actions and exogenous events. To achieve this we perform a diagnosis on the history of performed actions. Before a diagnosis can be calculated it is essential to detect if the world evolves as expected. To check if the world evolves as expected the belief of the robot and the sensing results are checked for consistency. We define the consistency of the world in Definition 1. The predicate $Invaria(s)$ denotes the background knowledge of the robot. We call an inconsistency an obvious fault if $SF(\alpha, \bar{\delta}) \not\equiv RealSense(\alpha, do(\alpha, \bar{\delta}))$.

Definition 1. *A history of performed actions σ is consistent iff $\mathcal{D}^* \models Cons(\sigma)$ with $Cons(\cdot)$ inductively defined as:*

1. $Cons(\epsilon) \doteq Invaria(S_0)$
2. $Cons(do(\alpha, \bar{\delta})) \doteq Cons(\bar{\delta}) \wedge Invaria(do(\alpha, \bar{\delta})) \wedge$
   $[SF(do(\alpha, \bar{\delta})) \wedge RealSense(\alpha, do(\alpha, \bar{\delta})) \vee \neg SF(\alpha, \bar{\delta}) \wedge \neg RealSense(\alpha, do(\alpha, \bar{\delta}))]$

To be able to perform the consistency check as defined above we use some additional assumptions about the modeled actions. One assumption states that the actions are modeled correctly. Otherwise we could end up in an inconsistent situation even all actions where executed correctly and no exogenous events happened.

We also assume that the initial state is modeled correctly and does not contradict the background knowledge.

Using these assumptions and the definition of consistency we can conclude that a contradiction can only be caused by a sensing action.

If the world does not evolve as expected a diagnosis is performed through a variation of the actions in the history of performed actions and insertions of possible exogenous events. The goal of the diagnosis is to find an alternative course of actions which is consistent with the observations. Please note that at the end even the actual sensing action might be altered. This would represent a perception error.

3 Use of Cyc as Background Model

To check the consistency we have to check if an obvious fault happened and if no invariant of the background knowledge is violated. In particular the last one is of interest because it needs the background knowledge ($Invaria(s)$). In order to avoid the necessity to hand-code this background knowledge we map a situation and the sensor readings into a common sense knowledge base (KB). Please note that we are not interested in mapping the situation itself. We define the mapping from the knowledge base of the situation calculus $\mathcal{L}_{SC}$ to some common sense knowledge base $\mathcal{L}_{KB}$ using Definition 2. We expect that both representations are logical complete. The mapping we define is a theory morphism and is not expected to be plain, conservative or axiom preserving. For a detailed description of these properties we refer the reader to [7].

Definition 2. *The function $\mathcal{M}_{\mathcal{L}_{SC},\mathcal{L}_{KB}} : \mathcal{L}_{SC} \times S \to \mathcal{L}_{KB}$ represents the transformation of a situation and an extended basic action theory to a theory in $\mathcal{L}_{KB}$.*

As we described above we are only interested to map the belief of the robot represented by a $\mathcal{D}^*$, an situation and the related fluents.

To make this mapping useful in our context we need to ensure an additional property as defined in Definition 3. This property states that the mapping preserves the consistency regarding to the background knowledge. A similar property is the query property [8] which is defined for mapping between OWL knowledge bases.

Definition 3. *The mapping $\mathcal{M}_{\mathcal{L}_{SC},\mathcal{L}_{KB}}$ is proper if the transferred situation s together with a possible empty set of additional sentences Δ lead to a contradiction if and only if the situation s together with the background knowledge lead to a contradiction:*

$$\forall s : \Delta \cup \mathcal{M}_{\mathcal{L}_{SC},\mathcal{L}_{KB}}(\mathcal{D}^*, s) \models_{\mathcal{L}_{KB}} \bot \Leftrightarrow \mathcal{D}^* \cup BK(s) \models_{\mathcal{L}_{SC}} \bot$$

The definition of the additional mapping property ensures that regardless if we use the situation in the situation calculus together with the background knowledge or if we use the mapped situation and the Δ we will always obtain the same result for the consistency. Thus we can use this mapping to answer if the situation is consistent with the background knowledge which is defined in the common sense knowledge base.

4   Implementation Using Cyc

Before we show an implementation of the mapping using Cyc we define two transformation functions. The first transformation function maps constants while the second transformation function maps fluents. To specify these functions we assume n constants $c_1, ..., c_n$ and m relational fluents $F_1(\bar{x}_1, s), ..., F_m(\bar{x}_m, s)$ used in the *extended basic action theory*. Additionally $\bar{x}_j$ represents the set of variables of the jth fluent.

Definition 4. *For a constant symbol c from the constants of the situation calculus we define a transformation function. $R_C : c \to \mathcal{L}_{KB}$.*

Definition 5. *For a positive fluent F_i we define a transformation function $R_{F_i} : T^{|\bar{x}_i|} \to \mathcal{L}_{KB}$. For a negative fluent F_i we define a transformation function $\bar{R}_{F_i} : T^{|\bar{x}_i|} \to \mathcal{L}_{KB}$. Where T denotes a term from $\mathcal{L}_{SC}$.*

The mapping $\mathcal{M}_{\mathcal{L}_{SC}, \mathcal{L}_{KB}}(\mathcal{D}^*, s)$ can now be defined as follows:

Definition 6. *The mapping* $\mathcal{M}_{\mathcal{L}_{SC}, \mathcal{L}_{KB}}(\mathcal{D}^*, s)$ *consists of a set of transformed fluents and constants such that:*

$$\mathcal{M}_{\mathcal{L}_{SC}, \mathcal{L}_{KB}}(\mathcal{D}^*, s) \equiv \{\forall i \forall \bar{x}_i . \mathcal{D}^* \models_{\mathcal{L}_{SC}} F_i(\bar{x}_i, s) | R_{F_i}(\bar{x}_i) \cup \bigcup_{x \in \bar{x}_i} R_C(x)\} \cup$$

$$\{\forall i \forall \bar{x}_i . \mathcal{D}^* \models_{\mathcal{L}_{SC}} \neg F_i(\bar{x}_i, s) | \bar{R}_{F_i}(\bar{x}_i) \cup \bigcup_{x \in \bar{x}_i} R_C(x)\}$$

In order to avoid problems caused by the different interpretation of unknown fluents (e.g. the situation calculus uses close world assumption while a potential target representation might use open world assumption) we define a complete mapping for all positive and negative fluents. We have implemented such a mapping between the situation calculus and Cyc. Cyc [9] is a common sense knowledge base which uses a first-order logic for predicates together with higher-order logic to represent special properties for predicates. Such a property is for instance the partly uniqueness. This property specifies that if all terms of a function except one are fixed the remaining term is uniquely defined. We will use such properties to specify object-place relations. It has to be mentioned that we are here only interested to map the state of the world from one representation into another and not in transferring reasoning capabilities.

The mapping maps each constant x from the situation calculus to a constant $\{x\}$ in CYC [1]. Each constant is mapped as it is depicted in Listing 1.1. The class *Partially Tangible* represents physical entity in Cyc.

Listing 1.1. Mapping for constants

```
R_C(x) → (isa {x} PartiallyTangible)
```

To map the fluents of our running example we use the functions depicted in Listing 1.2.

Listing 1.2. Mapping for fluents

```
R_{F_isAt}(x,y) → (isa storedAt_{x} AttachmentEvent)
    (objectAdheredTo storedAt_{x} {y})
    (objectAttached storedAt_{x} {x})
R̄_{F_isAt}(x,y) → (isa storedAt_{x} AttachmentEvent)
    (not (objectAdheredTo storedAt_{x} {y}))
    (objectAttached storedAt_{x} {x})
R_{F_holding}(x) → (isa robotsHand HoldingAnObject)
R̄_{F_holding}(x) → (not (isa robotsHand HoldingAnObject))
R_{F_carryAnything} → (isa robotsHand HoldingAnObject)
R̄_{F_carryAnything} → (not (isa robotsHand HoldingAnObject))
```

The first transformation functions handles the location of objects. Cyc represents the location of an object by an *AttachmentEvent* of an object x to a location y. Together

---

[1] $\{x\}$ denotes a unique string representation of the constant represented by variable x.

with the assertions specifying which object and location belongs to the event the location becomes uniquely defined for an object. Using this higher-order properties Cyc reports an inconsistency if an object is perceived at different locations. Please note that this property is already part of Cyc and was not specified by hand within the situation calculus. The second function is used if an object is not at a particular location.

The remaining functions map the belief about objects held by the robot and the related sensing. We use the Cyc event $HoldingAnObject$ to specify if the object $robotsArm$ is holding something. Cyc will report an inconsistency if the robot beliefs to hold something while the sensor reports the opposite. Please note that this rule was also not explicitly coded into the situation calculus representation. Please note that these invariants have to be hand-coded into the situation calculus representation if Cyc is not used.

Our implementation of the belief management system is based on the Prolog implementation of the IndiGolog interpreter [2]. This implementation allows to easily query if a fluent is true or false in a situation. Thus a snapshot of the situation's fluents can be simple created. Using the above mapping we transfer the semantics of the situation in the situation calculus to Cyc. To check if a situation is consistent using the mapping the algorithm depicted in Algorithm 1 is used.

Algorithm 1. $checkConsistency$

---

input : α ... action to perform in the situation,
        $\bar{\delta}$ ... situation before execution action α
output: true iff situation is consistent

1 **if** $sens\_act(\alpha)$ **then**
2   **if** $SF(\alpha, \bar{\delta}) \not\equiv RealSense(\alpha, do(\alpha, \bar{\delta}))$ **then**
3 | **return** $false$
4   **end**
5     $\bar{\delta}' = do(\alpha, \bar{\delta})$
6     $\mathcal{M} = \mathcal{M}_{\mathcal{L}_{SC}, \mathcal{L}_{KB}}(\mathcal{D}^*, \bar{\delta}')$
7   **if** $\neg checkCycConsistency(\mathcal{M})$ **then**
8 | **return** $false$
9   **end**
10 **end**
11 **return** $true$

The algorithm first checks if the actual sensing results leads to a direct contradiction with the expected ones. If not the algorithm checks if the current situation $\bar{\delta}$ is consistent with the background knowledge. Please note that this check is done within the Cyc framework (see line 7). If both checks are passed the sensor reading and the situation are consistent with the background knowledge. Here we can clearly see that background knowledge is important to be able to detect inconsistencies that do not lead to a direct contradiction to the sensing results.

5   Experiments

We performed three evaluations with the above presented implementation. All evaluations performed a simple delivery task in an simulated environment[2]. To solve this task the agent is able to perform three actions: pick up the object, go to destination place, and put down the object. All actions can fail. A pickup action can fail at all or picks up the wrong object located in the same room. The goto action can end up in a wrong room. The putdown action can fail in such a way that the robot is still holding the object. Additionally to these internal faults an exogenous event could happen which randomly moves an object to another room. The probabilities of the different action variations and the exogenous event can be seen in Table 1.

Table 1. Probabilities of fault of actions and the exogenous event. The probability denotes the chance that an action fails or an exogenous event occurs.

fault	probability
goto wrong location	0.05
pickup wrong object	0.2
pickup fails	0.2
putdown fails	0.3
random movement of object	0.02

The high-level control program for the agent is a Teleo-Reactive Program [10] which can be executed in every possible situation. The TR program allows to continue program execution even if the internal belief was altered by a diagnosis and repair step.

Each experiment ran for 20 different object settings (different object locations and goals). Each object setting was performed 5 times with different seeds for the random number generator. This ensures that the object setting or the triggered faults does not prefer a single setting.

The experiments use three different high-level control programs. The first program (HBD 0) uses no background knowledge and only directly detects contradictions. The second program (HBD 1) uses the hand coded background knowledge implemented in IndiGolog. The third program (HBD 2) uses the background knowledge in Cyc.

The agent moves three objects within a building with 20 rooms. The evaluations differ in the time the agent was granted to finish the task. For each evaluation the number of successful delivered objects, the time to deliver the object as well as the fact if a test run was finished in time was recorded.

We now postulate three hypotheses for our proposed mapping.

Hypothesis 1. *The programs which use the background knowledge have a higher success-rate compared to the program not using background knowledge.*

Hypothesis 2. *With an increasing timeout the success-rate increases too.*

---

[2] The tests were performed on an Intel quad-core CPU with 2.40 GHz and 4 GB of RAM running at 32-bit Ubuntu 12.04. The ROS version Fuerte was used for the evaluation.

Hypothesis 3. *With an increasing timeout the program which uses the common sense represented in Cyc shows the same success-rate as the program which uses the hand-coded background knowledge.*

The first evaluation used a timeout of 660 seconds. The results are depicted in Table 2. The second evaluation used a timeout of 6600 seconds. The results are depicted in Table 3. The third evaluation use a timeout of 66000 seconds. The results are depicted in Table 4. The relation of the success-rate to the timeout is depicted in Figure 5.

Table 2. Success-rate, timeout-rate (TIR) and runtime for a timeout of 660 seconds

	success-rate	TIR	average runtime/s	
			Mean	Std
HBD 0	48.47	15.00	9.32	46.20
HBD 1	89.49	7.00	11.79	47.18
HBD 2	66.44	44.00	423.29	141.50

Table 3. Success-rate, timeout-rate (TIR) and runtime for a timeout of 6600 seconds.

	success-rate	TIR	average runtime/s	
			Mean	Std
HBD 0	54.92	6.00	132.78	750.37
HBD 1	91.53	2.00	216.85	1188.02
HBD 2	77.29	22.00	903.02	1118.48

Table 4. Success-rate, timeout-rate (TIR) and runtime for a timeout of 66000 seconds.

	success-rate	TIR	average runtime/s	
			Mean	Std
HBD 0	46.78	0.0	704.21	2064.03
HBD 1	89.83	0.0	351.36	1513.69
HBD 2	78.31	17.00	2340.75	4789.59

The evaluations support the first hypothesis. In each evaluation the programs using background knowledge outperformed the program not using background knowledge. It has to be mentioned that the program without background knowledge never got close to the worst cases of the programs using background knowledge. The second hypothesis is not supported by the evaluations. While for HBD 1 the change of the success-rate in relation to the timeout is not significant there is a decrease for HBD 0 for larger timeouts after an initial increase. We suppose that the longer timeout led to longer action histories. Therefore, the chance for exogenous events that negatively affect the agent's performance increases. But it has to be noted that HBD 2 is able to benefit from the timeout increase. The mapping has a high runtime because we have to check for all

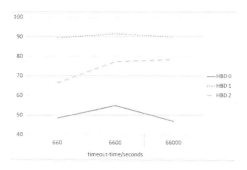

Fig. 1. Success-rate of the different approach in relation to the timeout

fluents if they hold in the current situation using regression. This leads to a maximum of $O(m \times n^k)$ fluent checks. Where m is the number of fluents, n the number of constants and k the maximum arity of the fluents. Moreover, the time needed for one check also depends on the length of the action history due to the used regression. Because of the above results it is plausible that HBD 2 will reach the same performance level as HBD 1. Therefore, the third hypothesis is supported by the evaluations. The evaluation also shows that even for a high timeout of more than 18 hours there are still some cases of timeout for HBD 2. Although, there is the advantage of being able to reuse common sense knowledge the runtime of the proposed approach remains a drawback.

6   Related Research

Before we conclude the paper we want to discuss related work on belief management, logic mapping and plan diagnosis. There are many different methods for belief management. Some of these methods have in common that they assign a measurement of trust to the predicates in the belief. The approach proposed in [11] uses weights for this measurement. In contrast possibilistic logic which can be related to believe revision techniques (see [12] for more details) uses a likelihood for the truth value of predicates. In [13] an approach was proposed which also takes the performed actions into consideration. The method we use (see [1] for details) does not individually weight predicates in the belief. Different beliefs are weighted instead.

If we consider work on logic mapping the approach presented in [14] has to be mentioned. The proposed method makes it possible to reuse reasoning of one logic in another one. In contrast we are not interested to transfer reasoning capabilities. We are interested in an oracle which tells us if a background knowledge is consistent with the internal belief instead.

Finally considering work on plan diagnose the approach proposed in [15,16] have to be mentioned. The approach use model based diagnosis on actions in a plan. Additionally partly observations are given to determine faulty actions. The approach differs not only due to usage of a plan instead of a history of executed actions. It also does not use a background knowledge to relate the observations to the internal belief.

7 Conclusion and Future work

Autonomous agents may fail to successfully finish their task because of an inconsistent belief about the world. In this paper we presented a mapping which allows to check if the internal belief of the robot is consistent with an external source of common sense knowledge. The idea is to map the internal belief to the same representation used in the common sense database and check the consistency in this framework. The advantage of this approach is that existing common sense knowledge can be reused for belief management without the necessity to hand-code this knowledge again. In an experimental evaluation it was shown that an agent using background knowledge to detect inconsistency significantly outperforms a simple agent. Moreover, it showed that the reuse of existing external common sense knowledge is almost as good as hand-coded knowledge in the original system. One current drawback of the proposed mapping is the high runtime preventing the system to utilize the full potential of the external knowledge source. In future work we will investigate if other knowledge bases could better serve our needs. One such knowledge base is [17] which is used to specify physical system and their behavior. Moreover, we will investigate if the mapping could be performed lazy such that only fluents are mapped which are essential to find an inconsistency.

References

1. Gspandl, S., Pill, I., Reip, M., Steinbauer, G., Ferrein, A.: Belief Management for High-Level Robot Programs. In: The 22nd International Joint Conference on Artificial Intelligence (IJCAI), Barcelona, Spain, pp. 900–905 (2011)
2. Giacomo, G.D., Lespérance, Y., Levesque, H.J., Sardina, S.: IndiGolog: A High-Level Programming Language for Embedded Reasoning Agents. In: Multi-Agent Programming: Languages, Tools and Applications, pp. 31–72. Springer (2009)
3. McCarthy, J.: Situations, Actions and Causal Laws. Technical report, Stanford University (1963)
4. Reiter, R.: Knowledge in Action. Logical Foundations for Specifying and Implementing Dynamical Systems. MIT Press (2001)
5. De Giacomo, G., Levesque, H.J.: An incremental interpreter for high-level programs with sensing. In: Logical Foundations for Cognitive Agents, pp. 86–102. Springer (1999)
6. Ferrein, A.: Robot Controllers for Highly Dynamic Environments with Real-time Constraints. PhD thesis, Knowledge-based Systems Group, RWTH Aachen University, Aachen Germany (2008)
7. Meseguer, J.: General logics. Technical report, DTIC Document (1989)
8. Arenas, M., Botoeva, E., Calvanese, D., Ryzhikov, V.: Exchanging OWL 2 QL Knowledge Bases. In: The 23rd International Joint Conference on Artificial Intelligence, pp. 703–710 (2013)
9. Ramachandran, D., Reagan, P., Goolsbey, K.: First-orderized researchcyc: Expressivity and efficiency in a common-sense ontology. In: AAAI Workshop on Contexts and Ontologies: Theory, Practice and Applications (2005)
10. Nilsson, N.J.: Teleo-reactive programs for agent control. Journal of Artificial Intelligence Research 1(1), 139–158 (1994)
11. Williams, M.A.: Anytime belief revision. In: The 15th International Joint Conference on Artificial Intelligence (IJCAI), pp. 74–81 (1997)

12. Prade, D.D.H.: Possibilistic logic, preferential models, non-monotonicity and related issues. In: Proc. of IJCAI, vol. 91, pp. 419–424 (1991)
13. Jin, Y., Thielscher, M.: Representing beliefs in the fluent calculus. In: The 16th European Conference on Artificial Intelligence (ECAI)
14. Cerioli, M., Meseguer, J.: May i borrow your logic? In: Borzyszkowski, A.M., Sokolowski, S. (eds.) MFCS 1993. LNCS, vol. 711, pp. 342–351. Springer, Heidelberg (1993)
15. Roos, N., Witteveen, C.: Diagnosis of plans and agents. In: Pěchouček, M., Petta, P., Varga, L.Z. (eds.) CEEMAS 2005. LNCS (LNAI), vol. 3690, pp. 357–366. Springer, Heidelberg (2005)
16. de Jonge, F., Roos, N., Witteveen, C.: Diagnosis of multi-agent plan execution. In: Fischer, K., Timm, I.J., André, E., Zhong, N. (eds.) MATES 2006. LNCS (LNAI), vol. 4196, pp. 86–97. Springer, Heidelberg (2006)
17. Borst, P., Akkermans, J., Pos, A., Top, J.: The physsys ontology for physical systems. In: 9th International Workshop on Qualitative Reasoning (QR), pp. 11–21 (1995)

Driving Behavior Analysis of Multiple Information Fusion Based on SVM

Jeng-Shyang Pan[1], Kaixuan Lu[1], Shi-Huang Chen[2], and Lijun Yan[1]

[1] Shenzhen Graduate School, Harbin Institute of Technology, Shenzhen, China
[2] Department of Computer Science and Information Engineering, Shu-Te University,
Kaohsiung County, 824, Taiwan, ROC
jengshyangpan@gmail.com, lukaixuan203@sina.com, shchen@stu.edu.tw,
yanlijun@126.com

Abstract. With the increase in the number of private cars as well as
the non-professional drivers, the current traffic environment is in urgent
need of driving assist equipment to timely reminder and to rectify the incorrect driving behavior. To meet this requirement, this paper proposes
an innovative algorithm of driving behavior analysis based on support
vector machine (SVM) with a variety of driving operation and traffic information. The proposed driving behavior analysis algorithm will mainly
monitor driver's driving operation behavior, including steering wheel angle, brake force, and throttle position. To increase the accuracy of driving
behavior analysis, the proposed algorithm also takes road conditions, including urban roads, mountain roads, and highways into account. The
proposed will make use of SVM to create a driving behavior model in various different road conditions, and then could determine whether the current driving behavior belongs to safe driving. Experimental results show
the correctness of the proposed driving behavior analysis algorithm can
achieve average 80% accuracy rate in various driving simulations. The
proposed algorithm has the potential of applying to real-world driver
assistance system.

Keywords: Driving behavior analysis, driver assistance system, SVM.

1 Introduction

Because the continuously increases of the global car ownership, it is unavoidable
to raise the traffic density and number of non-professional drivers. This also leads
to frequent traffic accidents which have become the first hazard of modern society
[1]. Among these traffic accidents, the improper driving behavior habits are an
important cause of crashes. Therefore the study of driving behavior analysis has
become extremely useful.

Thanks to the modern vehicle manufacturing technology, the safety factors
of the vehicle itself caused traffic accidents are smaller and smaller proportion.
However, the driver personal factors caused traffic accidents have become the
main reason for causing a traffic accident [2]. Toyota's Masahiro MIYAJI, Jiangsu

A. Moonis et al. (Eds.): IEA/AIE 2014, Part I, LNAI 8481, pp. 60–69, 2014.

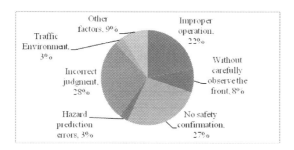

Fig. 1. Distribution of Traffic Accidents Statistics

Universitys Liang Jun [3][4] and other scholars count and analyze the specific reasons of traffic accidents and the results are shown in Fig. 1. From these results, it is observable that the individual factor account for the accidents are more than 90%. In addition to the driver and pedestrian essential training and education, it also needs to monitor and predict driving behavior of the drivers to prevent the traffic accidents and increase the traffic safety.

Many scholars have made a great of contributions in the study of driving behavior analysis. Yoshifumi Kishimoto and Koji Oguri propose a method of modeling driving behavior concerned with certain period of past movements by using AR-HMM (Auto-Regressive Hidden Markov Model) in order to predict stop probability [5]. Antonio Prez, M. Isabel Garca, Manuel Nieto, et.al present Argos, which is a complex and powerfully computerized car to help researchers in the study of car driver behavior [6]. Reza Haghighi Osgouei and Seungmoon Choi address a model-based objective measure for the evaluation of driving skills between different drivers. This metric is based on a stochastic distance between a pair of hidden Markov models (HMMs) each of which is trained for an individual driver [7]. Hong-mao Qin, Zhiqiang Liu and Peng Wang In order to overcome the limitations of single-channel information in the determination of drowsy driving behaviora method was proposed based on multi-channel information fusion [8]. The [9] is proposed for a realtime traffic surveillance system which can help the driver to get more surrounding traffic information.

On the basis of the above study, this paper proposed an algorithm to capture the driver operation and then to analysis the driver behavior and status. The proposed algorithm can assess whether the current driving behavior to make the car in a safe state. If necessary, it will provide the appropriate prompts or operation to make sure the vehicle from a dangerous state and back to a safe state.

The remaining sections are organized as follows. Section 2 brief introduces the SVM theorem. The details of the proposed driving behavior analysis algorithm are presented in Section 3. Section 4 shows experimental results and performance analysis. Finally, Section 5 concludes this paper.

2   Support Vector Machines

In this section we provide a brief review of the theory behind this type of algorithm; Support Vector Machines (SVMs) is developed from the optimal hyper-plane in the case of linearly separable, for more details we refer the reader to [10][11].

Assuming the set of training data (x_i, y_i), $i = 1, \cdots, n$, can be separated into two classes, where $x \in R^d$ is a feature vector and $y \in \{+1, -1\}$ its class label. they can be separated by a hyperplane $H : w \cdot x + b = 0$, and we have no prior knowledge about the data distribution, then the optimal hyperplane is the one which maximizes the margin[11]. The optimal values for w and b can be used to found by solving a constrained minimization problem, can be transform the optimal hyper-plane problem to its dual problem.Lagrange multipliers $\alpha_i\ (i = 1, \cdots, m)$

$$f(x) = \text{sgn}\left(\sum_{i=1}^{n} \alpha_i^* y_i K(x_i, x) + b^*\right) \tag{1}$$

where α_i^* and b^* are found by using an SVC learning algorithm [11]. Those x_i with nonzero α_i^* are the support vectors. For $K(x, y) = x \cdot y$, this corresponds to constructing an optimal separating hyperplane in the input space R^d.in this paper we use the kernel Radial basis function: $K(x_i, x_j) = \exp\left(-\gamma \|x_i - x_j\|^2\right)$, $\gamma > 0$ is an interval relaxation vector.

3   Driving Behavior Analysis System

The proposed driving behavior analysis system consists of driving operation acquisition module, data preprocessing module, the driving operation information fusion module, and SVM classification and recognition modules.

Driving behavior analysis data will be divided into training set and test set. Preprocessing and feature extraction are simultaneously applied to both sets. Classification makes the test samples into the driving model based on SVM algorithm to classify and determine the test sample category. The number of rightly or wrongly classified samples divided by the number of the test set samples is the classification correct rate or error rate.

For example, the following use the driving data in urban traffic road to illustrate the system processes. In the data acquisition module, at first, a good driving data and bad driving data should be collected as training set. Then collecting another data set includes good driving behavior data and bad driving behavior data as a test set using the same method. After data preprocess step, each time slice samples can be regarded as the rate of change the driving operation. We use the training set to establish the driving classification model in the city as a judge model by the SVM theory, and then we use the test se to judgment the accuracy of the model. Finally, judge the merits of driving behavior. Fig. 2 shows the flow of the entire system.

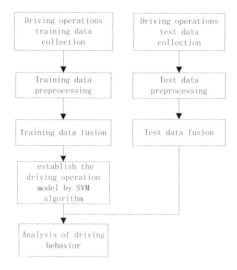

Fig. 2. The flow of the driving behavior analysis

4   Experimental results and analysis

4.1   Experimental Equipment

In this study, driving operation data is simulated through the YOYO driving training machine, produced by Guangzhou Great Gold East Network Technology Co. Driving operation data are read through the USB interface and combined with the simulation environment (city, mountains, and highway) to analysis the driving behavior. Fig. 3 show the data collection equipment and driving environment system.

4.2   Data Collection

In the PC terminal, we used the data collection program (Bus Hound 6.01) real-time recording the driving operation data when driver driving simulation. Each record driving time is about 5-15 minutes and driving environment are selected from the city, high-speed and mountainous road. Each of driving environment corresponds to simulate two driving data set and every set includes good driving behavior and bad driving behavior, respectively for training set and testing set. One of the driving data set as a training set is used to establish city driving model by SVM. The other driving set as the test set is used to test validity about the above driving model established by SVM. The method of collecting driving data is described as follows:

(1) Urban Roads Good driving record: maintain speed below 35 km/h, try to reduce the use of brake, deceleration in advance, try to keep a straight line driving, cornering speeds below 15 km/h, acceleration and braking are operated slowly. Bad driving record: Keep the car at speed driving state, quick step on

Fig. 3. Data collection equipment and driving environment system

and quick release the accelerator brake, Remain uncertain direction and curve driving.

(2) Highway Good driving record: maintain speed between 80 km/h and 100 km/h, Keep the car was going along a straight line with almost no brake operation, No big turning in high-speed. Bad driving record: speed is too high and instability, curve driving, quick step on and quick release the accelerator brake, often braking.

(3) Mountain road Good record: maintain speed below 40 km/h, turning speed below 20 km/h. Remained constant speed. Braking slowly and downshift in downhill. Bad driving record: speed is too high and instability, quick step on and quick release the accelerator brake, curve driving Finally obtain the following 12 groups data: Training set: good city driving, bad city driving, good mountain driving, bad mountain driving, good highway driving, bad highway driving Testing set: good city driving, bad city driving, good mountain driving, bad mountain driving, good highway driving, bad highway driveing

4.3   Data Preprocessing

he data record format is shown in Fig. 4.

(1) Extract data packet from the "data" column of the Fig. 4. The packet is converted from hex to decimal. These data include steering, brakes, and accelerator information. The first column is the steering wheel angle information, expressed by the number from 0 to 255. The second column is the turn left and turn right information: Digital 1 represents the first circle turn to the left; 0 represents the second circle turn to the left; 2 represents the first circle turn to the right; 3 represents the second circle turn to the right. In order to facilitate the data analysis, the steering angle information will be converted into a continuous angle data from -510 to 510, negative numbers indicate turn left, positive number indicate turn right, Where each number represents 1.4118 degree angle. The

Device	Phase	Data						Description	Cmd. Phase. Ofs(rep)
13.1	IN	00 02 80 80	00 19 10 55					.......U	1.1.0(90)
13.1	IN	00 02 80 88	00 19 10 4e					.......N	91.1.0
13.1	IN	00 02 80 9e	00 19 10 0e					....	92.1.0
13.1	IN	00 02 80 b5	00 19 10 1e					....	93.1.0
13.1	IN	00 02 80 ff	00 19 10 e3					....	94.1.0(48)
13.1	IN	00 02 80 ff	00 19 00 bf					....	142.1.0(24)
13.1	IN	00 02 80 ff	08 19 00 be					....	166.1.0(16)
13.1	IN	00 02 80 ff	00 19 00 bf					....	182.1.0(35)
13.1	IN	00 02 80 ff	00 19 02 57					.......W	217.1.0(79)
13.1	IN	00 02 80 ce	00 19 02 56					V	296.1.0
13.0	CTL	21 09 00 03	00 00 08 00					SET REPORT	297.1.0(2)
13.0	OUT	cc 00 00 00	00 00 00 00					....	297.2.0
13.0	CTL	21 09 00 03	00 00 08 00					SET REPORT	299.1.0
13.0	OUT	00 cc 00 00	00 00 00 00					....	299.2.0
13.0	CTL	21 09 00 03	00 00 08 00					SET REPORT	300.1.0(2)
13.0	OUT	00 0a 00 00	00 00 00 00					....	300.2.0
13.1	IN	00 02 80 cc	00 19 02 07					....	302.1.0

Fig. 4. Schematic of collection data text

third column is the brake and throttle Information, according the same method as changes of steering angle information, the throttle data is converted from -128 to 0, and the brake data is converted from 0 to 128.

(2) In the data record, the "Phase" column data indicates the data packet input or output state, "IN" represents the driving operation from driver training machine input information, "OUT, CTL" represents the pc control information output, we only need to extract the information that driver training machine input which the "IN" correspond to data packet from the "data" column.

(3) The time information processing, the "Cmd.Phase.Ofs (rep)" column represents the time series, where figures in brackets indicate the time of the operation remain unchanged. We will restore the driving operation information of each time slice and finally composite the driving operation record of the continuous time slice. The driving operation data of each time slice is a sample, including Attribute 1: steering wheel angle; Attribute 2: Brake throttle Information; and sample label, 1 represents a good driving, -1 represents a bad driving. According to the above preprocessing methods, we get a series of graph about city driving record data. Fig. 5 shows the city good driving record steering wheel data graph.

4.4   SVM Parameters Setting

In this paper, SVM algorithm with RBF kernel function is applied to convert the actually problem into a high dimensional space, where C and γ are punished coefficient parameters and intervals which are necessary two parameters of RBF. Its value directly will affect the classification accuracy. Here we adopt a cross-validation-based "grid search" method [12][13] to select the values of C and γ, that is selected the training samples divided into v parts, the part of v-1 as a training samples, the remaining as the test sample to determine model parameters, used to verify classification accuracy of the results classified by the v-1 part of the data, and constantly change the C and γ to obtain higher sample classification accuracy.

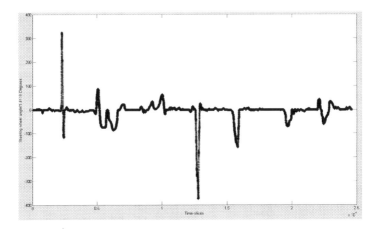

Fig. 5. The city good driving record data of steering wheel graph

Experience has shown, C and γ searched follow the exponential mode of growth $(C = 2^{-10}, 2^{-9}, \cdots, 2^{10}, \gamma = 2^{-10}, 2^{-9}, \cdots, 2^{10})$ is better way to quickly determine the parameters C and γ, where C and γ are independent growth.

4.5   Experimental Results

Use the processed data, select the good and bad driving data under different road as the SVM training and test sets, such as selecting good city driving and bad city driving in training set as training, good city driving in test set as good driving record test, bad city driving in test set as bad driving record test, using the method of parameters set on the section 4.4 described to find out the optimal parameters C and that the most suitable for the city, high-speed and mountain road.

The sample sets are obtained from the section 4.3 Data preprocessing, calculated the rate of change value of steering wheel angle, brake, and throttle to compose the new sample set to create the SVM driving model and test results are as shown in Table 1 SVM-based information fusion (steering wheel, brake throttle, road conditions) driving behavior data analysis, Table 2 SVM-based single information (steering wheel) driving behavior analysis, table 3 SVM-based single information (brake throttle) driving behavior analysis, table 4, 5, 6 are the result of cross-validation, respectively correspondence the SVM-based driving model is established in one of the three roads, and tested by driving data on the other road.

4.6   Experimental Analysis

From the data in Tables 1 and 2, one can clearly see where the testing process in the modeling, SVM comprehensive test accuracy rate can reach 80%, where

Table 1. SVM-based information fusion (steering wheel, brake throttle, road conditions) driving behavior data analysis

	Good driving behavior correct rate	Bad driving behavior correct rate	Comprehensive assessment
City road (C = 1, = 2)	95.5208%	68.7341%	82.12745%
Highway road (C = 1, = 4)	97.2902%	77.0102%	87.1502%
Mountain road (C = 1, = 4)	97.1227%	60.0445%	78.5836%

Table 2. SVM-based single information (steering wheel) driving behavior analysis

	Good driving behavior correct rate	Bad driving behavior correct rate	Comprehensive assessment
City road (C = 1, = 2)	95.7117%	60.882%	78.29685%
Highway road (C = 1, = 4)	93.722%	71.2691%	82.49555%
Mountain road (C = 1, = 4)	97.9687%	41.2564%	69.61255%

Table 3. SVM-based single information (brake throttle) driving behavior analysis

	Good driving behavior correct rate	Bad driving behavior correct rate	Comprehensive assessment
City road (C = 1, = 2)	99.9881%	2.17372%	51.08091%
Highway road (C = 1, = 4)	99.9403%	2.7111%	51.3257%
Mountain road (C = 1, = 4)	99.9739%	5.59072%	52.78231%

Table 4. The result of cross-validation SVM-based driving model is established in mountain

Mountain road driving training model (C = 1, = 4)	Good driving behavior correct rate	bad driving behavior correct rate	Comprehensive assessment
City road driving test data	95.4695%	62.3225%	78.896%
Highway road driving test data	97.8128%	62.7329%	80.27285%

Table 5. The result of cross-validation SVM-based driving model is established in highway

Highway driving training model (C = 1, = 4)	Good driving behavior correct rate	bad driving behavior correct rate	Comprehensive assessment
City road driving test data	88.764%	70.4949%	79.62945%
Mountain road driving test data	92.5274%	54.7351%	73.63125%

Table 6. The result of cross-validation SVM-based driving model is established in city

City road train driver model (C = 1, = 2)	Good driving behavior correct rate	bad driving behavior correct rate	Comprehensive assessment
Highway driving test data	98.8507%	62.1706%	80.51065%
Mountain road driving test data	97.9687%	45.4524%	71.71055%

the good driving behavior recognition rate is relatively higher, while the identification of bad driving behavior is lower, that mainly caused by the following points:

(1) Since in this test we used the test and training data set which composed with multiple time-slice sample set include the steering wheel angle gradient and brake throttle gradient to establish the driving model in different road. The driving behavior analysis system will judge the each of the time slice sample in real time. While there are so many good driving behavior sample in the bad driving data set, such as Uniform motion in a straight, slow start and so on. Thats the reason why the identification of bad driving behavior is lower.

(2) we can clearly see from Tables 1 and 2, the behavior analysis based on multi-information fusion driving is better than based on the single information in judgment the bad driving behavior. It is indicate that the single drive operation is difficult to correctly reflect the current driver's driving behavior, in real-time driving behavior analysis, the driving behavior analysis system need to combine multiple driving operation information to make driving behavior analysis to get more accurate results.

(3) Tables 4, 5, and 6 show that there are some major differences between the driving models which established in the different road based on SVM, resulting in lower correct rates in the driving behavior judgment. It is indicate that we need to create different driving behavior model to determine the correct analysis according to different road conditions.

5   Conclusion

With the real-time driving behavior analysis becomes more and more important, this paper utilizes a number of important driving operation data (brakes, throttle, steering wheel angle and road conditions) to comprehensive analysis the driving behavior. Driving behavior analysis model based on SVM can effectively achieve the correct judgment on driving behavior analysis; timely corrective driver's driving improperly. In future work We need to further study on combination with other data analysis methods to achieve more accurate and rapid analysis of driving behavior. Such as the method of Discrete Hidden Markov Model (DHMM) applied to driver behavior analysis [14] etc.

References

1. Leonard, E.: Traffic safety. Science Serving Society (2004)
2. Yan, X., Zhang, H., Wu, C., Jie, M., Hu, L.: Research progress and prospect of road traffic driving behavior. Traffic Information and Safety (1), 45–51 (2013)
3. Miyaji, M., Danno, M., Oguri, K.: Analysis of driver behavior based on traffic incidents for driver monitor systems. In: IEEE Intelligent Vehicles Symposium, pp. 930–935 (2008)
4. Liang, J., Cheng, X., Chen, X.: The research of car rear-end warning model based on mas and behavior. In: IEEE Power Electronics and Intelligent Transportation System, pp. 305–309 (2008)
5. Yoshifumi, K., Koji, O.: A modeling method for predicting driving behavior concerning with driver's past movements. In: IEEE International Conference on Vehicular Electronics and Safety, pp. 132–136. IEEE (2008)
6. Prez, A., Garcia, M.I., Nieto, M., Pedraza, J.L., Rodrguez, S., Zamorano, J.: Argos: an advanced in-vehicle data recorder on a massively sensorized vehicle for car driver behavior experimentation. IEEE Transactions on Intelligent Transportation Systems 11(2), 463–473 (2010)
7. Osgouei, R.H., Choi, S.: Evaluation of driving skills using an hmm-based distance measure. In: IEEE Haptic Audio Visual Environments and Games, pp. 50–55 (2012)
8. Qin, H., Liu, Z., Wang, P.: Research on drowsy driving behavior based on multichannel information fusion. China Safety Science Journal 2, 020 (2011)
9. Huang, D.Y., Chen, C.H., Hu, W.C., Yi, S.C., Lin, Y.F.: Feature-based vehicle flow analysis and measurement for a real-time traffic surveillance system. Journal of Information Hiding and Multimedia Signal Processing 3(3), 279–294 (2012)
10. Nello, C., John, S.: An introduction to support vector machines and other kernel-based learning methods. Cambridge University Press (2000)
11. Vapnik, V.: Statistical learning theory (1998)
12. Sathiya, K., ChihJen, L.: Asymptotic behaviors of support vector machines with gaussian kernel. Neural Computation 15(7), 1667–1689 (2003)
13. Joachims, Thorsten: Making large scale svm learning practical (1999)
14. Pan, S.T., Hong, T.P.: Robust speech recognition by dhmm with a codebook trained by genetic algorithm. Journal of Information Hiding and Multimedia Signal Processing 3(4), 306–319 (2012)

Constructing Support Vector Machines Ensemble Classification Method for Imbalanced Datasets Based on Fuzzy Integral

Pu Chen[1] and Dayong Zhang[2,*]

[1] Academy of Fundamental and Interdisciplinary Sciences, Harbin Institute of Technology,
Harbin 150001, China
{ No.92 ,West Da-zhi Street, Harbin, P.R. China 150001}
chenpu@live.com
[2] Department of New Media and Arts,Harbin Institute of Technology ,Harbin150001,China
{ No.92 ,West Da-zhi Street, Harbin, P.R. China 150001}
yonghit@163.com

Abstract. The problem of data imbalance have attract a lot of attentions of the researchers, and them of machine learning and data mining recognized this as a key factor in data classification. Ensemble classification is a excellent method that used in machine learning and has demonstrated promising capabilities in improving classification accuracy. And Support vector machines ensemble has been proposed to improve classification performance recently. In this paper we used the fuzzy integral technique in SVM ensemble to evaluate the output of SVM in imbalanced data. And we compared this method with SVM, neural network and the LDA. The results indicate that the proposed method has better classification performance than others.

Keywords: SVM Ensemble, imbalanced data, fuzzy integral.

1   Introduction

Data imbalance is a special and famous known problem in data mining as everyone knows, that class imbalance exists in a large number of real-world, such as financial risks, medical diagnosis and text classification[1], and the researcher of machine learning and data mining recognized this as a key factor in data classification. A lot of work has been done to deal with the class imbalance problem in recent years[2] .

The unbalanced data problem is when the dataset is dominated by a major class or classes that have significantly more data than the rest. Because the data sets in real life usually have class imbalance problems, due to the fact that one class is represented by a much larger number of instances than other classes. [3]Consequently, algorithms tend to be overwhelmed by the large classes and ignore the small classes. For example, Figure 1 shows two graphs. Graph (a) shows how there are significantly have unbalanced in

---

* Corresponding author.

A. Moonis et al. (Eds.): IEA/AIE 2014, Part I, LNAI 8481, pp. 70–76, 2014.

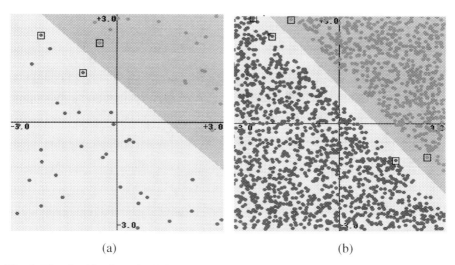

(a)   (b)

Fig. 1. The classification of unbalanced datasets, in the graph (a) the scale is 4:1 while in the graph(b) the scale is 25:1

datasets while Graph (b) is far easier to classify since there is a better balance between the two sets. This problem is considered one the "greatest challenges" within machine learning and since Support Vector Machines(SVM) has a lot of applications in real life, there is a high interest in solving this problem caused by imbalanced data sets. [4]

In recent years, a lot of research has shown that there are some algorithms that can handle largely unbalanced data sets. However each of them has their own drawbacks. In this paper we use the fuzzy integral and the support vector machines to deal with this problem. Section 2 introduces the basic theory of SVM classifier. In Section 3 we provided the fuzzy integral that used in this paper. The ensemble methods are presented in section 4. Section 5 presents experiment results applied to the actual problems . Finally, conclusions and future work are given in section 6.

2   A Brief Review on Support Vector Machines

SVMs provide a machine learning algorithm for classification Gene expression vectors can be thought of as points in an n-dimensional space. The SVM is then trained to discriminate between the data points for that pattern (positive points in the feature space) and other data points that do not show that pattern (negative points in the feature space). [5]Specifically, SVM chooses the hyperplane that provides maximum margin between the plane surface and the positive and negative points. The separating hyperplane is optimal in the sense that it maximizes the distance from the closest data points, which are the support vectors.

SVMs construct a classifier from a set of labeled pattern called training examples.Let $\{(x_i, y_i) \in R^d \times \{-1,1\}, i = 1,2,...l\}$ 1 be such a set of training examples. The SVMs try to find the optimal separating hyperplane $w^T x + b = 0$ that maximizes the margin of the nearest examples from two classes. To the nonlinear classification problem, the original data are projected into a high dimension feature space F via a nonlinear map $\Phi : R^d \rightarrow F$ so that the problem of nonlinear classification is transferred into that of linear classification in feature space F . By introducing the kernel function $K(x_i, x_j) = \Phi(x_i) \cdot \Phi(x_j)$, it is not necessary to explicitly know $\Phi(\cdot)$, and only the kernel function $K(x_i, x_j)$ is enough for training SVMs. The corresponding optimization problem of nonlinear classification can be obtained by

$$\min J(w, \xi) = \frac{1}{2} w^T w + C \sum_{i=1}^{l} \xi_i \tag{1}$$

s.t. $y_i(w^T \Phi(x) + b) \geq 1 - \xi_i, i = 1,2,...,l$

where C is a constant and ξ_i is the slack factor.

Eq. (1) can be solved by constructing a Lagrangian equality and transformed into the dual:

$$\max W(a) = \sum_{i=1}^{l} \alpha_i - \frac{1}{2} \sum_{i,j=1}^{l} \alpha_i \alpha_j y_i y_j K(x_i, x_j) \tag{2}$$

s.t. $\sum_{i=1}^{l} \alpha_i y_i = 0, 0 \leq \alpha_i \leq C, i = 1,2,...,l$

By solving the above problem (2), we can get the optimal hyperplane

$$\sum_i \alpha_i y_i K(x, x_i) + b = 0 \tag{3}$$

Then we can get the decision function of the form:

$$f(x) = sgn(\sum_i \alpha_i y_i K(x, x_i) + b) \tag{4}$$

In equation (4) the function $k(x_i, x)$, known as the kernel function. Kernel functions enable dot product to be performed in high-dimensional feature space using low dimensional space data input without knowing the transformation Φ . All kernel functions must satisfy Mercer's condition that corresponds to the inner product of some feature space.

3   The Method of Fuzzy Integral in SVM Ensemble

In 1974, Sugeno introduced the concept of fuzzy measure and fuzzy integral, generalizing the usual definition of a measure by replacing the usual additive property with a weak requirement, i.e. [6]the monotonic property with respect to set inclusion. In this section, we give a brief to some notions about fuzzy integral that used in this paper.

In a fuzzy measure space(X, β, g), let h be a measurable set function defined in the fuzzy measurable space. Then the definition of the fuzzy integral of h over A with respect to g is

$$\int_A h(x)dg = \sup_{\alpha \in [0,1]} [\alpha \wedge g(A \cap H_\alpha) \tag{5}$$

where $H_\alpha = \{x | h(x) \geq \alpha \}$.$A$ is the domain of the fuzzy integral. When $A=X$, then A can be taken out.

Next, the fuzzy integral calculation is described in the following. For the sake of simplification, consider a fuzzy measure g of $(X, \aleph)$ where X is a finite set. Let $h : x \rightarrow [0,1]$ and assume without loss of generality that the function $h(x_j)$ is monotonically decreasing with respect to j, i.e., $h(x_1) \geq h(x_2) \geq \cdots \geq h(x_n)$. To achieve this, the elements in X can be renumbered. With this, we then have

$$\int h(x)dg = \bigvee_{i=1}^{n} [f(x_i) \wedge g(x_i)] \tag{6}$$

where $X_i = \{x_1, x_2, \cdots, x_i\}$, $i = 1, 2, \cdots, n$.

In practice, h is the evaluated performance on a particular criterion for the alternatives, and g represents the weight of each criterion. The fuzzy integral of h with respect to g gives the overall evaluation of the alternative. In addition, we can use the same fuzzy measure using Choquet's integral, defined as follows (Murofushi and Sugeno, 1991).

$$\int hdg = h(x_n)g(X_n) + [h(x_{n-1}) - h(x_n)]g(X_{n-1}) + \cdots + [h(x_1) - h(x_2)]g(X_1) \tag{7}$$

The fuzzy integral model can be used in a nonlinear situation since it does not need to assume the independence of each criterion.

4   The Method of Fuzzy Integral in SVM Ensemble

In order to resolve the unbalanced data, a support vector machines ensemble strategy based on fuzzy integral is proposed in this paper. Usually, this presented method consists of four phases. firstly, we use bagging technique to construct the component

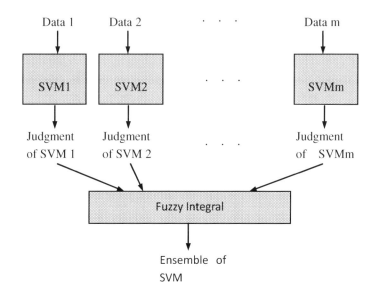

Fig. 2. The general architecture of an SVM ensemble based on fuzzy integral

SVM. In bagging methods, several SVMs are trained independently using training sets generated by a bootstrap method from the original training set. Furthermore, posterior class probabilities are required when using fuzzy integral to combine the component SVMs classification outputs for the overall decision. [7] So we obtain probabilistic outputs model of each component SVM in the second step. Thirdly, we assign the fuzzy densities, the degree of importance of each component SVM, based on how good these SVMs performed on their own training data. Finally, we aggregate the component predictions using fuzzy integral in which the relative importance of the different component SVM is also considered. [8] Fuzzy integral nonlinearly combines objective evidence, in the form of a SVM probabilistic output. Fig.2 shows the general architecture of an SVM ensemble based on fuzzy integral.

5   Numerical Experiments

In order to verify the method that we proposed, we select the sample from the real world. The sample includes 1124 data, and the major class has 980 data and the minor class has 144 data. And we used the bagging technique to construct the component SVM. This Experiments are implemented based on libsvm software. The Fig.3 is the parameter selection of support vector machine. And the final result is shown in Fig.4.

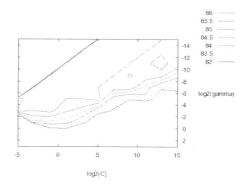

Fig. 3. Parameter selection of Support Vector Machine

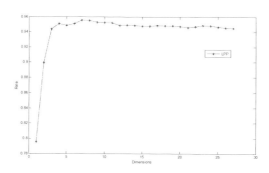

Fig. 4. Assessment results of ensemble svm

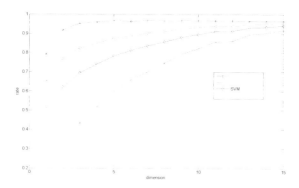

Fig. 5. Comparative of evaluation models

6   Conclusions

This paper proposes a support vector machines ensemble strategy based on fuzzy integral on classification of imbalanced data. The most important advantage of this approach is that not only are the classification results combined but that the relative importance of different class in imbalanced data is also considered. The simulating results show the effectiveness and efficiency of our method. Future research will be focus on how to set fuzzy densities more reasonably and how we can decreased the influence of the number of the different data in imbalance problems.

Acknowledgments. This work was partly supported by the NSF of PRC under Grant No. 70903016 and the national science & technology support program under Grant No. 2012BAH81F03.

References

1. Evgeniou, T., Pontil, M., Elisseeff, A.: Leave-one-out error, stability, and generalization of voting combinations of classifiers. Machine Learning (2002)
2. Cherkassky, V., Yunqian, M.: Practical selection of SVM parameters and noise estimation for SVM regression. Neural Networks 17(1), 113–126 (2004)
3. Le, X.H., Quenot, G., Castelli, E.: Speaker-Dependent Emotion Recognition for Audio Document Indexing. In: International Conference on Electronics, Information, and Communications, ICEIC 2004 (2004)
4. Sedaaghi, M.H., Kotropoulos, C., Ververidis, D.: Using Adaptive Genetic Algorithms to Improve Speech Emotion Recognition. In: IEEE 9th Workshop on Multimedia Signal Processing, MMSP 2007, pp. 461–464 (2007)
5. Joachims, T.: Making Large-Scale SVM Learning Practical. In: Schölkopf, B., Burges, C., Smola, A. (eds.) Advances in Kernel Methods - Support Vector Learning. MIT-Press, Cambridge (1999)
6. Danisman, T., Alpkocak, A.: Speech vs. Nonspeech Segmentation of Audio Signals Using Support Vector Machines. In: Signal Processing and Communication Applications Conference, Eskisehir, Turkey (2007)
7. Smola, A.J., Schölkopf, B.: A Tutorial on Support Vector Regression. NeuroCOLT, Technical Report NC-TR-98–030, Royal Holloway College, University of London, UK (1998)
8. Dehzangi, A., Phon Amnuaisuk, S., Ng, K.H., Mohandesi, E.: Protein Fold Prediction Problem Using Ensemble of Classifiers. In: Leung, C.S., Lee, M., Chan, J.H. (eds.) ICONIP 2009, Part II. LNCS, vol. 5864, pp. 503–511. Springer, Heidelberg (2009)

Distortion-Based Heuristic Sensitive Rule Hiding Method – The Greedy Way

Peng Cheng[1, 2], Shu-Chuan Chu[3], Chun-Wei Lin[2], and John F. Roddick[3]

[1] School of Computer and Information Science, Southwest University, Chongqing, P.R. China
[2] Harbin Institute of Technology Shenzhen Graduate School, Shenzhen, Guangdong, P.R. China
[3] School of Computer Science, Engineering and Mathematics,
Flinders University, Adelaide, South Australia, Australia
surf_mailbox@163.com

Abstract. Today, people can use various database techniques to discover useful knowledge from large collections of data. However, people also face the risk of disclosing sensitive information to competitor when the data is shared between different organizations. Thus, there is a balance between the legitimate mining need and protection of confidential knowledge when people release or share data. In this paper, we study the privacy preserving in association rule mining. A new distortion-based method was proposed which hides sensitive rules by removing some items in database so as to reduce the support or confidence of sensitive rules below specified thresholds. Aimed at minimizing side effects, the number of sensitive rules and the number of non-sensitive rules supported by each transaction are utilized to sort the transactions and the candidates which contain most sensitive rules and least non-sensitive rules are selected to modify. Comparative experiments on real datasets showed that the new method can achieve satisfactory results with fewer side effects and data loss.

Keywords: Association rule hiding, privacy preserving data mining, sensitive association rules, side effects.

1 Introduction

Nowadays, data often need to be shared among different organizations during business collaboration. People can utilize data mining techniques to extract useful knowledge from these large data collection. In particular, the discovery of association rules from large database can provide valuable information, such as customer purchasing patterns in supermarkets, fraudulent conduct detection in banking and so forth. However, it also might pose the threat of disclosing sensitive knowledge to other parties. In such circumstances it is necessary to ensure that sensitive knowledge behind database in the form of itemsets or association rules is not disclosed.

Hence, a problem arises how to balance the confidentiality of the disclosed data with the legitimate mining needs of the data users. To address this issue, the original database can be modified by adding new items or removing existing items so that the sensitive knowledge can't be mined out at some specified thresholds. However, such

A. Moonis et al. (Eds.): IEA/AIE 2014, Part I, LNAI 8481, pp. 77–86, 2014.

modification could lead to non-sensitive rules also to be lost or new ghost rules generated. So the challenge is how to protect the sensitive rules while the data utility of the database is maximized as much as possible. The transforming process from the original database into a released one to hide sensitive itemsets or rules is called association rules hiding/data sanitization.

By comparing the rules set mined from the original database and the rules set contained in the modified database, three criteria can be used to assess the side effects the hiding process brings. They are the number of the sensitive rules not hidden, the number of non-sensitive rules lost and the number of new ghost rules generated after sanitization. The ideal case is that all sensitive rules are hidden, and at the same time no non-sensitive rules from the original database are missing and no ghost rules are generated during the hiding process. However, in most real cases, it is difficult to achieve such an ideal goal without any side effects. It strongly depends on the dataset and sensitive rules user defined. Actually, there is a tradeoff between hiding sensitive ones and maintaining non-sensitive ones.

Some distortion-based methods have been proposed to solve the rule hiding or itemset hiding problems [7] [8]. Based on different heuristic information, these methods greedily chose transactions to modify so as to hide sensitive itemsets or rules. They tried to find an approximate solution close to the optimal one by minimizing the above three side effects. Atallah et al. [2] first proposed the protection algorithm for data sanitization to avoid the inference of association rules, and prove it as NP-hard problem. Under the assumption that sensitive rules are disjoint, Dasseni, Verikios and et al. [3] [1] first extended the itemset hiding to association rules and presented heuristic-based methods to hide sensitive rules by inserting or deleting items. Their methods use the size of transaction as the selection base and are not suitable to the database in which most transactions hold the same size. However, their methods run fast in most cases. Wu et al. [4] proposed a method aimed at avoiding all the side effects in rule hiding process instead of hiding all sensitive rules. Oliveira and Zaïane [5] introduced the multiple-rule hiding approach. One major novelty with their approach is to begin to take into account the impact on hiding legitimate non-sensitive patterns. They introduce three measures: hiding failures, missing costs, and artificial patterns (spurious patterns falsely generated). Amiri [6] proposed heuristic algorithms to hide itemset (not rules) by removing transactions or items, in terms of the sensitive and non-sensitive itemsets supported by each transaction. Borrowing the idea of TF-IDF (Term Frequency-Inverse Document Frequency) in the field of text mining, Hong et al. [9] devised a greedy-based hiding approach called SIF-IDF (Sensitive Items Frequency – Inverse Database Frequency). Each transaction in database is assigned a SIF-IDF value to evaluate the correlation degree of the transaction with the sensitive itemset.

In this paper, we proposed a new rule hiding approach. This approach hides rules by removing items in the identified transactions which support sensitive rules, so that sensitive rules can escape the mining in the modified database at some predefined thresholds, while the side effects and number of removed items are minimized. The modified transactions are selected based on the number of sensitive rules and the number of non-sensitive rules they support. The transactions which support more sensitive rules and support less non-sensitive rules are preferred, because the modification on them may hide sensitive ones effectively but impose fewer side effects on the non-sensitive ones. We empirically compared the proposed method with the SIF-IDF method [9] and demonstrated the new method can hide all sensitive rules with fewer side effects, and achieve such performance by modifying fewer items. In addition, the running time of the new method is not long.

2   The Problem Formulation

We first introduce some basic notions about frequent itemsets and association rules mining. Let $I = \{I_1, I_2, \ldots, I_m\}$ be a set of items available. An itemset X is a subset of I. A transaction t is characterized by an ordered pair, denoted as $t = <ID, X>$, where ID is a unique transaction identifier number and X represents a list of items making up the transaction. A transactional database D is a relation consisting of a set of transactions. For instance, in market basket data, a transactional database is composed of business transactions. Each transaction consists of items purchased in a store.

Absolute support of the itemset X is the number of transactions in D that contain X. Likewise, the relative support of X is the fraction (or percentage) of the transactions in database which contain itemset X, denoted as $Supp(X)$.

An itemset X is called frequent if $Supp(X)$ is at least equal to a minimum relative support threshold (denoted as MST) specified by user. The goal of frequent item set mining is to find all itemsets which are frequent with database D.

The notion of confidence is relevant to association rule. A rule has the form of $X \rightarrow Y$. It means that the condition X infers to the consequence Y. Here both X and Y are itemsets. $X \cap Y = \emptyset$. The confidence of a rule is computed as $Supp(X \cup Y)/Supp(X)$, and denoted as $Conf(X \rightarrow Y)$. It indicates a rule's reliability. Like MST, user also can define a minimum confidence threshold called MCT.

A rule $X \rightarrow Y$ is strong if it satisfies the following condition:

1) $Supp(X \cup Y) \geq MST$ and
2) $Conf(X \rightarrow Y) \geq MCT$.

Association rule mining usually includes two phases: (1) frequent itemsets are mined out with given MST. (2) Strong association rules are generated from the frequent itemsets obtained in phase 1 based on given MCT. In the following part of this paper, when the concept association rules are used, we refer to strong rules.

In order to ease the comprehensibility, we adopt the bit-vectors to express the transactional database, as indicated in Table 1. The database size in Table 1 is 6. The set of items available is $\{a, b, c, d, e, f, g\}$. Assume $MST = 50\%$ and $MCT = 0.6$. We may derive the following frequent itemsets and association rules using the improved Apriori algorithm [10-13], as indicated in Table 2.

The rules hiding problem can be formulated as follows.

Let D be a transaction database and R be the set of strong rules that can be mined from D with given MST, MCT. Let R_S denote a set of sensitive rules that need to be hidden, and $R_S \subset R$. R_N is the set of non-sensitive rules. $R_N \cup R_S = R$. The hiding problem is to transform D into a sanitized database D' such that only the patterns or rules which belong to R_N can be mined from D'. Let R' denote the strong rules mined from sanitized database D' with the same MST and MCT.

There are three possible side effects after transforming D into D'. Sensitive rules subset which is not hidden in modified database D' is called as S-N-H (Sensitive rules Not Hidden). S-N-H $= \{r \in R_s \mid r \in R'\}$. Some of non-sensitive rules are falsely hidden and lost in the modified database D', which is denoted as N-S-L (Non-Sensitive rules Lost). N-S-L $= \{r \in R_N \mid r \notin R'\}$. The rules falsely generated in sanitized database D' is marked as S-F-G (Spurious rules Falsely Generated). S-F-G $= \{r \in R' \mid r \notin R\}$.

Table 1. Conversion between transactional database and bit-vectors

TID	Trans
1	abcdefg
2	acefg
3	cdf
4	acfg
5	bcdfg
6	cdeg

⟺

TID	a	b	c	d	e	f	g
1	1	1	1	1	1	1	1
2	1	0	1	0	1	1	1
3	0	0	1	1	0	1	0
4	1	0	1	0	0	1	1
5	0	1	1	1	0	1	1
6	0	0	1	1	1	0	1

Table 2. Frequent itemsets and association rules generated from the Table 1. ($MST = 50\%$ and $MCT = 60\%$)

Frequent Itemset	Support
c, d	0.67
f, g	0.67
c, f	0.83
c, g	0.83

Rules	Confidence	Support
d→c	1	0.67
c→d	0.67	0.67
f→g	0.8	0.67
g→f	0.8	0.67
f→c	1	0.83
c→f	0.83	0.83
g→c	1	0.83
c→g	0.83	0.83

Assumption: a rule is considered to be hidden if its support is less than MST or its confidence is less than MCT. In other words, if a strong rule in D becomes not strong in D', we consider it to be hidden. The task of the rule hiding method is to transform D in some ways so that all sensitive rules of R_S become hidden in D'.

Table 3. Hiding sensitive rules by decreasing support and confidence

Before Hiding	After Hiding	Outcome
$Supp(r) \geq MST$ and $Conf(r) \geq MCT$ and r$\in R_S$	$Supp(r) < MST$ or $Conf(r) < MCT$	r is hidden

Table 3 indicates the means by which the sensitive rule can be hidden. The hiding process often may affect the non-sensitive rules in D or the pre-strong rules in D. The pre-strong rules refer to rules in D with support not less than MST and with confidence less than MCT. A pre-strong rule may become strong when its confidence is above MCT. A non-sensitive rule in D also may become not strong when its support is below MST or its confidence is below MCT in D' due to the item removing operation. Table 4 summarizes the possible side effects along with the hiding process

Table 4. Three possible side effects along with hiding process

Before Hiding	After Hiding	Side Effects
$Supp(r) \geq MST$ and $Conf(r) \geq MCT$ and $r \in R_S$	$Supp(r) \geq MST$ and $Conf(r) \geq MCT$	r is sensitive but not hidden
$Supp(r) \geq MST$ and $Conf(r) \geq MCT$ and $r \in R_N$	$Supp(r) < MST$ or $Conf(r) < MCT$	r is non-sensitive but falsely hidden
$Supp(r) < MST$ or $Conf(r) < MCT$ and r is not strong rules	$Supp(r) \geq MST$ and $Conf(r) \geq MCT$	r is a newly generated spurious rule

3   The Proposed Solution

3.1   The Hiding Process

The rule hiding process can be divided into two phases. In the first phase, an improved version of Apriori algorithm [12] [13] is used on the database to find all frequent patterns and corresponding association rules under the given threshold MST and MCT. The output of the first phase is a set of frequent item sets and association rules. Then user needs to select and specify the sensitive rules from the generated association rules. The definition of sensitive rules is based on the user's preference, policy enforcement or business benefit conflict. In the second phase, the hiding algorithm is performed on the original database and it identifies some candidate transactions to modify by removing items using heuristic information. The aim of the second phase is to reduce the support or confidence of sensitive rules below the minimum thresholds.

3.2   The Hiding Strategy

The basic rule hiding strategy is to modify the database by removing some items so that the support or confidence of all sensitive rules drop below the user-specified threshold: MST or MCT.

Based on the above strategy, the two further underlying problems need to be solved before the item removing operation.

1) Which transactions are identified to be modified?
2) Which item is selected to be removed in an identified transaction?

For the first problem, only the transactions which fully support sensitive rules need to be considered to modify, because the modification of other transactions bring no influences on the support of the generating itemset of sensitive rules. For the transactions which only support the antecedent part of the rule but not support the consequent part, if the removing operation is performed to reduce the support of the antecedent part, it will increase the confidence of the rule. This is opposite to the aim of rule hiding. For instance, we hope to hide the rule $A{\rightarrow}B$ by removing some items. If we select the transaction which contains the item A but not contains the item B, and remove the item A in this transaction, the support of item A will decreases but the support of the union $\{AB\}$ remains the same. So the confidence of the rule $A{\rightarrow}B$ is increased accordingly since $Conf(A{\rightarrow}B) = Supp(AB)/Supp(A)$. Therefore, what we are interested is the transactions which fully support one or more sensitive rules.

However, it is insufficient that the transactions which fully support any sensitive rule are filtered and randomly selected to modify, although the sensitive rules can be hidden by this way. The reason is that choosing different supporting transactions subset may lead to different side effects. So we need to find some measure to evaluate the relevance of different supporting transactions in order to select ones on which the modification can bring fewer side effects. A transaction can be evaluated according to the number of sensitive rules it support and the number of the non-sensitive rules it support. The transactions which contain more sensitive rules and less non-sensitive rules are preferred to be modified firstly. Assume $NUM_{sen}(t)$ is the number of sensitive rules supported by the current transaction t and $NUM_{non-sen}(t)$ is the number of non-sensitive rules supported by t. The relevance of the transaction t is calculated as

$$\text{Relevance}(t) = NUM_{sen}(t) / [1+NUM_{non-sen}(t)] \tag{1}$$

For each rule, all the supporting transactions are filtered out firstly, and then these supporting transactions are sorted by their relevance according to the formula (1). Then the foremost transactions which hold the highest relevance values are selected in turn to modify. The item corresponding to the consequent part of the sensitive rule and with highest support is selected to remove.

In order to minimize the damage to the database, we may calculate in advance how many transactions need to be modified at least in order to hide one sensitive rule. As indicated in Table 3, we may hide the sensitive rule by reducing its support below MST or its confidence below MCT. Thus, the following properties can be deduced.

Property 1. Let $\Sigma_{X \cup Y}$ be the set of all transactions which support sensitive rule $X{\rightarrow}Y$. In order to decrease the confidence of the rule below MCT, the minimal number of transactions which need to be modified in $\Sigma_{X \cup Y}$ is:

$$NUM_1 = \lceil (Supp(X \cup Y) - Supp(X) * MCT) * |D| \rceil + 1 \tag{2}$$

Proof. Removing one item from the transaction in $\Sigma_{X \cup Y}$ which corresponds to the consequent part will decrease the support of the rule $X{\rightarrow}Y$ by 1. Assume θ is the minimal number of transactions which need to be removed in $\Sigma_{X \cup Y}$ in order to reduce the confidence of the rule below MCT. Then we have:

$$(Supp(X \cup Y)*|D| - \theta) / (Supp(X)*|D|) < MCT$$
$$\rightarrow Supp(X \cup Y) *|D| - Supp(X)*|D|*MCT < \theta$$

Because θ is an integer and θ is the minimum number which is greater than $Supp(X \cup Y)*|D| - Supp(X)*|D|*MCT$, we can get:

$$\theta > Supp(X \cup Y)*|D| - Supp(X)*|D|*MCT$$
$$\rightarrow \quad \theta = \lceil (Supp(X \cup Y) - Supp(X)*MCT)*|D| \rceil + 1 \quad \square$$

Property 2. Let $\Sigma_{X \cup Y}$ be the set of all transactions which support sensitive rule $X \rightarrow Y$. In order to decrease the support of the generating itemset of $X \rightarrow Y$ below MST, the minimal number of transactions which need to be modified in $\Sigma_{X \cup Y}$ is:

$$NUM_2 = \lceil (Supp(X \cup Y) - MST)*|D| \rceil + 1 \tag{3}$$

Proof. Removing one item in a transaction belonging to $\Sigma_{X \cup Y}$ will decrease the support of the rule $X \rightarrow Y$ by 1. Assume θ is the minimal number of transactions which need to be removed in $\Sigma_{X \cup Y}$ in order to reduce the support of the rule below MST. Then we have:

$$(Supp(X \cup Y)*|D| - \theta) / |D| < MST \rightarrow \quad Supp(X \cup Y)*|D| - MST*|D| < \theta$$

Because θ is an integer and θ is the minimum number which is greater than $Supp(X \cup Y)*|D| - MST*|D|$, we can get:

$$\theta > Supp(X \cup Y)*|D| - MST*|D| \rightarrow \quad \theta = \lceil (Supp(X \cup Y) - MST)*|D| \rceil + 1 \quad \square$$

Based on Property 1 and Property 2, we can infer the minimum number of transactions to be modified to hide the sensitive rule is:

$Min\{ NUM_1, NUM_2 \}$
$= Min\{ \lceil (Supp(X \cup Y) - Supp(X)*MCT)*|D| \rceil + 1, \lceil (Supp(X \cup Y) - MST)*|D| \rceil + 1\}$ (4)

3.3   The Relevance-Sorting Algorithm

The transactions which support any sensitive rules are filtered out at the beginning. We denoted collection of all supporting transactions as Σ. In addition, in order to reduce the time for calculation of relevance value, the sensitive itemset list and non-sensitive itemset list supported by each transaction in Σ are determined in advance before the hiding procedure begins. In the hiding process, the supporting transactions for each sensitive rule are filtered out and sorted by their relevance values in descending order. The modification operations take place in the foremost supporting transactions with higher relevance values.

The algorithm flow is listed as below.

Step 1: Initialization

1) Use improved Apriori algorithm to discover all frequent itemsets and association rules contained in the original database D. User need specify the sensitive itemset R_S from R. $R = R_S \cup R_N$. (R is the set of all association rules in D, R_S is the set of sensitive itemsets, and R_N is the set of non-sensitive rules).
2) At the beginning, the sanitized database D' equals the original database D.
3) Filter out all transactions set Σ that support the sensitive rules.
4) **For each** sensitive transaction t in Σ

{
- Determine the sensitive rules list supported by t
- Determine non-sensitive rules list supported by t
  //These two lists are used to calculate relevance value of t
- Determine the relevance of the transaction t according to formula (1), i.e., Relevance(t) = $NUM_{sen}(t)$ / [$1+NUM_{non-sen}(t)$].
}

Step 2: Perform the hiding process.

For each sensitive rule r_i in R_S do
{
1) Filter out transactions which fully support r_i. We denote this transactions set as Σ_i. $\Sigma_i = \{t \in D \mid t$ fully support $r_i\}$
2) Sort the transactions in Σ_i by the relevance values in descending order.
3) Use the formula (2), formula (3) and formula (4) to calculate the minimum number of transactions which need modifications to hide the rule r_i, denoted as $N\_iterations$.
4) **For** i :=1 to $N\_iterations$ do
   {
   - Choose the transaction in Σ_i with the highest relevance value, i.e., $t = \Sigma_i[1]$.
   - Choose the item corresponding to the consequent of the rule r_i and with the highest support, i.e., $j=$ choose_item(r_i)
   - Remove the item j in the transaction t.
   - Update the support and confidence of r_i
   - Update the support and confidence of other rules which are originally supported by t and contain the item j.
   - $\Sigma_i = \Sigma_i - t$

   }
}

4   Performance Evaluations

We tested the proposed algorithm on three well-known real datasets: mushroom, BMS-WebView-1 and BMS-WebView-2. These datasets exhibit varying characteristics with respect to the number of transactions and items that they contain, as well as with respect to the average transaction length. They were summarized in Table 5. The experiments were carried out on an Intel Core(TM) i3 CPU with 2.53 GHz processor and with 2 GB of main memory. The proposed algorithm was implemented in C++.

The experiment results were measured according to three side effects and modification degree as following:

- $\alpha = |S\text{-}N\text{-}H|$: the number of sensitive rules not to be hidden.
- $\beta = |N\text{-}S\text{-}L|$: the number of non-sensitive lost rules.
- $\gamma = |S\text{-}F\text{-}G|$: the number of spurious rules newly generated.

- # of modified items: the number of items removed during the hiding process. It reflects the degree of data loss.

Table 6 shows the comparative experiments results. For each dataset, five rules were selected randomly, and the proposed method was compared with the SIF-IDF method [9] under various *MCTs*. Lower *MCT* values may produce more strong association rules. More strong rules increase the possibility of being affected by the hiding process and might result in more non-sensitive rules lost after sanitization. We can notice that both two algorithms can hide all sensitive rules in one time, i.e. $\alpha = 0$. On the values of β and γ, the relevance-sort algorithm can achieve better results (fewer side effects) apparently, especially on dataset mushroom and BMS-WebView-1. For the number of modified items, it is also clear that the proposed method can hide all sensitive rules with less modification on the datasets. It means less data loss.

Table 5. The characteristics of three datasets

Dataset	# of Trans.	# of Items	Average Trans. Len.
Mushroom	8124	119	23
BMS-WebView-1	59602	497	2.5
BMS-WebView-2	77512	3340	5.0

Table 6. Comparative results on real datasets

Dataset	MCT	|R|	Relevance-sorting		SIF-IDF	
			(α, β, γ)	# of modified items	(α, β, γ)	# of modified items
Mushroom (*MST*=0.05)	0.6	849	(0, 11, 1)	822	(0, 18, 0)	1398
	0.7	678	(0, 11, 0)	822	(0, 18, 0)	1398
	0.8	560	(0, 6, 0)	568	(0, 17, 0)	1398
	0.9	461	(0, 6, 0)	280	(0, 18, 0)	1398
BMS-1 (*MST*=0.001)	0.3	325	(0, 4, 0)	1034	(0, 13, 2)	2380
	0.4	131	(0, 1, 0)	766	(0, 7, 0)	2380
	0.5	34	(0, 0, 0)	497	(0, 6, 1)	2380
	0.6	11	(0, 0, 0)	226	(0, 1, 0)	2380
BMS-2 (*MST*=0.002)	0.3	482	(0, 6, 1)	493	(0, 12, 0)	526
	0.4	283	(0, 4, 1)	425	(0, 10, 0)	526
	0.5	112	(0, 3, 2)	311	(0, 7, 1)	526
	0.6	29	(0, 0, 0)	135	(0, 3, 0)	526

5   Conclusions

Privacy preserving in association rule mining is an important research topic in database security field. We proposed a new efficient method to solve the association rule hiding problem for data sharing. This method can hide all sensitive rules in one run and take short time. It is robust the database, parameter setting and sensitive rules user specifies. It hides sensitive rules by removing items in identified transactions. In order to hide each rule, the supporting transactions are sorted by their relevance degree to the sensitive rules and non-sensitive rules. The transactions which support more sensitive rules and less non-sensitive rules are selected firstly to be modified. The experiment result showed that the proposed method can hide all sensitive rules with fewer side effects and data loss compared with the method SIF-IDF. We will compare the new method with more existing distortion-based hiding methods and seek a way to improve its performance further.

References

1. Verykios, V.S., Elmagarmid, A.K., et al.: Association rule hiding. IEEE Transactions Knowledge and Data Engineering 16(4), 434–447 (2004)
2. Atallah, M.B.E., Elmagarmid, A., Ibrahim, M., Verykios, V.S.: Disclosure limitation of sensitive rules. In: Proceedings of IEEE Workshop on Knowledge and Data Engineering Exchange, Chicago, IL, pp. 45–52 (1999)
3. Dasseni, E., Verykios, V.S., Elmagarmid, A.K., Bertino, E.: Hiding association rules by using confidence and support. In: Proceedings of the 4th International Workshop on Information Hiding, pp. 369–383 (2001)
4. Wu, Y.H., Chiang, C.C., Chen, A.L.P.: Hiding sensitive association rules with limited side effects. IEEE Transactions Knowledge and Data Engineering 19(1), 29–42 (2007)
5. Oliveira, S.R.M., Zaïane, O.R.: Privacy preserving frequent itemset mining. In: Proceedings of IEEE International Conference on Privacy, Security and Data Mining, Australia, pp. 43–54 (2002)
6. Amiri, A.: Dare to share: Protecting sensitive knowledge with data sanitization. Decision Support Systems 43(1), 181–191 (2007)
7. Verykios, V.S.: Association rule hiding methods. Wiley Interdisc. Rew.: Data Mining and Knowledge Discovery 3(1), 28–36 (2013)
8. Sathiyapriya, K., Sadasivam, G.S.: A Survey on privacy preserving association rule mining. International Journal of Data Mining & Knowledge Management Process (IJDKP) 3(2) (March 2013)
9. Hong, T.P., Lin, C.W., Yang, K.T., Wang, S.L.: Using TF-IDF to hide sensitive itemsets. Applied Intelligence 38(4), 502–510 (2013)
10. Agrawal, R., Imielinski, T., Sawmi, A.: Mining association rules between sets of items in large databases. In: Proceedings of the ACM SIGMOD Conference on Management of Data (SIGMOD), pp. 207–216 (1993)
11. Agrawal, R., Srikant, R.: Fast algorithm for mining association rules. In: Proceedings of the International Conference on Very Large Data Bases (VLDB), pp. 487–499 (1994)
12. Bodon, F.: Surprising results of trie-based FIM algorithms. In: Proceedings of IEEE ICDM Workshop on Frequent Itemset Mining Implementations (FIMI 2004), Brighton, UK (2004)
13. Bodon, F.: A fast APRIORI implementation. In: IEEE ICDM Workshop on Frequent Itemset Mining Implementations (FIMI 2003), Melbourne, Florida, USA (2003)

Parallelized Bat Algorithm with a Communication Strategy

Cheng-Fu Tsai, Thi-Kien Dao, Wei-Jie Yang, Trong-The Nguyen, and Tien-Szu Pan

Department of Electronics Engineering,
National Kaohsiung University of Applied Sciences, Taiwan
jvnkien@gmail.com

Abstract. The trend in parallel processing is an essential requirement for optimum computations in modern equipment. In this paper, a communication strategy for the parallelized Bat Algorithm optimization is proposed for solving numerical optimization problems. The population bats are split into several independent groups based on the original structure of the Bat Algorithm (BA), and the proposed communication strategy provides the information flow for the bats to communicate in different groups. Four benchmark functions are used to test the behavior of convergence, the accuracy, and the speed of the proposed method. According to the experimental result, the proposed communicational strategy increases the accuracy of the BA on finding the near best solution.

Keywords: Bat algorithm; swarm intelligence; numerical optimization.

1   Introduction

Swarm intelligence has been paid more attention from researchers, who work in the related field. Many swarm intelligence based algorithms have been developed and been successfully used to solve optimization problems in the engineering, the financial, and the management fields for recently years. For instance particle swarm optimization (PSO) techniques have successfully been used to forecast the exchange rates, the optimizing, [1-3], to construct the portfolios of stock, human perception [4-6], ant colony optimization (ACO) techniques have successfully been used to solve the   routing problem of networks, the secure watermarking [7, 8], artificial bee colony (ABC) techniques have successfully been used to solve the lot-streaming flow shop scheduling problem [9], cat swarm optimization (CSO) [10] techniques have successfully been used to discover proper positions for information hiding [11].

Based on the algorithms in swarm intelligence, the idea of parallelizing the artificial agents by dividing them into independent subpopulations is introduced into the existing methods such as ant colony system with communication strategies [12], parallel particle swarm optimization algorithm with communication strategies [13], parallel cat swarm optimization [14], Island-model genetic algorithm [15], and parallel genetic algorithm [16]. The parallelized subpopulation of artificial agents increases the accuracy and extends the global search capacity than the original structure.

A. Moonis et al. (Eds.): IEA/AIE 2014, Part I, LNAI 8481, pp. 87–95, 2014.
© Springer International Publishing Switzerland 2014

The parallelization strategies simply share the computation load over several processors. The sum of the computation time for all processors can be reduced compared with the single processor works on the same optimum problem. In this paper, the concept of parallel processing is applied to Bat algorithm and a communication strategy for parallel BA is proposed.

The rest of this paper is organized as follows: a briefly   review of BA is given in session 2; our analysis and designs   for the parallel BA is presented in session 3; a series of experimental   results and the comparison between original BA and parallel BA are discussed   in session 4;   finally, the conclusion is summarized in session 5.

2   Metaheuristic Bat-Inspired Algorithm

In 2010, Xin-SheYang proposed a new optimization algorithm,   namely, Bat Algorithm or original Bat Algorithm (oBA), based on swarm intelligence   and the inspiration from observing the bats [17] . Original BA simulates parts of the echolocation characteristics of the micro-bat in the simplicity way.   It is potentially more powerful than particle swarm optimization and genetic algorithms as well as Harmony Search. The primary reason is that BA uses a good combination of major advantages of these algorithms in some way. Moreover, PSO and harmony search are the special cases of the Bat Algorithm under appropriate simplifications. Three major characteristics of the micro-bat are employed to construct the basic structure of BA. The used approximate and the idealized rules in Xin-SheYang's method are listed as follows:

All bats utilize the echolocation to detect their prey, but not all species of the bat do the same thing. However, the micro-bat, one of species of the bat is a famous example of extensively using the echolocation.   Hence, the first characteristic is the echolocation behavior. The second characteristic is the frequency that   the micro-bat sends a fixed frequency $fmin$ with a variable wavelength λ and the loudness A_0 to search for prey.

1. Bats fly randomly with velocity v_i at position x_i . They can adjust the wavelength (or frequency) of their emitted pulses and adjust the rate of pulse emission r ∈ [0, 1], depending on the proximity of their target;
2. There are many ways to adjust the loudness.   For simplicity, the loudness is assumed to be varied from a positive large A_0 to a minimum constant value, which is denoted by Amin.

In Yang's method, the movement of the virtual bat is simulated by equation (1) – equation (3):

$$f_i = f_{min} + (f_{max} - f_{min}) * \beta \tag{1}$$

$$v_i^t = v_i^{t-1} + (x_i^{t-1} - x_{best}) * f_i \tag{2}$$

$$x_i^t = x_i^{t-1} + v_i^t \tag{3}$$

where f is the frequency used by the bat seeking for its prey, f_{min} and f_{max},   represent the minimum   and maximum value, respectively. x_i denotes the location of the ith bat in the solution space, v_i represents the velocity of the bat, t indicates the current iteration, β is a random vector, which is drawn from a uniform distribution, and $\beta \in [0, 1]$, and x_{best} indicates the global near best solution found so far over the whole population.

In addition, the rate of the pulse emission from the bat is also taken to be one of the roles in the process. The micro-bat emits the echo and adjusts the wavelength depending on the proximity of their target. The pulse emission rate is denoted by the symbol r_i, and $r_i \in [0, 1]$, where the suffix i indicates the ith bat. In every iteration, a random number is generated and is compared with r_i. If the random number is greater than r_i, a local search strategy, namely, random walk, is detonated. A new solution for the bat is generated by equation (4):

$$x_{new} = x_{old} + \varepsilon A^t \tag{4}$$

where ε is a random number and $\varepsilon \in [-1, 1]$, and at represents the average loudness of all bats at the current time step. After updating the positions of the bats, the loudness A_i and the pulse emission rate r_i are also updated only when the global near best solution is updated and the random generated number is smaller than A_i. The update of A_i and r_i are operated by equation (5) and equation (6):

$$A_i^{t+1} = \alpha A_i^t \tag{5}$$

$$r_i^{t+1} = r_i^0[1 - e^{-\gamma t}] \tag{6}$$

where α and γ are constants. In Yang's experiments, $\alpha = \gamma = 0.9$ is used for the simplicity.

The process of oBA is depicted as follows:

Step 1. Initialize the bat population, the pulse rates, the loudness, and define the pulse frequency

Step 2. Update the velocities to update the location of the bats, and decide whether detonate the random walk process.

Step 3. Rank the bats according to their fitness value, find the current near best solution found so far, and then update the loudness and the emission rate.

Step 4. Check the termination condition to decide whether go back to step 2 or end the process and output the result.

3   Parallelized Bat Algorithm with a Communication Strategy

Several groups in a parallel structure are created from dividing the population into subpopulations to construct the parallel processing as having been presented in some previous methods, such as parallel Cat swarm optimization [14], parallel Particle swarm optimization algorithm with communication strategies [13], parallel Genetic algorithm [16], Island-model genetic algorithm [15], and Ant colony system with communication strategies [12]. Each of the subpopulations evolves independently in regular iterations. They only exchange information between subpopulations when the communication strategy is triggered. It results in the reducing of the population size for each subpopulation and the benefit of cooperation is achieved.

The parallelized BA is designed based on original BA optimization. The swarm of bats in BA is divided into G subgroups. Each subgroup evolves by BA optimization independently, i.e. the subgroup has its own bats and near best solution. The bats in one subgroup don't know the existence of other subgroups in the solution space. The total iteration contains R times of communication, where $R = \{R_1, 2R_1, 3R_1, ...\}$. Let g_p be the subgroup, where $g \in G$ and p is the index of the subgroup. If $t \cap R \neq \varphi$, k agents with the top k fitness in g_p will be copied to $g_{(p+1) \bmod G}$ to replace the same number of agents with the worst fitness, where t denotes the current iteration count, $p = 1, 2, 3, ..., G$ and k is a predefined constant. The diagram of the parallelized BA with communication strategy is shown in figure 1.

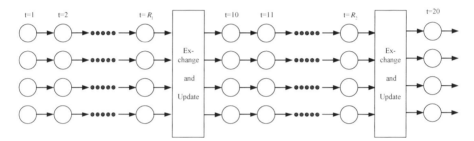

Fig. 1. The diagram of parallel BA with a communication strategy

1. **Initialization:** Generate bat population and divide them into G subgroups. Each subgroup is initialized by BA independently. Defined the iteration set R for executing the communication strategy. The N_j bats X_{ij}^T for the jth group, $i = 0, 1, ..., N_j$- 1, j = 0, 1, ..., S -1, where S is the number of groups, N_j is the subpopulation size for the jth group and t is the iteration number. Set $t = 1$.
2. **Evaluation:** Evaluate the value of $f(X_{ij}^T)$ for every bat in each group.
3. **Update:** Update the velocity and bat positions using Eqs. (1), (2) and (3).
4. **Communication Strategy:** Migrate the best bat among all the bats G^t to each group, mutate G^t to replace the poorer bats in each group and update G_j^t with G^t for each group every R_1 iterations.
5. **Termination:** Repeat step 2 to step 5 until the predefined value of the function is achieved or the maximum number of iterations has been reached. Record the best value of the function $f(G^t)$ and the best bat position among all the bats G^t.

4   Experimental Results

This section presents simulation results and compares the parallel BA with the original BA, both in terms of solution quality and in the number of function evaluations taken. Four benchmark functions are used to test the accuracy and the convergence of parallel BA. All the benchmark functions for the experiments are averaged over different random seeds with 25 runs. Let $X = \{x_1, x_2, ..., x_n\}$ be an n-dimensional real-value vector, the benchmark functions are listed in equation (4) to equation (6). The goal of the optimization is to minimize the outcome for all benchmarks. The population size is set to 30 for all the algorithms in the experiments. The detail of parameter settings of BA can be found in [17].

$$f_1(x) = \sum_{i=1}^{N}[10 + x_i^2 - 10cos2\pi x_i \tag{7}$$

$$f_2(x) = 1 + \sum_{i=1}^{N}\frac{x_i^2}{4000} + \prod_{i=1}^{N}cos\frac{x_i}{\sqrt{i}} \tag{8}$$

$$f_3(x) = 20 + e - 20e^{-0.2\sqrt{\frac{\sum_{i=1}^{n}x_i^2}{n}}} - e^{\frac{\sum_{j=1}^{n}cos(2\pi x_i)}{n}} \tag{9}$$

$$f_4(x) = \sum_{i=1}^{N}x_i^2 \tag{10}$$

The initial range and the total iteration number for all test functions are listed in Table I.

Table 1. The initial range and the total iteration of test standard functions

Function	Initial range $[x_{min}, x_{max}]$	Total iteration
$f_1(x)$	[-5.12,5.12]	4000
$f_2(x)$	[-100,100]	4000
$f_3(x)$	[-50,50]	4000
$f_4(x)$	[-100,100]	4000

The parameters setting for both parallel BA and original BA: are the initial loudness $A_i^0 = 0.25$, pulse rate $r_i^0 = 0.5$ the total population size $n = 30$ and the dimension of the solution space $M = 30$, frequency minimum $f_{min} = $ *the lowest of initial range function* and frequency minimum $f_{max} = $ *the highest of initial range function*. Each function contains the full iterations of 4000 is repeated by different random seeds with 25 runs. The final result is obtained by taking the average of the outcomes from all runs. The results are compared with the original BA.

Comparison Optimizing Performance Algorithms: Table II compares the quality of performance and time running for numerical problem optimization between parallel Bat algorithm and original Bat algorithm. It is clearly seen that, almost these cases of testing benchmark functions for parallel BA are faster than original BA in convergence. It is special case with test function $f_1(x)$, the Rastrigin has the mean of value function minimum of total 25 seed runs is 207.0142 with average time running equal 6.8426 seconds for parallel Bat algorithm evaluation. However, for original Bat algorithm this value of function minimum of total 25 seed runs is 230.2515 with time running equal 6.2785seconds in same executing computer. The average of four benchmark functions evaluation of minimum function 25 seed runs is 4.88E+04with average time consuming 24.6783 for original BA and 4.55E+04 with average time consuming 64.767 for parallel BA respectively.

Table 2. The comparison between oBA and cBA in terms of quality performance evaluation and speed

Function	Performance evaluation		Time running evaluation (seconds)	
	Original BA	*Parallel BA*	*Original BA*	*Parallel BA*
$f_1(x)$	230.2515	207.0142	6.2785	6.8426
$f_2(x)$	4.45E+00	3.94E+00	7.0421	45.0864
$f_3(x)$	19.9624	19.9531	6.2729	7.2432
$f_4(x)$	4.86E+04	4.53E+04	5.0848	5.5948
Average value	**4.88E+04**	**4.55E+04**	**24.6783**	**64.767**

Figure number from 2 to 5 show the experimental results of four benchmark functions in 25 seed runs output with the same iteration of 4000.

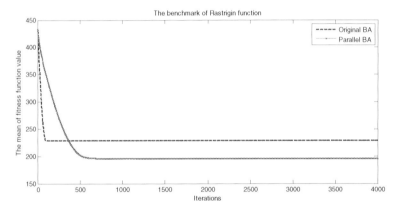

Fig. 2. The experimental results of Rastrigin function

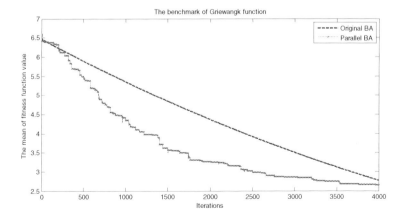

Fig. 3. The experimental results of Griewank function

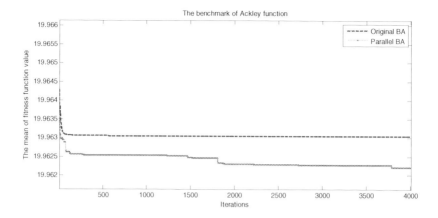

Fig. 4. The experimental results of Ackley function

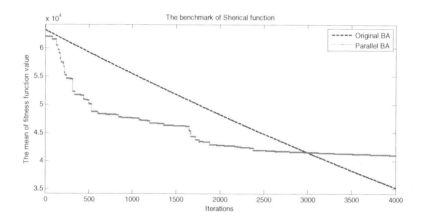

Fig. 5. The experimental results of Spherical function

According to the experimental result, parallel BA improves the convergence and accuracy of finding the near best solution about 8.3% than original BA.

5   Conclusion

In this paper, the parallelized Bat Algorithm (BA) optimization with a communication strategy is proposed for solving numerical optimization problems. The population bats are split into several independent groups based on the original structure of the BA, and the proposed communication strategy provides the information flow for the bats to communicate in different groups. In new proposed algorithm, the individual swarm poorer in among of subgroup of Bat algorithm is replaced with new individual swarm better from neighbor subgroups after each R_i iteration running. This feature is important for application problems characterized by parallel processing devices. The results of proposed algorithm on a set of various benchmark problems show that the proposed communicational strategy can be more convergence and increases the accuracy of the BA on finding the near best solution about 8.3%.

References

1. Chen, S.-M., Chien, C.-Y.: Solving the traveling salesman problem based on the genetic simulated annealing ant colony system with particle swarm optimization techniques. Expert Systems with Applications 38(12), 14439–14450 (2011)
2. Hsu, C.-H., Shyr, W.-J., Kuo, K.-H.: Optimizing Multiple Interference Cancellations of Linear Phase Array Based on Particle Swarm Optimization. Journal of Information Hiding and Multimedia Signal Processing 1(4), 292–300 (2010)
3. Chen, S.-M., Kao, P.-Y.: TAIEX forecasting based on fuzzy time series, particle swarm optimization techniques and support vector machines. Information Sciences 247, 62–71 (2013)
4. Jui-Fang, C., Shu-Wei, H.: The Construction of Stock_s Portfolios by Using Particle Swarm Optimization. In: Second International Conference on Innovative Computing, Information and Control, ICICIC 2007 (2007)
5. Parag Puranik, P.B., Abraham, A., Palsodkar, P., Deshmukh, A.: Human Perception-based Color Image Segmentation Using Comprehensive Learning Particle Swarm Optimization. Journal of Information Hiding and Multimedia Signal Processing 2(3), 227–235 (2011)
6. Ruiz-Torrubiano, R., Suarez, A.: Hybrid Approaches and Dimensionality Reduction for Portfolio Selection with Cardinality Constraints. IEEE Computational Intelligence Magazine 5(2), 92–107 (2010)
7. Pinto, P.C., et al.: Using a Local Discovery Ant Algorithm for Bayesian Network Structure Learning. IEEE Transactions on Evolutionary Computation 13(4), 767–779 (2009)
8. Khaled Loukhaoukha, J.-Y.C., Taieb, M.H.: Optimal Image Watermarking Algorithm Based on LWT-SVD via Multi-objective Ant Colony Optimization. Journal of Information Hiding and Multimedia Signal Processing 2(4), 303–319 (2011)
9. Pan, Q.-K., et al.: A discrete artificial bee colony algorithm for the lot-streaming flow shop scheduling problem. Inf. Sci. 181(12), 2455–2468 (2011)
10. Chu, S.-C., Tsai, P.-W.: Computational Intelligence Based on the Behavior of Cats. International Journal of Innovative Computing, Information and Control 3(1)(3), 8 (2006)
11. Wang, Z.-H., Chang, C.-C., Li, M.-C.: Optimizing least-significant-bit substitution using cat swarm optimization strategy. Inf. Sci. 192, 98–108 (2012)
12. Chu, S.-C., Roddick, J.F., Pan, J.-S.: Ant colony system with communication strategies. Information Sciences 167(1-4), 63–76 (2004)
13. Chang, J.F., Chu, S.C., Roddick, J.F., Pan, J.S.: A parallel particle swarm optimization algorithm with communication strategies. Journal of Information Science and Engineering 21(4), 9 (2005)
14. Pei-Wei, T., et al.: Parallel Cat Swarm Optimization. In: 2008 International Conference on Machine Learning and Cybernetics (2008)

15. Whitley, D., Rana, S., Heckendorn, R.B.: The Island Model Genetic Algorithm: On Separability, Population Size and Convergence. Journal of Computing and Information Technology 1305/1997, 6 (1998)
16. Abramson, D.A., Abela, J.: A parallel genetic algorithm for solving the school timetabling problem. In: Proc. of Appeared in 15 Australian Computer Science Conference, Hobart, Australia, p. 10 (1991)
17. Yang, X.-S.: A New Metaheuristic Bat-Inspired Algorithm. In: González, J.R., Pelta, D.A., Cruz, C., Terrazas, G., Krasnogor, N. (eds.) NICSO 2010. SCI, vol. 284, pp. 65–74. Springer, Heidelberg (2010)

Compact Artificial Bee Colony

Thi-Kien Dao[1], Shu-Chuan Chu[2], Trong-The Nguyen[1],
Chin-Shiuh Shieh[1], and Mong-Fong Horng[1]

[1] Department of Electronics Engineering,
National Kaohsiung University of Applied Sciences, Taiwan
jvnkien@gmail.com
[2] School of Computer Science, Engineering and Mathematics,
Flinders University, Australia

Abstract. Another version of Artificial Bee Colony (ABC) optimization algorithm, which is called the Compact Artificial Bee Colony (cABC) optimization, for numerical optimization problems, is proposed in this paper. Its aim is to address to the computational requirements of the hardware devices with limited resources such as memory size or low price. A probabilistic representation random of the collection behavior of social bee colony is inspired to employ for this proposed algorithm, in which the replaced population with the probability vector updated based on single competition. These lead to the entire algorithm functioning applying a modest memory usage. The simulations compare both algorithms in terms of solution quality, speed and saving memory. The results show that cABC can solve the optimization despite a modest memory usage as good performance as original ABC (oABC) displays with its complex population-based algorithm. It is used the same as what is needed for storing space with six solutions.

Keywords: Bee colony algorithm, Compact artificial bee colony algorithm, Optimizations, Swarm intelligence.

1 Introduction

Computational intelligence algorithms have also been successfully used to solve optimization problems in the engineering, the financial, and the management fields for recently years. For example, genetic algorithms (GA) have been successfully various applications including engineering, the financial, the security [1, 2], particle swarm optimization (PSO) techniques have successfully been used to forecast the exchange rates, the optimizing, [3, 4], to construct the portfolios of stock, human perception [2, 5], ant colony optimization (ACO) techniques have successfully been used to solve the routing problem of networks, the secure watermarking [6, 7], cat swarm optimization (CSO) [8] techniques have successfully been used to discover proper positions for information hiding [9]. Some applications require the solution of a complex optimization problem event though in limited hardware conditions. These conditions are to use a computational device due to cost and space limitations. For example, wireless

A. Moonis et al. (Eds.): IEA/AIE 2014, Part I, LNAI 8481, pp. 96–105, 2014.
© Springer International Publishing Switzerland 2014

sensor networks (WSN) are networks of small, battery-powered, memory-constraint devices named sensor nodes, which have the capability of wireless communication over a restricted area [10]. Due to memory and power constraints, they need to be well arranged to build a fully functional network. The other applications require a very fast solution of the optimization problem due to the communication time between a control/actuator devices and an external computer, real-time necessities within the control/actuator devices. For instance, in telecommunications[11] or in industrial plants for energy production[12]. The mentioned problem is not enough memory of computational devices to store a population composed of numerous candidate solutions of those computational intelligence algorithms.

Compact algorithms are a promise answer for this problem. An efficient compromise is used in compact algorithms to present some advantages of population-based algorithms but the memory is not required for storing an actual population of solutions.   Compact algorithms simulate the behavior of population-based algorithms by employing, instead of a population of solutions, its probabilistic representation. In this way, a much smaller number of parameters must be stored in the memory. Thus, a run of these algorithms requires much less capacious memory devices compared to their correspondent population-based structures.

The very first implementation of compact algorithms has been the compact Genetic Algorithm (cGA) [13]. The cGA simulates the behavior of a standard binary encoded Genetic Algorithm (GA). The compact Differential Evolution (cDE) algorithm has been introduced in [14]. The success of cDE implementation is the combination of two factors. The first is that a DE scheme seems to benefit from the introduction of a certain degree of randomization due to the probabilistic model. The compact Particle Swarm Optimization (cPSO) has been defined in [15]. The implementation of cPSO algorithm benefits from the same natural encoding of the selection scheme employed by DE and another "ingredient" of compact optimization.

In this paper, the behavior and the characteristic of the Bees are reviewed to improve the Artificial Bee Colony algorithms [16, 17] and to present the compact Artificial Bee Colony Algorithm (cABC) based on the framework of the original Artificial Bee Colony (oABC). According to the experimental results, our proposed cABC presents same result in finding original Artificial Bee Colony algorithm.

The rest of this paper is organized as follows: a briefly   review of ABC is given in session 2; our analysis and designs   for the cABC is presented in session 3; a series of experimental   results and the compare between oABC and cABC are discussed in session 4;   finally, the conclusion is summarized in session 5.

2   Related Works

A random walk is a mathematical formalization of a path that consists of a succession of random steps. This work is primarily inspired by the random walk model in [18, 19]. This model focused on building block for representing individual in warms. Compact algorithms is represented the population as probability distribution based on random steps over the set of solutions. The behavior of building blocks for solving to optimality could be simulated by the dynamics of the random walk model[19]. The Artificial   Bee Colony (ABC) algorithm was proposed Karaboga [16] in 2005, and on the performance of ABC was analyzed in 2008 [17] whose are based   on inspect-ing   the   behaviors   of   real   bees   on finding nectar and sharing the informa-tion of food sources to the bees in the nest. There is three kinds of bee was defined in ABC as being the artificial agent known as the employed bee, the onlooker, and the scout. Every kind of the bees plays different and important roles in the optimization process.

The process of ABC optimization is listed as follows:

Step 1. Initialization: Spray n_e percentage of the populations into the solution space randomly, and then calculate their fitness values, called the nectar amounts, where n_e represents the ratio of employed bees to the total population.

$$P_i = \frac{F(\theta_i)}{\sum_{k=1}^{S} F(\theta_k)} \tag{1}$$

Step 2. Move the Onlookers: Calculate the probability of selecting a food source by equation (1), where θ_i denotes   the   position   of   the ith employed   bee, $F(\theta_i)$ denotes   the fitness   function, S represents the number of employed bees, and P_i is the probability   of selecting the ith employed bee. The onlookers are moved by equation (2), where x_i denotes the position of the ith onlooker bee, t denotes the itera-tion number, θ is the randomly chosen employed bee, j represents the dimension of the solution and $\Phi(.)$ produces   a series of random variable in the range from -1 to 1.

$$x_{ij}(t+1) = \theta_{ij}(t) + \emptyset(\theta_{ij}(t) - \theta_{kj}(t)) \tag{2}$$

Step 3. Update the Best Food Source Found So Far: Memorize the best fitness value and the position, which are found by the bees.

Step 4. Move the   Scouts: If the   fitness   values   of   the employed   bees do not   be   improved   by   a   continuous predetermined   number   of   itera-tions, which is   called "Limit",   those   food   sources   are   abandoned,   and these employed bees become the scouts. The scouts are moved by equation (3), where r is a random number and r $\in$ range from 0 to 1.

$$\theta_{ij} = \theta_{jmin} + r \times (\theta_{jmax} - \theta_{jmin}) \tag{3}$$

Step 5. Termination Checking: Check if the amount of the iterations satisfies the ter-mination condition.   If the termination condition is satisfied, terminate the program and output the results; otherwise go back to Step 2.

3   Compact Artificial Bee Colony Algorithm

As mentioned above that compact algorithms process an actual population of solution as a virtual population.   This virtual population is encoded within a data structure, namely Perturbation Vector (PV) as probabilistic model of a population of solutions. The distribution of the individual in the hypothetical swarms must be described by a probability density function (PDF) [20] defined on the normalized interval is from -1 to +1. The distribution of the each Bee of swarms could be assumed as Gaussian PDF with mean μ and standard deviation σ [13]. A minimization problem is considered in an m-dimensional hyper-rectangle in Normalization of two truncated Gaussian curves (m is the number of parameters). Without loss of generality, the parameters assume to be normalized so that each search interval is rang from -1 to +1. Therefore PV is a vector of $m \times 2$ matrix specifying the two parameters of the PDF of each design variable being defined as:

$$PV^t = [\mu^t, \sigma^t] \qquad (4)$$

where μ and σ are mean and standard deviation values a Gaussian (PDF) truncated within the interval range from -1 to +1, respectively. The amplitude of the PDF is normalized in order to keep its area equal to 1. The apex t is time steps. The initialization of the virtual population is generated for each design variable i, $\mu_i^1 = 0$ and $\delta_i^1 = k$, where k is set as a large positive constant (e.g. $k = 10$). The PDF height normalization is obtained approximately sufficient in well the uniform distribution with a wide shape. The generating for a candidate solution x_i are produced from $PV(\mu_i, \delta_i)$. The value of mean μ and standard deviation δ in PV are associated equation of a truncated Gaussian PDF is described as following:

$$PDF\left(trucNormal(x)\right) = \frac{e^{\frac{(x-\mu_i)^2}{2\delta_i^2}}}{\delta_i(\mathrm{erf}\left(\frac{u_i+1}{\sqrt{2}\delta_i}\right) - \mathrm{erf}\left(\frac{u_i+1}{\sqrt{2}\delta_i}\right))} \qquad (5)$$

The PDF in formula (5) is then used to compute the corresponding Cumulative Distribution Function (CDF). The CDF is constructed by means of Chebyshev polynomials by following the procedure described in [21], the codomain of CDF is range from 0 to 1. The distribution function or cumulative distribution function (CDF) describes the probability that a real-valued random variable X with a given probability distribution will be found at a value less than or equal to x_i CDFs are also used to specify the distribution of multivariate random variables.

$$CDF = \int_0^1 PDF * dx \qquad (6)$$

The sampling of the design variable x_i from PV is performed by generating a random number rand $(0, 1)$ from a uniform distribution and then computing the inverse function of CDF in rand $(0, 1)$. The newly calculated value is x_i by the sampling mechanism as equation (7)

$$x_i = inverse(CDF) \qquad (7)$$

When the comparison between two design variables for individuals of the swarm (or better two individuals sampled from PV) is performed the winner solution biases the PV. Let us indicate with winner the vector that scores a better fitness value and with loser the individual losing the (fitness based) comparison. Regarding the mean values 1, the update rule for each of its elements is $\mu_i^t, \delta_i^t => \mu_i^{t+1}, \delta_i^{t+1}$.

$$\mu_i^{t+1} = \mu_i^t + \frac{1}{N_p}(winner_i - loser_i) \tag{8}$$

where N_p is virtual population size. Regarding δ values, the update rule of each element is given by:

$$\delta_i^{t+1} = \sqrt{(\delta_i^t)^2 + (\mu_i^t)^2 - (\mu_i^{t+1})^2 + \frac{1}{N_p}(winner_i^2 - loser_i^2)} \tag{9}$$

$$[winner, loser] = complete(x_{best}, x^{t+1}) \tag{10}$$

The construction of formulas (9) and (10) are persistent and non-persistent structures with tested results given in [22]. Similar to the binary cGA case, it was impossible to assess whether one or another elitist strategy was preferable. In elitist compact schemes, at each moment of the optimum performance is retained in a separate memory slot. If a new candidate solution is computed, the fitness based comparison between it and the elite is carried out. If the elite are a winner solution, it biases the *PV* as shown in formulas (9) and (10).

---

1) *Initialization probability vector $(PV(\mu, \delta))$*
   for $i=1:n$ **do** $\mu_i^t = 0$; $\delta_i^t = k = 10$;
2) *Initialization parameters:*   trial=0; limit=10;
   while *termination is not satisfied* **do**
3) Employed Bee Phase
   generate x^t from PV; Calculate $f(x^t)$;
   if $(f(x^t)<f(sol)$ then   sol=x^t; $f(sol)=f(x^t)$;
   else   trial=trial+1;   end if
   //*Update PV*
     $[winner, loser]=$**compete**$(x^{t+1}, best)$
     Equation (8), (9) and (1)
4) Onlooker bee phase
   $x^t=sol$;
   if (rand<Prob)
         for i=1:n do
         x^t (i)= x^t (i)+rand*($x^t(i)$-x(k)); with random $k=1,...,n$
   end if
     Calculate $f(x^t)$;   //Update local
     if $(f(x^t)<f(sol)$ then   sol=x^t; $f(sol)=f(x^t)$;
     else   trial=trial+1;   end if   //Update global
     if (f(sol)<fbest) then   best=sol;   fbest=f(sol);
5) Scout Bee Phase
   if (trial==limit) then
         x^t=generated from PV;   end if
   end while

Fig. 1. The pseudo code of compact Artificial Bee Colony algorithm

Figure 1 shows the pseudo code of algorithm working principles of cABC. The fitness value of the position x' is calculated and compared with *best* to determine a winner and a loser. Equation (9) and (10) are then applied to update the probability vector PV. If rand is smaller than *Prob* (probability equation (1) is calculated from employment bee phrase), x^t will be calculated by equation (2). Update local and update global are implemented in Onlooker bee phrase. If f(*sol*) < fbest the value of function is memorized the value of the global best is then updated.

4   Experimental Results

This section presents simulation results in running benchmark function tests and compares the cABC with the oABC, both in terms of solution quality and in the number of function evaluations taken. To evaluate the accuracy and the computational speed of the proposed cABC, four test standard functions are chosen to use in the experiments. All experiments are averaged over different random seeds with 10 runs. The used test standard functions are Rosenbrock, Griewank, Rastrigin and Sphere and are listed in equation (11) - equation (14).

$$f_1(x) = \sum_{i=1}^{n-1}(100(x_{i-1} - x_i^2)^2 + (1 - x_i)^2 \tag{11}$$

$$f_2(x) = 1 + \sum_{i=1}^{N} \frac{x_i^2}{4000} + \prod_{i=1}^{N} cos\frac{x_i}{\sqrt{i}} \tag{12}$$

$$f_3(x) = \sum_{i=1}^{N}[10 + x_i^2 - 10cos2\pi x_i] \tag{13}$$

$$f_4(x) = \sum_{i=1}^{N} x_i^2 \tag{14}$$

The initial range and the total iteration for all test functions are listed in Table 1.

Table 1. The initial range and the total iteration of test standard functions

Function	Initial range $[x_{min}, x_{max}]$	Total iteration
$f_1(x)$	[-30, 30]	1000
$f_2(x)$	[-100, 100]	1000
$f_3(x)$	[-5.12,5.12]	1000
$f_4(x)$	[-100, 100]	1000

The optimization goal for all of these test functions is to minimize the outcome. The parameters setting for both cABC and oABC: are the initial 'limit'=10 of food source the total population size $n = 20$ and the dimension of the solution space *dim = 10.* Each function contains the full iterations of 1000 is repeated by different random seeds with 10 runs. The final result is obtained by taking the average of the outcomes from all runs. The results are compared with the original ABC.

4.1   Comparison Optimizing Performance Algorithms

Table 2 compares the quality of performance and time running for numerical problem optimization between cABC and oABC. It is clearly seen that, the average cases of testing functions in compact Artificial Bee algorithm are faster convergence that original cases. The mean of four test functions evaluation of minimum function 10 seed runs is 6.78E+07 with average time consuming 1.174 s for oABC and 3.29E+07 with average time consuming 0.341 s for cABC respectively.

Table 2. The comparison between oABC and cABC in terms of quality performance evaluation and speed

Functions	Performance as mean of evaluation		Time running evaluation (seconds)	
	oABC	*cABC*	*oABC*	*cABC*
$f_1(x)$	1.7917e+008	1.6533e+007	0.7503	0.2118
$f_2(x)$	0.2387	0.3615	1.0895	0.2457
$f_3(x)$	307.8214	139.2473	0.6979	0.2075
$f_4(x)$	2.2582	6.6112	0.6241	0.1891
Average value	1.79E+08	1.65E+07	3.1618	0.8541

Figures 2 to 3 show the average of function minimum of four test functions in 10 seed of output with the same iteration of 1000.

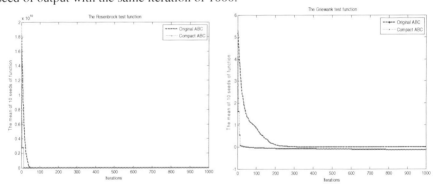

Fig. 2. The mean of 10 seeds function minimum curves in comparing cABC and oABC algorithms for the Rosenbrock and Griewank

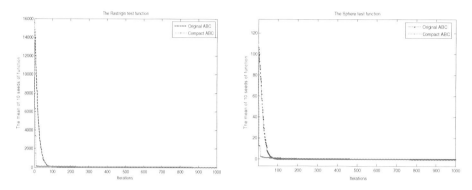

Fig. 3. The mean of 10 seeds function minimum curves in comparing cABC and oABC algorithms for the Rastrigin and Sphere

4.2   Comparison Saving-Memory and Time-Complexity Algorithms

Table 3 compares the saving-memory computations of two algorithms cABC and oABC.

Table 3. The saving-memory comparison between compact ABC and original ABC

Algorithms	Population size	Dimension	#Memory variable	# Equations	Computing complexity
Original ABC	N	D	$3{\times}N{\times}D$	*(1),(2), (3)*	$3{\times}T{\times}N{\times}D$ $\times iteration$
Compact BA	1	D	$6{\times}D$	*(1),(2), (3),(8, 9),(10)*	$6{\times}T{\times}D$ $\times iteration$

It can be clearly seen that the number memory variables of cABC is smaller than that of oABC in the same condition of computation such as iterations. The real numbers of population or population size and dimension space of food source of oABC are N and D, but that size for cABC is only one with dimension D. Even though, the number equations used for optimizing computation in cABC is six such as equations (1), (2), (3), (8), (9) and (10), and the number equations used for optimizing computation in oABC is only three of them such as equations (1), (2), and (3), the computing complexity of cABC is 6×T×D×iteration and it for oABC is 3×T×N×D×iteration. Thus, the rate of saving-memory equals the computing complexity of cABC per the computing complexity of oABC as given: rate = 2/N.

The computational times, for both the algorithms cABC and oABC, have been calculated by means of a PC Intel Core 2 Duo 2.4 GHz with 4 GB RAM employing in Windows7-OS, with Matlab (R2011b), version 7.13.0.564 32bits. Figure 4 illustrates the comparison of executing time between cABC and oABC in 10 seeds with iteration 1000 for four test functions. It is clearly seen that the most cases of test functions time executing in the proposed cABC (red colored bar) is smaller than that executing in oABC (blue colored bar).

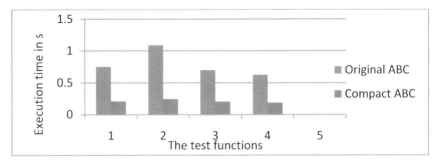

Fig. 4. Comparison two algorithms in term of time running for test functions

5   Conclusion

This paper, a novel proposed optimization algorithm is presented, namely compact Artificial Bee Colony algorithm (cABC). The implementation of compact for optimization algorithms could have important significance for the development of embedded devices small size, low price and being suitable with trend of ubiquitous computing today. In new proposed algorithm, the actual design variable of solutions search space of Artificial Bee Colony algorithm is replaced with a probabilistic representation of the population. This feature is important for application problems characterized by a limited memory since it allows the embedded implementation in small and cheap devices. The performance of cABC algorithm is as good as the other previous works in literature with respect to compact algorithms. The results of proposed algorithm on a set of various test problems show that cABC seems to be a valid alternative for optimization problems plagued by a limited memory. Experimental results on this real-world application also show the applicability of the proposed approach and highlight the good cABC performance within the category of memory-saving algorithms.

References

1. Wang, S., Yang, B., Niu, X.: A Secure Steganography Method based on Genetic Algorithm. Journal of Information Hiding and Multimedia Signal Processing 1(1), 8 (2010)
2. Ruiz-Torrubiano, R., Suarez, A.: Hybrid Approaches and Dimensionality Reduction for Portfolio Selection with Cardinality Constraints. IEEE Computational Intelligence Magazine 5(2), 92–107 (2010)
3. Chen, S.-M., Chien, C.-Y.: Solving the traveling salesman problem based on the genetic simulated annealing ant colony system with particle swarm optimization techniques. Expert Systems with Applications 38(12), 14439–14450 (2011)
4. Hsu, C.-H., Shyr, W.-J., Kuo, K.-H.: Optimizing Multiple Interference Cancellations of Linear Phase Array Based on Particle Swarm Optimization. Journal of Information Hiding and Multimedia Signal Processing 1(4), 292–300 (2010)
5. Parag Puranik, P.B., Abraham, A., Palsodkar, P., Deshmukh, A.: Human Perception-based Color Image Segmentation Using Comprehensive Learning Particle Swarm Optimization. Journal of Information Hiding and Multimedia Signal Processing 2(3), 227–235 (2011)

6. Pinto, P.C., et al.: Using a Local Discovery Ant Algorithm for Bayesian Network Structure Learning. IEEE Transactions on Evolutionary Computation 13(4), 767–779 (2009)
7. Khaled Loukhaoukha, J.-Y.C., Taieb, M.H.: Optimal Image Watermarking Algorithm Based on LWT-SVD via Multi-objective Ant Colony Optimization. Journal of Information Hiding and Multimedia Signal Processing 2(4), 303–319 (2011)
8. Chu, S.-C., Tsai, P.-W.: Computational Intelligence Based on the Behavior of Cats. International Journal of Innovative Computing, Information and Control 3(1)(3), 8 (2006)
9. Wang, Z.-H., Chang, C.-C., Li, M.-C.: Optimizing least-significant-bit substitution using cat swarm optimization strategy. Inf. Sci. 192, 98–108 (2012)
10. Akyildiz, I.F., et al.: A survey on sensor networks. IEEE Communications Magazine 40(8), 102–114 (2002)
11. Gene, C., Wai-tian, T., Yoshimura, T.: Real-time video transport optimization using streaming agent over 3G wireless networks. IEEE Transactions on Multimedia 7(4), 777–785 (2005)
12. Pourmousavi, S.A., et al.: Real-Time Energy Management of a Stand-Alone Hybrid Wind-Microturbine Energy System Using Particle Swarm Optimization. IEEE Transactions on Sustainable Energy 1(3), 193–201 (2010)
13. Harik, G.R., Lobo, F.G., Goldberg, D.E.: The compact genetic algorithm. IEEE Transactions on Evolutionary Computation 3(4), 287–297 (1999)
14. Mininno, E., et al.: Compact Differential Evolution. IEEE Transactions on Evolutionary Computation 15(1), 32–54 (2011)
15. Neri, F., Mininno, E., Iacca, G.: Compact Particle Swarm Optimization. Information Sciences 239, 96–121 (2013)
16. Hou, Y.T., et al.: On energy provisioning and relay node placement for wireless sensor networks. IEEE Transactions on Wireless Communications 4(5), 2579–2590 (2005)
17. El-Aaasser, M., Ashour, M.: Energy aware classification for wireless sensor networks routing. In: 2013 15th International Conference on Advanced Communication Technology, ICACT (2013)
18. Pearson, K.: The Problem of the Random Walk. Nature 72 (1905)
19. Pemantle, R.: A survey of random processes with reinforcement. Probability Surveys 4(2007), 9 (2007)
20. Billingsley, P.: Probability and Measure. John Wiley and Sons, New York (1979)
21. Cody, W.J.: Rational Chebyshev approximations for the error function. Mathematics of Computation 23(107), 631–637 (1969); 23(107), 6 (1965)
22. Mininno, E., Cupertino, F., Naso, D.: Real-Valued Compact Genetic Algorithms for Embedded Microcontroller Optimization. IEEE Transactions on Evolutionary Computation 12(2), 203–219 (2008)

A Genertic Algorithm Application on Wireless Sensor Networks

Haiyi Zhang and Yifei Jiang

Jodrey School of Computer Science, Acadia University
Wolfville, Canada
{haiyi.zhang,095997j}@acadiau.ca

Abstract. The Genetic Algorithm (GA) is performed in the Base Station (BS) to generate the optimal clusters and cluster heads for a given Wireless Sensor Networks. And then, Dijkstra's algorithm is used to generate the energy-efficient routing paths based on the optimal cluster heads produced by the GA. In addition, we use the concept of data aggregation to eliminate the redundant data. To demonstrate the feasibility of our approach, formal analysis and experimental results are presented.

Keywords: Genetic Algorithm, Dijkstra's algorithm, Greedy algorithm, Data Aggregation, Wireless Sensor Network.

1   Introduction

Using the GA in our proposed approach is used to create the approximately optimal clusters and cluster head (CH) for a given Wireless Sensor Network. Based on those optimal clusters and cluster heads (CHs), we utilize Dijkstra's algorithm to create a complete graph for the WSN, then form the single-source (the BS) shortest routing path, which is our expected energy-efficient routing path.

To the best of our knowledge, several attempts have been used by some researchers to propose many approaches to prolong nodes lifetime in [2] and to eliminate redundant sensory data in [4]. Recently, intelligent technology is widely used as a combined approach in WSNs in [6]. Moreover, there have also been plenty of attempts to reduce energy consumption based on clustering techniques in [5]. Currently, Using the GA, combined with a proper greedy algorithm is becoming the trend for designing an energy-efficient routing protocol in the research area of WSN. As for the work in [1], Muruganathan proposed another energy-efficient routing protocol named as Base-Station Controlled Dynamic Clustering Protocol (BCDCP). In BCDCP, the high-energy BS with a powerful processing ability is utilized to set up clusters and routing paths. The randomized CHs rotation is performed in BS as well. The core idea of BCDCP are the formation of clusters where every cluster has an approximately same number of member nodes to avoid CH overload, uniform placement the CHs through the whole WSN, and utilization of CH-to-CH routing pattern to transfer the gathered data to BS.

Hussain et al. [4] also proposed an energy-efficient routing protocol in 2007 that is based on BCDCP. One of the important contributions of their works is the

A. Moonis et al. (Eds.): IEA/AIE 2014, Part I, LNAI 8481, pp. 106–117, 2014.
© Springer International Publishing Switzerland 2014

introduction of GA used for cluster and CH creation. Their works are suitable for a large scale of WSN in industry. Although the protocol proposed in [5] shows a better experiment result based on the total data gathering rounds achieved and other aspects, it neither takes into account how to reduce the redundant data nor clearly explains when the member nodes should transmit their sensory data to the corresponding CH, when the new CHs should be re-generated and how to achieve the above procedures.

To deal with the aforementioned limitations in [4], we proposed a BSCICR protocol [8] by using an aggregation method to handle redundant data, and creating a scheduler to guide when the member and CHs nodes should gather and transfer data and when the new CHs should be reproduced. Moreover, we improved the fitness function of GA to produce better cluster and CHs, which is the foundation of generating the final energy-efficient routing path. We also introduced a repair function to consummate our GA. Furthermore, many optimizations related routing problems can be modeled by a Traveling Salesman Problem (TSP) for the WSN in a radio harsh environment. Compared with traditional TSP, the scalable and the dynamic topology of sensor networks may result in an incomplete graph. If there is an incomplete graph, it will turn out that the source nodes selected by the BS have no direct communication link with other selected source nodes. In order to allow the selected source nodes to communicate directly, generating a complete graph becomes a necessity. Therefore, our proposed approach utilizes Dijkstra's algorithm to achieve such conversion.

2   Intelligent Routing Approach

2.1   GA-Based BSCICR Operation Flow

In the BSCICR protocol [8], the sequence of the source nodes to be visited in the data transmission phase has a significant impact on the energy consumption. Therefore, to form an energy-efficient routing path, the BS utilizes the Dijkstra's algorithm to convert the network topology to a complete graph and create a single-source shortest path based on the optimal cluster-heads generated by the GA. Once the clusters, cluster-head nodes and routing path information have been identified, the BS forwards this information to all the source nodes. In the end, the source nodes accomplish the data transmission and aggregation phases based on their received packets. One thing to be highlighted is, during the clusters and cluster-heads formation phase, the BS also checks the Boolean flag LastRoundFlag within the sensory data packet gathered from the source nodes. If this Boolean flag is set to be true, the BS will utilize GA to update the cluster-heads and broadcast them to the source nodes. By doing this, it can guarantee that it always provides the optimal solution for the given WSN, which is extremely important for the WSNs used in a dynamic and radio harsh environment.

2.2   Efficient Routing Approach

Using GA to address energy-efficient routing, a BS requires knowing the whole topology of the network. It is not suitable to attempt to keep up-to-date topology information in a dynamic distributed environment as topology updates need to be broadcasted frequently [6]. These updated packets will consume a large portion of the network bandwidth and more energy. Moreover, our proposed technique is designed for a large-scale WSN and the CH for a cluster-based scheme needs to maintain the topology of the corresponding cluster to guide the data gathering, therefore, this large topology will increase the overload of each CH. For the above two reasons, in order to achieve the energy-efficient routing for a vast WSN used in a harsh environment, we need to address the two following issues.

(1) How to maintain a relatively small topology for each sub-cluster of a given WSN

(2) Due to the dynamic environment, how to minimize GA's computation interval and guarantee optimizations process not to converge quickly.

As for the first issue, it can be solved using the method discussed in the paper [8]. Regarding to the second issue, the scheduler, fitness parameter and fitness function of GA are defined to deal with it here.

Scheduler. Our proposed technique is designed for dynamic environment, which means lots of sensors may die by running out of their power, or many new sensors may join in to form a new cluster after several data gathering rounds. It turns out that the topology for the entire WSN is changed. In our approach, a GA is used for creating a globally optimal clusters and CHs as the initial inputs, which are utilized for generating the energy-efficient routing path (the single-source shortest path) according to Dijkstra's algorithm. Therefore, the changes for the topology of WSN will greatly influence the accuracy of the result of GA. According to the above reason, a scheduler is used in our proposed technique is to maintain relatively high accurate results of GA. Our scheduler generates schedules in the BS, where the topology for the entire WSN (the location information of all the source nodes) is available. In addition, the creation of the schedule packets is based on the cluster information (the formation of each sub-cluster) generated by the GA.

Genetic Algorithm. GA is one of the most powerful heuristics for solving optimization problems that is based on natural selection, the process that drives biological evolution [7]. With the constantly changed number of sensor devices for a WSN used in a radio harsh environment, the conventional search methods, such as calculus-based, enumerative and random methods, cannot meet our robustness requirement. The reasons proposed in [5] shows why the GA is different from the normal optimization and search procedures in four ways:

1. GA works with a coding of the parameter set, not the parameters themselves.
2. GA searches from a population of points, not a single point.
3. GA uses payoff (objective function) information, not derivatives or other auxiliary knowledge.
4. GA uses probabilistic transition rule, not deterministic rules.

In a word, the GA is a class of probabilistic optimization algorithms and is widely used in the applications where the accurate results are not very important and the search space is massive. The main advantage of a GA is that the process is completely automatic and avoids local minima. Therefore, we utilize a GA to generate an approximate optimal solution rather than using the conventional search methods mentioned above.

In a broad sense, a GA performs the fitness tests on the new structures to select the best population, where the fitness is used to determine the quality of the individual (chromosome) based on the defined criteria (fitness function). As in our proposed approach, a WSN is implemented as a population of chromosomes, where the candidate solutions (population) will be evaluated to obtain the better solutions for the further generations (populations). For example, as our proposed BSCICR protocol, a GA utilized in the BS, is performed on the given WSN to produce the optimal clusters and CHs as the best individual (the chromosome) by evaluating its fitness value. Then, a certain number of the best individuals from each generation are kept for a predefined number of data gathering round. These best individuals are also copied to the new generation by processing the following described operations, such as crossover, mutation and repair function, for further reproducing.

Before adopting GA to compute an approximate best solution, we have to consider encoding method, selection approach, fitness function, crossover and mutation operations, and the repair function. The main components and core operators of GA are as follows:

Chromosome. A chromosome is a collection of genes and represents a single cluster for a given WSN in our proposed technique. Each chromosome has the size of fixed length that is determined by the number of source nodes in the WSN. Here, the source nodes in the WSN are represented as the identification numbers (IDs) in each chromosome. For example, as every chromosome, it is the sequence of decimal numbers assigned by the BS. There are two main components consisted in each chromosome: gene index and gene value. A gene index represents the source node's identification number (ID). As for the gene value, it indicates the ID of the CH node for the corresponding cluster in the WSN. For the application of WSN, we choose the permutation method to encode a chromosome, since for our proposed approach, the cluster, in fact, is a data aggregation tree rooted at the CH node. It is an ordering problem considered the aforementioned scheduler method used to transmit the sensory data from all the member nodes to the CH node. In permutation encoding, every chromosome is a string of numbers that indicates the routing sequence, described as follows:

Chromosome A:

Gene index	3	1	4	6	2	5	7	0	8	9
Gene value	6									

Chromosome B:

Gene index	22	25	21	23	24	26	20	28	27	29
Gene value	26									

As for the above chromosome example, it represents a cluster with a fixed 10 source nodes. To be specific, for cluster A, it has the nodes from "0 to 9" (Gene index) with the node "6" (Gene value) as its CH. Similarly, as cluster B, it has the nodes from "20 to 29" (Gene index) with the node "2" (Gene value) as its CH. Here, the decimal number represents the corresponding ID for each source nodes (including the CH node) for their subordinate cluster. Although the gene index and gene value in the above example are represented as decimal format, they can also be denoted as binary format. For instance, as the gene index "26" in chromosome B can also be represented as "0001101". However, regardless of decimal or binary representation, for all genetic operations, the entire gene value is treated as atomic [10].**Selection.** The process of selection decides which chromosomes in the current population will mate (crossover) to produce a new chromosome. Then, these generated chromosomes will join the existing population. The combined population will also be the basis for the next selection. Here, the population is defined as a collection of chromosomes, where the size of the population, in fact, is the number of chromosomes, remains the same for all the generations. During the process of selection, a chromosome with a higher fitness value has a better chance of being selected. Several methods are used to select chromosomes to mate (crossover), such as: "Rank" selection, "Tournament" selection and "Roulette-Wheel" selection. According to Darwinian survival of the fittest principle, "Roulette-Wheel" selection, also called stochastic sampling with replacement, is a simple selection method of GA utilized in our proposed approach. As for this method, the individuals (chromosomes) are mapped to contiguous segments of a line, such that each individual's segment is equal in size to its fitness value. A random number is generated and the individual whose segment spans the random number is selected [15]. The process is repeated until the desired number of individuals is obtained, which is called mating population for crossover operation. This technique is analogous to a roulette wheel with each slice proportional in size to the fitness value, shown as the following table:

Table 1. Individual of Selection

Individual index	1	2	3	4	5	6	7	8	9	10
Fitness value	1.9	1.7	1.5	1.3	1.1	0.9	0.8	0.7	0.3	0.1
Probability of selection	0.17	0.15	0.14	0.12	0.10	0.08	0.06	0.05	0.02	0.01

Table1 shows the probability of selection for 10 individuals, linear ranking together with the fitness value. Individual 1 is the best fit individual and hence occupies the largest interval, whereas individual 10 as the least fit individual has the smallest interval on the line. For selecting the mating (crossover) population, the appropriate number of uniformly random numbers (uniform distributed between 0.0 and 1.0) is independently generated sample of 6 random numbers, for example: 0.75, 0.30, 0.90, 0.02, 0.61, and 0.41 as trail 1 to 6. Then the expression of "Roulette-Wheel" selection will be shown as below:

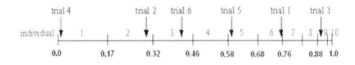

Fig. 1. "Roulette-Wheel" selection

After this selection of mating (crossover) population, the individuals (chromosomes) will consist: 1, 2, 3, 5, 6, and 9, these generated chromosomes will join the existing population and will be the basis for the next selection.

Fitness Parameters. In the GA, we use a defined function, called fitness function, to solve a given problem by evaluating the fitness of a chromosome, where a chromosome with a higher value has the better chance of survival. GAs solves design problems similar to that of natural solutions for biological design problems [11]. To be specific, in our proposed approach, the fitness of the chromosome is evaluated by the fitness function for every generation and each fitness parameter in that function is updated accordingly. In addition, the chromosome with a higher fitness function value will have a better chance to be selected for the selection operation. Moreover, that chromosome will also have the better chance to be selected as the best solution generated from the GA.

The fitness parameters listed below are designed to define the fitness function used to generate the optimal solutions of GA, which are the clusters and CHs in our proposed BSCICR protocol. Most important, these fitness parameters are the critical guidelines to minimize the energy consumption during the data transmission phase and prolong the network lifetime.

(i) The spatial transmission distance within the single cluster (*CD*): it denotes as the sum of the spatial transmission distances from all the non-CH nodes to the CH node. Since the source nodes may not be uniformly placed in the sensing field, the spatial transmission distance from each non-CH node to CH node is different. For this reason, we need to maintain a reasonable sum of the distances from all the non-CH nodes to the CH node. Recall the aforementioned network model, all the source nodes are stationary devices that are spatially deployed. Therefore, we know their exact positions, which are denoted as coordinates in our proposed approach. As described below, for one of the clusters in a given WSN, with the number of j source nodes (indexed from i to j), and the coordinates (x_{ch}, y_{ch}) and (x_{non-ch}, y_{non-ch}) for one of the CH node and non-CH node, respectively, the spatial transmission distance CD is:

$$CD = \sum_{i=1}^{j-1} \sqrt{(x_{ch} - x_{non-ch})^2 + (y_{ch} - y_{non-ch})^2}$$

(1)

(ii) The spatial transmission distance between a CH and the BS (*D*): Using the same principle in 1, since the BS is also a stationary device that is spatially deployed. Therefore, if the coordinates of CH and BS are (x_{ch}, y_{ch}) and (x_{bs}, y_{bs}), respectively. The spatial transmission distance between a CH and BS can be represented as the following equation.

$$D = \sqrt{(x_{bs} - x_{ch})^2 + (y_{bs} - y_{ch})^2}$$

(2)

The data transmission on the CH to BS is the core part for our cluster based routing protocol. In addition, according to the radio model discussed above, the energy consumed on transmit amplifier of a sensor node is proportional to the d^2, where d is the long range transmission happened on CH to BS or CH to CH. Therefore, for energy-efficient purpose, the value of D should be small.

(iii) The average spatial transmission distances between CHs (C_H): In our proposed technique, through executing the Dijkstra's algorithm in the BS, a shortest routing path is generated. The routing path here, in fact, is a routing sequence list, which contains the multi-hop transmission path among the CHs. After a certain number of data gathering rounds, a few new CHs will be generated by the GA. It turns out that the distances between CHs are changed. Therefore, take into account the radio model mentioned before, the average distance C_H should also be set as a small value to reduce the energy cost on routing the gathered data between CHs. With the same principle in 1 and 2, we can calculate the C_H by the equation as below:

$$C_H = \frac{\sum_{i=1}^{n} \sqrt{(x_n - x_i)^2 + (y_n - y_i)^2}}{n(n-1)/2} \tag{3}$$

where the (x_n, y_n) and (x_i, y_i) represent as the coordinates of two different CHs. As for the n ($n \geq 1$), it indicates the number of CH. In our knowledge, for a complete graph with n vertices, there are at most $n(n-1)/2$ edges.

(iv) Energy consumption (E_c): it represents the energy consumed on each cluster of the given WSN, which includes the energy expenditure on data gathering, data aggregation and data routing. For example, for a given cluster c with the number of j source nodes (indexed from i to j), the energy consumption can be defined as below:

$$E_c = \sum_{i=1}^{j} E_{T_{(i,CH)}} + j \times E_R + (j-1) \times E_{DA} + E_{T_{(CH,BS)}} \tag{4}$$

In Equation (4), the sum of the energy consumed on transmitting the sensory data from every non-CH node to the CH node is denoted as the first term. As for the second term, it shows the energy dissipation on receiving the gathering data from all the non-CH nodes within the cluster. Regarding to the third and fourth term, they represent the energy expenditure on executing the operation of data aggregation and transmitting the aggregation data from the CH to the BS, respectively. In order to design the energy-efficient routing protocol in WSN, obviously, on the precondition of ensuring a good data communication, the smaller value of E_c the better.

(v) The number of data gathering rounds (R): it is a predefined number dispatched by the BS. It is also a crucial parameter used in the scheduling phase that is discussed above. According to this parameter, the GA decides when to start reproducing the next generation of the population (the chromosomes) by using the aforementioned selection, crossover and mutation operations. Moreover, the value of R can be adjusted by the current energy status of all the source nodes in the WSN. Furthermore, if the R is assigned to be a larger value for the current population of GA, it indicates that this population has a better fitness value than others. It means this population will be used for a longer period of time. Considered that the quality of the best population is determined by the history of previous generations. Therefore, the R is a very important fitness parameter utilized by the GA to generate the optimal solution (population). In addition, a reasonable larger value of R will be good for the fitness function of the GA to generate the small variations in the best fitness value of the chromosomes.

(vi) The percentage of CHs for the given WSN (P): it is the ratio of the total number of CHs over the total number of participating source nodes (include non-CH and CH nodes) in the WSN, which is defined as follows:

$$P = \frac{N_{CH}}{N_{PS}} \times 100\%$$

(5)

In Equation (5), the N_{CH} and N_{PS} represent the total number of CH nodes and participating source nodes respectively. Here, only the source nodes that are alive during every data gathering round are considered to be the participating source nodes. Therefore, N_{PS} has the largest value during the first data gathering round R (in 4 as we discussed above), since the entire source nodes in the sensing region are alive during that round. As for N_{CH}, in our proposed approach, the whole WSN is separated uniformly by a certain number of clusters that have an approximately equal number of member nodes to avoid the CH nodes overload during the data transmission phase. In fact, N_{CH} is what we mentioned the certain number of clusters here. In order to minimize the spatial transmission distances between the CH and non-CH nodes within the individual cluster, the value of N_{CH} should be a relatively large number. Hence, P can be computed after the cluster setup and CH selection phase before the first data gathering round. In addition, P, at this moment, also means the minimum overload during each CH node. Although after certain number of data gathering round R, the value of N_{PS} may decrease due to running out of the power on the participating source nodes, we should keep the value of P the same for every data gathering round. By doing this, we maintain the lowest overload on every CH node for every data gathering round. Thereby extending the network lifetime by postponing the first dead node by prolonging its charge.

Fitness Function. The fitness function is defined over the above six fitness parameters and measures the quality of the represented solution (the chromosomes). Once we have the fitness parameters and the fitness function defined, GA proceeds to randomly select a population of solutions (the chromosomes), then, improve it through repetitive operations of GA, such as crossover, mutation, and repair function, to generate the optimal solution.

In most cases, we cannot come out a concrete equation for the fitness function of GA, due to its complexity and many uncertainties. Therefore, for the fitness function in our proposed approach, it is summarized as the following expression.

$$f(x) = \sum_{i=1}(\alpha_i \times f(x_i)), \forall f(x_i) \in \{CD, D, C_H, E_c, R, P\}$$

(6)

In expression 6, α_i is a set of arbitrarily assigned weights for the initial fitness parameters discussed above. Then, after every generation the best fit chromosome is evaluated and all the fitness parameters listed in $f(x_i)$ are updated as follows:

$$\Delta f = f(x_{i+1}) - f(x_i) \tag{7}$$

The Δf in Equation (7) represents the change in the fitness parameters' value and the index i represents the number of generations. Therefore, the Δf expression can be described as the subtraction of the fitness value for current population and the previous population. After every generation, the fitness parameters, such as the energy consumption (E_c), the percentage of CHs (P), the spatial transmission distances with each cluster (CD), between any CH and BS (D) and among the CHs (C_H), are evaluated to see the improvements. As for the initial weight α_i, a suitable range of it is assigned in my simulation part, as discussed in Section IV.

Based on the encoding, selection methods and the fitness function given above, we can select a group of chromosomes from a given population through evaluating each chromosome's fitness $f(x)$. For extending the search space and compute an approximate solution by producing the new generation, GA needs to accomplish crossover and mutation operations.

Crossover Operator. From the biological point of view, crossover operation simulates the transfer of genetic inheritance during the sexual reproductive process. In 1989, Goldberg [15] proposed a crossover method called partially matched crossover (PMX) to handle a traveling salesman problem (TSP) with permutation representation. In our proposed approach, we use the same PMX to achieve crossover operation, where two chromosomes (permutation and their associate alleles) are aligned, and two crossover sites are picked uniformly at random along a length of strings, which are defined as matching sections. These two points define a matching section that is used to affect a cross through position-by-position exchange operations. Finally, the alleles in the cross sites are moved to their new positions in the offspring. In other words, for crossover operation, we randomly select crossover point and the gene values of participating parents are flipped to produce a pair of child chromosomes. Consider the following chromosomes (Red color represents the two crossing sites):

Parent chromosome:									
A:									
12	30	45	26	70	16	10	99	36	18
B:									
30	16	10	45	18	99	36	12	70	26

PMX proceeds by position wise exchanges. First mapping chromosome B to chromosome A: the gene value 45 and 26, 18 and 70, 99 and 16 exchange places. Similarly, mapping chromosome A to chromosome B, the gene value 26 and 45, value 70 and 18, and value 16 and 99 exchange places. Following PMX two new offspring are generated:

Offspring chromosome:									
A*:									
2	0	6	5	8	9	0	6	6	0
B*:									
0	9	0	6	0	6	6	2	8	5

Where, each chromosome contains ordering information partially determined by its parent.

Mutation Operator. In biology, as a weak point of crossover operation, the new produced population may only have the traits of their parents. This could cause a problem that there would be no new genetic material introduced in the offspring. In order to avoid losing any feasible gene segments and maintain various populations, the mutation operation has to be implemented. Basically, the new genetic materials are allowed to introduce to the new chromosomes by the mutation operation. Although the mutation operation introduces a new sequence of genes into a chromosome, there is no guarantee that mutation will produce desirable features in the new chromosome. In addition, there exists the differences from the conventional mutation operators (like bit inversion) based on varied encoding methods. The mutation operation for the WSN in our proposed approach is adopted as an "order changing" method because a chromosome is encoded as a permutation method.

Although the crossover and mutation operations manipulate between two valid chromosomes, they may produce invalid chromosome(s), which would contain cycles. Since a chromosome represents a cluster for a given WSN in our proposed technique, which is also a data aggregation tree, there should not be a cycle or loop in it. Hence, repair function is used to identify and remove the inclusion of invalid chromosomes in the new generation..

In [12], a large of experiments show that, adding repair function to a GA, will generate much better solutions than the one without a repair function. Furthermore, on terms of the efficiency, GA with repair function also generates the best solution significantly faster than the one without a repair function.

3   Simulation

The main objectives of our simulation are to create a visible view of our proposed BSCICR protocol and generate the results based on network lifetime, energy dissipation, and total transmissions achieved (the amount of data gathering rounds) to compare with existing energy-efficient routing protocols discussed in Section II for further research. Two crucial parts are involved in our simulation. The first is to simulate the creation process of the optimal clusters and CHs for a given WSN, which is done by using GA. Secondly, simulating the process of producing the energy-efficient routing paths (the single-source shortest path) by utilizing Dijkstra's algorithm on the optimal CHs generated from the GA. The simulator is implemented using Java language under Eclipse development environment. Several modules are required in simulating a WSN environment, such as network configuration, communication models and data structures. The communication channel in our defined WSN is not ideal due to the radio harsh environment that we are trying to simulate. Therefore, we take into account the communication interference during the data transmission phases. However, we still assumed that there are no packet drops in consideration of collision, because the addressing scheme discussed in Section C is used to handle this communication interference. In addition, we do simulate the energy required to receive schedules from the BS. Moreover, we also assumed that the packet queries and schedules sent from BS are continuous and persistent.

4   Conclusions and Future Work

We proposed a genetic algorithm-based solution for a large-scale WSN used in a radio harsh environment. We described this wireless sensor network as a set of chromosomes (population), which is represented by an intelligent technology, the genetic algorithm. By utilizing Dijkstra's algorithm, we were able to transform a topology of the entire network to a dynamic TSP. However, the TSP is in the class of NP-problems. Thus, the genetic algorithm was used in order to compute the approximate optimal routing paths used during the data transmission phases.

More study of genetic algorithm with improved fitness function is our next step to improve our approach. One of the areas of focus will be on parameter tuning of the variables used in the genetic algorithm. In addition, how to properly predefine the number of data gathering rounds and CHs, is another our research direction.

References

1. Hou, Y.T., Shi, Y., Pan, J., Midkiff, S.F.: Maximizing the Lifetime of Wireless Sensor Networks Through Optimal Single-Session Flow Routing. IEEE Transactions on Mobile Computing 5(9), 1255–1266 (2006)
2. Hua, C., Yum, T.-S.P.: Optimal Routing and Data Aggregation for Maximizing Lifetime of Wireless Sensor Networks. Journal IEEE/ACM Transactions on Networking (TON) 16(4), 892–903 (2008)
3. Bhondekar, A.P., Vig, R., Singla, M.L., Ghanshyam, C., Kapur, P.: Genetic algorithm based node placement methodology for wireless sensor networks. In: Proceeding of the International MultiConference of Engineers and Computer Scientists (IMECS 2009), March 18-20 (2009)
4. Hussain, S., Matin, A.W., Islam, O.: Genetic algorithm for energy efficient clusters in wireless sensor networks. In: Fourth International Conference on Information Technology: New Generations, ITNG 2007 (April 2007)
5. Hussain, S., Matin, A.W., Islam, O.: Genetic Algorithm for Hierarchical Wireless Sensor Networks. Journal of Networks 2(5) (September 2007)

6. Nallusamy, R., Duraiswamy, K., Muthukumar, D.A.: Energy efficient clustering and shortest path routing in wireless ad hoc sensor networks (WASN) using approximation algorithms. Journal of Mathematics and Technology (February 2010) ISSN: 2078-0257
7. Mollanejad, A., Khanli, L.M., Zeynali, M.: DBSR: Dynamic base station repositioning using genetic algorithm in wireless sensor network. International Journal of Computer Science Issues 7(2(2)) (March 2010)
8. Jiang, Y., Zhang, H.: Base Station Controlled Intelligent Clustering Routing in Wireless Sensor Networks. In: Butz, C., Lingras, P. (eds.) Canadian AI 2011. LNCS, vol. 6657, pp. 210–215. Springer, Heidelberg (2011)

A New Fuzzy Support Vector Data Description Machine

Pei-Yi Hao

Department of Information Management,
National Kaohsiung University of Applied Sciences, Kaohsiung, Taiwan
haupy@cc.kuas.edu.tw

Abstract. Data domain description or one-class classification concerns the characterization of a data set. A good description covers all target data but includes no superfluous space. The boundary of a dataset can be used to detect novel data or outliers. One-class classification is important in many applications where one of the classes is characterized well, while no measurements are available for the other class. Tax et al. first introduced a method of adapting the support vector machine (SVM) methodology to the one-class classification problem, called support vector data description (SVDD). In this paper, we incorporate the concept of fuzzy set theory into the SVDD. We apply a fuzzy membership to each input point and reformulate the SVDD such that different input points can make different contributions to the learning of decision surface. Besides, the parameters to be identified in SVDD, such as the components within the spherical center vector and the radius, are fuzzy numbers. This integration preserves the benefits of SVM learning theory and fuzzy set theory, where the SVM learning theory characterizes the properties of learning machines which enable them to effectively generalize the unseen data and the fuzzy set theory might be very useful for finding a fuzzy structure in an evaluation system.

1 Introduction

In modeling some systems where available information is uncertain, we must deal with a fuzzy structure of the system considered. This structure is represented as a fuzzy function whose parameters are given by fuzzy sets. The fuzzy functions are defined by Zadeh's extension principle [1-2]. Basically, the fuzzy function provides an effective means of capturing the approximate, inexact natural of real world. Fuzzy theory appears very useful when the processes are too complex for analysis by conventional quantitative techniques or when the available source information are interpreted qualitatively, inexactly, or uncertainly.

The Support Vector Machines (SVMs) were developed at AT&T Bell Laboratories by Vapnik and co-works [3-5]. It is based on the idea of structural risk minimization, which shows that the generalization error is bounded by the sum of the training error and a term depending on the Vapnik–Chervonenkis dimension. By minimizing this bound, high generalization performance can be achieved. Due to this industrial context, SVM research is up to date and has a sound orientation towards real-world applications. In many applications, SVM has been shown to provide higher

A. Moonis et al. (Eds.): IEA/AIE 2014, Part I, LNAI 8481, pp. 118–127, 2014.
© Springer International Publishing Switzerland 2014

performance than traditional learning machines and has been introduced as a powerful tool for solving classification problems.

Much effort of SVM has been expended to solve classification and regression tasks. In these problems a mapping between objects represented by a feature vector and outputs (a class label or real valued outputs) is inferred on the basis of a set of training examples. In practice, another type of problem is of interest too: the problem of data description or One-Class Classification [6]. Here the problem is to make a description of a training set of objects and to detect which (new) objects resemble this training set. This is important in many applications [7-9]. Consider, for example, trying to classify sites of "internet" to a web surfer where the only information available is the history of the user's activities. One can envisage identifying typical positive examples by such tracking, but it would be hard to identify representative negative examples. Another example is that when we want to monitor a machine. A classifier should detect when the machine is showing abnormal, faulty behavior. Measurements on the normal operation of the machine are easy to obtain. In faulty situations, on the other hand, the machine might be destroyed completely.

In one-class classification, one class of data has to be distinguished from the rest of the feature space. In this type of classification problems, one of the classes is characterized well, while for the other class (almost) no measurements are available. It is assumed that we have examples from just one of the classes, called the target class and that all other possible objects, per definition of the outlier objects, are uniformly distributed around the target class. This one-class classification problem is often solved by estimating the target density [6], or by fitting a model to the data support vector classifier [10]. Schölkopf et al. [11] suggested a method of adapting the SVM methodology to the one-class classification, called one-class SVM. Instead of using a hyperplane, a hypersphere around the target set is used. This method is called the support vector data description, developed by Tax and Duin [8-9].

In this paper, we incorporate the concept of fuzzy set theory into the support vector data description (SVDD) proposed by Tax et al. [8-9]. Different from SVDD, the proposed fuzzy SVDD treat the training data points with different importance in the training process. Beside, the parameters to be identified in the fuzzy SVDD model, such as the components of spherical center vector and the radius term, are set to be the fuzzy numbers. In other words, we construct a fuzzy hypersphere in the feature space to distinguish the target class from the rest. Moreover, the decision function of the proposed fuzzy SVDD is based on a fuzzy partial ordering relation. This integrates the benefits of SVM learning theory and fuzzy set theory, where the VC theory characterizes properties of learning machines which enables them to generalize well the unseen data, whereas the fuzzy set theory might be very useful for finding a fuzzy structure in an evaluation system.

2   Support Vector Data Description

Tax and Duin introduced the use of a data domain description method [8-9], inspired by the SVM developed by Vapnik. Their approach is known as the support vector

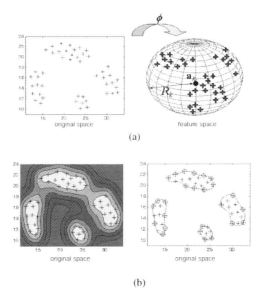

(a)

(b)

Fig. 1. Support vector domain description

domain description (SVDD). In domain description the task is to give a description of a set of objects. This description should cover the class of objects represented by the training set, and should ideally reject all other possible objects in the object space. To begin, let   denote a nonlinear transformation, which maps the original input space into a high-dimensional feature space. Data domain description gives a closed boundary around the data: a hypersphere. The sphere is characterized by center a and radius R>0. We minimize the volume of the sphere by minimizing R2, and demand that the sphere contains all training objects .

We now illustrate the support vector domain description method as shown in Figure 1. To begin, let   denotes a nonlinear transformation, which maps the original input space into a high-dimensional feature space. Data domain description gives a closed boundary around the data: a hypersphere. The SVDD may be viewed as finding a smallest sphere that encloses all the data points in the feature space, as shown in Figure 1 (a). The contour diagram shown in Figure 1 (b) can be obtained by estimating the distance between the spherical center and the corresponding input point in the feature space. The larger the distance, the darker is the gray level in the contour diagram. The algorithm then further identifies domain description using the data points enclosed inside the boundary curve. The mathematical formulation of the SVDD is as follows.

$$\underset{R, a, \xi_i}{\text{minimize}} \quad R^2 + C\sum_i \xi_i$$

$$\text{subject to} \quad \left\| \Phi(\mathbf{x}_i) - \mathbf{a} \right\|^2 \leq R^2 + \xi_i, \quad \xi_i \geq 0 \qquad i = 1, \ldots, n \tag{1}$$

where R is the radius and $\mathbf{a}$ is the center of the enclosing sphere; ξ_i is a slack variable; and C is a constant controlling the penalty of noise. Using the Lagrangian theorem, we can formulate the dual problem as

$$\underset{\beta_i}{\text{maximize}} \quad L = \sum_i \Phi(\mathbf{x}_i) \cdot \Phi(\mathbf{x}_i)\alpha_i - \sum_{i,j} \alpha_i \alpha_j \Phi(\mathbf{x}_i) \cdot \Phi(\mathbf{x}_j)$$

$$\text{subject to} \quad \sum_i \alpha_i = 1 \text{ and } 0 \le \alpha_i \le C , \forall i \tag{2}$$

where $\{\alpha_i\}_{i=1,...,n}$ are the Lagrange multipliers.

3   Fuzzy Support Vector Data Description Machine

In this section, inspired by the support vector data description (SVDD), we propose a novel fuzzy support vector data description machine.

3.1   The Quadratic Programming Problem

Now, we incorporated the concept of fuzzy set theory into the support vector data description machine. We construct a "fuzzy" hypersphere that tightly enclosed the target class. The parameters to be identified, such as the components of spherical center vector and radius term, are fuzzy numbers. The fuzzy parameters studied in this work are restricted to a class of "triangular" membership functions. To do this, we need the following preliminaries.

Preliminary 1 [12]: For any fuzzy number A, B and $\alpha \in (0,1]$, where $A^\alpha = [a_1, a_2]$ and $B^\alpha = [b_1, b_2]$ denoted the α-cuts of A and B, respectively. If we define the partial ordering of closed intervals in the usual way, that is
$$[a_1, a_2] \ge [b_1, b_2] \text{ iff } a_1 \ge b_1 \text{ and } a_2 \ge b_2$$
then for any fuzzy number A, B, we have $A \underset{f}{\ge} B$ iff $A^\alpha \ge B^\alpha$ for all $\alpha \in (0,1]$, where "$\underset{f}{\ge}$" denotes the *fuzzy* larger than.

Let $X=(m,c)$ be a symmetric triangular fuzzy number where m is the center and c is the width. From Preliminary 1, for any two symmetric triangular fuzzy numbers $A = (m_A, c_A)$ and $B = (m_B, c_B)$ in $T(R)$, we have
$A \underset{f}{\ge} B$ iff $m_A + c_A \ge m_B + c_B$ and $m_A - c_A \ge m_B - c_B$.

Moreover, the components in the spherical center vector and radius term used in the hypersphere are symmetric triangular fuzzy numbers. Given the fuzzy spherical center vector $\mathbf{A}=(\mathbf{a}, \mathbf{c})$ and fuzzy spherical radius term $\mathbf{R}=(r, d)$, The fuzzy spherical center vector $\mathbf{A}=(\mathbf{a}, \mathbf{c})$ is denoted in the vector form of $\mathbf{a} = [a_1,...,a_n]'$ and

$\mathbf{c} = [c_1,...,c_n]'$, which means "approximation $\mathbf{a}$", described by the center $\mathbf{a}$ and the width $\mathbf{c}$. Similarly, $\mathbf{R} = (r, d)$ is the fuzzy spherical radius, which means "approximation r", described by the center r and the width d.

Preliminary 2: Given the fuzzy data vector $\mathbf{X}_i = (\mathbf{x}_i, \mathbf{e}_i)$ where $\mathbf{x}_i = [x_{i1},...,x_{in}]'$ and $\mathbf{e}_i = [e_{i1},...,e_{in}]'$, which means "approximation $\mathbf{x}_i$", described by the center $\mathbf{x}_i$ and the width $\mathbf{e}_i$. The distance between $\mathbf{X}_i$ and $\mathbf{A}$, which is denoted as $\|\mathbf{X}_i - \mathbf{A}\|_f$, are also a symmetric triangular fuzzy number with center $\|\mathbf{x}_i - \mathbf{a}\|$ and width $\|\mathbf{e}_i\| + \|\mathbf{c}\|$.

From the above two preliminaries, our fuzzy support vector data description (fuzzy SVDD) task here is therefore to

$$\min \ \|\mathbf{R}\|_f^2 + M\left(\frac{1}{2}\|\mathbf{c}\|^2 + d^2\right) + C\sum_{i=1}^N \mu_i \xi_i \qquad (3)$$

$$\text{subject to} \quad \|\mathbf{X}_i - \mathbf{A}\|_f \underset{f}{\leq} \mathbf{R} + \xi_i \ \text{ for all } i=1,..,N,$$

where the minimization of $\|R\|_f^2$ means to seek fuzzy hypersphere with minimal spherical radius to enclosed all training data vectors. $\frac{1}{2}\|\mathbf{c}\|^2 + d$ is the term which characterizes the vagueness of the model. The more vagueness in the fuzzy SVDD model means the more inexactness in the result, and M is a trade off parameter chosen by the decision-maker. In this case, the vector $\mathbf{c}$ is a user-specified parameter which controls the vagueness of the result model. The $\{\xi_i\}_{i=1,...,N}$ are sets of slack variables that measure the amount of variation of the constraints for each point where C is a fixed penalty parameter chosen by the user. The fuzzy membership μ_i is the attitude of the corresponding point $\mathbf{x}_i$ toward the target class.

More specifically, from the above preliminaries, our problem is to find out the fuzzy spherical center vector $\mathbf{A}^* = (\mathbf{a}, \mathbf{c})$ and fuzzy spherical radius $\mathbf{R}^* = (r, d)$, which is the solution of the following quadratic programming problem:

$$\min \quad r^2 + M\left(\frac{1}{2}\|\mathbf{c}\|^2 + d^2\right) + C\sum_{i=1}^N \mu_i(\xi_{1i} + \xi_{2i})$$

$$\text{subject to} \quad \|\mathbf{x}_i - \mathbf{a}\|^2 + \left(\|\mathbf{c}\|^2 + \|\mathbf{e}_i\|^2\right) \leq r^2 + d^2 + \xi_{1i} \qquad (4)$$

$$\|\mathbf{x}_i - \mathbf{a}\|^2 - \left(\|\mathbf{c}\|^2 + \|\mathbf{e}_i\|^2\right) \leq r^2 - d^2 + \xi_{2i}$$

$$\text{and } \ d \geq 0, \ \xi_{1i}, \xi_{2i} \geq 0 \ \text{ for } i=1,...,N.$$

We can find the solution of this optimization problem in dual variables by finding the saddle point of the Lagrangian:

$$L = r^2 + M\left(\frac{1}{2}\|\mathbf{c}\|^2 + d\right) + C\sum_{i=1}^{N}\mu_i(\xi_{1i} + \xi_{2i}) + \sum_{i=1}^{N}\alpha_{1i}\left(\|\mathbf{x}_i - \mathbf{a}\| + \|\mathbf{c}\|^2 + \|\mathbf{e}_i\|^2 - r^2 - d^2 - \xi_{1i}\right)$$

$$+ \sum_{i=1}^{N}\alpha_{2i}\left(\|\mathbf{x}_i - \mathbf{a}\| - \|\mathbf{c}\|^2 - \|\mathbf{e}_i\|^2 - r^2 + d^2 - \xi_{2i}\right) - \sum_{i=1}^{N}\eta_{1i}\xi_{1i} - \sum_{i=1}^{N}\eta_{2i}\xi_{2i} - \mu d$$

where $\alpha_{1i}, \alpha_{2i}, \eta_{1i}, \eta_{2i}$ and μ are the nonnegative Lagrange multipliers. This function has to be minimized with respect to the primal variables $\mathbf{a}$, r, d, ξ_{1i}, ξ_{2i} and maximized with respect to the dual variables $\alpha_{1i}, \alpha_{2i}, \eta_{1i}, \eta_{2i}$ and μ. To eliminate the former, we compute the corresponding partial derivatives and set them to zero, obtaining the following conditions:

$$\partial L\big/\partial\mathbf{a} = 0 \quad\Rightarrow\quad \mathbf{a} = \sum_{i=1}^{N}(\alpha_{1i} + \alpha_{2i})\mathbf{x}_i \tag{5}$$

$$\partial L\big/\partial r = 0 \quad\Rightarrow\quad \sum_{i=1}^{N}(\alpha_{1i} + \alpha_{2i}) = 1 \tag{6}$$

$$\partial L\big/\partial d = 0 \quad\Rightarrow\quad \sum_{i=1}^{N}(\alpha_{1i} - \alpha_{2i}) = M - \mu \le M \tag{7}$$

$$\partial L\big/\partial\xi_{1i} = 0 \quad\Rightarrow\quad \alpha_{1i} = C\mu_i - \eta_{1i} \text{ and } \alpha_{1i} \le C\mu_i \tag{8}$$

$$\partial L\big/\partial\xi_{2i} = 0 \quad\Rightarrow\quad \alpha_{2i} = C\mu_i - \eta_{2i} \text{ and } \alpha_{2i} \le C\mu_i. \tag{9}$$

Substituting Eqs. (5)-(9) into L, we obtain

$$L = -\sum_{i=1}^{N}\sum_{j=1}^{N}(\alpha_{1i} + \alpha_{2i})(\alpha_{1j} + \alpha_{2j})\mathbf{x}_i \cdot \mathbf{x}_j + \sum_{i=1}^{N}(\alpha_{1i} + \alpha_{2i})\mathbf{x}_i \cdot \mathbf{x}_i$$

$$+ \left(\frac{M}{2} + \sum_{i=1}^{N}(\alpha_{1i} - \alpha_{2i})\right)\|\mathbf{c}\|^2 + \sum_{i=1}^{N}(\alpha_{1i} - \alpha_{2i})\|e_i\|^2$$

using $\alpha_{1i}, \alpha_{2i}, \eta_{1i}, \eta_{2i}, \mu \ge 0$ leaves us with the following quadratic optimization problem:

$$\max_{\alpha_{1i}, \alpha_{2i}} \begin{cases} -\sum_{i=1}^{N}\sum_{j=1}^{N}(\alpha_{1i} + \alpha_{2i})(\alpha_{1j} + \alpha_{2j})\mathbf{x}_i \cdot \mathbf{x}_j \\ +\sum_{i=1}^{N}(\alpha_{1i} + \alpha_{2i})\mathbf{x}_i \cdot \mathbf{x}_i + \left(\frac{M}{2} + \sum_{i=1}^{N}(\alpha_{1i} - \alpha_{2i})\right)\|\mathbf{c}\|^2 + \sum_{i=1}^{N}(\alpha_{1i} - \alpha_{2i})\|e_i\|^2 \end{cases} \tag{10}$$

$$\text{subject to} \begin{cases} \sum_{i=1}^{N}(\alpha_{1i} + \alpha_{2i}) = 1, \quad \sum_{i=1}^{N}(\alpha_{1i} - \alpha_{2i}) \le M \\ 0 \le \alpha_{1i} \le C\mu_i, \quad 0 \le \alpha_{2i} \le C\mu_i, \quad i = 1, \dots, N \end{cases}$$

As for the spherical center $\mathbf{a}$, from Eqs. (5) we derive $\mathbf{a} = \sum_{i=1}^{N}(\alpha_{1i} + \alpha_{2i})\mathbf{x}_i$.

Knowing the fuzzy spherical center vector $\mathbf{A}^*=(\mathbf{a}, \mathbf{c})$, we can subsequently determine the fuzzy radius $\mathbf{R}^*=(r, d)$ by exploiting the Karush-Kuhn-Tucker (KKT) conditions:

$$\alpha_{1i}\left(\|\mathbf{x}_i - \mathbf{a}\| + \|\mathbf{c}\|^2 + \|\mathbf{e}_i\|^2 - r^2 - d^2 - \xi_{1i}\right) = 0 \tag{11}$$

$$\alpha_{2i}\left(\|\mathbf{x}_i - \mathbf{a}\| - \|\mathbf{c}\|^2 - \|\mathbf{e}_i\|^2 - r^2 + d^2 - \xi_{2i}\right) = 0 \tag{12}$$

$$\left(C\mu_i - \alpha_{1i}\right)\xi_{1i} = 0, \tag{13}$$

$$\left(C\mu_i - \alpha_{2i}\right)\xi_{2i} = 0, \tag{14}$$

$$\mu d = 0. \tag{15}$$

For some $\alpha_{1i} \in (0, C\mu_i)$ and $\alpha_{2j} \in (0, C\mu_j)$, we have $\xi_{1i} = \xi_{2j} = 0$ (using Eqs. (13) and (14)) and moreover the second factor in Eqs. (11) and (12) has to vanish. Hence, for some $\alpha_{1i} \in (0, C\mu_i)$ and $\alpha_{2j} \in (0, C\mu_j)$, r and d can be computed as

$$r^2 = \frac{1}{2}\left(\|\mathbf{x}_i - a\| + \|\mathbf{x}_j - a\| + \|\mathbf{e}_i\|^2 - \|\mathbf{e}_j\|^2\right), \tag{16}$$

$$d^2 = \frac{1}{2}\left(\|\mathbf{x}_i - a\| - \|\mathbf{x}_j - a\| + 2\|c\|^2 + \|\mathbf{e}_i\|^2 + \|\mathbf{e}_j\|^2\right) \tag{17}$$

For any fuzzy data vector $\mathbf{X}_i=(\mathbf{x}_i, \mathbf{e}_i)$, the distance between $\mathbf{X}_i$ and spherical center $\mathbf{A}$, $\|\mathbf{X}_i - \mathbf{A}\|_f$, are also a symmetric triangular fuzzy number with center $\|\mathbf{x}_i - \mathbf{a}\|$ and width $\|\mathbf{e}_i\| + \|\mathbf{c}\|$. The fuzzy spherical radius $\mathbf{R}=(r, d)$ a symmetric triangular fuzzy number with center r and width d. For a new point $\mathbf{X}_i$, we evaluate which side of the hypersphere it falls on by defining the following fuzzy partial ordering relation. For any two symmetric triangular fuzzy numbers $A = (m_A, c_A)$ and $B = (m_B, c_B)$ in $T(R)$, the degree that A is smallerer than B (i.e. A falls on the left side of B) is defined by the following membership function:

$$Rank_{\leq B}(A) = \begin{cases} 1 & \text{if } \alpha < 0 \text{ and } \beta < 0 \\ 0 & \text{if } \alpha > 0 \text{ and } \beta > 0 \\ 0.5\left(1 + \dfrac{\alpha + \beta}{\max(|\alpha|, |\beta|)}\right) & \text{o.w.} \end{cases} \tag{18}$$

where $\alpha = (m_A + c_A) - (m_B + c_B)$ and $\beta = (m_A - c_A) - (m_B - c_B)$.

Notice that $R_{\leq B}(A) = 0.5$ if $m_A = m_B$, $R_{\leq B}(A) < 0.5$ if $m_B < m_A$, and $R_{\leq B}(A) > 0.5$ if $m_B > m_A$. And the decision function of the proposed fuzzy SVDD model is $f(\mathbf{X}_i) = Rank_{\leq \mathbf{R}}\left(\|\mathbf{X}_i - \mathbf{A}\|_f\right)$.

This decision function takes a value within a specified range that indicates the membership grade of the new point **x** belongs to the target class. A vague, fuzzy boundary exists between members and nonmembers of the set.

3.2   Extension to the Nonlinear Case

To extend the proposed fuzzy SVDD to nonlinear one-class classification, we will use the idea of SVM for crisp-nonlinear classification [5,10]. The basic idea is to simply map the input patterns $\mathbf{x}_i$ by Φ: $R^n \rightarrow F$ into a higher dimensional feature space F. Note that the only way in which the data appears in the algorithm for the model is in the form of inner products $\langle \mathbf{x}_i \cdot \mathbf{x}_j \rangle$. The algorithm would only depend on the data through inner products in F, i.e. on functions of the form $\langle \Phi(\mathbf{x}_i) \cdot \Phi(\mathbf{x}_j) \rangle$. Hence it suffices to know and use $k(\mathbf{x}_i, \mathbf{x}_j) = \langle \Phi(\mathbf{x}_i) \cdot \Phi(\mathbf{x}_j) \rangle$ instead of $\Phi(\bullet)$ explicitly. By replacing $\langle \mathbf{x}_i \cdot \mathbf{x}_j \rangle$ with $k(\mathbf{x}_i, \mathbf{x}_j)$, we obtain the dual quadratic optimization problem given by Eq. (20). Here, we should note that the constraints are not changed.

$$
\max_{\alpha_{1i}, \alpha_{2i}} \begin{cases} -\sum_{i=1}^{N}\sum_{j=1}^{N}(\alpha_{1i}+\alpha_{2i})(\alpha_{1j}+\alpha_{2j})k(\mathbf{x}_i,\mathbf{x}_j) \\ +\sum_{i=1}^{N}(\alpha_{1i}+\alpha_{2i})k(\mathbf{x}_i,\mathbf{x}_i)+\left(\dfrac{M}{2}+\sum_{i=1}^{N}(\alpha_{1i}-\alpha_{2i})\right)\|\mathbf{c}\|^2+\sum_{i=1}^{N}(\alpha_{1i}-\alpha_{2i})\|e_i\|^2 \end{cases} \tag{19}
$$

$$
\text{subject to} \begin{cases} \sum_{i=1}^{N}(\alpha_{1i}+\alpha_{2i})=1, \qquad \sum_{i=1}^{N}(\alpha_{1i}-\alpha_{2i})\leq M \\ 0\leq\alpha_{1i}\leq C\mu_i, \quad 0\leq\alpha_{2i}\leq C\mu_i, \quad i=1,...,N \end{cases}
$$

4   Experiments

We now apply the proposed fuzzy SVDD approaches to the handwritten digits problem. The data set used for this experiment consists of isolated binary handwritten digits partially extracted from the BR digit set of the SUNY CDROM-1[13] and the ITRI database [14]. We used two categories of confusion group, class "2" and class "7", with 600 samples per class. Each digit has different size, and a sample of some of the digits is shown in Figure 9. We utilize the 60-dimensional transition feature set [15] obtained from the original digit patterns as input for the algorithm to construct the hyperplane. For handwritten digit patterns, most of the digit classes consist of different styles of writing, and it may not always be feasible to construct an adequate pattern distribution for each individual class due to the fact of various shapes of patterns. In addition, the inherent problem of outliers or noise patterns makes the digit classification even more difficult.

We fed our algorithms with the training instances of digit 2 only (using the first 400 samples). The training instances were fuzzified by consulting an expert's domain knowledge. Testing was done on both digit 2 (using the remaining 200 samples) and digit 7 (600 samples). Table I presents the result of comparing these methods. We report the optimal model parameters and the corresponding classification errors in the training and test set. As illustrated in Table I, associating a fuzzy membership to each data point such that different data points can have different effects in the learning of the separating hypersphere can effectively reduce the effects of outliers, and incorporating the concept of fuzzy set theory into the fuzzy SVDD might be very useful for finding a fuzzy structure in an evaluation system. As a whole, the experimental results show that the proposed method performs fairly well on the handwritten digit datasets.

Table 1. A comparison of classification performance

algorithm		SVDD	Fuzzy-SVDD
Num. of misclassified patterns	Digit "2" (Training set)	8	9
	Digit "2"(Test set)	6	5
	Digit "7"(Test set)	21	18

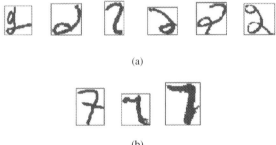

(a)

(b)

Fig. 2. The 9 worst outliers (a) digit 2 and (b) digit 7, that are located in the fuzzy boundary for the handwritten digit dataset

Although this experiment leads to satisfactory results, it did not really address the actual task the fuzzy SVDD algorithm was designed for. Therefore, we next focused on a problem of novelty detection. Figure 2 shows the 9 worst outliers for the handwritten digit dataset that are located in the fuzzy boundary constructed by the proposed fuzzy SVDD. In other words, the values of the fuzzy decision function for all of these patterns are in the range of (0, 1), which represent the maximum degree of fuzziness for corresponding category. It can be seen that the number of "fuzzy digit pattern" in class "2" is more than in class "7". This cause by the various writing styles in training set of class "2". It can be seen that those patterns are significantly different from normal writing styles.

5   Conclusion

In this paper, we incorporate the concept of fuzzy set theory into the support vector data description (SVDD). We apply a fuzzy membership to each input point such that different input points can make different contributions to the learning of decision surface. Besides, the parameters to be identified in the proposed fuzzy SVDD, such as the components within the spherical center vector and the radius, are fuzzy numbers. The decision function is also based on a fuzzy partial ordering relation. A vague, fuzzy boundary exists between members and nonmembers of the set. The fuzzy set theory is demonstrated to be very useful for finding a fuzzy structure in an evaluation system.

References

1. Yager, R.R.: On solving fuzzy mathematical relationships. Inform. Contr. 41, 29–55 (1979)
2. Zadeh, L.A.: The concept of linguistic variable and its application to approximate reasoning—I. Inform. Sci. 8, 199–249 (1975)
3. Guyon, B.B., Vapnik, V.: Automatic capacity tuning of very large VC-dimension classifier. Advances in Neural Information Processing Systems 5, 147–155 (1993)
4. Vapnik, V.N.: The Nature of Statistical Learning Theory. Springer, New York (1995)
5. Cortes, C., Vapnik, V.N.: Support vector network. Machine Learning 20, 1–25 (1995)
6. Moya, M.R., Koch, M.W., Hostetler, L.D.: One-class classifier networks for target recognition applications. In: Proceedings of World Congress on Neural Networks, Portland, OR, pp. 797–801. International Neural Network Society (1993)
7. Manevitz, L.M., Yousef, M.: One-class SVMs for document classification. Journal of Machine Learning Research 2, 139–154 (2001)
8. Tax, D.M.J., Duin, R.P.W.: Support vector domain description. Pattern Recognition Letters 20, 1191–1199 (1999)
9. Tax, D.M.J., Duin, R.P.W.: Support vector data description. Machine Learning 54, 45–66 (2004)
10. Vapnik, V.N.: Statistical Learning Theory. Wiley (1998)
11. Schölkopf, B., Platt, J.C., Shawe-Taylor, J., Smola, A.J., Williamson, R.C.: Estimating the support of a high-dimensional distribution. Neural Computation 13, 1443–1471 (2001)
12. Klir, G.J., Yuan, B.: Fuzzy Sets and Fuzzy Logic: Theory and Applications. Prentice-Hall, New Jersey (1995)
13. Hull, J.J.: A database for handwritten text recognition research. IEEE Trans. on Pattern Analysis and Machine Intelligence 16, 550–554 (1994)
14. Chiang, J.-H., Gader, P.: Recognition of handprinted numerals in VISA card application forms. Machine Vision and Applications 10, 144–149 (1997)
15. Gader, P., Mohamed, M., Chiang, J.-H.: Handwritten word recognition with character and inter-character neural networks. IEEE Trans. on Systems, Man, and Cybernetics 27(1), 158–164 (1997)

Answer Set Programs with One Incremental Variable

Junjun Deng and Yan Zhang

Abstract. In the past decade, Answer Set Programming (ASP) has emerged as a popular paradigm for declarative problem solving, and a number of answer set solvers have been developed. However, most existing solvers require variables occurring in a logic program to be bounded by some finite domains, which limits their applications in one way or another. In this paper, we introduce answer set programs with one incremental variable to overcome this limitation. Based on existing ASP solving techniques, an approach to solve answer set programs with one incremental variable is proposed, and a prototype solver is then developed based on this approach. By conducting some experiments, our approach is shown to be comparable to *iClingo*'s modular domain description approach in the incremental problem solving setting.

1 Introduction

Answer Set Programming (ASP) [1] is a declarative language for problem solving. Normally a problem is described in an ASP program containing bounded variables, and given an instance of that problem, an ASP grounder will output one propositional ASP program, which is handed to a propositional ASP solver to search for answer sets. This process is intuitive and straightforward for many classical search problems. However, some real-world applications may call for unbounded variables.

The Graph Coloring Problem (GCP) is a classical NP hard problem. Given an undirected graph, the goal is to find the least number of colors such that there exists at least one legal graph coloring scheme which assigns every vertex of the graph a color such that no two adjacent vertices have the same color.

Example 1. If the number of available colors k is provided, we can encode this problem into an ASP program such that each answer set of that program corresponds to a legal coloring scheme.

$$col(1..k).$$
$$color(V, C) \leftarrow node(V), col(C), not\ ncolor(V, C).$$
$$ncolor(V, C) \leftarrow node(V), col(C), col(C1), color(V, C1), C \neq C1.$$
$$\leftarrow edge(V1, V2), col(C), color(V1, C), color(V2, C).\ \square$$

A. Moonis et al. (Eds.): IEA/AIE 2014, Part I, LNAI 8481, pp. 128–137, 2014.

However, if k is unknown and to be minimized, then the domain of predicate *col* is unbounded, thus above program is not valid. Another example is the planning problem, where we search for a sequence of actions to reach the goal state, and the length of that sequence is also not known precisely.

To solve the problems comprising a parameter which reflects its solution size, an incremental approach to both grounding and solving in ASP is proposed in [2]. This approach introduces a *(parametrized) domain description* as a triple (B, P, Q) of logic programs, among which P and Q contain a single parameter k ranging over the natural numbers: *a*) B is meant to describe static knowledge; *b*) $P[k]$ capture knowledge accumulating with increasing k; *c*) $Q[k]$ is specific for each value of k. The goal is to decide if the program $R[k] = B \cup \bigcup_{1 \leq i \leq k} P[i] \cup Q[i]$ has an answer set for some integer k [2].

Based on that domain description concept, an incremental ASP solver called *iClingo* [2] was developed, which can solve problems encoded in modular domain description. However, this incremental ASP approach requires programmers to split the whole encoding into three parts, and ensure that their domain description is modular. In addition, some intuitive encodings are excluded from modular domain description for some problems, e.g. GCP in Example 1.

Our goal is to develop an approach that is not only easy and intuitive for users to encode their problems, but also allows solvers to compute answer sets efficiently. In this paper, we propose a new approach for answer set programming, where programs are armed with one incremental variable. This incremental variable is ranging over the natural numbers, thus can model problems with unbounded variables. Based on the Clark's completion and loop formula [7], we propose a dynamic transformation for answer set programs with one incremental variable. With that we can bypass grounding to propositional rules, and directly construct dynamic transformation formula which is valid even when value of the incremental variable increases, and feed it to a propositional ASP solver repetitively until answer sets are found.

The rest of the paper is organized as follows. In Section 2 we define answer set programs with one incremental variable (ivASP). Then safe ivASP programs are introduced in Section 3. Section 4 presents the dynamic transformation of ivASP programs. Based on that transformation, we then develop a prototype solver for safe ivASP programs, and report its performance in Section 5. Finally we conclude this paper with related works in Section 6.

2   ASP with One Incremental Variable

In this section, we introduce answer set programs with one incremental variable (ivASP). The language of ivASP is defined over an alphabet including the following classes of symbols: (*a*) constants, (*b*) function symbols $\mathcal{F}$ including builtin $+$ and $-$; (*c*) variable symbols $\mathcal{V}$; (*d*) predicate symbols $\mathcal{P}$ including builtin $\leq, \geq$ and $=$; and (*e*) a special variable k ranged over natural numbers, which is called the *incremental variable*. A *term* is inductively defined as follows:

- A variable (including k) is a term.
- A constant is a term.

– If f is an n-ary function symbol and $t_1, \ldots, t_n$ are terms then $f(t_1, \ldots, t_n)$ are terms.

A term is said to be *ground* if no variable occurs in it. An *atom* is of the form $p(t_1, \ldots, t_n)$ where p is a n-ary predicate symbol and each t_i is a term. An atom is ground if all t_i $(1 \leq i \leq n)$ in the atom is ground. A *literal* is either an atom or an atom preceded by the symbol "not".

A logic program is a finite set of rules of the form

$$a \leftarrow b_1, \ldots, b_m, not\ c_{m+1}, \ldots, not\ c_n$$

where a is an atom or $\perp$, and b_i, c_i are atoms.

A rule is *ground* if all of its atoms are ground, and a logic program is *ground* if every rule in it is ground. The head atom (a here) of a rule r is denoted as $head(r)$. $\{b_1, \ldots, b_m\}$ is called the positive body, denoted by $pos(r)$. Similarly, $\{not\ c_{m+1}, \ldots, not\ c_n\}$ is called the negative body of a rule, denoted by $neg(r)$, and each $not\ c_i$ is called *negative literal*. All atoms of a rule $atom(r) = head(r) \cup pos(r) \cup neg(r)$.

The *Herbrand Universe* of a language $\mathcal{L}$, denoted by $HU_{\mathcal{L}}$, is the set of all ground terms formed with the functions and constants (no variables) in $\mathcal{L}$. Similarly, the *Herbrand Base* of a language $\mathcal{L}$, denoted by $HB_{\mathcal{L}}$, is the set of all ground atoms formed with predicate $\mathcal{P}$ from $\mathcal{L}$ and terms from $HU_{\mathcal{L}}$.

Let r be a rule in the ivASP language $\mathcal{L}$. The grounding of r in $\mathcal{L}$ on level L where $L \in \mathbf{N}$, denoted by $ground(r, \mathcal{L}, L)$, is the set of all rules obtained from r by all possible substitutions of elements of $HU_{\mathcal{L}} \cup \{1..L\}$ for the variables in r except k, and the substitution of L for the special variable k. For any ivASP logic program Π, we define

$$ground(\Pi, \mathcal{L}, L) = \cup_{r \in \Pi} ground(r, \mathcal{L}, L) \tag{1}$$

and we use $ground(\Pi, L)$ as a shorthand for $ground(\Pi, \mathcal{L}(\Pi), L)$.

A *Herbrand interpretation* of a logic program Π is any subset of its Herbrand Base. A Herbrand interpretation I of Π is said to satisfy a ground rule, if

i) if $a \neq \perp$ then $\{b_1, \ldots, b_m\} \subseteq I \wedge \{c_{m+1}, \ldots, c_n\} \cap I = \emptyset$ implies that $a \in I$;
ii) otherwise $a = \perp$, $\{b_1, \ldots, b_m\} \not\subseteq I \vee \{c_{m+1}, \ldots, c_n\} \cap I \neq \emptyset$

A *Herbrand model A* of a logic program Π is a Herbrand interpretation I of Π such that it satisfies all rules in Π.

Definition 1. *Given a ground logic program Π, for any set S of atoms from Π, let Π^S be the program obtained from Π by deleting*

i) each rule that has a negative literal not b in its body with $b \in S$, and
ii) all negative literals in the bodies of the remaining rules.

The set S is a *stable model (answer set)* of Π if S is a minimal Herbrand model of Π^S [6].

Definition 2. *Given an ivASP program Π, a pair of a set of atoms $S \subseteq HB_{\mathcal{L}(\Pi)}$ and an integer L, (S, L) is a stable model (answer set) of Π if S is a stable model of ground(Π, L).*

Two ivASP programs are *equivalent* if their answer sets are the same.

Example 2. Following is an ivASP program:

$$p(1..k).$$
$$q(X + 1) \leftarrow p(X), not\ p(X + 1).$$

The pair $(\{p(1), q(2)\}, 1)$ and $(\{p(1), p(2), q(3)\}, 2)$ are two answer sets of it. □

3   Safe ivASP Programs

Given an ivASP program Π, its grounding program may consist of infinite number of rules. To make the problem of finding answer sets feasible, we should consider ivASP programs that is equivalent to ground programs with finite rules. This calls for the definition of safe ivASP programs.

Definition 3. *Given a logic program Π, the dependency graph $DG(\Pi) = (V_\Pi, E_\Pi)$ of Π is a graph where each node corresponds to a predicate in Π and there is an edge from node p to node q iff there exists a rule in Π where p is the predicate symbol in head and q is a predicate symbol in the body.*

In a logic program Π, a predicate p *depends on* a predicate q iff there is a path from p to q in the dependency graph $DG(\Pi)$. A predicate p is an *extensional predicate* iff it does not depend on any predicate including itself.

Definition 4. *In an ivASP program Π, a predicate $p \in \Pi$ is a domain predicate iff it holds that every path in $DG(\Pi)$ starting from the node corresponding to p is cycle-free.*

Since the domain (extension) of a domain predicate can be computed without search, domain predicates are able to serve as the basis for an efficient grounding.
    Given a variable X in a rule of a logic program, it is *range bounded* if there are at least a lower bound atom of the form $X \geq Y$ and an upper bound atom of the form $X \leq Z$ in the positive body where Y and Z are integer constants or range bounded variables.

Definition 5. *A rule r in an ivASP program is safe if it holds that: if the head predicate of r is an extensional predicate, then every variable except k occurs in the head is also range bounded; otherwise, every variable except k that occurs in the rule also appears in a positive domain predicate. An ivASP program is safe if every rule in it is safe.*

In practice, a rule of the form $p(a..k)$ is commonly used as a shorthand of $p(X) \leftarrow a \leq X, X \leq k$, thus is safe by definition. Example 1 is a safe ivASP program.

In Equation (1), the grounding of an ivASP program instantiates rules based on the Herbrand Universe. This naive grounding method may generate too many unnecessary rules in practice. Given a safe ivASP program Π and an integer L, a more concise ground program $grd(\Pi, L)$ which is equivalent to $ground(\Pi, L)$ can be constructed as follows:

1. Build the dependency graph $DG(\Pi)$, and identify a set of domain predicates $D(\Pi)$. Then predicates in $D(\Pi)$ are sorted topologically from bottom to top in $DG(\Pi)$, resulting an ordered set $D(\Pi) = (p_1, p_2, \ldots p_i, \ldots p_n)$ such that for any two predicates $p_i, p_j \in D(\Pi)$, if p_i depends on p_j then $i > j$.
2. For any predicate p that is a leaf node in $DG(\Pi)$ (extensional predicate), its ground instances are determined by fact atoms, or instantiated from range bound variables of the program. These ground atoms are added to $grd(\Pi, L)$, and the domain of this predicate, $Dom(p)$ is populated.

$$Dom(p) = \{\bar{c}|p(\bar{c}). \in \Pi\} \cup$$
$$\bigcup \{(c_1, \ldots, c_n)|p(\bar{X}) \leftarrow \bigwedge_{X_i \in \bar{X}} lb \leq X_i, X_i \leq ub. \in \Pi, lb_i \leq c_i \leq ub_i (1 \leq i \leq n)\}$$

where $\bar{c}$ is a tuple of constants, $\bar{X}$ is a tuple of variables and lb and ub are expressions which can be evaluated to integers.

3. For each domain predicate $d \in D(\Pi)$ which is not extensional, compute its domain $Dom(d)$ based on set operations for domains in Table 1.
4. Given the occurrence of a domain predicate of this form $d(X, Y, ...)$, let its variable binding $Bind(d) = \{(X/x, Y/y, ...)|(x, y, ...) \in Dom(d)\}$. For all rules $r \in \Pi$, the variable binding of r, B_r is the natural join of variable bindings of all positive domain predicates, $B_r =\bowtie_{d \in D(\Pi) \cap pos(r)} Bind(d)$. Then some elements in B_r are filtered out if they do not satisfy any relation test in the body of r. Finally instantiations of r is obtained through substitutions of variables to bindings in B_r, and $grd(\Pi, L) = grd(\Pi, L) \cup \{r/\theta|\theta \in B_r, r \in \Pi\}$.

Table 1. Set operations for domains

Rules	Operations
$p(X) \leftarrow q(X)$ $p(X) \leftarrow r(X)$	$Dom(p) = Dom(q) \cup Dom(r)$
$p(X) \leftarrow q(X), r(X)$	$Dom(p) = Dom(q) \cap Dom(r)$
$p(X) \leftarrow q(X), not\ r(X)$	$Dom(p) = Dom(q) \setminus Dom(r)$
$p(X, Y) \leftarrow q(X), r(Y)$	$Dom(p) = Dom(q) \times Dom(r)$
$p(X, Y, Z) \leftarrow q(X, Y), r(Y, Z)$	$Dom(p) = Dom(q) \bowtie Dom(r)$

Proposition 1. *Given a safe ivASP program and an integer $L \geq 0$, $grd(\Pi, L)$ is equivalent to $ground(\Pi, L)$.*

4   Dynamic Transformation

In the incremental problem solving setting, one of the most important techniques is computation reuse. To do this, we regard each logic program as a propositional formula by employing Lin and Zhao's loop formulas [7]. Our main idea is to revise the completion formula and loop formula so that they are valid when the value of the incremental variable increases. To do this, we distinguish propositional clauses in the completion formula and loop formula that may be changed and annotate them with auxiliary propositional variables such that those clauses are satisfied whenever they become invalid.

To define the dynamic transformation, we first need some notations. Suppose Π is a safe ivASP program, a a ground atom, and L a natural number. We define (a) the supporting rules of an atom on level L, $sup_L(a, \Pi) = \{r \in grd(\Pi, L) \mid head(r) = a\}$; (b) the ground rules firstly instantiated on level L, $rule_L(\Pi) = grd(\Pi, L) \setminus \cup_{0 \leq l < L} grd(\Pi, l)$; and (c) the atoms firstly instantiated on level L, $atom_L(\Pi) = atom(grd(\Pi, L)) \setminus \cup_{0 \leq l < L} atom(grd(\Pi, l))$.

A ground rule r is *cumulative* wrt Π if for any two integers $1 \leq L_1 < L_2$ it holds that $r \in grd(\Pi, L_1)$ implies $r \in grd(\Pi, L_2)$. Let $cm_L(\Pi)$ be the set of rules in $rule_L(\Pi)$ which are cumulative wrt Π, and $ncm_{\leq L}(\Pi)$ be the set of rules $grd(\Pi, L) \setminus \cup_{0 \leq l \leq L} cm_l(\Pi)$. A ground atom $a \in atom_L(\Pi)$ is *level restrained* if $sup_l(a, \Pi) = sup_L(a, \Pi)$ for all integers $l > L$. Given a natural number L, let $lr_L(\Pi)$ denotes the sets of all level restrained atoms in $atom_L(\Pi)$, $lr_{\leq L}(\Pi)$ be the set of atoms $\cup_{0 \leq l \leq L} lr_l(\Pi)$, and $nlr_{\leq L}(\Pi)$ be the set of atoms $atom(grd(\Pi, L)) \setminus lr_{\leq L}(\Pi)$.

Definition 6 (Dynamic Completion Formula). *Given a safe ivASP program Π and an integer $L \geq 0$, the dynamic completion formula of Π on level L, denoted by $DCF(\Pi, L)$, is the conjunction of the static part*

$$\bigwedge_{r \in cm_L(\Pi)} \left[\widehat{body}(r) \supset head(r) \right] \wedge \bigwedge_{a \in lr_L(\Pi)} \left[a \supset \bigvee_{r \in sup_L(a,\Pi)} \widehat{body}(r) \right], \quad (2)$$

and the dynamic part

$$\bigwedge_{r \in ncm_{\leq L}(\Pi)} \left[\widehat{body}(r) \wedge k_L \supset head(r) \right] \wedge \bigwedge_{a \in nlr_{\leq L}(\Pi)} \left[a \wedge k_L \supset \bigvee_{r \in sup_L(a,\Pi)} \widehat{body}(r) \right],$$
$$(3)$$

where $\widehat{body}(r)$ denotes the conjunction of all literals in $body(r)$.

For each natural number L, an auxiliary variable k_L is introduced to annotate precedents of clauses in the dynamic part. When these clauses become invalid, the assignment of false to k_L make them satisfied. As we will see later, this technique is also applied to the dynamic loop formula.

Next, let us define the dynamic loop formula for ivASP programs. Suppose P is a ground logic program. The *positive dependency graph* of P is the directed graph whose vertices are atoms appearing in P and that consists of all edges from p to q such that p and q positively occur in the head and the body of a rule in P respectively. A nonempty set ℓ of atoms occurring in P is called a *loop* of P if for each pair of atoms a and b, there is a path from a to b in the positive dependency graph of P such that each vertex in the path belongs to ℓ.

Definition 7 (Dynamic Loop Formula). *Given an ivASP program Π and a natural number L, the dynamic loop formula, denoted by $DLF(\Pi, L)$, is the conjunction of the static part*

$$\bigwedge_{\ell \in \mathcal{L}_L(\Pi) \wedge \ell \subseteq lr_{\leq L}(\Pi)} \bigwedge_{a \in \ell} [a \supset ES(\ell, grd(\Pi, L))], \tag{4}$$

and the dynamic part

$$\bigwedge_{\ell \in loop_L(\Pi) \wedge \ell \not\subseteq lr_{\leq L}(\Pi)} \bigwedge_{a \in \ell} [a \wedge k_L \supset ES(\ell, grd(\Pi, L))], \tag{5}$$

where $loop_L(\Pi)$ is the set of all loops of $grd(\Pi, L)$, and $\mathcal{L}_L(\Pi) = loop_L(\Pi) \setminus \cup_{0 \leq l < L} loop_l(\Pi)$, $ES(\ell, grd(\Pi, L))$ is the formula $\vee_{a \in \ell} \vee_{r \in sup_L(a, \Pi) \wedge pos(r) \cap \ell = \emptyset} \widehat{body}(r)$.

Theorem 1. *Let Π be a safe ivASP program and L a natural number. Then a set A of ground atoms is an answer set of $grd(\Pi, L)$ iff $A \cup \{k_L\}$ is a model of*

$$\bigwedge_{0 \leq l \leq L} [DCF(\Pi, l) \wedge DLF(\Pi, l)]. \tag{6}$$

Proof. Let φ be the formula obtained from (6) by substituting $\perp$ for each k_l where $0 \leq l < L$ and by substituting $\top$ for k_L. Let $\varphi_1 = \bigwedge_{0 \leq l \leq L} DCF(\Pi, l)$ and $\varphi_2 = \bigwedge_{0 \leq l \leq L} DLF(\Pi, l)$ then $\varphi = \varphi_1 \wedge \varphi_2$.

$$\varphi_1 = \bigwedge_{0 \leq l \leq L} \bigwedge_{r \in cm_l(\Pi)} \left[\widehat{body}(r) \supset head(r) \right] \wedge \bigwedge_{0 \leq l \leq L} \bigwedge_{a \in lr_l(\Pi)} \left[a \supset \bigvee_{r \in sup_l(a, \Pi)} \widehat{body}(r) \right]$$

$$\wedge \bigwedge_{r \in ncm_{\leq L}(\Pi)} \left[\widehat{body}(r) \supset head(r) \right] \wedge \bigwedge_{a \in nlr_{\leq L}(\Pi)} \left[a \supset \bigvee_{r \in sup_L(a, \Pi)} \widehat{body}(r) \right],$$

Note that $grd(\Pi, L) = \bigcup_{0 \leq l \leq L} cm_l(\Pi) \cup ncm_{\leq L}(\Pi)$, and for a level restrained atom a, $sup_l(a, \Pi) = sup_L(a, \Pi)$ as $l < L$, thus conjuncts in φ_1 can be merged, resulting

$$\varphi_1 = \bigwedge_{r \in grd(\Pi, L)} \left[\widehat{body}(r) \supset head(r) \right] \wedge \bigwedge_{a \in atom(grd(\Pi, L))} \left[a \supset \bigvee_{r \in sup_L(a, \Pi)} \widehat{body}(r) \right],$$

which is exactly the Clark's completion of $grd(\Pi, L)$. In addition,

$$\varphi_2 = \bigwedge_{0 \leq l \leq L} \bigwedge_{\ell \in \mathcal{L}_l(\Pi) \wedge \ell \subseteq lr_{\leq l}(\Pi)} \bigwedge_{a \in \ell} [a \supset ES(\ell, grd(\Pi, l))]$$

$$\wedge \bigwedge_{\ell \in loop_L(\Pi) \wedge \ell \not\subseteq lr_{\leq L}(\Pi)} \bigwedge_{a \in \ell} [a \supset ES(\ell, grd(\Pi, L))],$$

Algorithm 1. IVASPSOLVE

Require: A safe ivASP program Π.
Ensure: Returns an answer set of Π or "no answer set".
1: $l \leftarrow 1$
2: $\Delta \leftarrow \emptyset$
3: **while** $l \leq UB$ **do**
4:      $P \leftarrow \text{GROUND}(\Pi, l)$
5:      $\Delta \leftarrow \Delta \wedge DCF(\Pi, l)$
6:      $(res, Ans, \Delta') \leftarrow \text{ASPSOLVE}(P, \Delta, k_l)$
7:      $\Delta \leftarrow \Delta'$
8:      **if** $res = $ true **then**
9:          **return** Ans
10:     **else**
11:          $l \leftarrow l + 1$
12:     **end if**
13: **end while**
14: **return** no answer set

For level restrained atoms, once they are grounded on a level, their supporting rules do not change on higher levels, so for a loop ℓ whose elements are all level restrained atoms, its external support does not change as well. Therefore $ES(\ell, grd(\Pi, l)) = ES(\ell, grd(\Pi, L))$.

$$\varphi_2 = \bigwedge_{\ell \in loop_L(\Pi)} \bigwedge_{a \in \ell} [a \supset ES(\ell, grd(\Pi, L))],$$

which equals to the loop formula of $grd(\Pi, L)$. By Theorem 1 of [7], we then obtain the desired result.   □

5   Implementation and Experiment

Theorem 1 provides an approach to solve ivASP programs, as we can determine whether a set of ground atoms is the answer set by checking if it satisfies Equation 6.

Algorithm 1 shows the high level structure of our prototype solver called *ivASP*. After initialization, it repetitively grounds the program, computes the dynamic completion formulas, and then call ASPSOLVE to do actual searching, until answer sets are found or global configured upper bound UB is reached. GROUND is a grounding procedure that outputs rules without variables, and marks each atom as level restrained or not and each rule as cumulative or not. Then dynamic completion formulas can be constructed. Δ is the constraint (clause) array, and stores all the constraints transformed from the input program and constraints learnt during search.

ASPSOLVE is a conflict-driven constraint learning ASP solving procedure, based on Algorithm 1 in [4]. The differences are: (a) the last argument k_l is an assumption that forces k_l be in the answer set. (b) for any detected unfounded

Table 2. Running time of *ivASP* and *iClingo* in selected benchmarks

instance	n	ivASP	iClingo	instance	n	ivASP	iClingo
1_FullIns_3	4	0.010	0.000	LCL661+1.015	2	0.150	0.170
1_Insertions_4	5	38.270	51.760	LCL669+1.015	2	0.180	0.260
2_FullIns_3	5	0.010	0.000	LCL671+1.015	2	2.520	2.780
2_Insertions_3	4	0.020	0.010	LCL677+1.015	2	0.160	0.200
3_Insertions_3	4	0.170	0.140	LCL689+1.010	2	0.150	0.140
queen5_5	5	0.010	0.000	LCL689+1.020	2	0.270	0.340
queen6_6	7	1.810	2.460	NLP160+1	2	1.360	1.260
queen7_7	7	0.060	0.050	NLP161+1	2	2.670	1.370
queen8_8	9	–	–	NLP162+1	2	1.310	1.260
queen8_12	12	97.020	183.450	NLP164+1	2	2.090	2.300
queen9_9	10	–	–	NLP165+1	2	2.040	2.370
	33	36.810	8.350	NLP190+1	2	46.870	22.330
	34	27.460	9.450	NLP192+1	2	13.360	13.430
Towers of Hanoi	36	112.750	21.200	NLP193+1	2	21.100	21.960
	39	213.750	52.030	NLP194+1	2	13.800	13.980
	41	393.200	99.070	NLP195+1	2	13.220	13.080
LCL651+1.010	2	0.090	0.120	NLP196+1	2	20.830	24.880
LCL651+1.020	2	0.360	0.370	NLP211+1	7	1.450	0.540
LCL653+1.005	2	0.110	0.090	NLP212+1	7	1.480	0.570
LCL653+1.015	2	0.310	0.340	NLP213+1	7	1.440	0.560
LCL655+1.010	2	5.790	3.460	SYN330+1	8	0.360	0.230
LCL659+1.015	2	270.580	59.690	SYN335+1	11	106.950	59.930
LCL661+1.001	2	0.140	0.120	SYN519+1	3	0.080	0.170

set, its dynamic loop formula (instead of normal loop formula) is added to Δ; and (c) if the input ground logic program has a answer set *Ans*, then it returns (*true*, *Ans*, Δ); otherwise it return (*false*, $\emptyset$, Δ). Each call of ASPSOLVE would probably add more learnt constraints to Δ, thus constraints learnt from current levels can be utilized on later levels.

To evaluate the performance of our proposed approach to ivASP solving, we compare *ivASP* to *iClingo* in some benchmarks. The first series of instances are from the GCP problem. The encoding of GCP for *ivASP* is like Example 1 with the addition of two meta-statement declaring k as the incremental variable and atoms instantiated from *ncolor* as non-level restrained. By treating GCP as a finite model computation problem, we derive its domain description [5], which is the encoding of GCP for *iClingo*. For the Towers of Hanoi problem, we use the encoding and benchmarks from [2]. The rest of benchmarks are from the FNT (First-order form syntactically non-propositional Non-Theorem) division of the CASC-23 competition, also requiring *iclingo* to solve in at least 2 steps and more than 0.1 second. Similar to *fmc2iasp*, which compiles finite model computation (FMC) problems to *iClingo* programs, we also implement a tool that can convert FMC problems to ivASP programs.

The experiments are all done in a linux desktop with an Intel Core i7-3520M CPU and 4 GB memory. In Table 2, total times of *ivASP* and *iClingo* running each selected instances are listed, with "−" denoting timeout in 500 seconds. The second column n denotes the least value of the incremental variable when answer sets are found.

As a conclusion, when a problem is natural to express by an ivASP program, our approach has advantages in both the language and efficiency, as shown in the GCP instances. In general, the performance of *ivASP* in these benchmarks is comparable to that of *iClingo*, though there indeed some instances *iClingo* shows over performance to *ivASP*. Nevertheless, we need to emphasize that this is mainly due to the shortage of optimization of ivASP at this stage, that we will seriously take into account in our next research.

6 Conclusion and Related Works

In this paper, an incremental variable was introduced to ASP language for incremental problem solving, and a dynamic transformation, which is based on the techniques of loop formulas [7] and conflict-driven learning [3], was then proposed to accelerate the solving of answer sets for the new language. With this transformation, a prototype system, *ivASP*, was developed, and the effectiveness of our system was demonstrated by experiments on some benchmarks.

Another system, *iClingo* [2], was also developed for incremental answer set solving. However, the problems to be solved in this system are required being encoded in three parts, which challenges the programmer on both domain knowledge and programming techniques. Instead our system only requires the problems being encoded in a natural way. Surprisingly, our system is still comparable to their system with regard to computation efficiency.

References

1. Baral, C.: Knowledge Representation, Reasoning, and Declarative Problem Solving. Cambridge University Press, New York (2003)
2. Gebser, M., Kaminski, R., Kaufmann, B., Ostrowski, M., Schaub, T., Thiele, S.: Engineering an incremental ASP solver. In: Garcia de la Banda, M., Pontelli, E. (eds.) ICLP 2008. LNCS, vol. 5366, pp. 190–205. Springer, Heidelberg (2008)
3. Gebser, M., Kaufmann, B., Neumann, A., Schaub, T.: Conflict-driven answer set solving. In: Proceedings of the 20th International Joint Conference on Artifical Intelligence, pp. 386–392 (2007)
4. Gebser, M., Kaufmann, B., Schaub, T.: Conflict-driven answer set solving: From theory to practice. Artificial Intelligence (2012)
5. Gebser, M., Sabuncu, O., Schaub, T.: An incremental answer set programming based system for finite model computation. AI Commun. 24(2), 195–212 (2011)
6. Gelfond, M., Lifschitz, V.: The stable model semantics for logic programming. In: Proceedings of the 5th International Conference on Logic Programming, vol. 161 (1988)
7. Lin, F., Zhao, Y.: Assat: Computing answer sets of a logic program by sat solvers. Artificial Intelligence 157(1), 115–137 (2004)

Programmable Managing of Workflows in Development of Software-Intensive Systems

P. Sosnin, Y. Lapshov, and K. Svyatov

Ulyanovsk State Technical University, Severny Venetc str. 32,
432027 Ulyanovsk, Russia
sosnin@ulstu.ru

Abstract. The paper focuses on the improvement of management processes in developments of software-intensive systems (SIS). An improvement can be achieved if a team of designers will use specialized programming of own actions in implementing of workflows. The offered approach is based on modelling of question-answer reasoning used in personal and collective solutions of project tasks. One type of such models is pseudo-code programs. Developed means are realized in a specialized instrumental environment, which provides conceptual designing of SIS.

Keywords: Conceptual designing, pseudo-code programming, question-answering, reasoning, software intensive systems, workflow.

1   Introduction

Four years ago, the group of well-known researchers and developers in software engineering has initiated a process of innovations [1], which have been named SEMAT (Software Engineering Methods And Theory). The main reason of this initiative was extremely low degree of success in developments of Software-Intensive Systems. This problem is steadily registered in statistics (success approximately 35%) that are presented in reports [2] of the company "Standish Group" last twenty years.

It is necessary to notice that, in normative documents of SEMAT, a way of working used by a team of designers is marked as a very important essence [3]. There "way-of-working" as a notion is defined as "the tailored set of practices and tools used by the team to guide and support their work." Therefore, researches aimed at the search of effective ways-of-working are perspective and can lead to increase of success in developments of SISs. In software engineering, such researches allocate a specialized subject domain solving the tasks of which it is necessary to consider the specificity of personal and collective activity of designers.

Development of SISs is based on the following features of designers' activity:

1. The joint work is fulfilled by a team in collaboration and coordination.
2. Typical units of the work are or a solution of a typical project task or creative task with using the technological tasks.

A. Moonis et al. (Eds.): IEA/AIE 2014, Part I, LNAI 8481, pp. 138–147, 2014.

3. Designers have a deal with a huge amount of tasks the greater part of which is arisen in front of them operatively.
4. Tasks being solved are combined in workflows, the amount of which is also very huge.
5. Any task, which should be solved, is appointed only to one designer under its responsibility.
6. At any moment of time, tasks of the definite workflow can be solved by a group of designers in parallel.
7. At any moment of time, any designer can solve a number of tasks in pseudo-parallel.

The indicated features directly concern the problem of managing the designers' activity under complicated circumstances in the real time. Therefore, investigations of such type of management can lead to innovative ways-of-working that will provide increasing the degree of success in designing of SISs.

This paper will focus on positive effects that can be achieved when programming of designers' actions is used in workflows' management. Methods and means that are suggested in the paper are oriented on pseudo-code programming of workflows implemented in instrumentally technological environment WIQA (Working In Questions And Answers [4].)

2   Preliminary Bases

The choice of the environment WIQA (for managing of workflows) has been caused by its question-answer memory (QA-memory) which supports the real time work with question-answer reasoning in interactions of designers with tasks being solved in designing of SIS.

Any cell of QA-memory is an interactive object which is intended and specified for storing of s simple question or answer on it. By other words, the specification of the cell includes a set of attributes with a corresponding set of operations that provide registering, visualizing and using models of the simple question or corresponding answer. We shall count that the question has a simple type if subordinated questions are not embedded to its structure.

It is necessary to note that tasks also are questions, a type of which will be designated as a "Z-type" in this paper. Let us clarify our understanding that will be below bound with Z-type.

Tasks are naturally artificial phenomena, which are arisen when   people apply an accessible experience in definite conditions to find and implement a composition of actions for achieving the definite aims. People interact with any task by using natural language and other language means. In such interactions, a statement of the task as its textual (sign) model plays a very important role.

The task statement as a communicative construction can be divided on "theme" (predictable starting point) and "rheme" (new information) that should receive additional meaning in answering on corresponding Z-question. The answer on Z-question is the composition of actions that provide an achievement of the necessary aims.

Usually, the creation of the answer on Z-question is beginning with attempts to understand the task statement. In this work, extraction of the subordinate questions is useful. Thus, the necessary level of understanding of the task can be achieved by question-answer reasoning (QA-reasoning) the registered form of which can be used as a question-answer model (QA-model) of the task.

In the general case, QA-model of the task includes subordinated questions of different types including Z-type, but always, the textual expression of any question has the predicative meaning with an obvious accent on its rheme. Decreasing uncertainty of the rheme till necessary level is the main aim of answering which adds the registered text of QA-reasoning to the task statement.

A very important kind of questions is bound with indicating in a clause the relation between a name and its meaning. Similar clauses can bind, for example, the name of a variable with its value or the name of an object with its characteristic. The rheme is also presented in such clauses.

It is necessary to note that presented understanding of questions and answers have been taking into account in the specification of QA-memory and its cells. The generalized scheme of this memory is presented in Fig. 1 where a diversity of design tasks is also reflected.

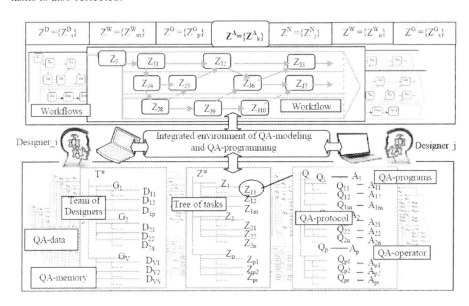

Fig. 1. QA-memory of WIQA

The team competence should be sufficient for real-time work with the following sets of tasks: subject tasks $Z^S = \{Z^S_i\}$ of the SIS subject area; normative tasks $Z^N = \{Z^N_j\}$ of the technology used by designers; adaptation tasks $Z^A = \{Z^A_k\}$ that provide an adjustment of the tasks $\{Z^N_j\}$ for solving the tasks $\{Z^N_j\}$; workflow tasks $\{Z^W_m\}$ that provide work with the tasks of Z^S-type in workflows $\{W_m\}$ in SIS; workflow tasks $\{Z^W_n\}$ that provide work with Z^N-type tasks in corresponding workflows $\{W_n\}$ in the used technology; and workflow tasks $\{Z^G_p\}$ and $\{Z^G_r\}$, any of which correspond to the specific group of workflows in SIS or the technology.

The environment WIQA with its QA-memory has been chosen for programming of workflows because this instrumental complex supports [5]:

- Modeling of the designers' team which can have any organizational structure:
- Dynamic registering and visualizing of a tasks' tree for the SIS being developed;
- Question–answer modeling of reasoning used in the solution process of any task being solved by members of the team.

The named constructions have hierarchical structures, which are demonstrated in general in Fig. 1. Let us notice that the task being solved can be presented with different QA-models. It depends on a used viewpoint on the task. One of such versions is a program of actions being implemented in the solution of the task. This version in a pseudo-code form is indicated in the scheme as QA-program because its operators (QA-operators) inherit the features of memory cells.

Indicated hierarchical constructions were used in a number of WIQA-applications among which we mark "Conceptual Designing" and "Multi-Agent System for Monitoring of Surrounding of the Sea Vessel". QA-programming of the work with tasks was undertook for several plug-ins of WIQA among which we mark "Controlling of Assignment", "Subsystem of Interruption" and "Kanban board".

3   Related Works

A well-known approach to programming of the human actions was realized in the "Model Human Processor" (MH-processor) as an engineering model of human performance in solving different tasks in real-time regime [6].

The EPIC version [7] of MH-processor uses programs that are written in the specialized command language Keystroke Level Model (KLM). A set of basic KLM actions includes the following operators: K – key press and release (keyboard), P – point the mouse to an object on screen, B – button press or release (mouse), H – hand from the keyboard to mouse, or vice versa, and other commands. Operators of the KLM-language and their values are used to estimate temporal characteristics of human interactions for alternative schemes of interfaces. KLM-programs do not correspond exactly to the used reasoning, and; therefore, they do not reflect interactions with the accessible experience.

It should be noted that formal languages are widely used in applications that involve workflows. Therefore, a separate group of related research includes publications, for example, papers [8] and [9] that open solutions that are connected with the use of workflows in collaborative designing. In interactions with workflows, the graphical means are widely used. Among such means, we mark Kanban [10] and Scrum [11] which help to organize the real time interactions in rational forms. It is necessary to note that in this group of related works, means of pseudo-code programming were not suggested for managing of workflows.

The offered version of workflows' managing is coordinated with basic principles of the SEMAT Kernel, which is described in [2]. "The process is what the team does. Such processes can be formed dynamically from appropriate practices in accordance with current situations. The informational cards and queue mechanisms are being used for managing of ways-of-working [2]."

One more group of related publications concerns the use of question-answering in computerized mediums, for example, papers [12] and [13]. In this group, the closest research presents the experience-based methodology "BORE" [13], in which question-answering is applied as well, but for the other aims, this methodology does not support programming of the creative designer activity.

4   Programmable Managing

4.1   Operational Conditions

The first attempt of programmable managing of workflows has been realized with an orientation on visualizing the state of designing in a Kanban board [10]. In developing the second rationalized version, all changes and modifications were bound with a Scrum approach [11] to an organization of collective activity. Let us notice that Scrum method is used as an example of applying the SEMAT idea in the description of its kernel [2]. The scheme of the second version is generally presented in Fig. 2. In the scheme, the added and modified components are marked by the grey colour.

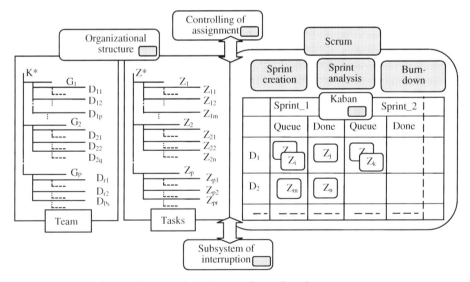

Fig. 2. Operational conditions of workflows' management

The tasks' tree is a dynamic construction, and its typical units of evolving are task and workflow. As interests of this paper are focused at ways-of-working for managing of workflows, the work of designers in showed conditions will be explained only for this case. It is necessary to note that all on the scheme has explicit relation to pseudo-code means of WIQA (embedded pseudo-code algorithmic language L^{WIQA} (specialized editor, interpreter, compiler, a number of QA-programs).

4.2   Preparation to Managing

Let us assume that a technology workflow task Z^W_n is included to the tree of tasks and a group of designers $G_p(\{D_{ps}\})$ should implement the task Z^W_n with a subordinated set of tasks $\{Z^N_j\}$. After that, the following programmed procedures will be executed:

1. The leader of the group distributes tasks $\{Z^N_j\}$ among members $\{D_{ps}\}$ of the group in accordance with them competences.
2. A specialized (pseudo-code) programmed agent registers all appointments in tables of "Controlling of Assignments". A fragment of the agent source code has the following view:

```
&OrgProject& := QA_GetProjectId ("Organizational structue")
&ProjectsTask& := QA_GetQaId(&OrgProject&, "          ")
&cnt& := 0
&max& := QA_GetDirectChildsCount (&OrgProject&,
&ProjectsTask&)
LABEL &L1&
IF &cnt& >= &max& THEN GOTO &L2&
&Pid& := QA_GetDirectChildId (&OrgProject&, &ProjectsTask&,
&cnt&)
&Pname& := QA_GetQAText(&Pid&)
&npid& := QA_GetProjectId(&Pname&)
if &npid& == &current_project& THEN GOTO &L3&
&cnt& := &cnt& + 1
GOTO &L1&
LABEL &L2&
```

This code is only presented for a demonstration of syntax L^{WIQA}. The code of this agent is processing by the WIQA-compiler.

3. The leader of the group uploads the necessary typical QA-program for the task Z^W_n from a library of QA-programs. The uploaded pattern is adjusted to the specificity of the executed work and, after that, QA-program of the task Z^W_n is executed. Results of the execution register in "Controlling of Assignments" for each task Z^N_j in the form of "pseudo-code of condition opening the opportunity to start work with this task".
   If the pattern for the task Z^W_n is absent, then the necessary QA-program is created and processed in accordance with points (p.1 - p.3).
4. Additionally, the leader of the group sets the estimated characteristics of time for planned work with each task Z^N_j.
5. The responsible member of the group specified a next sprint (list of tasks being solved, values of story-points for tasks, real estimates, initial velocity of the team and other characteristics).

Fig. 3. Tasks' queues on the board

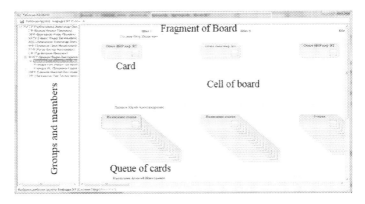

Fig. 4. Tasks' queues on the board

This part of the work is implemented in the operational conditions presented in Fig. 3, where some labels are used because all interface forms in Russian.

6. The other programme agent, that is responsible for the current state of the Kanban board visualizes a number of characteristics for each task Z^N_j in two-side cards in corresponding cells of the board. An interface of the visualization is presented in Fig. 3.

4.3   Real Time Work

The real time work of designers with tasks of any workflow is based on the following actions:

1. Informational interactions with queues of tasks that are located in the board of cards includes possibilities of visual and touchable navigation (on groups, members of groups, previous sprints), sorting and selecting of cards, their dynamic visualizing and choosing of the necessary card.

2. Informational interactions with the choosing card include choosing of the necessary side of the card, analyzing of information on the visual side and activating of the necessary actions from the card.
3. Programmable interactions of the designer with the personal queue of the appointed tasks the scheme of which is figuratively presented in Fig 5.

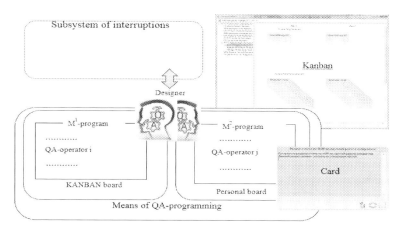

Fig. 5. Operational conditions of programmable managing

In the discussed case, the queues with which the designer works is expediently interpreted as a special type of program (M-programs), which manages the activity of the designer. Any queue in M-program includes the names of the tasks and attributes that indicate conditions in which the work with the corresponding task can be begun or interrupted. Any unit of such a queue is interpreted as an operator of an M-program.

A program character of such an interpretation is clarified by an abstract example with a workflow contains a sequence of three tasks:

```
Z0: &IsEnabled& := 1
Z1:IF &idZ0&.STATE == &running& THEN &IsEnabled& := 1 ELSE
&IsEnabled& := 0
Z2:IF &idZ1&.STATE == &done& THEN &IsEnabled& := 1 ELSE
&IsEnabled& := 0Z.
```

There is a possibility of using two types of M-programs in the WIQA-environment. The first type M1 provides the pseudo-parallel solving of tasks (QA-programs) by the designer playing a role of I-processor [14]. Such an opportunity is supported by the plug-in "System of interruption".

Any M2-program manages the execution of tasks by the group of designers in the workflows, which are processed collectively with the use of other means presented in Fig. 2. In this case, it is necessary to take into account an inheritance among conditions that are opening the possibilities for the work with tasks in the group. In both cases, the access to the queues includes priorities of their units.

4.4   Analysis of Sprint Execution

The analysis (inspection) of the sprint execution is the principal part of a Scrum process. This part of the work helps to define the characteristics of the definite team activity that will allow more adequately planning of the next sprint. The most part of data for analysis is gathered from daily Scrums.

The one of important results of the analysis is a burn-down diagram that helps to manage by the sprint work during its execution. In the described case, this type of diagrams is visualized as it is shown in Fig. 6.

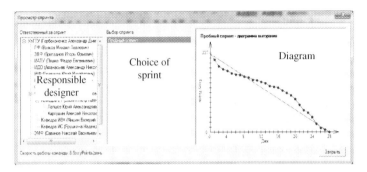

Fig. 6. Burn-down diagram

There are a number of typical trends of a watched diagram curve, which show evolving of events in the execution of the current sprint.

5   Conclusion

The following conclusions can be drawn from the present study of programmable managing of workflows in the development of software-intensive systems. The offered way-of-working for such managing allows including in the management process a number of additional effects. These effects are caused by the automation of a number of designers' actions in them collaborative activity. Moreover, the activity of the designers' team is estimated with using of a set of metrics, helping to adjust the executed work on specificity of the team competence and its power.

The offered approach correlates with innovations declared in SEMAT for program engineering. The positive effects of managing are achieved by using: the means of pseudo-code programming oriented on the memory that is specified for the registration of QA-reasoning of designers; the programmable activity of designers; the visualized cards in the frames of Kanban and Scrum means. In conceptual designing, all workflows, including of programmable managing, can be implemented in WIQA environment.

References

1. Jacobson, I., Meyer, B., Soley, R.: The SEMAT Initiative: A Call for Action, http://www.drdobbs.com/architecture-and-design/the-semat-initiative-a-call-for-action/
2. Reports of the Standish Group, https://secure.standishgroup.com/reports/
3. Jacobson, I., Ng, P.-W., McMahon, P., Spence, I., Lidman, S.: The essence of software engineering: the SEMAT kernel. Queue 10(10), 1–12 (2012)
4. Sosnin, P.: Conceptual solution of the tasks in designing the software intensive systems. In: Proc. MELECON, Ajaccio, France, pp. 293–298 (2008)
5. Sosnin, P.: A Scientifically Experimental Approach to the Simulation of Designer Activity in the Conceptual Designing of Software Intensive Systems. IEEE Access 1, 488–504 (2013)
6. Card, S.K., Thomas, T.P., Newell, A.: The Psychology of Human-Computer Interaction. Lawrence Erbaum Associates, London (1983)
7. Kieras, D.: Using the Keystroke-Level Model to Estimate Execution Times (2001)
8. Held, M., Blochinger, W., Structured, M.: collaborative workflow design. Future Generation Computer Systems 25(6), 638–653 (2009)
9. Van der Aalst, W.M.P., Hofstede, A.H.M.: Workflow Patterns Put into Context. Software and Systems Modeling 11(3), 319–323 (2012)
10. Wang, J.X.: Lean Manufacturing Business Bottom-Line Based. In: Kanban: Align Manufacturing Flow with Demand Pull, pp. 185–204. CRC Press (2010)
11. The State of Scrum: Benchmarks and Guidelines. Scrum Alliance (2013), http://www.scrumalliance.org/scrum/media/ScrumAllianceMedia/Files%20and%20PDFs/State%20of%20Scrum/2013-State-of-Scrum-Report_062713_final.pdf
12. Webber, B., Webb, N.: Question Answering. In: Clark, Fox, Lappin (eds.) Handbook of Computational Linguistics and Natural Language Processing. Blackwells (2010)
13. Henninger, S.: Tool Support for Experience-based Software Development Methodologies. Advances in Computers 59, 29–82 (2003)
14. Sosnin, P.: Pseudo-code Programming of designer activity in development of software intensive systems. In: Jiang, H., Ding, W., Ali, M., Wu, X. (eds.) IEA/AIE 2012. LNCS, vol. 7345, pp. 457–466. Springer, Heidelberg (2012)

An Answer Validation Concept Based Approach for Question Answering in Biomedical Domain

Wen-Juan Hou and Bing-Han Tsai

Department of Computer Science and Information Engineering
National Taiwan Normal University
No.88, Section 4, Ting-Chou Road, Taipei 116, Taiwan, R.O.C
{emilyhou,60047039S}@csie.ntnu.edu.tw

Abstract. With the continuously growing literatures in the biomedical domain, it is not feasible for researchers to manually go through all information for answering questions. The task of making knowledge contained in texts in forms that machines can use for automated processing is more and more important. This paper describes a system to answer multiple-choice questions for the biomedical domain while reading a given document. In this study, we use the data from the pilot task "machine reading of biomedical texts about Alzheimer's disease" which is a task of the Question Answering for Machine Reading Evaluation (QA4MRE) Lab at CLEF 2012. We adapt the concept of answer validation that assumes the over-generation hypotheses will be checked in the validation step. In the following, the query expansion technique "global analysis" is applied. The best result is 0.51 $c@1$ score which is clearly above the baseline at CLEF 2012 and shows an exhilarating performance.

Keywords: Question answering system, Answer validation, Query expansion, Question answering for machine reading evaluation.

1   Introduction

Question Answering (QA) is the task whereby an automated system receives a question in natural language and returns small snippets of texts that contain an answer to the question [1]. Nowadays many scientific researchers have investigated practiced QA systems, and some mature QA systems have been widely applied. The recent successes of IBM's Watson on Jeopardy highlight the possibilities and potential power of QA [2]. However, few researchers are doing research on the biomedical domain with the complexity of questions and abundant domain knowledge [3]. The amount of the current biomedical text data and research papers is very huge and it grows at a very high and unprecedented rate. Obviously, it is not feasible for researchers to manually go through all documents stored in the MEDLINE digital library to extract the best answers to their questions.[1] Hence, the task of making knowledge contained in texts available in forms that machines can use for automated

---

[1] http://www.ncbi.nlm.nih.gov/pubmed

A. Moonis et al. (Eds.): IEA/AIE 2014, Part I, LNAI 8481, pp. 148–159, 2014.
© Springer International Publishing Switzerland 2014

processing is more and more important. The study on QA systems oriented to the biomedical domain is still under development and needs further investigation.

The question answering task has two reference inputs: the corpora to be used to extract the relevant answers and the question itself. In general, the QA processing in a QA system consists of three main processing phases, namely, question processing, document processing, and answer processing phases [4]. For the biomedical QA systems, MEDLINE abstracts were referenced in order to retrieve relevant documents, e.g., AskHERMES [5], EAGLi [6] and MedQA [7]. To help the QA in the biomedical domain, many techniques were applied. In [8], the PICO (Problem/Population, Intervention, Comparison and Outcome) framework and statistical techniques were used. Several machine learning based systems were proposed [5, 7]. Semantic-based approaches were studied by many researchers such as in [5, 9-10]. Pattern-based methods were often applied in the biomedical QA systems [11-12]. [10, 13] used the clustering techniques to help with the task. Accompanying with the semantic-based methods, some researchers also used the summarization or IR approaches to finding answers [14-15]. In [12, 16], researchers explored the logic-based approach. The inference-based approach to the biomedical QA was investigated in [17].

Machine reading of biomedical texts about Alzheimer's disease is a pilot task of the Question Answering for Machine Reading Evaluation (QA4MRE) Lab at CLEF 2012.[2] The task follows the same set up and principles as the QA4MRE, with the difference of focusing on the biomedical domain [18]. The task focuses on the reading of single documents and the identification of the answers to a set of questions about information that is stated or implied in the text. Questions are in the form of multiple choices, each having five options, and only one correct answer. The task is like the reading comprehension test, and it is the problem we address in this paper.

For solving the task of machine reading of biomedical texts about Alzheimer's disease, researchers have applied several methods. Attardi *et al.* [19] used the index expansion technique, adding variants of terms and relations to a specialized sentence retrieval engine. Bhattacharya and Toldo [20] developed several information retrieval (IR) and semantic web-based strategies. Some researchers [21-22] adapted the existing question answering systems and developed methods to select answers. In [23-24], similarity matching scores of answer/question pairs in a document were calculated to select the best answer. Patel *et al.* [25] introduced the Configuration Space Exploration framework for building and exploring configurations for intelligent information systems.

The study aims at exploring the ability of a machine reading system to answer questions about Alzheimer's disease. We adapt the concept of answer validation that assumes the over-generation hypotheses will be checked in the validation step. We follow the query expansion technique [26] in finding the most suitable answer to a given question.

The rest of the paper is organized as follows. Section 2 presents the overview of our system architecture. We describe the proposed method in details in Section 3. The

---

[2] http://celct.fbk.eu/QA4MRE

experimental data used to evaluate the method and the results achieved by the proposed method are shown and discussed in Section 4. Finally, we express our main conclusions and outline the future work.

2   Architecture Overview

Figure 1 shows the overall architecture of our method for answering questions about Alzheimer's disease. As we mentioned before, the study focuses on the reading of single documents and the identification of the answers to a set of questions about information that is stated or implied in the text. Questions are in the form of multiple choices, each having five options, and only one correct answer. Hence, the test data are first separated to three parts: documents, questions and answers. We then preprocess the data and get the stemmed documents, query words/phrases and answer words/phrases, which will be stated in details in Section 3.1. After that, the answer validation concept for solving the QA problem is used. In the hypothesis generation phase, query words/phrases and answer words/phrases are combined together. For assessing which combination is the best, we give the hypothesis words/phrases weights by referencing to the background knowledge collection. Next, we retrieve the hypothesis related sentences from the document with the weighting values. Finally, the selected answer is obtained according to a hypothesis scoring scheme.

3   Methods

The methods are divided into five parts: (1) data preprocessing, (2) hypothesis word/phrase generation, (3) hypothesis word/phrase weighting, (4) hypothesis related sentence retrieval and weighting, and (5) hypothesis scoring and answer selection. The details for each part are explained in Sections 3.1-3.5. Furthermore, we add the query expansion concept and design other experiments. It is elucidated in Section 3.6.

3.1   Data Preprocessing

Exclusion of Stop Words and Punctuations. After turning uppercase letters to lowercase letters, we first remove stop words. Stop words are common English words that frequently appear in the text but are not helpful in discriminating special classes. The stop word list in this study is collected with reference to a text retrieval toolkit website.[3] The wordlist contains 429 words. At this step, we also remove punctuation symbols like (, -, and so on.

Stemming. Stemming is a procedure of transforming an inflected form to its root form. For example, "inhibited" and "inhibition" will be mapped into the root form "inhibit" after stemming. Porter's stemmer is applied in this study.[4]

---

[3] http://www.lextek.com/manuals/onix/stopwords1.html
[4] http://tartarus.org/martin/PorterStemmer/

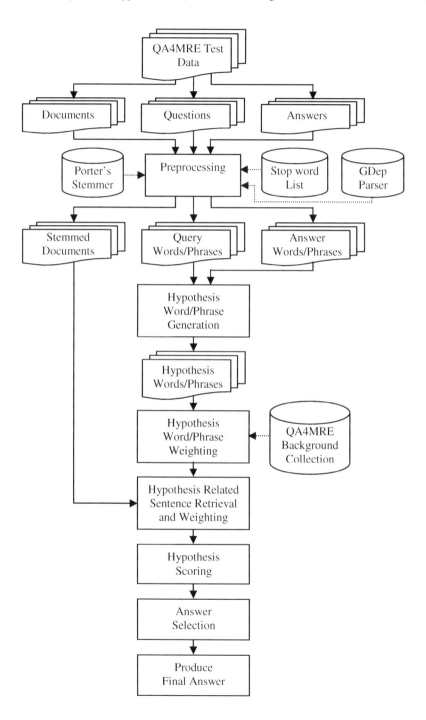

Fig. 1. System architecture of answering questions about Alzheimer disease

Chunking. Chunking can parse sentences into partial syntactic structures such as noun, verb, or adjective phrases. For the documents and questions, GDep is utilized to chunk the noun phrases since most named entities in biomedical texts are contained within noun phrases.[5] But the noun phrases that include interrogatives (e.g., what, which, …) are rejected. For the answers, we remain all phrases because the answer items usually contain important information that we cannot discard them arbitrarily.

3.2   Hypothesis Word/Phrase Generation

Because we want to compare the results between using single words and using phrases, we generate hypothesis words and hypothesis phrases in this step. A hypothesis word/phrase is composed of query words/phrases and answer words/phrases. An example is shown in Figure 2. For briefness, we give only two choices in the example.

Question: Which technique was used to determine the cellular locations of the CLU1 and CLU2 gene products?
(a) intracellular and secreted   (b) intracellular localization
Query words: techniqu, determin, cellular, locat, clu1, clu2, gene, product
Answer words: (a) intracellular, secret   (b) intracellular, local
Hypothesis words:
H_1: techniqu, determin, cellular, locat, clu1, clu2, gene, product, intracellular, secret
H_2: techniqu, determin, cellular, locat, clu1, clu2, gene, product, intracellular, local
Query phrases: the cellular location, the clu1 and clu2 gene product
Answer phrases: (a) intracellular, secrete   (b) intracellular localization
Hypothesis phrases:
H_1: the cellular location, the clu1 and clu2 gene product, intracellular, secrete
H_2: the cellular location, the clu1 and clu2 gene product, intracellular localization

---

Fig. 2. An example for hypothesis word/phrase generation

3.3   Hypothesis Word/Phrase Weighting

Hypothesis Word Weighting. For deciding the importance of hypothesis words to each sentence, we need to give the words weighting schemes. They are Term Frequency (TF) and Term Frequency-Inverse Document Frequency (TF-IDF).

TF Weighting. If some word occurs frequently in the document, it means the word may play an important role. The term frequency weight TF_{H_i} is defined as follows.

$$TF_{H_i} = 1 + \frac{f_{H_i}}{\max_{\forall H_i} f_{H_i}} \tag{1}$$

---

[5] http://people.ict.usc.edu/~sagae/parser/gdep/index.html

where f_{H_i} is the number of occurrences of the hypothesis word H_i in the document, and the denominator is the max number of occurrences of all hypothesis words in the document. The summand 1 means each word has a weighting value at least one.

TF–IDF Weighting. Equations (2) and (3) compute for TF-IDF weighting values.

$$TF - IDF_{H_i} = TF_{H_i} \times IDF_{H_i}, \quad \text{and} \tag{2}$$

$$IDF_{H_i} = \begin{cases} \log_2 \dfrac{N}{n_{H_i}} & \text{if } n_{H_i} \neq 0 \\[2mm] 0.1 & \text{if } n_{H_i} = 0 \text{ and } f_{H_i} \neq 0 \\[2mm] 0 & \text{otherwise} \end{cases} \tag{3}$$

where N is the total number of documents in the background collection, and n_{H_i} is the number of documents where the hypothesis word H_i appears in the background. f_{H_i} is the number of occurrences of H_i in the test document. If H_i is not in the background but occurs in the test document, we will use the value of 0.1 instead.

Hypothesis Phrase Weighting. For deciding the importance of hypothesis phrases to each sentence, we give a weighting scheme called Phrase Frequency (PF).

PF Weighting. The formula is defined as follows.

$$PF_{P_i} = \frac{f_{P_i}}{\max\limits_{\forall P_i} f_{P_i}} \tag{4}$$

where f_{P_i} is the number of the hypothesis phrase P_i in the test document.

3.4 Hypothesis Related Sentence Retrieval and Weighting

In this step, we first extract the sentences where a hypothesis word or phrase appears. Then the weighting values are computed as follows.

Weighting with Hypothesis Words. A sentence weighting is the sum of all matching hypothesis word weightings in the sentence. Since we have TF and TF-IDF weighting values to hypothesis words, the corresponding weights for sentence S_j are as below.

$$SHW\_TF_j = \sum_{H_i \in S_j} TF_{H_i} \tag{5}$$

$$SHW\_TFIDF_j = \sum_{H_i \in S_j} (TF_{H_i} \times IDF_{H_i})$$

(6)

Weighting with Hypothesis Phrases. The formula for sentence S_j is as follows.

$$SHW\_PF_j = \sum_{P_i \in S_j} PF_{P_i}$$

(7)

3.5   Hypothesis Scoring and Answer Selection

We rank the hypothesis related sentences according to its weighting values. If there is only one sentence with the highest score, the answer is selected. Otherwise, no answer is proposed.

3.6   Query Expansion

Query expansion is the process of reformulating a seed query to improve retrieval performance in information retrieval operations. That is, it adds terms to the query in order to achieve a better recall. In this paper, we do query expansion using the global document analysis technique.

The basic idea of global analysis is that the global context can be used to determine similarities between concepts [26]. The context for a word (i.e., concept) is all the words that co-occur in documents with that word. The context in the study comes from QA4MRE background collections. It includes 1,041 full articles. We call the words extracted from the background collection as "background words." After doing lowercase turning, removal of stop words and punctuations, and stemming, four steps are applied. The methods are introduced as follows.

Building of a Background Word-Document Matrix. A background word-document matrix is based on how background words are indexed by the background documents. Figure 3 shows the matrix. In Figure 3, $\overrightarrow{K_i} = (w_{i,1}, w_{i,2}, \cdots, w_{i,N})^T$ stands for a weight vector of a background word K_i in all background documents. N is the total number of documents. By reference to [26], the weight is computed as Equation (8).

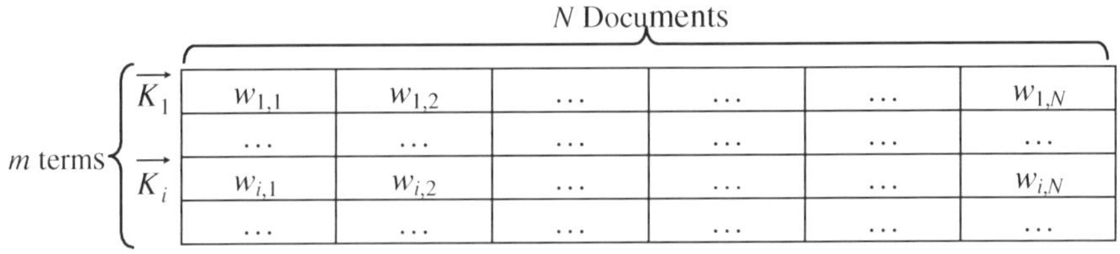

		N Documents				
	$\overrightarrow{K_1}$	$w_{1,1}$	$w_{1,2}$	...	...	...	$w_{1,N}$
		...	...	...	...	...	...
m terms	$\overrightarrow{K_i}$	$w_{i,1}$	$w_{i,2}$	...	...	...	$w_{i,N}$
		...	...	...	...	...	...

Fig. 3. A background word-document matrix example

$$w_{i,j} = \frac{(0.5 + 0.5 \frac{f_{i,j}}{\max_g f_{i,g}}) \times ITF_j}{\sqrt{\sum_{l=1}^{N} [(0.5 + 0.5 \frac{f_{i,l}}{\max_g f_{i,g}}) \times ITF_l]^2}} \tag{8}$$

where $f_{i,j}$ is the frequency of word k_i in the document d_j. ITF_j stands for the inverse term frequency (ITF) of document d_j. ITF can evaluate the importance of a document. If a document contains more different words, it means the topic of the document is more unclear, thus less important. The formula of ITF_j is shown as follows.

$$ITF_j = \log \frac{m}{m_j} \tag{9}$$

where m is the number of words in the document collection and m_j is the number of words indexed by the document d_j.

Computing of a Virtual Query Vector. A query vector $\vec{q} = (q_1, q_2, \cdots, q_m)^T$ represents a query q. Here, q_i is the weight of the query word K_i contained in the query q; m is the total number of words in the collection. We define q_i as term frequency in the query. We then map the query vector to the background word-document matrix, and call it "virtual query vector."

$$\vec{q_v} = \begin{cases} \sum_{K_i \in q} q_i \cdot \vec{K_i} & \text{if } K_i \text{ is in background words} \\ \sum_{K_i \in q} q_i \cdot \overrightarrow{ZeroVector} & \text{otherwise} \end{cases} \tag{10}$$

where q means the set of query words. Because the query word K_i may exist outside the background documents, we define $\overrightarrow{ZeroVector}$ to solve this problem.

$$\overrightarrow{ZeroVector} = (\frac{0.5 \cdot ITF_1}{\sqrt{\sum_{l=1}^{N}(0.5 \cdot ITF_l)^2}}, \frac{0.5 \cdot ITF_2}{\sqrt{\sum_{l=1}^{N}(0.5 \cdot ITF_l)^2}}, \cdots, \frac{0.5 \cdot ITF_N}{\sqrt{\sum_{l=1}^{N}(0.5 \cdot ITF_l)^2}}) \tag{11}$$

Similarity Computing of Words and Queries. The similarity between a background word K_i and the query q is denoted by $sim(q, K_i)$.

$$sim(q, K_i) = (\vec{q_c})^T \cdot \vec{K_i} = (\sum_{K_j \in q} q_j \cdot \vec{K_j})^T \cdot \vec{K_i} = \sum_{K_j \in q} q_j \cdot (\vec{K_j}^T \cdot \vec{K_i}) \tag{12}$$

Selection of Expanded Words. According to $sim(q, K_i)$, we select some top words as the expanded words. Then they are merged with the query words to form a new query.

4   Experiments and Results

4.1   Evaluation Metric

The evaluation metric we employed in this paper is $c@1$ which is used in QA4MRE. The formula of $c@1$ is stated as follows. In Equation (13), n_R is the number of questions correctly answered, n_U is the number of questions unanswered, and n is the total number of questions.

$$c@1 = \frac{1}{n}(n_R + n_U \frac{n_R}{n}) \tag{13}$$

4.2   Experimental Results and Discussions

Table 1 presents the evaluation results where we make different combination of techniques in the study.

Table 1. Experimental results

No.	Experiment Name	C1	C2	$c@1$
1	Hypothesis Words+TF	0	7	0.18
2	Hypothesis Words+TFIDF	0	17	0.43
3	Top4+Hypothesis Words+TF	2	10	0.26
4	Top4+Hypothesis Words+TFIDF	1	16	0.41
5	Top5+Hypothesis Words+TFIDF	2	17	0.45
6	Top5+Hypothesis Words+TFIDF+PF	2	17	0.45
7	QE250+Top5+Hypothesis Words+TFIDF+PF	2	18	0.47
8	QE350+Top5+Hypothesis Words+TFIDF+PF	2	19	0.50
9	Top5+Hypothesis Words+TFIDF+Hypothesis Phrases+PF	1	18	0.46
10	QE350+Top5+Hypothesis Words+TFIDF+ Hypothesis Phrases+PF	1	20	0.51

In the test data, there are four test documents with 10 questions for each document. Thus, there are 40 questions in total. In Table 1, "C1" is the number of questions unanswered; "C2" is the number of questions answered correctly.

In the "Experiment Name" column, "Hypothesis Words" means that the answer validation concept with hypothesis words is used. "TF" and "TFIDF" mean weighting schemes are TF and TF-IDF, respectively. "Top4" and "Top5" stand for selecting top 4 or 5 related sentences from the test document. "PF" means the phrase frequency is utilized. "QE250" and "QE350" represent that the number of expanded words are 250 and 350. "Hypothesis Phrases" means that the answer validation concept with hypothesis phrases is applied.

From experiments 1-2 and 3-4, it proves that TF-IDF weighting scheme is better than TF weighting scheme. It is reasonable in the natural language process domain.

Comparing with experiments 6-8 and 9-10, it illustrates that the query expansion technique helps well. We also make a baseline experiment that only uses TFIDF and gets 0.20 of $c@1$. Comparing with the baseline and No.2, it elucidates that the answer validation concept can increase the performance a lot. Utilizing both the answer validation concept and the query expansion shows the best score that is shown in No.10.

From the report of [18], the top 3 highest scores of participating in the task are 0.55, 0.47 and 0.30, respectively. The baseline is 0.20, too. Comparing with those teams, our best score is less than 0.55 but higher than others. The results are clearly above the baseline and thus show an exhilarating performance.

5   Conclusion

This study aims at exploring the ability of a machine reading system. The reading comprehension documents focus on Alzheimer's disease and the questions are multiple-choice forms. We adapt the concept of answer validation and the query expansion technique "global analysis" is applied. The best result gets 0.51 $c@1$ score which is clearly above the baseline at CLEF 2012 and shows an exhilarating performance. In the future, we can try the parsing-based technique to understand the structure of the sentence. Proposing some strategies to retrieve the related sentences from the background collection is the other direction. Anaphora resolution should be helpful in the study. Furthermore, using more background collections is a feasible way worthy of investigation.

Acknowledgements. Research of this paper was partially supported by National Science Council, Taiwan, under the contract NSC 102-2221-E-003-027.

References

1. Voorhees, E.M., Tice, D.M.: The TREC-8 Question Answering Track Evaluation. In: Proceedings of Text Retrieval Conference TREC-8, pp. 83–105 (1999)
2. IBM Watson: IBM (2011), http://www-03.ibm.com/innovation/us/watson/
3. Xu, B., Lin, H., Liu, B.: Study on Question Answering System for Biomedical Domain. In: Proceedings of IEEE 2009 International Conference on Granular Computing (GrC 2009), pp. 626–629 (2009)
4. Hirschman, L., Gaizauskas, R.: Natural Language Question Answering: the View from Here. Nat. Lang. Eng. 7, 275–300 (2001)
5. Cao, Y., Liu, F., Simpson, P., Antieau, L., Bennett, A., Cimino, J.J., Ely, J., Yu, H.: AskHERMES: an Online Question Answering System for Complex Clinical Questions. J. Biomed. Inform. 44(2), 277–288 (2011)
6. Gobeill, J., Patsche, E., Theodoro, D., Veuthey, A.-L., Lovis, C., Ruch, P.: Question Answering for Biology and Medicine. In: Proceedings of the 9th International Conference on Information Technology and Applications in Biomedicine (ITAB 2009), pp. 1–5 (2009)

7. Yu, H., Lee, M., Kaufman, D., Ely, J., Osheroff, J.A., Hripcsak, G., Cimino, J.: Development, Implementation, and a Cognitive Evaluation of a Definition Question Answering System for Physicians. J. Biomed. Inform. 40(3), 236–251 (2007)
8. Demner-Fushman, D., Lin, J.: Answering Clinical Questions with Knowledge-based and Statistical Techniques. Comput. Linguist. 33(1), 63–103 (2007)
9. Delbecque, T., Jacquemart, P., Zweigenbaum, P.: Indexing UMLS Semantic Types for Medical Question-Answering. Stud. Health Technol. Inform. 116, 805–810 (2005)
10. Weiming, W., Hu, D., Feng, M., Wenyin, M.: Automatic Clinical Question Answering Based on UMLS Relations. In: Proceedings of the Third International Conference on Semantics, Knowledge and Grid (SKG 2007), pp. 495–498 (2007)
11. Slaughter, L.A., Soergel, D., Rindflesch, T.C.: Semantic Representation of Consumer Questions and Physician Answers. Int. J. Med. Inform. 75, 513–529 (2006)
12. Terol, R.M., Martinez-Barco, P., Palomar, M.: A Knowledge Based Method for the Medical Question Answering Problem. Comput. Biol. Med. 27, 1511–1521 (2007)
13. Yu, H., Lee, M.: Accessing Bioscience Images from Abstract Sentences. Bioinformatics 22(14), e547–e556 (2006)
14. Demner-Fushman, D., Few, B., Hauser, S.E., Thoma, G.: Automatically Identifying Health Outcome Information in MEDLINE Records. J. Am. Med. Inform. Assoc. 13(1), 52–60 (2006)
15. Shi, Z., Melli, G., Wang, Y., Liu, Y., Gu, B., Kashani, M.M., Sarkar, A., Popowich, F.: Question Answering Summarization of Multiple Biomedical Documents. In: Proceedings of the 20th Conference of the Canadian Society for Computational Studies of Intelligence on Advances in Artificial Intelligence (CAI 2007), pp. 284–295 (2007)
16. Rinaldi, F., Dowdall, J., Schneider, G., Persidis, A.: Answering Questions in the Genomics Domain. In: Proceedings of the ACL-2004 Workshop Question Answering in Restricted Domains, pp. 46–53 (2004)
17. Kontos, J., Lekakis, J., Malagardi, I., Peros, J.: Grammars for Question Answering Systems Based on Intelligent Text Mining in Biomedicine. In: Proceedings of the 7th Hellenic European Conf. Computer Mathematics and Its Applications (HERCMA (2005), http://www.aueb.gr/pympe/hercma/proceedings2005/H05-FULL-PAPERS-1/KONTOS-LEKAKIS-MALAGARDI-PEROS-1.pdf
18. Morante, R., Krallinger, M., Valencia, A., Daelemans, W.: Machine Reading of Biomedical Texts about Alzheimer's Disease. In: CLEF 2012 Evaluation Labs and Workshop - Working Notes Papers (2012)
19. Attardi, G., Atzori, L., Simi, M.L.: Index Expansion for Machine Reading and Question Answering. In: CLEF 2012 Evaluation Labs and Workshop - Working Notes Papers (2012)
20. Bhattacharya, S., Toldo, L.: Question Answering for Alzheimer Disease Using Information Retrieval. In: CLEF 2012 Evaluation Labs and Workshop - Working Notes Papers (2012)
21. Grau, B., Pho, V.M., Ligozat, A.L., Abacha, A.B., Zweigenbaum, P., Chowdhury, F.: Adaptation of LIMSI's QALC for QA4MRE. In: CLEF 2012 Evaluation Labs and Workshop - Working Notes Papers (2012)
22. Vishnyakova, D., Gobeill, J., Ruch, P.: Using a Question-Answering in Machine Reading Task of Biomedical Texts About the Alzheimer Disease. In: CLEF 2013 Evaluation Labs and Workshop - Working Notes Papers (2013)
23. Martinez, D., MacKinlay, A., Molla-Aliod, D., Cavedon, L., Verspoor, K.: Simple Similarity-based Question Answering Strategies for Biomedical Text. In: CLEF 2012 Evaluation Labs and Workshop - Working Notes Papers (2012)

24. Tsai, B.H., Hou, W.J.: Biomedical Text Mining about Alzheimer's Diseases for Machine Reading Evaluation. In: CLEF 2012 Evaluation Labs and Workshop - Working Notes Papers (2012)
25. Patel, A., Yang, Z., Nyberg, E., Mitamura, T.: Building an Optimal Question Answering System Automatically Using Configuration Space Exploration (CSE) for QA4MRE 2013 Tasks. In: CLEF 2013 Evaluation Labs and Workshop - Working Notes Papers (2013)
26. Qiu, Y., Frei, H.P.: Concept Based Query Expansion. In: Proceedings of the 16th Annual International ACM SIGIR Conference on Research and Development in Information Retrieval, pp. 160–169 (1993)

Implementation of Question Answering System Based on Reference-Based Ranking Algorithm*

Takayuki Ito and Yuta Iwama

Master Course of Techno-Business Administration, Nagoya Institute of Technology,
Gokiso-cho, Showa-ku, Nagoya-city, Aichi, Japan
ito.takayuki@nitech.ac.jp
iwama.yuta@itolab.nitech.ac.jp
http://www.itolab.nitech.ac.jp/

Abstract. The importance of support centers continues to increase. They enhance user satisfaction by accurately answering user questions and increase user satisfaction in products or services. The responses required by support centers are handled by hand. However, responding by hand to many user questions is very time consuming. It is also difficult for support centers to quickly and exactly respond to user questions. Therefore, we implemented a question answering system that retrieves documents that correspond to user queries. Our system also scores answers and provides them to users. Scoring functions are implemented by a reference-based ranking algorithm. Finally, we evaluated the effectiveness of our prototype system.

Keywords: question answering system, web application, ranking algorithm, search engine.

1 Introduction

The importance of support centers continues to increase. They enhance user satisfaction by accurately answering user questions and increase user satisfaction in products or services. By developing information technology, enterprises provide various products. As services increase, questions from users also increase. The person in charge of a support center responds by hand to the above support questions. Questions to support centers are also increasing due to service diversification problems. Operating a support center is becoming more time-consuming. A support center searches for answers based on previous support or frequently asked questions (FAQs). Therefore, most user questions can be dealt with by FAQs. Spending much time and manpower to deal with frequently asked questions places a huge burden on businesses.

Users also have more problems because services are becoming more diverse and more complex and the size of support manuals is becoming unwieldy. Users

---

* Please note that the LNCS Editorial assumes that all authors have used the western naming convention, with given names preceding surnames. This determines the structure of the names in the running heads and the author index.

A. Moonis et al. (Eds.): IEA/AIE 2014, Part I, LNAI 8481, pp. 160–169, 2014.
© Springer International Publishing Switzerland 2014

have difficulty finding the information they actually want. Users might stop using products if they cannot find the information they need. To deal with such problems, support centers require an application whose interface is easily viewable and a lot of work which a support center done in the past.

This paper presents our implementation of a support system that retrieves answers wanted by users based on frequently asked questions and past specific support about the domain. The application results calculate scores retrieved by the system with a ranking algorithm that evaluated the system results. A trial of our system has been used by KDDI Web Communications, Inc.

The remainder of this paper consists of the following. Section 2 describes the implementation and the functions of our system. Section 3 describes the data, the retrieval method, and the scoring algorithm. Section 4 describes the usefulness of our system and its performance evaluation. Section 5 provides a conclusion and describes future work.

2   Question Answering System

2.1   System Function

Our system user interface of a retrieval function is depicted in Fig. 2. In the search form (①), user input questions (Fig. 2). Our system shows answers that match the user questions by the order of highest to lowest under the search form (②). If user clicks on the links in the system, it shows detailed pages of that link. We implemented our retrieval function for support center staff and users as described below.

Retrieval Function. We explain the overview of our retrieval function in Fig. 1. (1) In it, the user questions are input into the system. (2) They are analyzed. (3) Our system retrieves data that match the results of the analyzed questions from the database data. (4) Our system evaluates the data that match the results of the analyzed questions using the reference based scoring functions that we proposed. Finally, our system shows data in descending order scores to users. We analyze user questions with MeCab, which is an open source morphological analysis engine. We describe the details of our search method in Section 3.3. We evaluate the data that matches the results of analyzed question using scoring functions and describe the scoring method in Section 3.3.

2.2   System Implementation

Our system is a web application that consists of server and client programs. The server program was developed by Ruby on Rails (Rails). Generally, a web application needs a screen transition when communicating between server and client programs on a web browser. Rails adopts Model View Controller (MVC) as design pattern to achieve this process. However, recently users prefer dynamic UIs that allow operation without screen transitions by an Ajax technique to

Fig. 1. Overview of Question Answering System

provide asynchronous communication on client programs. Users do not need to worry about data communication between server and client programs. In our system design, Rails has a Model and a Controller, and the Client has a View. Our system is implemented on client programs using CoffeeScript, which is a program language that becomes Javascript after it is compiled. CoffeeScript has ruby-like syntax.

3   Ranking Algorithm Method

3.1   Data

We used two types of real world data (from KDDI Web Communication, Inc.): template and mail. Template data are the FAQ data of a pair of corresponding answers and questions. Mail data include e-mails that were exchanged between end users and the support center.

Template Data. There are 310 kinds of template data, which are constructed with question and answer data. A piece of question data is the FAQ data on a support page. An piece of answer data is a solution to a question. Template data, which are summarized as question and answer data, have category attributes. A category is an attribute that divides template data into several groups. (e.g., such statements as "I want to change my account name" can be categorized as "Accounts.") We classified the data by hand into 73 categories because the category attributes were not done in advance. Template data are crucial for our system. We improved the answer accuracy using template data because of the user question frequency. We designed our system to suggest higher answers than the e-mail data to improve the answer accuracy.

Mail Data. There are 4935 kinds of mail data, which are constructed by question and answer data. Question data include e-mails sent by users to the support

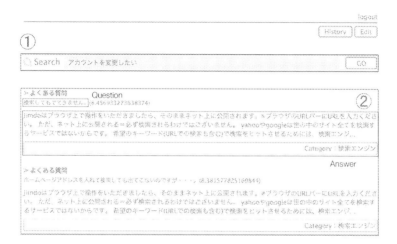

Fig. 2. Our Systemfs User Interface

center. Answer data include e-mail responses from the person in charge of the support center to user questions. As explained above, a pair of questions and answers is called mail data. They are given a label that refers to a particular template datum that is employed to calculate the ranking weight.

We manually did the labeling and found many similar questions. A support centerfs main task is to find adequate templates for certain questions. Thus, since many questions can be answered by the same template, template data can be helpful data because they are referred to by much mail data. By labels, we can easily find helpful template data and cope with the special questions that cannot be expressed by the template data.

3.2   Retrieval Method

Our system has three retrieval methods (Fig. 3)D

Retrieving Template Data. Method 1 retrieves template data. When retrieving template data, we use the analysis results of the user questions. If template data exist that fit such analysis, the template data will be shown in the search result data. This method retrieves template data based on the analysis of user questions (Fig. 3-①).

Retrieving Mail Data. Our second method retrieves mail data. When retrieving them, we use the analysis results of user questions. If mail data exist that fit the analysis of the user questions, the mail data are shown in the search result

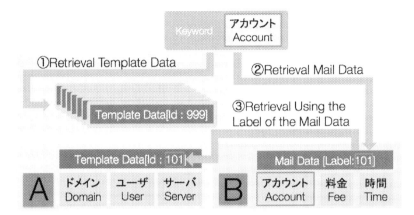

Fig. 3. Overview of Retrieval Function

data. This second method retrieves mail data based on the analysis results of user questions (Fig. 3-②).

Retrieving Mail Data Using Labels. Our third method uses the labels of the mail data. A label refers to template data referred by some mail data (Section 3.1). Some template data are not included in the search results in the first retrieval method (Fig. 3-A)D There are mail data in the search results in the first retrieval method (Fig. 3-B)D This method retrieves template data A using labels that refer to mail data B.

3.3 Reference-Based Scoring Method

We propose a new scoring method for our system. The scoring assesses how appropriate the answers are for a given question from a user. In our system, we combine four indexes into one scoring function: reference-based (the noun frequency), the occurrence in the documents, the consistency degree to the categories, and the reference weight of the template data.

Noun Frequency. This index assesses how many analysis results of user questions are located in the document data. We apply this index to the template and mail data and defined it as a $frequency$ function and number of nouns in a document as N. Then we can define the $frequency$ function: (1):

$$frequency_i = \sum_{j=1}^{k} \frac{n_j}{N_i},$$ (1)

where i is the id of document datum, j is the analysis result of the user questions, and k is the number of nouns in the analysis results.

Occurrence in Documents. This index assesses where the analysis results of the user questions are found in the document data. We evaluate how quickly a word in the analysis result appears and apply this index to the template and mail data. We define this index as the *location* function, the number of nouns in a document as N, and the occurrence position of the analysis results of the user questions as *index*. Then we can define *location* function as (2):

$$location_i = \sum_{j=1}^{k} \left(1 - \frac{index_j}{N_i} \right),$$ (2)

where i is the id of a document datum, j is the analysis results of the user questions, and k is number of nouns in the analysis result.

Consistent Degree to Category. This index evaluates how consistent the analysis results are with the category attributes. A category attribute is a category given to the template data. We apply this index to the template data and define it as *category* function (3):

$$category_i = \begin{cases} 1 & \text{if itfs in the consistent category} \\ 0 & \text{otherwise out of the consistent category} \end{cases},$$ (3)

where i is the id of a document datum.

Reference Weight to Template Data. Mail data have labels that refer to the template data. This index, which evaluates how often the template data are referred to by the mail data, is applied to the template data. We define this index as the *label* function, the score of the mail data that refers to the template data as $MailScore$, and the amount of mail data, which refer to the template data, as $RefN$. Then we can define *label* function as (4):

$$label_i = \frac{\sum_{j=1}^{RefN} MailScore_j}{RefN},$$ (4)

where i is the id of a document datum and $MailScore_i$ is the mail data that refer to the template data.

4   System Evaluation

4.1   Evaluation Setting

We experimentally compared our system and a simple keyword search system. We randomly chose 30 questions from actual mail data. We input questions into both systems and compared the results using the following three evaluation indexes. The first index is the average of the reciprocal of the rank. The second index is the correct answer rate in the top 5. The third index is the occupancy rate of the correct data in the top 5. We describe our results for each index below.

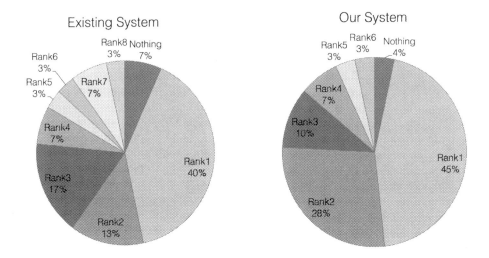

Fig. 4. Percentage of Rank

Average of Reciprocal of the Rank. The average of the reciprocals of the highest ranks is the average of the inversed values of the highest ranks of the correct data proposed by our system. Fig.5 shows an example. Let us assume that we have 3 questions to the system. For Question 1, our sytem proposes two correct answers (data) which ranked at 2nd and 3rd. Similary, for Quetion 2 and Question 3, our system proposes two correct answers ranked at 3rd and 5th, and three correct answers ranked at 2nd, 4th, and 5th, respectively. We can select the highest rank (top rank) 2nd, 3rd, and 2nd for Question 1, Question 2, and Question 3, respectively, We calculate the average of the reciprocals, $\frac{\frac{1}{2}+\frac{1}{3}+\frac{1}{2}}{3} = 0.666...$, and this is the index of the averateg of the reciprocals of the highest ranks. We show the percentage of the ranks of both systems (Fig.4). Our systems results are shown on the right and the existing systemfs results on the left. The average reciprocal of the rank is calculated by the rankfs percentage. The score of our system is 0.625, and the score of the current system is 0.531. Our system outperforms the current system. One reason is because most of the data are around rank 4, meaning that our system's proposed answer order is better.

Correct Answer Rate in Top 5. The correct answer rate is the number of the questions that our system can propose the correct answers. Fig. 6 shows an example where we have 3 questions. Here, our system proposes the correct answers for Question 1 and Question 3. Thus, the correct answer rate becomes $\frac{1+1}{3}$. Our systemfs score is 86.6%, and the current systemfs score is 80.0%. We confirmed that our system works better than the existing system in the correct

	Question1	Quesiton2	Question3	
Rank1				Calculate Value
Rank2	✓		✓	✓ :Correct Data
Rank3	✓	✓		
Rank4			✓	
Rank5		✓	✓	
Top Rank	2	3	2	

$$\frac{\frac{1}{2}+\frac{1}{3}+\frac{1}{2}}{3}=0.666...$$

Fig. 5. Example of Reciprocal Average of the Rank

	Question1	Quesiton2	Question3	
Rank1				Calculate Value
Rank2	✓		✓	✓ :Correct Data
Rank3	✓			
Rank4			✓	
Rank5			✓	
Correct	T	F	T	

$$\frac{1+1}{3}=0.666...$$

Fig. 6. Example of Correct Answer Rate in Top 5

answer rate in the top 5. This result means that our system's scoring function outperforms the existing system.

Occupancy Rate of Correct Data in Top 5. The average occupancy rate of the correct data in the top 5 is the percentage of correct answers in the top 5 candidates for each question. Fig. 7 shows an example with 3 questions. In this example, our system can propose 2 correct answers within top 5-ranked answers for qeustion 1, 2 correct answers within top 5-ranked answers for question 2, and 3 correct answers within top 5-ranked answers for question 3. Thus, we can calculate the occupancy rate, i.e., the percentage of the currect answers in the 5 answers from the top rank, for each question. They are $\frac{2}{5}$, $\frac{2}{5}$, $\frac{3}{5}$, for Question 1, Question 2, and Question 3, respectively. The average occupancy rate of the correct data in the top 5 is $\frac{\frac{2}{5},\frac{2}{5},\frac{3}{5}}{3} = 4.666....$ For example, if there are two correct answers in the top 5 candidates of the search results, this index is 40%). We show the amount of correct data in both our system and the current system (Fig. 8). The score of our system was 38.6%, and the score of the current system was 36.6%. We confirmed that our system works better than the current system in

	Question1	Quesiton2	Question3	
Rank1				
Rank2	✓		✓	
Rank3	✓	✓		
Rank4			✓	
Rank5		✓	✓	
Percentage of Correct answer	$\dfrac{2}{5}$	$\dfrac{2}{5}$	$\dfrac{3}{5}$	

▨ Calculate Value
✓ :Correct Data

$$\frac{\dfrac{2}{5}+\dfrac{2}{5}+\dfrac{3}{5}}{3}=4.666...$$

Fig. 7. Example of Occupancy Rate of Correct Data in Top 5

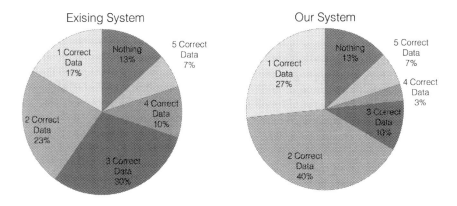

Fig. 8. Number of Correct Data of Top 5

the occupancy rate of the correct data in the top 5. This result means that our system can provide various answers to questions from users.

5   Conclusion and Future Work

Even though the number of questions received by support centers is increasing, most can be answered in the FAQs. The heaviest burden on support centers could be reduced by automatically answering FAQs. In this paper, we developed a question answering system using a reference-based ranking algorithm to solve this problem. We evaluated the performance of our prototype and confirmed its usefulness. Our prototype system works better than current systems in all indicators. There are two factors underlying our results. The first is using new mail data, which can compensate for the template data. Our system can respond to a wide range of questions. The second factor is that compared to current systems, our systemfs scoring function is superior because it has been tuned. Future work will enhance the scoring and search accuracy. To enhance the former, we will tune the weights. To enhance the latter, we will employ SVM.

References

1. Zhang, D., Lee, W.S.: A web-based question answering system. In: The SMA Annual Symposium 2003, Singapore (2003)
2. Joachims, T., Granka, L., Pan, B., Hembrooke, H., Gay, G.: Accurately Interpreting Clickthrough Data as Implicit Feedback. In: The 28th Annual International ACM SIGIR Conference on Research and Development in Information Retrieval (2005)
3. Ishioroshi, M., Kano, Y., Kando, N.: An Analysis of the Questions of the University Entrance Examination to Answer Using the Question Answering System. In: The 27th Annual Conference of the Japanese Society for Artificial Intelligence (2013)
4. Cha, J., Nabeshima, K., Mizuno, J., Okazaki, N., Inui, K.: Why-question answering using document structure. In: The 27th Annual Conference of the Japanese Society for Artificial Intelligence (2013)
5. Nasukawa, T.: Text Mining Application for Call Centers. Japanese Society for Artificial Intelligence 16(2), 219–225 (2001)
6. Segaran, T.: Programming Collective Intelligence. O'Reilly Japan
7. Okumura, M., Isozaki, H., Higashinaka, R., Nagata, M., Kato, T.: Question Answering System. Corona Publishing Co., Ltd. (2009) (in Japanese)
8. Nishida, K.: Technology of support Google. Gijutsu-Hyohron Co., Ltd. (March 28, 2008) (in Japanese)

An Optimized Approach Based on Neural Network and Control Theory to Estimate Hydro-geological Parameters

Chih-Yung Hung and Cheng-Haw Lee

Department of Resources Engineering
National Cheng Kung University
Tainan, Taiwan 701, ROC

Abstract. In the past, typical approaches simulating the groundwater flow are based on continuous trials to approach to the targeted measurement accuracy. In this study, we propose a new neural network based on feedback observer technique to estimate the hydro-geological structure and hydraulic parameters of a large-scale alluvial fan in Taiwan. We develop an under-ground water level observer (UGW-LO) based on feedback control theory to simulate the dynamics of groundwater levels and estimate water levels of wells in the large area. In the proposed observer system, a large-scale back-propagation neural network (BPNN) is proposed to simulate water levels dynamics of multiple wells. The simulation results are fed back as a reference for BPNN to approach to refine estimation. Based on that model, a groundwater flow is simulated correctly by software MODFLOW. Experimental results indicate that the innovative method works better than conventional regression estimations. The learning ability of BPNN also contributes to overcome the gap between legacy dynamics UGW equations and real UGW dynamics. The applicability and precision are verified in a large scale experiment that is beneficial to the management of under-ground waters and reduce the risk of ground–sink.

Keywords: Under-ground Water Resource, Feedback Control System, Observer System, Back-Propagation Neural Networks, Water-level Estimation.

1 Introduction

Under-ground water resource, due to its purity and low cost has become a popular solution to the water-demanding problem. Thus utilization of under-ground water has a rapid growth as the development of industrial and population. However, as unlimited utilization of underground water, the problems of ground-sink and salt water have serious impact on environment. To solve such problems, a study of the factors affecting the flow of underground water is initialized. In other words, the modeling of the underground water flow will be explored in this paper. In fact, the factors affecting the flow dynamics of underground water are very complicated. In the past, there were various approaches to derive the model of underground water. In previous work, simulation models were established for small-scale under-ground water systems such as sandbox water. Mathematic models [1, 2] were proposed to describe the flow

A. Moonis et al. (Eds.): IEA/AIE 2014, Part I, LNAI 8481, pp. 170–177, 2014.
© Springer International Publishing Switzerland 2014

system in terms of initial condition, boundary condition and dynamic equations. In general, there are two categories of mathematic models; analytical equation-based (AEB)[1-2] and numeric equation-based (NEB). In AEB model, the dynamic model is expressed in terms of analytical solutions after certain simplifications of selected factors. For examples, the geological characteristics of ground are uniform and the flow of underground water is the first/second order dynamics. But, in real world, due to the deposit environment and forming duration, water layer usually is non-uniform with irregular boundary conditions and various thicknesses. Thus the assumptions narrow down the scope and accuracy of the AEB model.

The simplified model is too rough to present the real condition of underground water. Thus, numeric equation-based (NEB) approach is more common for the modeling of underground water. A typical NEB is designed with the inputs of field data to emulate the real physical system. The emulated results should be verified by the patterns collected form the real field. Finite Element Method (FEM) [3] is a typical approach of this category. In this paper, we propose a new approach of water level observer based on feedback control theory to model the dynamic of underground water system. Through feedback control theory, the underground water system is manipulated as a controlled plant. Based on that, the flow model of underground water between neighboring wells is modeled as a physical system. The relationship of water levels in two neighboring wells is derived as an input/output relationship. Through the mathematic model, the geological and hydrological parameters of underground watering systems are obtained.

Filters, compensators and observers are common components used in control systems. Filter is a signal processing component. Compensator is activated to change the system behavior. And observers are used to estimate the state variables existed in a control system. In this paper, we propose an observing system to derive the geological and hydrological parameters of underground watering systems. The proposed system composed of state observers, is tuned by a training process of neural networks to obtain the optimal parameters.

The rest of this paper is organized as follows. The literature review of related work is given in Section 2. A neural network model to evaluate hydro-geological parameters is developed in Section 3. A series of experiments and their results are illustrated in Section 4. Finally, we conclude this work in Section 5

2   Related Work

Neural Networks have been suggested as a method of offline adapting and optimizing observers to increase simulated performance. It provides a simple method to empower computer systems with knowledge and reasoning ability that improve the accuracy of handle imprecision. There are many studies about the application of neural networks on water resource. Chang *et. al.* [4] applied a recurrent neural network to evaluate the water level of rivers. Liu [5] estimated the quality of underground water through a back-propagation neural network. Dawson *et. al.* [6] proposed a RBF neural network and multi-layer feed forward neural network to model hydrological systems. The proposed approach models the relationship between rainfall and water level of rivers to demonstrate a better accuracy than the conventional statistics approach.

Lin *et. al.* [7] presented a neural-network approach to tune the observer parameters in an optimal control system. This approach confirms a superior performance contributed by a closed-loop control system. Chang *et. al.*[8] proposed an approach based on grey-fuzzy technique to a real-time dam operation . Shiao *et. al* [9] presented a genetic algorithm with constrained differential dynamic programming technique to develop an integrated observer system to manage the utilization and the pollution of underground water. In the related studies of flow modeling in underground water systems, MODFLOW is a popular simulation tool. Ding *et. al.* [10] studied the modeling of hydrological parameters in Ping-Tung plain through the simulation on MODFLOW. Based the obtained parameters, a geological model of underground water system had been developed.

In this paper, a set of dynamic equations of underground water and a neural network model are interacted in parallel to simulate the dynamics of underground water system and to tune the control system. The proposed approach features with (1) fast response in tracking the desired output (2) reliable stability (3) high immunity to noise and interference. Thus the proposed system will detect the level dynamics of the underground water through the observation of the monitored well. A case study of underground water in A large alluvial fan is investigated. In this case study, the dynamic of underground water and geological model are presented. The results are also compared to the performance from MODFLOW to illustrate the superior performance of this proposed approach

3   Estimation of Under-Ground Water Level Based on Back Propagation Neural Networks Emulating Dynamics Equations

3.1   Design of Level Estimation Based on Neural Networks

In this section, we will develop a level observer (LO) for under-ground water (UGW) system. The proposed approach is designed for a basic estimation. In this basic estimation, the water level of the target well is derivable from the water levels of the neighboring wells. The estimation model decouples large-area water level estimation to multiple small-scale water level estimation. That is contributive to the scalability of the water-level estimation for large area. The architecture of the developed observer is presented in Fig. 1. The estimation is realized by a controller and an observer. The presented level observer employees a back-propagation neural network (BPNN) to learn the dynamics equations of UGW systems. The controller is realized by a back propagation neural network is to learn the relationship of water levels between two neighboring wells. The inputs of controller are water level (h_t) and estimation deviation (e_{t1}) of two successive estimations. The observer is designed to search the optimal solution of the modeling parameters. The composition of an observer is a set of water flow dynamic equations. The inputs of the equation set are the controller output (y_m) and the water level (h_t). The observer output is the estimation value of water level in the next time instance.

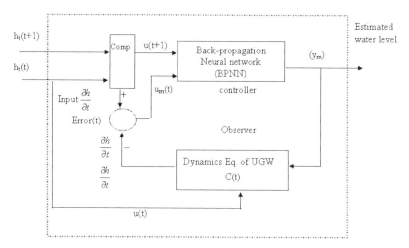

Fig. 1. Diagram of the observer system for the modeling of underground water system

The controller is trained by a training pattern set. The training procedure is by batch. In each learning cycle, all training data sets are fed to the BPNN controller to derive the most-fit weight set of this BPNN. The trained controller will generate the water-level estimation (y_m) of the neighboring well for the observer (discussion in 3.2). The observer based on the estimation will derive the fitting parameters of the estimated hydrological and geological structure model, including conductive parameter (k), storactivity coefficient (S), pumping rate (Q) and recharge ratio (R). The learning pattern will be applied to the controller continuously. The learning procedure will not finish until the error deviation convergence.

The learning of controller is contributed by the involvement of BPNN. BPNN is a kind of supervise learning. Its advantages are high recognition on the training pattern. The network structure developed in this work is as shown in Fig. 2. The first layer (P) stands for water levels and deviation error. The second layer (H) is a hidden layer. And the third layer is output layer (T) to deliver the water level estimation to the observer. The weights of the connections between layers indicated as IW_{ij}. To derive the estimation values as an output. First, the inputs of the first layer are summed with the corresponding weights to feed to the hidden layer as follows.

$$net_j(t) = \sum_{i=1}^{n} IW_{ij}(t) * P_i(t) + b_i \tag{1}$$

where net_j is the input of the j-th neuron in the hidden layer. P_i and b_i are the input value and the input bias of the i-th neuron in the first layer. Thus the neuron output of the hidden layer is given by

$$H_j(t) = tansig\ (net_j(t)) \tag{2}$$

where tansig is the sigmoid function expressed as $(x) = 1/(1+e^{-x})$

Then the output value of the k-th neuron in the hidden layer is obtained as

$$net_c(t) = \sum_{i=1}^{m} IP_{ic}(t) * H_i(t) \qquad (3)$$

Then the outputs of the hidden layer are delivered to the output layer to generate the estimation of the underground water level as expressed in Eq. 4

$$R_c(t+1) = purelin \ (net_c(t) + b_c) \qquad (4)$$

and

$$purelin \ (x) = ax \qquad (5)$$

The developed estimator is to derive the mapping relationship between two consecutive wells. Consider the error between the estimation value $R_c(t)$ and the measurement value $T_c(t)$ given by

$$e_c(t) = \sum_{c=1}^{C} \ [(R_c \ (t) - T_c \ (t))^2/2] \qquad (6)$$

The total error of T neighboring wells yields to

$$E = \sum_{t=1}^{T} e_c(t) \qquad (7)$$

Certainly, the BPNN should be trained to minimize E by adapting the connection weights according to the steepest algorithm to derive the modification of each connection weights $\triangle \varepsilon_{ij}$ as

$$\Delta \varepsilon (t-1) = -\eta \frac{\partial E(t)}{\partial \varepsilon(t-1)} \qquad (8)$$

where η is the learning rate. Besides we also have

$$\frac{\partial E(t)}{\partial \varepsilon_{cj}(t-1)}$$

$$= \frac{\partial E(t)}{\partial T_c(t)} \frac{\partial T_c(t)}{\partial net_j} \frac{\partial net_j}{\partial \varepsilon_{cj}(t-1)} \qquad (9)$$

$$= -(Rc(t) - Tc(t)) \frac{\partial [f(net)]}{\partial net} \frac{\partial (\sum_{j=1}^{j} \varepsilon_{cj} * P_j + b_1)}{\partial \varepsilon_{cj}(t-1)}$$

$$= -\sum_{j=1}^{j} e_c(t) f'(net(t)) P_j$$

Finally we have

$$\Delta\varepsilon(t-1) = -\eta\frac{\partial E(t)}{\partial\varepsilon(t-1)} = \eta\sum_{j=1}^{j}e_C(t)f'(net(t))P_j \tag{10}$$

and

$$W_{ij}(t)=W_{ij}(t-1)+\Delta\varepsilon(t-1) \tag{11}$$

According to Eq. 6-9, the controller is continuously trained to a convergent status. The correction of each connection weights that are calculated by Eq. 10, and the connection weights are modified by Eq.11 in train stage.

4   Experimental Results and Analysis

A series of experiments are designed and conducted to verify the performance of the proposed algorithm. The largest alluvial fan, in Taiwan, is selected to be the target to explore the hydrological and geological structure of the underground water. There are 34 observing wells built by Water Resources Agency (WRA), Ministry of Economic Affairs (MOEA), Taiwan. The training and testing data sets are collected from the WRA. The large alluvial fan spreads from the east to the west in the west area of Taiwan. According the geological study from Central Geological Survey (CGS), MOEA, this area is divided to 7 hydrological sections. According to the slope of the geological structure, we divided the groundwater well into 32 groups on the 7 hydrological sections. We had finished the exploration hydro-geological parameter as shown in Table 1, for these clusters of a large alluvial fan. Now, we select randomly wells from 32 clusters to show the results for the study case.

All the data sets are the historic data of the water level measurements from WRA. Erhshui station is the main well and Gan-yuan station is the neighboring well in the lower reach. The entire data set is divided training data and testing data according to

Table 1. The hydro-geological parameters of clusters

Well No.	Pumping (m³/day)	Recharge (m³/day)	K(m/day)	S
1-1	3345	27258	102	0.00007
1-2	1879	6874	82	0.00004
1-3	23115	10467	49	0.0005
1-4	42378	11159	145	0.0002
2-2	23775	30569	316.3	0.03
2-3	21367	18649	7.4	0.03
3-1	88	24792	653	0.21
3-2	13432	27430	187	0.15
3-3	11990	16012	150	0.05
3-4	16906	10221	74	0.02
3-5	24857	14571	6	0.01

the portions of 50%, and 50%, respectively. The proposed BPNN-based level estimator is trained iteratively. The comparison of both the proposal and MODFLOW is show in Fig. 2 and 3. First, the estimated water level is depicted in Fig. 2. During the duration of 12 months, the water level varies with reasons. The proposal estimator effectively indicated the level variation of the underground water. In the conventional approach based on MODFLOW, the derived water level is not accurate with the maximum error of 0~7.4 meters. Comparing to the result from MODFLOW, the proposed approach has better performance. The proposed approach is fast in tracking the target variation. The error is also smaller than before. In most case the error approached to zero. The error tolerance for a convergence of learning is set to 2% as depicted in Fig. 3. After the training, the testing data set is fed to derive the output value. According to the comparison of estimation and actual value, the accuracy of this estimator is evaluated fairly. The proposed approach demonstrates little error during the estimation period of 24 months. In contrast, the estimations from MODFLOW are varying and obvious. The main reason is that the estimation is generated by the designed dynamic equations in MODFLOW and this tool lacks of the ability to modify the estimation from the feedback error. The proposed approach based on a feedback system. The estimation is continuously modified to generate according to feedback error. Thus, the proposed approach demonstrates a better estimation performance.

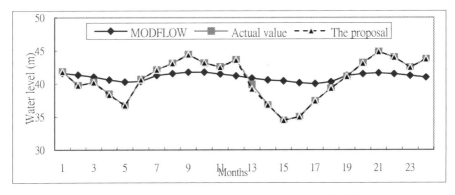

Fig. 2. Comparison of underground water level estimation

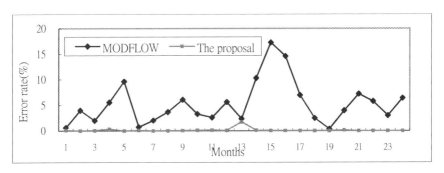

Fig. 3. Rate comparison of estimation error

5   Conclusion

Global warming is a critical issue to human and environment. In future, more and more abnormal climates are expected. The variation of rainfall definitely impacts on the resource of under-ground water. When the rainfall is varying heavily, the management of groundwater resource will become very important without doubt. Furthermore, because the geology of soil layer is usually non-uniform with irregular boundary conditions and various thicknesses in the underground water system, the previous numeric method did not present a satisfying result. In this work, a case study of a large alluvial fan, the application of neural network and observer system is applicable for the monitoring and estimation of underground water level in a large-scale and scalable methodology. The obtained accuracy is better than the previous approach based on MODFLOW. The proposed approach offers the feature of fast response, close tracking ability and stability. Thus it is proved as a good solution to the modeling of large-scale underground water systems.

References

[1] Ildoromi, A.: Analysis and Modeling of Landslide Surface with Geometrical Hydrology and Slope Stability: Case Study Ekbatan Watershed Basin, Hamedan, Iran 6(3), 920–924 (2012)

[2] Rimmer, A., Hartmann, A.: Simplified Conceptual Structures and Analytical Solutions for Groundwater Discharge Using Reservoir Equations. In: Water Resources Management and Modeling. InTech (2012) ISBN 978-953-51-0246-5

[3] Wang, H., Anderson, M.P.: Introduction to groundwater modeling finite differences and finite element methods. Elsevier (1995) ISBN-10: 012734585X

[4] Chang, F.C., Huang, H.L., Chang, L.T.: Application of recurrent neural networks on flow estimation of rivers. Journal of Argi. Eng. 49(2), 32–39 (2001)

[5] Liu, C.Y., Kuo, Y.M.: Variation analysis of underground water in Yulin area based on back-propagation neural networks. Journal of Taiwan Water Conservancy 48(1), 9–25 (2000)

[6] Dawson, C.W., Wilby, R.L.: Hydrological modeling using artificial neural networks. Progress in Physical Geography 25(1), 80–108 (2001)

[7] Lin, S.W., Chou, S.Y., Chen, S.C.: Irregular shapes classification by back-propagation neural networks. Journal of Advance Manufacturing Technology 34, 1164–1172 (2007)

[8] Chang, F.C., Hsu, R.T.: Application of grey theory on real-time dam operation. Journal of Taiwan Water Conservancy 47(1), 44–53 (1999)

[9] Shiao, J.T., Chang, L.C.: Application of dynamic control and genetic algorithms on the management and operation of underground water resource. In: Proceeding of the Tenth Conference on Water Conservancy, pp. 40–45 (1990)

[10] Ding, T.S., Cao, C.M., Chen, C.Y.: A study of pollution management on underground water. Journal of Technology 11(3), 159–169 (1997)

Heuristic Search for Scheduling Flexible Manufacturing Systems Using Multiple Heuristic Functions

Bo Huang, Rongxi Jiang, and Gongxuan Zhang

School of Computer Science and Engineering, Nanjing University of Science and Technology,
Nanjing, P.R. China
huangbo@njust.edu.cn

Abstract. To cope with the complexities of flexible manufacturing system (FMS) scheduling, this paper proposes and evaluates an improved search strategy and its application to FMS scheduling in a Petri net framework. Petri nets can concisely model multiple lot sizes for each job, the strict precedence constraint, multiple kinds of resources and concurrent activities. On the execution of the Petri nets, our algorithm can use both admissible heuristic functions and nonadmissible heuristic functions having the upper supports of the relative errors in A^* heuristic search algorithm. In addition, the search scheme can ensure the results found are optimal and invokes quicker termination conditions. To demonstrate it, the scheduling results are derived and evaluated through a simple FMS with multiple lot sizes for each job. The algorithm is also applied to a set of randomly-generated FMSs with such characteristics as multiple resources and alternative routings.

Keywords: Petri nets, Flexible manufacturing, Scheduling, Heuristics.

1 Introduction

Nowadays, in order to provide wide product variety and quick response to changes in marketplace, FMSs have been adopted broadly in modern production environments [1,2]. An FMS is an automated production environment where there may exist multiple concurrent flows of processes. Different products may be manufactured at the same time, and shared resources are often exploited to reduce the production cost. In an FMS, a high-level control system must decide what resources are to be assigned to what job and at what time, so as to optimize some criteria, e.g. makespan, cost, etc.

In this kind of systems, scheduling is a typical combinatorial optimization problem, which decides starting times and allocations of jobs to be processed. A desirable scheduling method must have both easy formulation of the problem and quick identification of solutions (with small computational efforts). So, many industry and research communities are now focusing on developing methods for solving real-world FMS scheduling problems.

The performance of FMS has been recently studied by the PN community. It is appropriate to select Petri net-based algorithms for optimization purposes. Recent approaches have attempted to use artificial intelligence techniques [3,4] to selectively search the Petri net reachability graph using the well-known A^* search algorithm [5]. A^* algorithm is a general search procedure that explores the search space in a Best-First manner.

A. Moonis et al. (Eds.): IEA/AIE 2014, Part I, LNAI 8481, pp. 178–187, 2014.

The A^* search algorithms presented in [3,6,7,8,9] developed for minimizing the objective function of flow time of parts in the system are all derived from the original algorithm presented in [10]. These A^* heuristic algorithms use some admissible heuristic functions and they can guarantee that the results obtained are optimal, but they limit the selection of heuristic functions to only those that never overestimate the optimal completion cost.

In this paper, we present an A^* scheduling strategy which can simultaneously use both admissible heuristic functions and nonadmissible heuristic functions and the search scheme can also ensure the results found are optimal and invokes quicker termination conditions. Then we model a simple FMS using Petri nets. The scheduling results are derived and evaluated through the modelled system with multiple lot sizes for each job. The algorithm is also applied to a set of randomly-generated FMSs with such characteristics as operations with multiple resources and jobs with alternative routings.

2   Modelling and Scheduling of FMS Based on P-timed Petri Net

In prior works, P-timed Petri net model was widely used because its markings are well defined in each firing [7,9,11,12,13,14], so we adopt P-timed Petri net in this paper. The definition of the general P-timed Petri net is represented as below.

Definition 1. *A general P-timed Petri net is a six-tuple PPN= (P, T, I, O, m, d) where:*

$P = \{p_1, p_2, \ldots, p_n\}, n > 0$, *is a finite set of places;*

$T = \{t_1, t_2, \ldots, t_s\}, s > 0$, *is a finite set of transitions with* $P \cup T \neq \emptyset$ *and* $P \cap T = \emptyset$;

$I : P \times T \rightarrow \{0, 1\}$ *is an input function or direct arcs from P to T ;*

$O : T \times P \rightarrow \{0, 1\}$ *is an output function or direct arcs from T to P;*

$m : P \rightarrow \{0, 1, 2, \ldots\}$ *is a* $|P|$ *dimensional vector with* m_p *being the token count of place p.* m_0 *is an initial marking;*

$d : P \rightarrow R^+ \cup \{0\}$ *is a delaying function that associates the time delay with each places. Note that* R^+ *is a set of positive real numbers.*

In this paper, a place represents a resource status or an operation, a transition represents either start or completion of an event or operation process, and the stop transition for one activity will be the same as the start transition for the next activity following. Token(s) in a resource place indicates that the resource is available and no token indicates that it is not available. A token in an operation place represents that the operation is being executed and no token shows no operation is being performed. A certain time may elapse between the start and the end of an operation. This is represented by associating timing with the corresponding operation place.

The modelling is briefed as follows. First, construct a Petri net model for each job based on their sequence and the use of resources. Then merge these models to obtain a complete Petri net model through the shared resource places which model the availability of resources [14].

A simple example is used to demonstrate the modeling capability of P-timed Petri nets. Table 1 shows the requirements of each job. The example system consists of three types of resource R_1, R_2, R_3 and four types of job J_1, J_2, J_3, J_4. Each job has three stages of operations and some stages may have alternative routings. For example, the 3rd stage of J_1 can be performed by using resource R_1 or resource R_1 and R_3 alternatively and the corresponding operating times are 80 and 108 (57 + 51) respectively. Figure 1 shows the P-timed Petri net model for this system, where the resource places with a same name represent the same resource place.

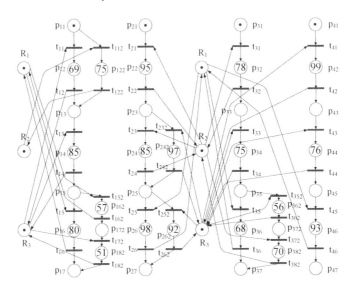

Fig. 1. P-timed Petri Net model for the example

Table 1. Job requirements

Tasks/Jobs	J1	J2	J3	J4
1	R3(69) or R2(75)	R2(95)	R2(78)	R3(99)
2	R3(85)	R2(85) or R3(97)	R3(75)	R3(76)
3	R1(80) or R1(57) and R3(51)	R1(98) or R3(92)	R1(68) or R3(56) and R2(70)	R1(93)

Since we employ deterministic P-timed Petri nets by associating time delays with places, the transitions can be fired without duration. In the P-timed Petri net model of a system, firing an enabled transition changes the token distribution (marking). A sequence of firings results in a sequence of markings, and all possible behaviours of the system can be completely tracked by the reachability graph of the net. The search space for the optimal event sequence is the reachability graph of the net, and the problem is to find a firing sequence of the transitions in the Petri net model from the initial marking to the final one. [7,9,10,11] adopted the well-known A^* search algorithm to Petri net dynamics and structures to perform scheduling. A^* is an informed search algorithm that expands only the most promising branches of the reachability graph of a Petri net. Given a Petri net model, A^* expands the reachability graph of the model from the initial marking until the generated portion of the reachability graph touches the final marking. Once the final marking is found, the optimal path is constructed by tracing the pointers that denote the parenthood of the markings, from the final marking to the initial marking. Then, the transition sequence of the path provides the order of the initiations of the activities, i.e., the schedule. The basic algorithm is as follows:

Algorithm 1. *A* algorithm*

1. *Put the initial marking m_0 on the list OPEN.*
2. *If OPEN is empty, terminate with failure.*
3. *Remove the first marking m from OPEN and put m on the list CLOSED.*
4. *If m is the final marking, construct the scheduling path from the initial marking to the final marking and terminate.*
5. *Find the enabled transitions of the marking m.*
6. *Generate the next marking, or successor, for each enabled transition, and set pointers from the next markings to m. Compute $g(m')$ for every successor m'.*
7. *For every successor m' of m:*
   (a) If m' is already on OPEN, direct its pointer along the path yielding the smallest $g(m')$.
   (b) If m' is already on CLOSED, direct its pointer along the path yielding the smallest $g(m')$. if m' requires pointer redirection, move m' to OPEN.
   (c) Calculate $h(m')$ and $f(m')$, and put m' on OPEN.
8. *Reorder OPEN in the increasing magnitude of f.*
9. *Go to Step 2.*

The function $f(m)$ in Algorithm 1 is calculated from the following expression: $f(m) = g(m) + h(m)$. $g(m)$ represents the cost (makespan) of the partial schedule determined so far. On the other hand, $h(m)$, called heuristic function, represents an estimate of the remaining cost to reach the marking that represents the goal state m_f. The purpose of the heuristic function $h(m)$ is to guide the search process in the most profitable direction by suggesting which transition to fire first (i. e. which marking to process next).

Definition 2. *If $h(m)$ is a lower bound to all complete solutions descending from the current marking, i.e., $h(m) \leq h^*(m), \forall m$, where $h^*(m)$ is the optimal cost of paths going from the current marking m to the final one, the $h(m)$ is admissible, which guarantees for an optimal solution.*

Definition 2 motivates naming this class of estimates: admissible heuristics. At each step of the A^* search process with an admissible heuristic function, the most promising of the markings generated so far is selected. This is done by applying the heuristic function to each of them. Then, it expands the chosen marking by firing all enabled transitions under this marking. If one of successor markings is a final marking, the algorithm quits. If not, all those new markings are added to the set of markings generated so far. Again the most promising marking is selected and the process continues. Once the Petri net model of the system is constructed, given initial and final markings, an optimal schedule can be obtained [10].

3 Scheduling with Admissible Functions and Nonadmissible Functions

A^* algorithms with admissible heuristic functions can guarantee for optimal solutions, but admissible search strategies limit the selection of heuristic functions to only those that never overestimate the optimal completion cost. In some cases, we may have some admissible functions and nonadmissible functions or only nonadmissible functions. Then how should one aggregate the estimates provided by several different heuristic functions? This section will provide answers to these questions.

Definition 3. *A heuristic function h_1 is said to be more informed than h_2 if both are admissible and $h_1(m) > h_2(m)$ for every nongoal marking node m.*

The pruning power of A^* is directly tied to the accuracy of the estimates provided by h. A^* expands every OPEN marking node satisfying the inequality: $g(m) + h(m) < C^*$, where C^* is the cheapest cost of paths from the start marking to the goal marking. Clearly, the higher the value of h, the fewer nodes will be expanded by A^*, as long as h remains admissible. Therefore, if some heuristics are all admissible, the highest heuristic, which is the closest approximation to h^*, would expand the least nodes and naturally lead to a most efficient search.

However, the scheme of taking the maximum of the available heuristics may not work as well in the nonadmissible cases. This is because overestimations of nonadmissible heuristics reduce the search complexity when they occur at off-track nodes and increase it when they take place along the solution path. The exact effect of overestimations on the mean run-time of A^* depends on a delicate balance between these two effects [15]. These difficulties disappear if the heuristics available are known to induce proportional errors. In such a case, the availability of a debiasing technique offers a procedure of combing several heuristics, admissible as well as nonadmissible.

Definition 4. *A random variable X is said to have an upper support r if $X \leq r$ and $P(X \leq x) < 1$ for all $x < r$.*

Proposition 1. *Given the set of heuristic functions $h_1, h_2, \ldots, h_k$ for which the upper supports of the relative errors are $r_1, r_2, \ldots, r_k$, the combination rule: $h_m = \max[\frac{h_1}{1+r_1}, \frac{h_2}{1+r_2}, \ldots, \frac{h_k}{1+r_k}]$ is more informed than any of its constituents.*

Proof. If r_i is the upper support of the relative errors of h_i, the division by $1 + r_i$ removes the bias from h_i and renders it admissible, with $r = 0$; the "max" operation further improves the heuristic estimate by rendering h_m larger than each of its constituents, while also maintaining $r = 0$. Therefore, h_m is more efficient than each of its constituents according to Definition 3.   □

To demonstrate it, we employ several admissible functions and nonadmissible functions in A^* algorithm. The first heuristic function comes from [7]: $h_a = \max_i \{\xi_i(m), i = 1, 2, \ldots, N.\}$, where $\xi_i(m)$ is the sum of operation times of those remaining operations for all jobs which are definitely to be processed with the ith resource when the current system state is represented by marking m; N is the total number of resources. The second heuristic function comes from [14]: $h_b = \max_{i \in SM}(RWT_i + Mr_i)$, where RWT_i is the sum of costs (associated with places), along the shortest path, that a part token located at the ith place would require to reach an end place of its job; Mr_i is the remaining processing time at the ith place; $SM = \{$marked places$\}$. These two heuristic functions have been proved admissible, so their upper supports of the relative errors are zero. To compare the effects of different functions, the third is a nonadmissible one: $h_c = \sum_{i \in SM}(RWT_i + Mr_i) \times M_i$, where M_i is the number of tokens at the ith place. This heuristic function is nonadmissible, and the following proposition will give the upper support of the relative errors of h_c.

Proposition 2. *The upper support of the relative errors of h_c is $n_{part} - 1$ (n_{part} is the number of part tokens in the system).*

Proof. Adopting the following notation:

h^* = the cheapest cost of paths from the current marking to the goal marking;

Obviously, $h_c = \sum_{i \in SM}(RWT_i + Mr_i) \times M_i$ is the sum of remaining processing time of all parts in the system, and $\dfrac{\sum_{i \in SM}(RWT_i + Mr_i) \times M_i}{n_{part}}$ is the minimum processing time required from the current marking to the goal marking when all parts in the system are processed and finish simultaneously.

So, we have: $\dfrac{\sum_{i \in SM}(RWT_i + Mr_i) \times M_i}{n_{part}} \leq h^*$.

Hence: $\dfrac{h_c - h^*}{h^*} = \dfrac{\sum_{i \in SM}(RWT_i + Mr_i) \times M_i}{h^*} - 1 \leq \dfrac{n_{part} \times h^*}{h^*} - 1 = n_{part} - 1$.

In addition, if and when all parts can be processed and finish at the same time, we can get $\dfrac{\sum_{i \in SM}(RWT_i + Mr_i) \times M_i}{n_{part}} = h^*$ and $\dfrac{h_c - h^*}{h^*} = n_{part} - 1$.

Therefore, $\dfrac{h_c - h^*}{h^*} \leq n_{part} - 1$ and $P(\dfrac{h_c - h^*}{h^*} \leq x) < 1$ for all $x < n_{part} - 1$, and $n_{parts} - 1$ is the upper support of the relative errors of h_c. $\square$

Then, according to Proposition 1, the combinational heuristic function $h_m = \max[\dfrac{h_a}{1+r_a}, \dfrac{h_b}{1+r_b}, \dfrac{h_c}{1+r_c}] = \max[h_a, h_b, \dfrac{h_c}{n_{part}}]$ will outperform its constituents, i.e. $\max_i\{\xi_i(m), i = 1, 2, \ldots, N.\}$, $\max_{i \in SM}(RWT_i + Mr_i)$ and $\dfrac{\sum_{i \in SM}RWT_i + Mr_i \times M_i}{n_{part}}$.

The FMS instances from [7] with five sets of lot size (1, 1, 1, 1), (2, 2, 1, 1), (5, 5, 2, 2), (8, 8, 4, 4), and (10, 10, 6, 6) are tested. We employ A^* algorithms with h_a, h_b, h_c/n_{part}, and h_m respectively. The code was written in C# in its entirety. The tests were performed on personal computer having an Intel Core microprocessor at a speed of 3.2 GHz with 4 GB of memory. The scheduling results of makespan, number of expanded markings (EM), and CPU time (TM) are shown in Table 2. From the testing results, we can see that in these problems, A^* with h_m yields a significant drop in search efforts (number of expanded markings and CPU time are low), while the solutions found (makespan) at termination are still the optimal results.

Table 2. Scheduling results of the instances from [7]

Lot sizes	makespan	h_a from [7]		h_b from [14]		h_c/n_{part}		h_m	
		EM	$TM(sec.)$	EM	$TM(sec.)$	EM	$TM(sec.)$	EM	$TM(sec.)$
1 1 1 1	17	41	0.062	598	0.134	2177	0.775	34	0.057
2 2 1 1	25	394	0.031	58127	610	64413	876	189	0.014
5 5 2 2	58	974	0.124	–	–	–	–	868	0.118
8 8 4 4	100	1838	0.454	–	–	–	–	1685	0.420
10 10 6 6	134	2567	0.768	–	–	–	–	2507	0.756

(– means the result can not be got in a reasonable time)

4 Application to a Set of Randomly Generated Cases

In this section, we test the algorithms with some randomly generated problems which have such characteristics as (1) jobs with alternative routings and (2) operations with multiple resources. These characteristics are illustrated in Fig. 2 where $O_{i,j,k}$ represents the jth operation of the ith job type being processed with the kth resource. In Fig. 2(a), the jth operation of job i can be performed by alternative routings, i.e., by using either resource k or resource r. In Fig. 2(b), the performance of the operation needs multiple resources, resource k and resource r.

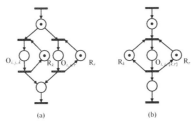

(a)   (b)

Fig. 2. (a) A model for a job having alternative routings (b) A model for an operation with dual resources

We generated a set of 40 random problems to test the algorithms. These problems, which were generated by the method of randomly selecting and linking predefined Petri net modules [11], had the following characteristics. The system had three resources and three different jobs with four operations each. The 75% of jobs had two alternative routings and 40% of operations had dual resources. Each operation was assigned a random cost from a uniform distribution (1-100). The lot size of each job was one.

We solved these problems using A^* algorithm with h_a, h_b, h_c/n_{part}, and h_m respectively, and we compare the results obtained using h_m with the results obtained using other heuristics. The scheduling results are summarized in Table 3. It represents the comparisons between the performances of A^* algorithm with h_m and A^* algorithms with h_a, h_b and h_c/n_{part} respectively.

Table 3. Computational results for randomly generated problems with alternative routings and multiple resources

Test	makespan	$\frac{h_m-h_a}{h_a}$			$\frac{h_m-h_b}{h_b}$			$\frac{h_m-h_c/n_{part}}{h_c/n_{part}}$		
		RDMS	RDEM	RDTM	RDMS	RDEM	RDTM	RDMS	RDEM	RDTM
1	326	0%	-20.51%	-16.38%	0%	-82.54%	-86.22%	0%	-86.79%	-93.34%
2	294	0%	-10.95%	-4.76%	0%	-87.75%	-96.06%	0%	-91.62%	-98.08%
3	246	0%	-41.35%	-58.58%	0%	-35.36%	-40.17%	0%	-70.30%	-84.95%
4	257	0%	-13.64%	-17.76%	0%	-43.67%	-50.14%	0%	-50.70%	-63.56%
5	344	0%	0.00%	-11.11%	0%	-93.81%	-97.80%	0%	-93.78%	-98.00%
6	326	0%	-63.64%	-78.03%	0%	-51.88%	-64.15%	0%	-80.24%	-92.84%
7	338	0%	0.00%	7.69%	0%	-83.63%	-90.00%	0%	-89.61%	-96.19%
8	394	0%	-0.08%	0.00%	0%	-18.24%	-31.88%	0%	-9.31%	-23.40%
9	214	0%	-4.68%	2.77%	0%	-61.25%	-81.02%	0%	-72.93%	-92.31%
10	237	0%	-41.61%	-58.33%	0%	-50.00%	-54.55%	0%	-81.98%	-90.91%
11	186	0%	-67.23%	-82.96%	0%	-71.46%	-76.04%	0%	-91.17%	-97.39%
12	343	0%	-95.20%	-99.49%	0%	-26.28%	-25.00%	0%	-85.80%	-95.70%
13	302	0%	-9.87%	-13.33%	0%	-93.99%	-98.50%	0%	-88.88%	-96.67%
14	236	0%	-66.00%	-78.99%	0%	-55.76%	-71.48%	0%	-77.00%	-91.10%
15	258	0%	-10.14%	-15.70%	0%	-44.06%	-61.32%	0%	-54.05%	-73.92%
16	214	0%	-14.95%	-14.81%	0%	-53.23%	-53.06%	0%	-69.75%	-79.09%
17	244	0%	149.07%	344.44%	0%	-46.97%	-51.22%	0%	-80.37%	-91.03%
18	288	0%	6.65%	11.76%	0%	-20.80%	-21.49%	0%	-22.75%	-37.09%
19	221	0%	-4.26%	9.52%	0%	-74.31%	-83.39%	0%	-83.03%	-93.26%
20	318	0%	-21.85%	-32.39%	0%	-51.12%	-57.14%	0%	-73.75%	-85.84%
21	329	0%	-25.60%	-38.80%	0%	-33.85%	-42.48%	0%	-71.12%	-88.02%
22	230	0%	-9.62%	-15.38%	0%	-92.47%	-97.30%	0%	-94.09%	-98.50%
23	323	0%	-16.13%	-50.00%	0%	-97.09%	-99.26%	0%	-98.19%	-99.71%
24	346	0%	24.12%	51.38%	0%	-47.97%	-62.67%	0%	-53.01%	-74.65%
25	224	0%	-0.68%	-14.29%	0%	-11.12%	-5.26%	0%	-37.60%	-53.04%
26	377	0%	-35.41%	-48.28%	0%	-86.00%	-92.98%	0%	-81.76%	-93.15%
27	414	0%	-8.59%	-12.50%	0%	-90.40%	-96.32%	0%	-91.75%	-97.39%
28	306	0%	-6.87%	-9.09%	0%	-93.05%	-97.23%	0%	-93.70%	-98.07%
29	400	0%	17.15%	34.78%	0%	-60.90%	-72.57%	0%	-74.15%	-86.92%
30	319	0%	-14.23%	-22.10%	0%	-49.29%	-62.40%	0%	-54.86%	-72.78%
31	324	0%	-44.94%	-57.14%	0%	-70.88%	-79.45%	0%	-85.07%	-94.34%
32	220	0%	-23.44%	-35.29%	0%	-49.55%	-58.88%	0%	-70.50%	-84.78%
33	271	0%	39.34%	40.00%	0%	-65.02%	-66.67%	0%	-95.01%	-98.38%
34	249	0%	-9.84%	-14.58%	0%	-90.02%	-95.49%	0%	-89.16%	-96.74%
35	295	0%	0.00%	33.33%	0%	-81.40%	-88.46%	0%	-90.88%	-96.58%
36	335	0%	-1.10%	5.77%	0%	-76.58%	-86.32%	0%	-81.33%	-92.48%
37	269	0%	0.00%	0.00%	0%	-95.52%	-98.74%	0%	-96.82%	-99.41%
38	289	0%	-87.54%	-92.99%	0%	-93.68%	-97.22%	0%	-96.47%	-99.29%
39	232	0%	-2.20%	-0.66%	0%	-66.91%	-80.82%	0%	-82.71%	-93.87%
40	329	0%	-88.90%	-95.20%	0%	-88.00%	-94.09%	0%	-97.02%	-99.53%
AVG		0%	-15.62%	-13.69%	0%	-64.64%	-71.63%	0%	-77.22%	-87.31%

The following phenomena were observed: First, the scheduling results (makespan) of these algorithms are all optimal. This phenomenon is to be expected, considering while the constituents of h_m are admissible, the maximum of them (h_m) is also admissible. Second, although A^* with h_m has a more complex heuristic function to compute, it has the smallest search efforts. This is because that A^* guarantees that no marking with $f(m) = g(m) + h(m) > C^*$ (where C^* denotes the optimal cost) would ever be expanded, so the higher the value of h, the fewer markings will be expanded, as long as h remains admissible. In Table 3, the comparison of the results of A^* algorithm with h_m and that of A^* algorithm with other heuristic function is $RDMS = \frac{makespan(h_m) - makespan(h_a|h_b|h_c/n_{part})}{makespan(h_a|h_b|h_c/n_{part})} \times 100\%$, and the percentages of computation complexity reduced by using h_m, which are the comparisons of the storage (number of expanded markings) and the computational time, are equal to: $RDEM = \frac{NEM(h_m) - NEM(h_a|h_b|h_c/n_{part})}{NEM(h_a|h_b|h_c/n_{part})} \times 100\%$ and $RDTM = \frac{T(h_m) - T(h_a|h_b|h_c/n_{part})}{T(h_a|h_b|h_c/n_{part})} \times 100\%$. We can see that A^* algorithm with h_m is more efficient in terms of number of expanded markings and computational time than that of using other heuristics and the research results of using h_m are still optimal. For example, on average, A^* algorithm with h_m explores 64.64% of markings less and it executes 71.63% faster than A^* algorithm with h_b, while its results found are the same as the results of A^* algorithms with the admissible heuristic function h_b.

From the above results, we found that it is promising to use h_m as candidates for heuristic function. h_m can simultaneously use admissible heuristic functions and nonadmissible heuristic functions having an upper support of relative errors, and h_m is still admissible and more informed than its constituents.

5   Conclusions

In this paper, we proposed a heuristic scheduling strategy for FMSs in a P-timed Petri net framework. Timed Petri nets provide an efficient method for representing concurrent activities, shared resources and precedence constraints encountered frequently in FMSs. On the Petri net reachability graph, the method can use both admissible heuristic functions and nonadmissible heuristic functions having the upper supports of the relative errors to search the schedules. It can guarantee the results found are optimal and invoke quicker termination conditions. We use it to search for scheduling of a simple manufacturing system with multiple lot sizes for each job type considered. The algorithm is also used for a set of randomly generated FMSs with alternative routings and dual resources.

Further work will be conducted in setting different performance indices such as minimization of tardiness, and developing multi criteria heuristic functions for Petri-net based scheduling problems. We will also investigate the robustness of the resulting systems.

Acknowledgments. This work was supported in part by National Natural Science Foundation of China (Grant No. 61203173, 61272420), National and Jiangsu Planned Projects for Postdoctoral Research Funds, Doctoral Fund of Ministry of Education of China and the Zijin Intelligent Program of Nanjing University of Science and Technology.

References

1. Agnetis, A., Alfieri, A., Nicosia, G.: Part batching and scheduling in a flexible cell to minimize setup costs. Journal of Scheduling 6(1), 87–108 (2003)
2. Li, Z.W., Zhou, M.C., Wu, N.Q.: A survey and comparison of Petri net-based deadlock prevention policies for flexible manufacturing systems. IEEE Transactions on Systems, Man, and Cybernetics, Part C: Applications and Reviews 38(2), 173–188 (2008)
3. Yu, H., Reyes, A., Cang, S., Lloyd, S.: Combined petri net modelling and ai based heuristic hybrid search for flexible manufacturing systems part 1. Petri net modelling and heuristic search. Computers and Industrial Engineering 44(4), 527–543 (2003)
4. Tuncel, G., Bayhan, G.M.: Applications of Petri nets in production scheduling: a review. The International Journal of Advanced Manufacturing Technology 34(7-8), 762–773 (2007)
5. Russell, S.J., Norvig, P., Canny, J.F., Malik, J.M., Edwards, D.D.: Artificial intelligence: a modern approach, vol. 74. Prentice Hall, Englewood Cliffs (1995)
6. Yim, S.J., Lee, D.Y.: Multiple objective scheduling for flexible manufacturing systems using Petri nets and heuristic search. In: IEEE International Conference on Systems, Man, and Cybernetics, pp. 2984–2989 (1996)
7. Xiong, H.H., Zhou, M.: Scheduling of semiconductor test facility via petri nets and hybrid heuristic search. IEEE Transactions on Semiconductor Manufacturing 11(3), 384–393 (1998)
8. Reyes, A., Yu, H., Kelleher, G., Lloyd, S.: Integrating Petri nets and hybrid heuristic search for the scheduling of fms. Computers in Industry 47(1), 123–138 (2002)
9. Huang, B., Sun, Y., Sun, Y.: Scheduling of flexible manufacturing systems based on Petri nets and hybrid heuristic search. International Journal of Production Research 46(16), 4553–4565 (2008)
10. Lee, D.Y., DiCesare, F.: Scheduling flexible manufacturing systems using Petri nets and heuristic search. IEEE Transactions on Robotics and Automation 10(2), 123–132 (1994)
11. Mejia, G.: An intelligent agent-based architecture for flexible manufacturing systems having error recovery capability. Thesis, Lehigh University (2002)
12. Lee, J., Lee, J.S.: Heuristic search for scheduling flexible manufacturing systems using lower bound reachability matrix. Computers and Industrial Engineering 59(4), 799–806 (2010)
13. Huang, B., Sun, Y., Sun, Y.M., Zhao, C.X.: A hybrid heuristic search algorithm for scheduling fms based on Petri net model. The International Journal of Advanced Manufacturing Technology 48(9-12), 925–933 (2010)
14. Huang, B., Shi, X.X., Xu, N.: Scheduling fms with alternative routings using Petri nets and near admissible heuristic search. The International Journal of Advanced Manufacturing Technology 63(9-12), 1131–1136 (2012)
15. Pearl, J.: Heuristics: intelligent search strategies for computer problem solving. Addison-Wesley, MA (1984)

Developing Data-driven Models to Predict BEMS Energy Consumption for Demand Response Systems

Chunsheng Yang, Sylvain Létourneau, and Hongyu Guo

Information and Communication Technologies
National Research Council Canada, Ottawa, Ontario, Canada
{Chunsheng.Yang,Sylvain.Letourneau,Hongyu.Guo}@nrc.gc.ca

Abstract. Energy consumption prediction for building energy management systems (BEMS) is one of the key factors in the success of energy saving measures in modern building operation, either residential buildings or commercial buildings. It provides a foundation for building owners to optimize not only the energy usage but also the operation to respond to the demand signals from smart grid. However, modeling energy consumption in traditional physic-modeling techniques remains a challenge. To address this issue, we present a data-mining-based methodology, as an alternative, for developing data-driven models to predict energy consumption for BEMSs. Following the methodology, we developed data-driven models for predicting energy consumption for a chiller in BEMS by using historic building operation data and weather forecast information. The models were evaluated with unseen data. The experimental results demonstrated that the data-driven models can predict energy consumption for chiller with promising accuracy.

Keywords: building energy management systems (BEMS), data-driven models, air handling unit (AHU), machine learning algorithms, modeling, demand response systems.

1 Introduction

Today it is estimated that building industry contributes over 40% of total energy consumption. The owners of building aim at reducing the cost of energy consumption by developing efficient technologies to optimize or reduce the energy consumptions. With the quick increase of the energy price, developing new technology for building energy management systems (BEMS) or applying new philosophy of operating the systems is becoming more and more important and necessary. In utility industry, one of philosophy to change the existing operation policy is called demand response (D/R) program. It is defined as the incentive program to promote the lower electricity use at times of high wholesale market prices [1]. To avoid the high prices or the peak time of electricity consumption, BEMS must be able to adjust the use of electricity by changing the usage of energy from their normal consumption pattern or routine operation temporally. To achieve the response, the operators can just reduce the electricity usage during peak hours when the price is high without changing their

A. Moonis et al. (Eds.): IEA/AIE 2014, Part I, LNAI 8481, pp. 188–197, 2014.

consumption pattern during other period. However, this will temporally lose the comfort. It is not expected from the customers. It is desirable that the BEMS can change the energy consumption pattern to avoid the peak hour without any lose of the comfort. For instance, the BEMS can use ice bank for air conditioning (A/C) during peak hour, and make and story ice during the idle hours such as midnight. To address this issue, it is necessary and urgent to develop a novel technology which allows BEMS automatically to modify the building operation pattern/condition or change control policy to respond to D/R signal.

BEMS mainly consists of heating, ventilation, and air-conditioning (HVAC) systems and lighting systems. HAVC contains many subsystems such as AHU (Air Handling Units), chiller, ice bank, etc. To automatically respond to the D/R signal from smart grid, these subsystems must be integrated as a smart energy consumption system to perform optimized operation by changing control policies or energy usage pattern. Such a smart system is capable of making decision on when to change the operation pattern, where to optimize the usage of energy for each subsystem, and how to set up the control policy without losing the comfort based on D/R signal and weather forecast. To this end, one of the fundamental issues is how we can predict the peak usage of energy consumption, the short term /long term energy consumption, and the energy consumption of each subsystem.   In general, four kinds of models are required: (1) to predict the expected total building energy consumption; (2) to predict energy consumption for each sub-systems such as chiller, AHU, ice bank, lighting etc.; (3) to express control policy by exploiting the relationship between consumption of a given subsystem and its variables such as speed, water temperature, and valves; and (4) to optimize the energy usage to respond to D/R single from grid or to avoid the peak hours based on the prediction results from former models, weather forecast information, and constrains of comfort settings. Unfortunately due to time varying characteristic and nonlinearity in existing BEMS, it is difficult to build traditional mathematic/physic models to meet those needs. To our best knowledge there do not exist publicly available models that could predict the complete performance and energy consumption for subsystems [9].

Recently, many research efforts have been invested on the development of data-driven models for long/short term energy consumption prediction by applying machine learning algorithm such as neural network, decision tree and regression analysis [2-6]. There have been some research focusing on development of advanced AHU control methods [7], or accessing the building occupancy for energy load prediction [8]. There are, however, few works in developing the models for predicting energy consumption of each subsystem or optimize the usage of energy consumption. To address the issues in developing predictive models for energy consumption estimation and provide an alternative for modeling energy consumption, we propose to develop data-driven models from historical building operation data by using the techniques from machine learning and data mining. This work focused on the development of energy consumption prediction for BEMS subsystems. In this paper we first introduce the developed methodology. Then we present the development of data-driven models for predicting energy consumption of subsystems. In particular, the modeling results for a chiller subsystem will be presented and discussed as a case study.

The rest of this paper is organized as follows. Section 2 briefly describes the methodology. Section 3 presents the development of data-driven models for a chiller subsystem. Section 4 provides some preliminary experimental results. Section 5 discusses the results and limitation. The final section concludes the paper.

2   Data-driven Methodology

Modern building operation has generated massive data from BEMS, including energy consumption history, building characteristics, operating condition, occupancy, and control records for subsystems, amongst others. These data are a valuable resource for developing data-driven predictive models. To build the models from these historic operation data, we develop a data-driven methodology by using machine learning and data mining techniques. Figure 1 illustrates the methodology which consists of four steps: data gathering, data transformation, modeling, and model evaluation. The following is a brief description of each step.

2.1   Data Gathering

Most data mining algorithms require as input a dataset, which contains instances consisting of vectors of attribute values. Modern building operation often generates many such datasets. The first problem is to select the best dataset(s) to use to build models for a particular subsystem. Advice from subject matter experts and reliable documentation can simplify this choice and help avoid a lengthy trial and error process. Not only must a dataset be selected, but a subset of instances must be selected to use in the analysis. The datasets are typically very large so it is inefficient to build models using all instances. Simple solutions, such as random sampling, are also inappropriate. To build the desired predictive models, a much more focused approach was required. In this work, 4 types of data are used, including energy

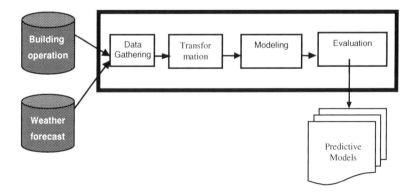

Fig. 1. The data-driven methodology

consumption history, control variables, weather forecast, building parameters, and ambient conditions. Data gathering is to obtain a dataset for subsystems from a big building operation database. This dataset will combine all types of operation data into a set of vectors.

2.2   Data Transformation

In modeling energy consumption, the target variable is numeric. Therefore, this methodology focuses on numeric modeling by applying machine algorithms. In other words, the main goal is to develop repressors by using regression algorithms. To improve the initial, as measured, representation, and remove outliers from the original data, data transformation is necessary. The main task is to generate some new features such as moving average, standard deviation, pattern expressions by using methods from process physics, signal processing, time series analysis, etc. The generated new features will enhance model performance significantly.

2.3   Modeling

After updating the initial dataset incorporating data representation enhancements, machine learning algorithms were used to build the predictive models. Dataset are separated into training and testing datasets. The training dataset were used for developing the models and the remaining data were kept for testing for evaluation. Any regression learning algorithm can be used. In early experiments, simple algorithms such as regression decision trees and support vector machine were preferred over more complex ones such as ANN because of their efficiency and comprehensibility. The same algorithm was applied several times with varying attribute subsets and a range of cost functions.   Therefore feature selection was also applied on the augmented data representation to automatically remove redundant or irrelevant features.

In order to build high-performance energy prediction models, another method is to perform the model fusion. Model fusion can be used for two reasons. First, when more than one data set is relevant for a given component, we can build a model for each dataset and then use model fusion to combine predictions from the various models. Second, we can apply model fusion for performance optimization regardless of the number of datasets selected. In this case, we learn various models using various techniques or parameter settings and combine them to obtain better performance than using any single model. Bagging and boosting [10] are two popular techniques to combine models but they are only applicable when there is a single data set and one kind of models (a single learning algorithm). For heterogeneous models or multiple data sets, we apply methods based on voting or stacking strategy [11], [12]. These techniques are globally referred to as multiple model systems.

2.4   Model Evaluation

To evaluate the performance of the developed models, the most important measure of performance is the prediction accuracy achieved by the models after development.

The accuracy is often defined using the forecast error which is the difference between the actual and predicted values [13]. Several criteria are available from statistics. The most widely used are the mean absolute error (*MAE*), the sum of squared error (*SSE*), the mean squared error (*MSE*), the root mean squared error (*RMSE*), and the mean absolute percentage error (*MAPE*). They are defined as follows.

$$MAE = \frac{1}{N} \sum_{i=1}^{N} |e_i| \tag{1}$$

$$SSE = \sum_{i=1}^{N} e_i^2 \tag{2}$$

$$MSE = \sum_{i=1}^{N} \sqrt{e_i} \tag{3}$$

$$RMSE = \sqrt{MSE} \tag{4}$$

$$MAPE = \frac{1}{N} \sum_{i=1}^{N} |\frac{e_i}{y_i}| \times 100 \tag{5}$$

where e_i is the individual prediction error; y_i is the actual value; and N is the number of examples in the test data.

The general method is to apply the testing dataset (unseen data) on the models by computing the error terms as mentioned above. Each measure term has advantages and limitations. It is not necessary to compute all accuracy measures. In this work we focused on three widely used error criteria, namely MAE, MSE, and MAPE.

3   Model Development for Chiller Energy Consumption

In this section, we demonstrate the development of models for predicting the energy consumption of BEMS subsystems by using the data-driven methodology. We focus

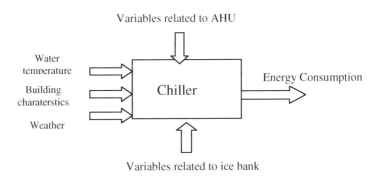

Fig. 2. The diagram of chiller energy consumption

on one of the main subsystems, namely the chiller in a HVAC system. Chiller is a key subsystem in the HAVC system, which provides the chilled water or cooled air to the coil of the AHU for air-conditioning the building. For this work, the chiller is used together with ice bank to provide A/C for the building. It is possible to increase the water temperature of chiller to reduce energy consumption by using ice bank to provide the cooled air for substituting chiller function during the peak hour. On the other hand, chiller can be used to charge ice bank during valley time such as midnight. There is a great potential for optimizing the energy consumption to avoid peak time or to respond the D/R signal if we can accurately predict the energy consumption of chiller and set up an optimized control policy.

3.1   Model Development

For development of chiller energy consumption models, we use building operation data from a modern building which uses chiller and ice bank for A/C in the summer. The data were collected from 2009 to 2010. The database contains over one thousand variables for all sensors or control points on the building. There is only one chiller in this building. The first task is to get data related to the chiller from this database over thousand of variables. After analyzing the signals of chiller control system and consulting with building operator, the four groups of data are identified and their variables are extracted from the database. These data are energy consumption (output of chiller), building occupancy, water temperatures at different points, and variables related to ice bank. In order to build high-performance models, we also collected weather forecast data from a local weather station. The weather data consists of temperature, wind speed, wind direction, and humidity. In the end, we can describe the chiller diagram as shown in Figure 2.

After gathering data relevant to Chiller from the database and combining with weather forecast data, data transformation was performed to generate some new features following the developed methodology. The new features mainly include time-series features such as weekday, seasons, moving averages, and the energy consumption for selected past days. These new features enhance the prediction accuracy for the models.

Using dataset with the new features, the modeling experiments were conducted by carefully choosing machine learning algorithms. As mentioned in Section 2.3, many regression-based algorisms were first evaluated by trial-error approach. Then we decide to use two algorithms, i.e., DecisionRegTree and SVM, to model energy consumption for chiller. The next subsection presents the experimental results for these selected algorithms.

3.2   Experiment Results

Using the dataset with new features, we conducted modeling experiments. The dataset contains 12,734 instances. We separated the dataset into training and testing sets. Training dataset contains 8000 instances and testing dataset has 4734 instances. Against the training dataset, the predicted models are built with two selected learning

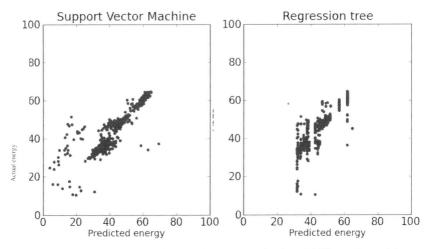

Fig. 3. The results of predicted values vs. actual values (chilling only mode)

Table 1. The performance of the built models (chilling only mode)

	SVM	DecisionRegressionTree
MAE	1.72	1.28
MSE	13.25	8.15
MAPE(%)	5	4

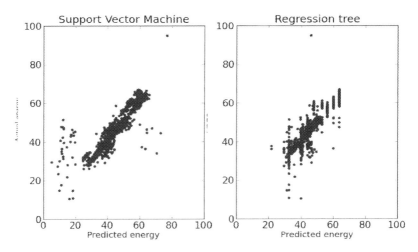

Fig. 4. The results of predicted values vs. actual values (chilling and charging mode)

Table 2. The performance of the built models (chilling and charging mode)

	SVM	DecisionRegressionTree
MAE	1,40	1.36
MSE	9.64	6.63
MAPE(%)	3	3

algorithms. Then we run the trained models on the testing dataset. Using evaluation criteria we computed the performance for each model. In building operation, the chill has two working modes: chilling only mode, and chill and charging mode. Therefore we conducted two kind of modeling experiments corresponding two chiller working modes. Figure 3 shows the modeling result of chilling only mode. The x-axis is the predicted value and y-axis is actual value of energy consumption. The performance of models, as measured by the MAE, MSE, and MAPE, are shown in Table 1.

The Figure 4 shows modeling results of chilling and charging mode, and the performance of the model is shown in Table 2.

4   Discussions

The experimental results above demonstrated that the data-driven modeling methodology is useful and effective for developing predictive models to predict the energy consumption for BEMS subsystems such as chiller and AHU. From the Table 1 and Table 2, it is obvious that non-linear models show high accuracy results for predicting energy consumptions. In many building operations, it is difficult to build traditional math models to predict energy consumption because of operation complexity and mutual interaction and constrains among subsystems. The proposed data-driven modeling method provides a feasible alternative for modeling BEMS energy consumption.

It is worth to noting that the energy predictions for two chiller modes are very similar This suggested that the developed models are very robust and transferable for different chiller working modes.

It is also worth to point out that the predictive accuracy for low energy consumption is not as good as that for high energy consumption. The reasons need to be further investigated. But it also suggested that there is a possibility to improve the model performance by focusing on low energy consumption of the BEMS subsystems.

One limitation for the proposed method is lack of ability of dealing with the sensor data noisy. In this work, we found that some data is not accurate under some operation condition. For example, when chiller is operated at low energy consumption the sensor readings are not correct or noised. We removed those data based on domain experts' input. Another problem, some data were missing in the database. Therefore, it is better to investigate an effective mechanism to deal with noisy and missing values in the data.

The proposed modeling techniques mainly focused on energy consumption prediction for reactive building operation. For proactive or long term building operation, it is more important to predict energy consumption ahead in a period of time, say 24 hours beyond. To meet such a requirement, we have to develop a predictive model which is able to predict energy load for whole building or all BEMS 24 hours ahead or longer period of time. This will be more challenge. We are working on this issue. The results will be reported in other papers.

5   Conclusions and Future Works

In this paper, we proposed a data-driven methodology for developing predictive models to generate energy consumption predictions for BEMS subsystems such as chiller and AHU. To attain this goal, we deployed some state-of-the-art machine learning and data mining techniques. We applied the proposed method to build predictive models for chiller in a given commercial building. The developed predictive models were evaluated using real data from the building owner. The experimental results show that the data-driven method is a useful and effective alternative for modeling BEMS energy consumptions.

As presented in this paper, we have only developed the predictive models for chiller subsystems. For our future work, we will continue working on modeling other subsystems such as AHU. In order to perform optimized energy consumption management, we have to also develop data-drive models for tailoring control policies based on the energy consumption information for specified subsystems, and for projecting short term energy loads for the whole building. Consequently, such predictive models can be used to make decision on energy consumption for each subsystem, resulting in reaching the final goal of energy saving for building operations.

Acknowledgment. Many people at the National Research Council Canada have contributed to this work. Special thanks go to Elizabeth Scarlett for technical support. Thanks to National Resource of Canada for providing the building data to conduct the modeling experiments and domain knowledge to exploit the database.

References

[1] Albadi, M.H., El-Saadany, E.F.: A summary of demand response in electricity markets. Electric Power Systems Research 78, 1989–1996 (2008)
[2] He, M., Cai, W.J., Li, S.Y.: Multiple fuzzy model-based temperature predictive control for HVAC systems. Information Science 169, 155–174 (2005)
[3] Yalcintas, M., Akkurt, S.: Artificial neural networks applications in building energy predictions and a case study for tropical climates. International Journal of Energy Research 29, 891–901 (2005)

[4] Catalina, T., Virgone, J., Blanco, E.: Development and validation of regression models to predict monthly heating demand for residential buildings. Energy and Buildings 40, 1825–1832 (2008)

[5] Olofsson, T., Andersson, S., Ostin, R.A.: Energy load prediction for buildings based on a total demand perspective. Energy and Buildings 28, 109–116 (1998)

[6] Tso, G.K.F., Yau, K.K.W.: Predicting electricity energy consumption: A comparison of regression analysis, decision tree and neural networks. Energy 32, 1761–1768 (2007)

[7] Bi, Q., Cai, W.J., Wang, Q.G., et al.: Advanced Controller auto-tuning and its application in HAVC systems. Control Engineering Practices 8, 633–644 (2000)

[8] Kwok, S.S.K., Yuen, R.K.K., Lee, E.W.M.: An intelligent approach to assessing the effect of building occupancy on building cooling load prediction. Building and Environment 46, 1681–1690 (2011)

[9] Bendapudi, S., Braun, J.E., Groll, E.A.: A Dynamic model of a vapor compression liquid chiller. In: The Proceedings of International Refrigeration and Air Conditioning Conference (2002)

[10] Dietterich, T.: An experimental comparison of three methods for constructing ensembles of decision trees: Bagging, boosting and randomization. IEEE Intelligent Systems and their Applications 40, 139–158 (2000)

[11] Tsoumakas, G., Katakis, I., Vlahavas, I.P.: Effective voting of heterogeneous classifiers. In: Boulicaut, J.-F., Esposito, F., Giannotti, F., Pedreschi, D. (eds.) ECML 2004. LNCS (LNAI), vol. 3201, pp. 465–476. Springer, Heidelberg (2004)

[12] Džeroski, S., Ženko, B.: Stacking with multi-response model trees. In: Roli, F., Kittler, J. (eds.) MCS 2002. LNCS, vol. 2364, pp. 201–211. Springer, Heidelberg (2002)

[13] Zhang, G., Patuwo, B.E., Hu, M.Y.: Forecasting with artificial neural networks: the state of the art. International Journal of Forecasting 14, 35–62 (1998)

Particle Filter-Based Method for Prognostics with Application to Auxiliary Power Unit

Chunsheng Yang[1], Qingfeng Lou[2], Jie Liu[2], Yubin Yang[3], and Yun Bai[4]

[1] National Research Council Canada, Ottawa ON, Canada
[2] Department of Mechanical & Aerospace Engineering,
Carleton University, Ottawa ON, Canada
[3] State Key Laboratory for Novel Software Technology, Nanjing University, Nanjing, China
[4] School of Computing, Engineering & Mathematics, University of Western Sydney, Australia

Abstract. Particle filter (PF)-based method has been widely used for machinery condition-based maintenance (CBM), in particular, for prognostics. It is employed to update the nonlinear prediction model for forecasting system states. In this work, we applied PF techniques to Auxiliary Power Unit (APU) prognostics for estimating remaining useful cycle to effectively perform APU health management. After introducing the PF-based prognostic method and algorithms, the paper presents the implementation for APU Starter prognostics along with the experimental results. The results demonstrated that the developed PF-based method is useful for estimating remaining useful cycle for a given failure of a component or a subsystem.

Keywords: Particle filter (PF), data-driven prognostics, remaining useful cycle (RUC), condition-based maintenance (CBM), APU Starter prognostics.

1 Introduction

Condition-based maintenance(CBM) is an emerging technology that recommends maintenance decisions based on the information collected through system condition monitoring (or system state estimation) and equipment failure prognostics (or system state forecasting), in which prognostics still remains as the least mature element in real-world applications [1]. Prognostics entail the use of the current and previous system states (or observations) to predict the likelihood of a failure of a dynamic system and to estimate remaining useful life (RUL). Reliable forecast information can be used to perform predictive maintenance in advance and provide an alarm before faults reach critical levels so as to prevent system performance degradation, malfunction, or even catastrophic failures [2].

In general, prognostics can be performed using either data-driven methods or physics-based approaches. Data-driven prognostic methods use pattern recognition and machine learning techniques to detect changes in system states [3, 4]. The classical data-driven methods for nonlinear system prediction include the use of stochastic models such as the autoregressive (AR) model [5], the threshold AR model [6], the bilinear model [7], the projection pursuit [8], the multivariate adaptive

A. Moonis et al. (Eds.): IEA/AIE 2014, Part I, LNAI 8481, pp. 198–207, 2014.
© Springer International Publishing Switzerland 2014

regression splines [9], and the Volterra series expansion [10]. Since the last decade, more interests in data-driven system state forecasting have been focused on the use of flexible models such as various types of neural networks (NNs) [11, 12] and neural fuzzy (NF) systems [13, 14]. Data-driven prognostic methods rely on past patterns of the degradation of similar systems to project future system states; their forecasting accuracy depends on not only the quantity but also the quality of system history data, which could be a challenging task in many real applications [2, 15]. Another principal disadvantage of data-driven methods is that the prognostic reasoning process is usually opaque to users [16]; consequently, they sometimes are not suitable for some applications where forecast reasoning transparency is required. Physics-based approaches typically involve building models (or mathematical functions) to describe the physics of the system states and failure modes; they incorporate physical understanding of the system into the estimation of system state and/or RUL [17-19]. Physics-based approaches, however, may not be suitable for some applications where the physical parameters and fault modes may vary under different operation conditions [20]. On one hand, it is usually difficult to tune the derived models *in situ* to accommodate time-varying system dynamics. On the other hand, physics-based approaches cannot be used for complex systems whose internal state variables are inaccessible (or hard) to direct measurement using general sensors. In this case, inference has to be made from indirect measurements using techniques such as particle filtering (PF). Recently the PF-based approaches have been widely used for prognostic applications [21-25], in which the PF is employed to update the nonlinear prediction model and the identified model is applied for forecasting system states. It is proven that FP-based approach, as a Sequential Monte Carlo (SMC) statistic method [26, 27], is affective for addressing the issues that data-driven and physic-based approach face.   In this work, we apply the PF method to Auxiliary Power Unit (APU) Starter prognostics by updating the models of the performance-monitoring parameters. This paper presents the developed PF-based methods for prognostics along with the experimental results from APU Starter prognostic application.

The rest of this paper is organized as follows. Section 2 briefly describes the background of the APU. Section 3 presents the PF-based method for prognostics; Section 4 provides some experimental results. Section 5 discusses the results and future work. The final section concludes the paper.

2   APU Overview

2.1   APU and APU Data

The APU engines on commercial aircrafts are mostly used at the gates. They provide electrical power and air conditioning in the cabin prior to the starting of the main engines and also supply the compressed air required to start the main engines when the aircraft is ready to leave the gate. APU is highly reliable but they occasionally fail to start due to failures of components such as the Starter Motor. APU starter is one of the most crucial components of APU. During the starting process, the starter accelerates APU to a high rotational speed to provide sufficient air compression for self-sustaining operation. When the starter performance gradually degrades and its output power decreases, either the APU combustion temperature or the surge risk will increase significantly. These consequences will then greatly shorten the whole APU life and even result in an immediate thermal damage. Thus the APU starter degradation can result in unnecessary economic losses and impair the safety of airline operation. When Starter fails, additional equipment such as generators and compressors must be used to deliver the functionalities that are otherwise provided by the APU. The uses of such external devices incur significant costs and may even lead to a delay or a flight cancellation. Accordingly, airlines are very much interested in monitoring the health of the APU and improving the maintenance.

For this study, we considered the data produced by a fleet of over 100 commercial aircraft over a period of 10 years. Only ACARS (Aircraft Communications Addressing and Reporting System) APU starting reports were made available. The data consists of operational data (sensor data) and maintenance data. The maintenance data contains reports on the replacements of many components which contributed the different failure modes. Operational data are collected from sensors installed at strategic locations in the APU which collect data at various phases of operation (e.g., starting of the APU, enabling of the air-conditioning, and starting of the main engines). The collected data for each APU starting cycle, there are six main variables related to APU performance: ambient air temperature (T_1), ambient air pressure (P_1), peak value of exhaust gas temperature in starting process (EGT_{peak}), rotational speed at the moment of EGT_{peak} occurrence (N_{peak}), time duration of starting process (t_{start}), exhaust gas temperature when air conditioning is enable after starting with 100% N (EGT_{stable}). There are 3 parameters related to starting cycles: APU serial number (S_n), cumulative count of APU operating hours (h_{op}), and cumulative count of starting cycles (cyc). In this work, in order to find out remaining useful cycle, we define a remaining useful cycle (RUC) as the difference of cyc_0 and cyc. cyc_0 is the cycle count when a failure happened and a repair was token. When RUC is equal to zero (0), it means that APU failed and repair is needed. RUC will be used in PF prognostic implementation in the following.

2.2   APU Data Correction

The APU data collected in operation covers a wide range of ambient temperatures from -20° to 40° and ambient pressures relevant to the airport elevations from sea level to 3557ft. Since the ambient conditions have a significant impact on gas turbine engine performance, making the engine parameters comparable requires a correction from the actual ambient conditions to the sea level condition of *international standard atmosphere* (ISA). To improve the data quality, the data correction is performed based on the March number similarity from gas turbine engine theory. Two main parameters (EGT_{peak} (noted as EP) and N_{peak} (noted as NP)) related APU performance are corrected using Equation 1 and 2.

$$EP = EGT_c = \frac{EGT_{peak}}{\Theta^{a_{EGT}}} \tag{1}$$

$$\text{NP} = N_c = \frac{N_{peak}}{\Theta^{a_N}} \tag{2}$$

Here the empirical exponents a_{EGT} and a_N are normally determined by running a calibrated thermodynamic computer model provided by engine manufacturers.

3   PF-Based Prognostics

3.1   PF-Based Prognostic Approach

In forecasting the system state, if internal state variables are inaccessible (or hard) to direct measurement using general sensors, inference has to be made from indirect measurements. Bayesian learning provides a rigorous framework for resolving this issue. Given a general discrete-time state estimation problem, the unobservable state vector $X_k \in R^n$ evolves according to the following system model

$$X_k = f(X_{k-1}) + w_k , \tag{3}$$

where $f : R^n \to R^n$ is the system state transition function and $w_k \in R^n$ is a noise whose known distribution is independent of time. At each discrete time instant, an observation (or measurement) $Y_k \in R^p$ becomes available. This observation is related to the unobservable state vector via the observation equation

$$Y_k = h(X_k) + v_k , \tag{4}$$

where $h : R^n \to R^p$ is the measurement function and $v_k \in R^p$ is another noise whose known distribution is independent of the system noise and time. The Bayesian learning approach to system state estimation is to recursively estimate the probability density function (*pdf*) of the unobservable state X_k based on a sequence of noisy measurements $Y_{1:k}$, $k = 1, \dots, K$. Assume that X_k has an initial density $p(X_0)$ and the probability transition density is represented by $p(X_k \mid X_{k-1})$. The inference of the probability of the states X_k relies on the marginal filtering density $p(X_k \mid Y_{1:k})$. Suppose that the density $p(X_{k-1} \mid Y_{k-1})$ is available at step k-1. The prior density of the state at step k can then be estimated via the transition density $p(X_k \mid X_{k-1})$,

$$p(X_k \mid Y_{1:k-1}) = \int p(X_k \mid X_{k-1}) p(X_{k-1} \mid Y_{1:k-1}) \, dX_{k-1} . \tag{5}$$

Correspondingly, the marginal filtering density is computed via the Bayes' theorem,

$$p(X_k \mid Y_{1:k}) = \frac{p(Y_k \mid X_k) p(X_k \mid Y_{1:k-1})}{p(Y_k \mid Y_{1:k-1})} , \tag{6}$$

where the normalizing constant is determined by

$$p(Y_k \mid Y_{1:k-1}) = \int p(Y_k \mid X_k) p(X_k \mid Y_{1:k-1}) \, dX_k . \tag{7}$$

Equations (5)-(7) constitute the formal solution to the Bayesian recursive state estimation problem. If the system is linear with Gaussian noise, the above method simplifies to the Kalman filter. For nonlinear/non-Gaussian systems, there are no closed-form solutions and thus numerical approximations are usually employed [28].

The PF or so-called sequential important sampling (SIS), is a technique for implementing the recursive Bayesian filtering via Monte Carlo simulations, whereby the posterior density function $p(X_k \mid Y_{1:k})$ is represented by a set of random samples (particles) $x_k^i (i = 1,2, \dots, N)$ and their associated weights $w_k^i (i = 1,2, \dots, N)$.

$$p(x_k \mid Y_{1:k}) \approx \sum_{i=1}^{N} w_k^i \delta(x_k - x_k^i), \quad \sum_{i=1}^{N} w_k^i = 1. \tag{8}$$

The w_k^i, normally known as importance weight, is the approximation of the probability density of the corresponding particle. In a nonlinear/non-Gaussian system where the state's distribution cannot be analytically described, the w_k^i of a dynamic set of particles can be recursively updated through Equation 9.

$$w_k^i \propto w_{k-1}^i \frac{p(y_k \mid x_k^i) p(x_k^i \mid x_{k-1}^i)}{q(x_k^i \mid x_{k-1}^i, y_k)}, \tag{9}$$

where $q(x_k^i \mid x_{k-1}^i, y_k)$ is a proposal function called *importance density function*. There are various ways of estimating the importance density function. One common way is to select $q(x_k^i \mid x_{k-1}^i, y_k) = p(x_k^i \mid x_{k-1}^i)$ so that

$$w_k^i \propto w_{k-1}^i p(y_k \mid x_k^i). \tag{10}$$

3.2   Implementation for APU Prognostics

This section presents an implementation of PF-based prognostics for APU starter. As we mentioned in Section 2, two key parameters related to APU starter degradation are *NP* and *EP* from data correction. We conducted statistic analysis of these two parameters using data collected during 10 years' operation. It is clear that these two parameters are identical to show two phases: normal operation and degraded operation. In order to apply PF-based prognostic methods to these parameters, we take *EP* as an example to demonstrate the implementation. Figure 1 shows an example of *EP* moving average during evolution of APU degradation. It shows that the moving average $\mu_{X_{RUC}}$ and the moving standard deviation $\sigma_{X_{RUC}}$ are relatively stable in the normal phase, but increase dramatically in the degraded phase. In the normal operation phase, *EP* measurements satisfy a stationary Gaussian $\mathcal{N}(\mu_{nor}, \sigma_{nor}^2)$. The starter is healthy in this phase, and this healthy state is indicated by the starter signal which is a relative constant value equivalent to μ_{nor}. Meanwhile, the noise signal is a stationary white noise with variance of σ_{nor}^2. In the degraded phase, *EP* measurements satisfy a non-stationary distribution that cannot be analytically described. The starter is experiencing degradation in this phase, and the degradation level is indicated by the starter signal which is the estimation of the measurements. Meanwhile, the noise signal is a non-stationary white noise with a variance that varies with the degradation level of starter.

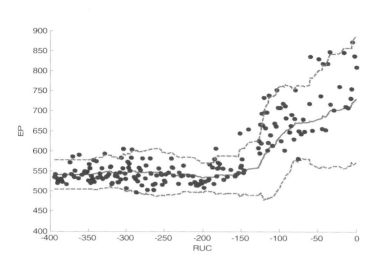

Fig. 1. an example of moving average for EP statistic analysis

Therefore, we can apply PF method to filter out the white noise and identify the degradation trend. To this end, we developed APU states estimation models for *EP*. These models are as follows:

$$\overline{EP}_k: \quad x_{1_k} = x_{1_{k-1}} \left(\frac{x_{3_k}}{x_{3_{k-1}}} \right) \exp\left[x_{2_k} (RUC_k - RUC_{k-1}) \right], \tag{11}$$

$$\lambda_k: \qquad\qquad x_{2_k} = x_{2_{k-1}} + \omega_{2_k}, \tag{12}$$

$$C_k: \qquad\qquad x_{3_k} = x_{3_{k-1}} + \omega_{3_k}, \tag{13}$$

$$EP_k: \qquad\qquad y_k = x_{1_k} + v_k. \tag{14}$$

where the subscript k represents the kth time step and RUC_k represents the starting cycle in this kth time step. There are three system states, $\overline{EP}$, λ, C, and one measurement, EP, in this system state model. These states and measurement are also denoted as x_1, x_2, x_3 and y respectively. ω_2 and ω_3 are independent Gaussian white noise processes, the v is approximate by the standard deviation of RUC in the collected dataset.

The first system state, $\overline{EP}$, represents the starter signal. As described in Equation 11, its value at time step k is determined from the system states at the previous time step. The second system state λ represents the starter degradation rate. It is located in the exponential part of Equation 11. Therefore, the starter degradation rate between two adjacent starting cycles is indicated by e^{λ}. The higher λ is, the faster a starter degrades along an exponential growth. When $\lambda = 0$, no degradation develops between two starting cycles. The third system state C represents a discrete change of the starter degradation between two adjacent starting cycles. During the PF iterations, the systems states are estimated in the framework of recursive Bayesian by constructing their conditional *pdf* based on the measurements.   Consequently, APU starter prognostic is implemented by λ estimation. Once the measurements stops, both λ and C are fixed with their most recent values. Thus the future degradation trend is expressed as an exponential growth of e^{λ}. The implementation of PF techniques is executed in an MATLAB environment.

4   Experimental Results

By implementing PF technique for **EP**, we can use λ to perform APU starter prognostics. The idea is that λ is fixed at its most recent values updated by the available measurements. Then the future degradation trend is expressed as an exponential growth, e^{λ}, started from the latest $\overline{EP}$ estimations. The experiments were mainly conducted to learn the weight parameters for PF methods and to predict or estimate the **EP** using learned parameters. The triggering point for prediction is determined based on the statistic analysis given a failure mode.   Figure 2 and 3 show the PF results when the prognostics is triggered at 650C and 750C for **EP** prediction corresponding to RUC at -100 and -50 starting cycles prior to the failure or replacement respectively. In these figures, we use "negative" numbers to represent the remaining cycles to failure event. Zero (0) represents the timing of failure event.

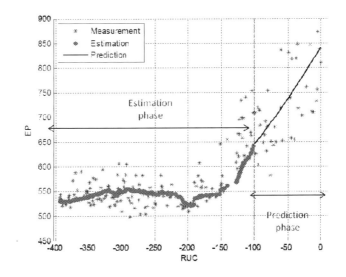

Fig. 2. PF prognostics result for EP (Triggered RUC= -100)

From the results, the APU starter prognostics can be easily performed by setting up a threshold for $\overline{EP}$.   From the Figure 2, $\overline{EP}$ threshold is set at 840C. In other words, when **EP** estimation from the learnt PF model reached 650C, it starts to use **EP** prediction to perform prognostics and the RUC corresponding to triggered EP will be used as onset point of RUC estimation. If the **EP** prediction is reaching 840C, it means APU starter should be changed or replacement within 100 RUCs.   In the Figure 2, the "star dots" represents the measurements; the red points are estimation from PF model during learning phase; and black line is *EP* prediction from the learnt PF models.

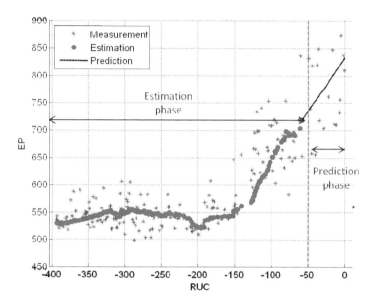

Fig. 3. PF prognostic result for EP (Triggered RUC= -50)

Similarly, Figure 3 shows the result of prediction triggered at *EP* = 750c correspond RUC = -50. From that point, the trained PF model starts to predict the *EP* for   APU Starter prognostics.

5   Discussions

The experimental results above demonstrated that PF-based method is useful and effective for performing APU starter prognostics. If the threshold value is decided correctly, the RUC can be predicted precisely by monitoring APU EGT or engine speed (N) with the developed techniques. This result is useful for developing onboard prognostic systems, which makes the prognostic decision transparent and simplified. In turn, it promotes largely application of prognostic technique to real-world problems.

Since there is existing a large variance in the different failure models, the precise RUC prediction for a particular APU Starter is really challenged. However, our PF-based prognostic techniques suggested clearly that once the estimated $\overline{EP}$ is $50^{o}C$ higher than μ_{nor}, the APU starts the degradation phase. This information is also useful for helping make decision on predictive maintenance.

It is worth to note that the results of *NP* for APU starter prognostic are similar to *EP*. It is also assume that the APU starter degradation follows a certain exponential growth pattern when we implemented PF-based prognostic for APU starter. This may not be effective for repetitive fluctuations of the starter degradation. In the future we should integrate data-driven prognostic techniques with PF-based prognostics to develop a hybrid framework for prognostics.

The results in this work only demonstrated one failure mode, "Inability to Start". The threshold value described above is determined only for this failure mode. For other failure modes, the corresponding statistics analysis is needed and the threshold values may vary. However, the developed PF-based method is still useful and applicable.

6   Conclusions

In this paper we developed a PF-based method for prognostics and applied it to APU Starter prognostics. We implemented the PF-based prognostic algorithm by using sequential importance sampling, and conducted the experiments with 10 years historic operational data provided by an airline operator. From the experimental results, it is obvious that the developed PF-based prognostic technique is useful for performing predictive maintenance by estimating relative precise remaining useful life for the monitored components or machinery systems.

Acknowledgment. Many people at the National Research Council Canada have contributed to this work. Special thanks go to Sylvain Letourneau, Jeff Bird and Craig Davison. We also thank for Air Canada to provide us APU operation data.

References

[1] Jardine, A., Lin, D., Banjevic, D.: A review on machinery diagnostics and prognostics implementing condition-based maintenance. Mechanical Systems and Signal Processing 20, 1483–1510 (2006)

[2] Liu, J., Wang, J.W., Golnaraghi, F.: A multi-step predictor with a variable input pattern for system state forecasting. Mechanical Systems and Signal Processing 2315, 86–99 (2009)

[3] Gupta, S., Ray, A.: Real-time fatigue life estimation in mechanical structures. Measurement Science and Technology 18, 1947–1957 (2007)

[4] Yagiz, S., Gokceoglu, C., Sezer, E., Iplikci, S.: Application of two non-linear prediction tools to the estimation of tunnel boring machine performance. Engineering Applications of Artificial Intelligence 22, 808–814 (2009)

[5] Groot, C.D., Wurtz, D.: Analysis of univariate time series with connectionist nets: a case study of two classical cases. Neurocomputing 3, 177–192 (1991)

[6] Tong, H., Lim, K.S.: Threshold autoregression, limited cycles and cyclical data. Journal of the Royal Statistical Society 42, 245–292 (1991)

[7] Subba, R.T.: On the theory of bilinear time series models. Journal of the Royal Statistical Society 43, 244–255 (1981)

[8] Friedman, J.H., Stuetzle, W.: Projection pursuit regression. Journal of the American Statistical Association 76, 817–823 (1981)

[9] Friedman, J.H.: Multivariate adaptive regression splines. Annals of Statistics 19, 1–141 (1981)

[10] Brillinger, D.R.: The identification of polynomial systems by means of higher order spectra. Journal of Sound and Vibration 12, 301–313 (1970)

[11] Atiya, A., El-Shoura, S., Shaheen, S., El-Sherif, M.: A comparison between neural-network forecasting techniques-case study: river flow forecasting. IEEE Transactions on Neural Networks 10, 402–409 (1999)

[12] Liang, Y., Liang, X.: Improving signal prediction performance of neural networks through multi-resolution learning approach. IEEE Transactions on Systems, Man, and Cybernetics-Part B: Cybernetics 36, 341–352 (2006)

[13] Husmeier, D.: Neural networks for conditional probability estimation: forecasting beyond point prediction. Springer, London (1999)

[14] Korbicz, J.: Fault Diagnosis: Models, Artificial Intelligence, Applications. Springer, Berlin (2004)

[15] Wang, W., Vrbanek, J.: An evolving fuzzy predictor for industrial applications. IEEE Transactions on Fuzzy Systems 16, 1439–1449 (2008)

[16] Tse, P., Atherton, D.: Prediction of machine deterioration using vibration based fault trends and recurrent neural networks. Journal of Vibration and Acoustics 121, 355–362 (1999)

[17] Adams, D.E.: Nonlinear damage models for diagnosis and prognosis in structural dynamic systems. In: Proc. SPIE, vol. 4733, pp. 180–191 (2002)

[18] Luo, J., et al.: An interacting multiple model approach to model-based prognostics. System Security and Assurance 1, 189–194 (2003)

[19] Chelidze, D., Cusumano, J.P.: A dynamical systems approach to failure prognosis. Journal of Vibration and Acoustics 126, 2–8 (2004)

[20] Pecht, M., Jaai, R.: A prognostics and health management roadmap for information and electronics-rich systems. Microelectronics Reliability 50, 317–323 (2010)

[21] Saha, B., Goebel, K., Poll, S., Christophersen, J.: Prognostics methods for battery health monitoring using a Bayesian framework. IEEE Transactions on Instrumentation and Measurement 58, 291–296 (2009)

[22] Liu, J., Wang, W., Golnaraghi, F., Liu, K.: Wavelet spectrum analysis for bearing fault diagnostics. Measurement Science and Technology 19, 1–9 (2008)

[23] Liu, J., Wang, W., Golnaraghi, F.: An extended wavelet spectrum for baring fault diagnostics. IEEE Transactions on Instrumentation and Measurement 57, 2801–2812 (2008)

[24] Liu, J., Wang, W., Ma, F., Yang, Y.B., Yang, C.: A Data-Model-Fusion Prognostic Framework for Dynamic System State Forecasting. Engineering Applications of Artificial Intelligence 25(4), 814–823 (2012)

[25] García, C.M., Chalmers, J., Yang, C.: Particle Filter Based Prognosis with application to Auxiliary Power Unit. In: The Proceedings of the Intelligent Monitoring, Control and Security of Critical Infrastructure Systems (September 2012)

[26] Doucet, A., Godsill, S., Andrieu, C.: On sequential Monte Carlo sampling methods for Bayesian filtering. Statistics and Computing, 197–208 (2000)

[27] Arulampalam, M., Maskell, S., Gordon, N., Clapp, T.: A tutorial on particle filters for online nonlinear/non-gaussian bayesian tracking. Trans. Sig. Proc. 50(2), 174–188 (2002), http://dx.doi.org/10.1109/78.978374

[28] Simon, D.: Optimal State Estimation: Kalman, and Nonlinear Approaches. Wiley Interscience (2006)

A Branch-and-Bound Algorithm for the Talent Scheduling Problem

Xiaocong Liang[1], Zizhen Zhang[1,*],
Hu Qin[2], Songshan Guo[1], and Andrew Lim[3]

[1] Sun Yat-Sen University, Guangdong, China
[2] Huazhong University of Science and Technology, Wuhan, China
[3] Nanjing University, Nanjing, China
zhangzizhen@gmail.com

Abstract. The talent scheduling problem is a simplified version of the real-world film shooting problem, which aims to determine a shooting sequence so as to minimize the total cost of the actors involved. We devise a branch-and-bound algorithm to solve the problem. A novel lower bound function is employed to help eliminate the non-promising search nodes. Extensive experiments over the benchmark instances suggest that our branch-and-bound algorithm performs better than the currently best exact algorithm for the talent scheduling problem.

Keywords: Branch-and-bound, talent scheduling, dynamic programming.

1 Introduction

The scenes of a film are not generally shot in the same sequence as they appear in the final version. Finding an optimal sequence in which the scenes are shot motivates the investigation of the talent scheduling problem, which is formally described as follows. Let $S = \{s_1, s_2, \ldots, s_n\}$ be a set of n scenes and $A = \{a_1, a_2, \ldots, a_m\}$ be a set of m actors. All scenes are assumed to be shot on a given location. Each scene $s_j \in S$ requires a subset $a(s_j) \subseteq A$ of actors and has a duration $d(s_j)$ that commonly consists of one or several days. Each actor a_i joins a subset $s(a_i) \subseteq S$ of scenes. We denote by Π the permutation set of the m scenes and define $e_i(\pi)$ (respectively, $l_i(\pi)$) as the earliest day (respectively, the latest day) in which actor i is required to be present on location in the permutation $\pi \in \Pi$. Each actor $a_i \in A$ has a daily wage $c(a_i)$ and is paid for each day from $e_i(\pi)$ to $l_i(\pi)$ regardless of whether they are required in the scenes. The objective of the talent scheduling problem is to find a shooting sequence (i.e., a permutation $\pi \in \Pi$) of all scenes that minimizes the total paid wages.

Table 1 presents an example of the talent scheduling problem, which is reproduced from [1]. The information of $a(s_j)$ and $s(a_i)$ is determined by the $m \times n$ matrix M shown in Table 1(a), where cell $M_{i,j}$ is filled with an "X" if actor

---

* Corresponding author.

A. Moonis et al. (Eds.): IEA/AIE 2014, Part I, LNAI 8481, pp. 208–217, 2014.

a_i participates in scene s_j and with a "·" otherwise. If the shooting sequence is $\pi = \{s_1, s_2, s_3, s_4, s_5, s_6, s_7, s_8, s_9, s_{10}, s_{11}, s_{12}\}$, the cost of each scene is presented in the second-to-last row and the total cost is 604. The cost incurred by the waiting status of the actors is called *holding cost*, which is shown in the last row of Table 1(b). Since minimizing the total cost is equivalent to minimizing the total holding cost, we only focus on minimizing the total holding cost in this paper.

Table 1. An example of the talent scheduling problem reproduced from [1]

	s_1	s_2	s_3	s_4	s_5	s_6	s_7	s_8	s_9	s_{10}	s_{11}	s_{12}	$c(a_i)$
a_1	X	·	X	·	·	X	·	X	X	X	X	X	20
a_2	X	X	X	X	X	·	X	·	X	·	X	·	5
a_3	·	X	·	·	·	·	X	X	·	·	·	·	4
a_4	X	X	·	·	X	X	X	·	·	·	·	·	10
a_5	·	·	·	X	·	·	·	X	X	·	·	·	4
a_6	·	·	·	·	·	·	·	·	·	X	·	·	7
$d(s_j)$	1	1	2	1	3	1	1	2	1	2	1	1	
cost	35	39	78	43	129	43	33	66	29	64	25	20	604
holding cost	0	20	28	34	84	13	24	10	0	10	0	0	223

The talent scheduling problem was originated from [2]. The paper introduced an orchestra rehearsal scheduling problem, which can be viewed as a restricted version of the talent scheduling problem with all actors having the same cost. [3] studied another restricted talent scheduling problem in which all scenes have identical duration. [4] proposed to use constraint programming to solve the both problems. In her subsequent work ([5]), she accelerated this constraint programming approach by catching search states. The talent scheduling problem studied in this work is the same as the one in [1], which proposed a dynamic programming algorithm to solve the problem. To the best of our knowledge, the dynamic programming algorithm is currently the best exact algorithm for the talent scheduling problem. In literature, there also exist several meta-heuristic approaches developed for the talent scheduling problem and its related problems. These approaches include hybrid genetic algorithms by [6], simulated annealing and tabu search algorithm by [7].

In the paper, we propose a branch-and-bound approach, which integrates a novel lower bound and the memoization technique, to find the optimal solution to the problem. The experimental results show that our proposed algorithm is superior to the current best exact algorithm by [1].

2   A Branch-and-Bound Approach

Branch-and-bound is a general technique for optimally solving various combinatorial optimization problems. The basic idea of the branch-and-bound algorithm is to systematically and implicitly enumerate all candidate solutions, where non-promising solutions are discarded by using upper and lower bounds.

2.1   Double-Ended Search

The solutions of the talent scheduling problem can be easily presented in a branch-and-bound search tree. Suppose we aim to find an optimal permutation $\pi^* = (\pi^*(1), \pi^*(2), \ldots, \pi^*(n))$. We can employ a *double-ended search* strategy that alternatively fixes the first and the last undetermined positions in the permutation. Therefore, the double-ended search determines a scene permutation following the order $\pi(1), \pi(n), \pi(2), \pi(n-1)$ and so on. When using the double-ended search strategy, a node in some level of the search tree corresponds to a partially determined permutation with the form $(\hat{\pi}(1), \ldots, \hat{\pi}(k-1), \pi(k), \ldots, \pi(l), \hat{\pi}(l+1), \ldots, \hat{\pi}(n))$, where $1 \leq k \leq l \leq n$ and the value of $\pi(h)$ $(k \leq h \leq l)$ is undetermined. We denote by $\overrightarrow{B}$ (respectively, $\overrightarrow{E}$) the partial permutation at the beginning, namely $\overrightarrow{B} = (\hat{\pi}(1), \hat{\pi}(2), \ldots, \hat{\pi}(k-1))$ (respectively, $E = (\hat{\pi}(l+1), \hat{\pi}(l+2), \ldots, \hat{\pi}(n)))$, and B (respectively, E) the corresponding set of scenes. The remaining scenes are put in a set Q, i.e., $Q = S - B - E$. If an actor is required by scenes of both B and E, this actor is called *fixed* since the number of days on which he/she is present on location is fixed and his/her cost in the final schedule becomes known. Let $a(Q) = \cup_{s \in Q} a(s)$ be the set of actors required by at least one scene in $Q \subseteq S$. The set of fixed actors can be represented by set $F = a(B) \cap a(E)$.

The double-ended branch-and-bound framework is given in Algorithm 1. In this algorithm, the operator "∘" in lines 2, 11 and 15 indicates the concatenation of two partially determined sequences. The function **search**$(\overrightarrow{B}, Q, \overrightarrow{E})$ returns the optimal solution to the talent scheduling problem with known $\overrightarrow{B}$ and $\overrightarrow{E}$, denoted by the subproblem $P(\overrightarrow{B}, Q, \overrightarrow{E})$. The optimal solution of the talent scheduling problem can be achieved by invoking **search**$(\overrightarrow{B}, Q, \overrightarrow{E})$ with $B = E = \emptyset$ and $Q = S$. The function **lower_bound**$(\overrightarrow{B} \circ s, Q - \{s\}, \overrightarrow{E})$ provides a valid lower bound to $P(\overrightarrow{B} \circ s, Q - \{s\}, \overrightarrow{E})$, where the set B of scenes is scheduled before scene s and the set $S - B - \{s\}$ of scenes is scheduled after scene s.

2.2   The Subproblem $P(\overrightarrow{B}, Q, \overrightarrow{E})$

The subproblem $P(\overrightarrow{B}, Q, \overrightarrow{E})$ corresponds to a node in the search tree. Its lower bound **lower_bound**$(\overrightarrow{B}, Q, \overrightarrow{E})$ can be expressed as:

$$\textbf{lower\_bound}(\overrightarrow{B}, Q, \overrightarrow{E}) = cost(\overrightarrow{B}, \overrightarrow{E}) + \textbf{lower}(B, Q, E),$$

where $cost(\overrightarrow{B}, \overrightarrow{E})$, called *past cost*, is the cost incurred by the path from the root node to the current node, and **lower**(B, Q, E) provides a lower bound to *future cost*, namely the holding cost to be incurred by scheduling the scenes in Q. It is irrelevant to the orders of scenes in B and E. Note that the holding costs of the fixed actors are determined, i.e., the fixed actors will not contribute to the holding cost in the later stages of the search. We use set A_F to contain all *fix actors*, namely $A_F = \{a_i \in A : a_i \in a(B) \cap a(E)\}$. We also use A_N to contain

---

Algorithm 1. The double-ended branch-and-bound search framework

---

Function: search$(\overrightarrow{B}, Q, \overrightarrow{E})$

1. **if** $Q = \emptyset$ **then**
2. $current\_solution = \overrightarrow{B} \circ \overrightarrow{E}$;
3. $z = \textbf{evaluate}(current\_solution)$;
4.   **if** $z < UB$ **then**
5. $UB := z$;
6. $best\_solution := current\_solution$;
7.   **end if**
8.   **return** ;
9. **end if**
10. **for** each $s \in Q$ **do**
11. $LB := \textbf{lower\_bound}(\overrightarrow{B} \circ s, Q - \{s\}, \overrightarrow{E})$;
12.   **if** $LB \geq UB$ **then**
13.   continue;
14.   **end if**
15.   **search**$(\overrightarrow{E}, Q - \{s\}, \overrightarrow{B} \circ s)$;
16. **end for**

---

all *non-fixed actors*, i.e., $A_N = \overline{A_F}$. We discuss the past cost $cost(\overrightarrow{B}, \overrightarrow{E})$ in this subsection and leave the description of **lower**(B, Q, E) in Subsection 2.3.

When $\overrightarrow{B}$ and $\overrightarrow{E}$ have been fixed, a portion of holding cost, namely $cost(\overrightarrow{B}, \overrightarrow{E})$, is determined regardless of the schedule of the scenes in Q. $cost(\overrightarrow{B}, \overrightarrow{E})$ is incurred by the holding days that can be confirmed by the following three ways.

1. For the actor $a_i \in a(B) \cap a(E)$, the number of his/her holding days in any complete schedule can be fixed.
2. For the actor $a_i \in a(B) \cap a(Q) - a(E)$, the number of his/her holding days in the time period for completing scenes in B can be fixed.
3. For the actor $a_i \in a(E) \cap a(Q) - a(B)$, the number of his/her holding days in the time period for completing scenes in E can be fixed.

Furthermore, we use $cost(s, B, E)$ to represent the newly confirmed holding cost incurred by placing scene $s \in Q$ at the first unscheduled position, namely the position after any scene in B and before any scene in $S - B - \{s\}$. Obviously, we have $cost(\overrightarrow{B} \circ \{s\}, \overrightarrow{E}) = cost(\overrightarrow{B}, \overrightarrow{E}) + cost(s, B, E)$. The lower bound function in line 11, Algorithm 1 can be rewritten as:

$$\textbf{lower\_bound}(\overrightarrow{B} \circ s, Q - \{s\}, \overrightarrow{E}) = cost(\overrightarrow{B}, \overrightarrow{E}) + cost(s, B, E) + \textbf{lower}(B \cup \{s\}, Q - \{s\}, E).$$

The value of $cost(s, B, E)$ is incurred by the following two types of actors:

Type 1. If actor a_i is included in neither $a(B) \cap a(E)$ nor $a(s)$ but is still present on location during the days of shooting scene s (i.e., $a_i \notin a(B) \cap a(E)$, $a_i \notin a(s)$ and $a_i \in a(B) \cap a(Q - \{s\})$), he/she must be held during the shooting days of scene s.

Type 2. If actor a_i is not included in $a(B) \cap a(E)$ but is included in $a(E)$, and scene s is his/her first involved scene (i.e., $a_i \notin a(B)$ and $a_i \in a(s)$ and $a_i \in a(E)$), the shooting days of those scenes in $Q - \{s\}$ that do not require actor a_i can be confirmed as his/her holding days.

To demonstrate the computation of $cost(\overrightarrow{B}, \overrightarrow{E})$ and $cost(s, B, E)$, let us consider a partial schedule presented in Table 2, where $\overrightarrow{B} = (s_1, s_2)$, $\overrightarrow{E} = (s_5, s_6)$ and $Q = S - B - E = \{s_3, s_4\}$. In the columns "$cost(\overrightarrow{B}, \overrightarrow{E})$", "$cost(s_3, B, E)$" and "$cost(s_4, B, E)$", we present the corresponding holding cost associated with each actor. Since actor a_1 is a fixed actor, his/her holding cost must be $c(a_1)(d(s_2) + d(s_4))$ no matter how the scenes in Q are scheduled. Actor a_2 is involved in B and Q but is not involved in E, so we can only say that the holding cost of this actor is at least $c(a_2)d(s_2)$. Similarly, actor a_3 has an already incurred holding cost $c(a_3)d(s_5)$. For actors a_4 and a_5, we cannot get any clue on their holding costs from this partial schedule and thus we say their already confirmed holding costs are 0. Suppose scene s_4 is placed at the first unscheduled position. Since actors a_2 and a_4 must be present on location during the period of shooting scene s_4, the newly confirmed holding cost is $cost(s_4, B, E) = (c(a_2) + c(a_4))d(s_4)$. If we suppose scene s_3 is placed at the first unscheduled position, the newly confirmed holding cost is only related to actor a_3, namely, $cost(s_3, B, E) = c(a_3)d(s_4)$.

Table 2. An example for computing $cost(\overrightarrow{B}, \overrightarrow{E})$ and $cost(s, B, E)$

	$\overrightarrow{B}$		Q	$\overrightarrow{E}$		$cost(\overrightarrow{B}, \overrightarrow{E})$	$cost(s_3, B, E)$	$cost(s_4, B, E)$
	s_1	s_2	$\{s_3, \ s_4\}$	s_5	s_6			
a_1	X	.	$\{$X, $\cdot\}$	X	X	$c(a_1)(d(s_2) + d(s_4))$	0	0
a_2	X	.	$\{$X, $\cdot\}$	.	.	$c(a_2)d(s_2)$	0	$c(a_2)d(s_4)$
a_3	.	.	$\{$X, $\cdot\}$	.	X	$c(a_3)d(s_5)$	$c(a_3)d(s_4)$	0
a_4	.	X	$\{$X, $\cdot\}$	.	.	0	0	$c(a_4)d(s_4)$
a_5	.	.	$\{$X, $\cdot\}$	.	.	0	0	0

Define $o(Q) = a(S - Q) \cap a(Q)$ as the set of on-location actors just before any scene in Q is scheduled ([1]). That is, $o(Q)$ includes a set of actors required by scenes in both Q and $S - Q$. Then, $cost(s, B, E)$ can be mathematically computed by:

$$cost(s, B, E) = d(s) \times c\big(o(B) - o(E) - a(s)\big)$$
$$+ \sum_{s' \in Q - \{s\}} d(s') \times \left(c\big((a(s) - o(B)) \cap o(E) \big) - c\big((a(s) - o(B)) \cap o(E) \cap a(s') \big) \right)$$

$$\tag{1}$$

where $c(G)$ is the total daily cost of all actors in $G \subseteq A$, i.e., $c(G) = \sum_{a \in G} c(a)$. Due to space limitations, we omit the explanation of how Equation (1) is derived.

2.3 Lower Bound to Future Cost

In [1], the authors proposed a lower bound to the future cost. They generated two lower bounds on the set of actors $o(B) - F$ and $o(E) - F$, respectively.

They claimed that the sum of these two lower bounds is still a lower bound to the future cost, i.e., **lower**(B,Q,E). The reader is encouraged to refer to [1] for the details of this lower bound. This lower bound is denoted by L_0 in our implementation.

In the remaining of this subsection, we propose a new implementation of **lower**(B,Q,E). Suppose σ is an arbitrary permutation of the scenes in Q. We denote by x_i the holding cost of the non-fixed actor a_i during the period of shooting the scenes in Q with the order specified by permutation σ. If **lower**(B,Q,E) $= \min_\sigma\{\sum_{i \in A_N} x_i\}$, we get the optimal future cost. However, it is impossible to achieve the equality except that all σ are enumerated. In the following context, we present a method for generating a lower bound to $\min_\sigma\{\sum_{i \in A_N} x_i\}$.

For any two different actors $a_i, a_j \in (o(B) - F) \cup (o(E) - F)$, we can derive a constraint $x_i + x_j \geq c_{i,j}$, where $c_{i,j}$ is a number computed based on the following four cases.

Case 1: $a_i, a_j \in o(B) - F$. Let $a_i(s) =$ "X" if actor a_i is required by scene s and $a_i(s) =$ "·" otherwise. For any scene $s \in Q$, the tuple $(a_i(s), a_j(s))$ must have one of the following four patterns: (X, X), (X, ·), (·, X), (·, ·). First, we schedule all scenes with pattern (X, X) immediately after the scenes in B and schedule all scenes with pattern (·, ·) immediately before the scenes in E. Second, we group the scenes with (X, ·) and the scenes with (·, X) into two sets. Third, we schedule these two set of scenes in the middle of the permutation, creating two schedules as shown in Table 3. If only actors a_i and a_j are considered, the optimal schedule must be either one of these two schedules. The value of $c_{i,j}$ is set to the holding cost of the optimal schedule. For the schedule in Table 3(a), if we define $S_1 = \{s \in Q|(a_i(s), a_j(s)) = (X, \cdot)\}$, then the holding cost is $c(a_j) \times d(S_1)$, where $d(S_1) = \sum_{s \in S_1} d(s)$. Similarly, for the schedule in Table 3(b), we have a holding cost $c(a_i) \times d(S_2)$, where $S_2 = \{s \in Q|(a_i(s), a_j(s)) = (\cdot, X)\}$. As a result, we have $c_{i,j} = \min\{c(a_j) \times d(S_1), c(a_i) \times d(S_2)\}$.

Table 3. Two schedules in Case 1

(a) The first schedule

	B				Q						E
		s_1	s_2	s_3	s_4	s_5	s_6	s_7	s_8		
a_i	X	X	X	X	X	·	·	·	·		·
a_j	X	X	X	·	·	X	X	·	·		·

(b) The second schedule

	B				Q						E
		s_1	s_2	s_5	s_6	s_3	s_4	s_7	s_8		
a_i	X	X	X	·	·	X	X	·	·		·
a_j	X	X	X	X	X	·	·	·	·		·

Case 2: $a_i, a_j \in o(E) - F$. We schedule all scenes with pattern (X, X) immediately before the scenes in E and schedule scenes with pattern (·, ·) immediately after the scenes in B. The remaining analysis is similar to that in Case 1.

Case 3: $a_i \in o(B) - F$ and $a_j \in o(E) - F$. We schedule all scenes with pattern (X, ·) immediately after the scenes in B and schedule all scenes with pattern

$(\cdot, X)$ immediately before the scenes in E. If there does not exist a scene with pattern (X, X), the holding cost must be 0 and $c_{i,j}$ is set to 0; otherwise $c_{i,j}$ is set to $\min\{c(a_i), c(a_j)\} \times d(S_0)$, where $S_0 = \{s \in Q | (a_i(s), a_j(s)) = (\cdot, \cdot)\}$, which can be observed from Table 4.

Table 4. Two schedules in Case 3

(a) The first schedule

a_i	B	s_1	s_2	s_3	s_4	s_5	s_6	s_7	s_8	E
					Q					
a_i	X	X	X	X	X	$\cdot$	$\cdot$	$\cdot$	$\cdot$	$\cdot$
a_j	$\cdot$	$\cdot$	$\cdot$	X	X	$\cdot$	$\cdot$	X	X	X

(b) The second schedule

a_i	B	s_1	s_2	s_5	s_6	s_3	s_4	s_7	s_8	E
					Q					
a_i	X	X	X	$\cdot$	$\cdot$	X	X	$\cdot$	$\cdot$	$\cdot$
a_j	$\cdot$	$\cdot$	$\cdot$	$\cdot$	$\cdot$	X	X	X	X	X

Case 4: $a_i \in o(E) - F$ and $a_j \in o(B) - F$. This case is the same as Case 3.

A valid lower bound to the future cost can be obtained by solving the following linear programming model:

$$(LB) \quad z^{LB} = \min \sum_{a_i \in A_N} x_i \tag{2}$$

$$\text{s.t. } x_i + x_j \geq c_{i,j}, \ \forall \, a_i, a_j \in A_N, i < j \tag{3}$$

$$x_i \geq 0, \ \forall \, a_i \in A_N \tag{4}$$

The value of z^{LB} must be a valid lower bound to $\min_\sigma\{\sum_{i \in A_N} x_i\}$. If the daily holding cost of actor a_i is an integral number, decision variable x_i should be integer. When all variables x_i are integers, the model (LB) is an NP-hard problem since it can be easily reduced to the *minimum vertex cover problem* [8]. If all variables x_i are treated as real numbers, this model can be solved by a liner programming solver. For some instances, the (LB) model needs to be solved a million times. Therefore, to save computation time, we apply the following two heuristic approaches to rapidly produce lower bounds (L_1 and L_2) to z^{LB}. Obviously, L_1 and L_2 are also valid lower bounds to the future cost.

Approach 1: Sum up left-hand-side and righ-hand-side of Equations (3), generating $(|A_N| - 1) \sum_{a_i \in A_N} x_i \geq \sum_{a_i \in A_N, a_j \in A_N, i<j} c_{i,j}$. The valid lower bound L_1 is defined as:

$$L_1 = \sum_{a_i \in A_N, a_j \in A_N, i<j} c_{i,j}/(|A_N| - 1).$$

Approach 2: Sort $c_{i,j}$ in descending order. If we select a $c_{i,j}$, we call the corresponding x_i and x_j *marked*. Beginning from the largest $c_{i,j}$, we select all $c_{i,j}$ whose x_i and x_j are not marked until all x_i are marked. The valid lower bound L_2 equals the sum of all selected $c_{i,j}$. This approach was termed the *greedy matching algorithm* [9]. To demonstrate the process of computing L_2, we consider the following constraints:

$$x_1 + x_2 \geq 2, \ x_1 + x_3 \geq 7, \ x_1 + x_4 \geq 6,$$
$$x_2 + x_3 \geq 12, \ x_2 + x_4 \geq 8, \ x_3 + x_4 \geq 5$$

We first select $c_{2,3} = 12$ and mark x_2 and x_3. Then, we can only select $c_{1,4} = 6$ since x_1 and x_4 have not been marked. Now all x_i are marked and the value of L_2 equals 18.

2.4 State Memoization

In [1], the talent scheduling problem was solved by a double-ended dynamic programming (DP) algorithm, where a DP state is represented by $\langle B, E \rangle$. The DP algorithm stores the best value of each examined state, which equals the minimum past cost of the explored search nodes associated with sets B and E. A better state representation for the DP algorithm is $\langle o(B), o(E), Q \rangle$, where $Q = S - B - E$. Readers are referred to [1] for more details.

We embed the DP process in the branch-and-bound framework by use of the *memoization* technique ([10]). Specifically, when the search reaches a tree node $P(\vec{B}, Q, \vec{E})$, it first checks whether the value of $cost(\vec{B}, \vec{E})$ is less than the current value in the state $\langle o(B), o(E), Q \rangle$. If so, it updates the state value by $cost(\vec{B}, \vec{E})$; otherwise, the current node must be dominated by some node and therefore can be safely discarded.

3 Computational Experiments

Our algorithm was coded in C++ and compiled using the g++ compiler. All experiments were run on a Linux server equipped with an Intel Xeon E5430 CPU clocked at 2.66 GHz and 8 GB RAM. In this section, we present our results for the benchmark instances provided by [1] and then compare them with the results obtained by the dynamic programming.

In the benchmark data, 100 instances were randomly generated for each combination of $n \in \{16, 18, 20, \ldots, 64\}$ and $m \in \{8, 10, 12, \ldots, 22\}$, so there are 200 instance groups and 200,000 instances in total. [1] tried to solve these instances using their dynamic programming algorithm with a memory bound of 2GB on a machine with Xeon Pro 2.4 GHz processors. For each instance, if the execution did not run out of memory, they recorded the running time and the number of subproblems generated. They reported the average running time and the average number of subproblems with more than 80 optimally solved instances in a group.

We tried to solve these benchmark instances using our branch-and-bound algorithm with a limit of 1,000,000 memoized states, which consume the memory no greater than 1GB. Table 5 gives the number of instances optimally solved in each instance group, where an underline sign ("_") is added to the cell associated with the instance group with less than 80 optimally solved instances. For an instance group, if our algorithm optimally solved 80+ instances while the dynamic programming algorithm failed to achieve so, the number in its corresponding cell is marked with an asterisk (*). From this table, we can see that our algorithm can solve 80+ out of 100 instances with the number of scenes (n) not greater than 30. Comparatively, the dynamic programming algorithm by [1] only optimally solved 80+ out of 100 instances for the instance groups with $n \leq 26$. Thus, we can conclude that the performance of our algorithm is better than the dynamic programming approach in terms of the optimal solutions found. Tables 6 show the average running times over all optimally solved instances for each instance group, in which the average running times are no greater than 30 seconds. This demonstrates the efficiency of our algorithm in solving these instances.

Table 5. The number of optimally solved instances

m \ n	16	18	20	22	24	26	28	30	32	34	36	38	40	42	44	46	48	50	52	54	56	58	60	62	64
8	100	100	100	100	100	100	100	100	100	100	100	100	100	100	100	100	99	99	100	95	95	92	92	93	82*
10	100	100	100	100	100	100	100	100	100	100	100	98	100	98	95*	91*	95*	95*	82*	84*	81*	76	74	63	59
12	100	100	100	100	100	100	100	100	100	100	97	98*	93*	94*	88*	80*	82*	70	66	52	59	34	46	37	32
14	100	100	100	100	100	100	100	99	99	96*	94*	89*	87*	76	71	61	68	58	42	41	31	27	24	24	17
16	100	100	100	100	100	100	100	99	99	96*	93*	89*	79	75	69	52	45	35	34	37	24	29	14	13	8
18	100	100	100	100	100	100	100	95*	93*	89	78	64	71	54	36	43	29	28	24	17	14	8	9	9	9
20	100	100	100	100	100	100	98	97*	87*	67	61	50	38	33	27	14	16	14	11	5	7	3	4	5	1
22	100	100	100	100	100	100	98*	88*	72	65	40	29	29	15	14	10	8	4	9	3	5	1	2	0	2

Table 6. The average running time (in seconds)

m \ n	16	18	20	22	24	26	28	30	32	34	36	38	40
8	0.02	0.02	0.03	0.03	0.04	0.05	0.06	0.09	0.14	0.15	0.20	0.17	0.32
10	0.02	0.02	0.03	0.04	0.06	0.08	0.14	0.17	0.19	0.42	0.57	1.11	1.40
12	0.02	0.03	0.04	0.05	0.09	0.15	0.53	0.54	0.75	0.98	1.38	2.28	1.87
14	0.02	0.03	0.06	0.10	0.15	0.26	0.78	0.85	1.68	2.53	3.32	4.47	4.92
16	0.03	0.05	0.07	0.14	0.25	0.54	0.95	1.86	4.26	4.65	5.93	7.04	6.34
18	0.04	0.08	0.14	0.26	0.43	1.49	2.45	3.67	5.95	6.30	10.48	11.46	9.74
20	0.08	0.12	0.20	0.42	1.04	4.22	9.09	9.33	11.82	11.78	16.34	16.07	14.82
22	0.17	0.23	0.37	0.69	2.23	3.58	6.59	12.21	12.43	16.34	16.07	16.32	19.74

m \ n	42	44	46	48	50	52	54	56	58	60	62	64
8	0.44	0.72	0.94	1.01	1.00	1.45	1.96	1.84	2.22	2.43	2.63	3.20
10	1.35	2.30	2.22	3.18	3.35	4.00	3.91	3.45	3.75	5.25	5.19	5.65
12	3.67	4.08	4.01	5.07	4.87	4.73	7.16	6.06	6.36	7.29	8.11	6.63
14	5.49	4.97	6.97	6.79	8.30	7.02	9.72	8.65	9.26	5.17	11.15	9.37
16	7.63	7.70	9.81	9.70	7.95	11.20	11.23	10.11	11.08	9.47	4.57	11.10
18	11.81	11.04	12.54	9.62	11.49	8.54	11.04	10.65	19.68	14.63	15.65	12.87
20	14.29	13.59	9.91	12.58	15.81	13.70	8.98	17.07	15.78	12.86	21.92	4.25
22	20.23	13.49	13.21	18.33	18.44	10.11	18.23	25.33	28.49	15.37	0.00	8.80

In order to make further comparisons between the performance of our lower bound and the lower bound provided by [1], we recorded the following statistics for those instances which are optimally solved.

- W_1: The number of search nodes with $\max\{L_1, L_2\} > L_0$.
- W_2: The number of search nodes with $\max\{L_1, L_2\} < L_0$.

W_1 indicates the number of times that our LB functions win (provide tighter LB than) the LB function given in [1], while W_2 indicates the number of times that our LB functions lose. Table 7 summarizes the rate between W_1 and W_2 on each instance group. The table clearly reveals that our LB functions are much better, as the rate between W_1 and W_2 is greater than 1 for almost all of the cases.

Table 7. The rate between W_1 and W_2

m \ n	16	18	20	22	24	26	28	30	32	34	36	38	40	42	44	46	48	50	52	54	56	58	60	62	64
8	3.3	3.2	2.6	3.5	2.9	2.2	2.8	2.0	2.2	2.2	2.0	2.0	2.1	1.8	1.5	1.7	1.1	1.3	1.3	1.4	1.4	1.3	1.0	1.1	1.4
10	2.7	3.1	3.0	2.8	2.8	2.2	2.5	1.8	1.7	2.4	2.0	2.2	2.0	2.2	1.6	1.9	1.7	1.4	1.8	1.8	1.4	1.4	1.2	0.9	0.8
12	3.2	2.5	3.0	2.4	2.8	2.6	3.2	2.2	1.8	2.2	2.2	2.2	2.1	1.9	2.0	1.7	1.8	1.4	1.6	1.5	1.1	1.0	1.3	1.2	1.3
14	3.0	3.0	2.9	2.9	2.5	3.1	2.3	2.8	1.9	2.1	2.1	2.0	2.3	2.3	1.9	2.1	1.8	1.5	1.7	1.6	1.3	1.6	1.2	1.5	1.1
16	2.7	2.9	2.6	2.8	2.7	2.6	2.7	2.5	2.2	2.4	2.3	2.3	2.4	1.9	1.7	1.7	2.1	1.7	1.5	1.9	1.2	1.5	1.3	0.8	1.5
18	2.8	3.0	2.6	2.8	2.7	2.7	2.8	2.5	2.8	2.3	2.5	2.0	2.6	1.9	1.6	1.5	1.7	1.7	1.5	2.3	2.0	2.1	1.3	1.2	1.4
20	3.0	2.6	3.1	2.8	3.1	2.8	2.1	2.7	2.3	2.8	2.5	2.3	2.2	2.0	1.9	2.8	1.7	1.5	1.8	2.3	1.4	1.1	1.1	2.3	0.8
22	3.3	3.1	3.0	2.9	3.1	2.7	2.8	2.9	2.5	2.2	2.5	3.0	1.9	2.4	1.5	2.1	3.6	2.3	1.9	1.1	1.9	3.0	1.2	-	1.6

4   Conclusions

In this paper, we propose a branch-and-bound algorithm to solve the talent scheduling problem, which is a very challenging combinatorial optimization problem. This algorithm use a new lower bound and memoization techniques to reduce the search nodes. The experimental results clearly show that our algorithm performs better than the currently best known approach and achieved the optimal solutions for considerably more benchmark instances.

References

1. de la Banda, M.G., Stuckey, P.J., Chu, G.: Solving talent scheduling with dynamic programming. INFORMS Journal on Computing 23(1), 120–137 (2011)
2. Adelson, R.M., Norman, J.M., Laporte, G.: A dynamic programming formulation with diverse applications. Operational Research Quarterly, 119–121 (1976)
3. Cheng, T.C.E., Diamond, J.E., Lin, B.M.T.: Optimal scheduling in film production to minimize talent hold cost. Journal of Optimization Theory and Applications 79(3), 479–492 (1993)
4. Smith, B.M.: Constraint programming in practice: Scheduling a rehearsal. Research Report APES-67-2003, APES group (2003)
5. Smith, B.M.: Caching search states in permutation problems. In: van Beek, P. (ed.) CP 2005. LNCS, vol. 3709, pp. 637–651. Springer, Heidelberg (2005)
6. Nordström, A.L., Tufekci, S.: A genetic algorithm for the talent scheduling problem. Computers & Operations Research 21(8), 927–940 (1994)
7. Fink, A., Voß, S.: Applications of modern heuristic search methods to pattern sequencing problems. Computers and Operations Research 26(1), 17–34 (1999)
8. Karp, R.M.: Reducibility among combinatorial problems. Springer (1972)
9. Drake, D.E., Hougardy, S.: A simple approximation algorithm for the weighted matching problem. Information Processing Letters 85(4), 211–213 (2003)
10. Michie, D.: "memo" functions and machine learning. Nature 218(5136), 19–22 (1968)

An Improved Multi-Gene Genetic Programming Approach for the Evolution of Generalized Model in Modelling of Rapid Prototyping Process

Akhil Garg and Kang Tai

School of Mechanical and Aerospace Engineering, Nanyang Technological University,
50 Nanyang Avenue, Singapore 639798, Singapore
{akhil1,mktai}@ntu.edu.sg

Abstract. The Rapid prototyping (RP) processes are widely used for the fabrication of complex shaped functional prototypes from the 3-D design. Among the various RP processes, fused deposition modeling (FDM) is widely known among researchers. The working mechanism behind the FDM process is governed by multiple input and output variables, which makes this process complex and its implementation costly. Therefore, the highly generalized mathematical models are an alternative for the practical realization of the process. Artificial intelligence methods such as multi-gene genetic programming (MGGP), artificial neural network and support vector regression can be used. Among these methods, MGGP evolves explicit models and its coefficients automatically. Since MGGP uses a multiple sets of genes for the formulation of model and is population based, it suffers from the problem of *over-fitting*. *Over-fitting* is caused due to inappropriate procedure of formation of MGGP model and the difficulty in model selection. To counter *over-fitting*, the present paper proposes an improved MGGP (I-MGGP) approach by embedding the statistical and classification algorithms in the paradigm of MGGP. The proposed I-MGGP approach is tested on the wear strength data obtained from the FDM process and results show that the I-MGGP has performed better than the standard MGGP approach. Thus, the I-MGGP model can be deployed by experts for understanding the physical aspects as well as optimizing the performance of the process.

Keywords: Wear strength prediction, *over-fitting*, FDM, Rapid prototyping modeling, FDM modeling.

1   Introduction

Fused deposition modelling (FDM) process makes use of rapid prototyping technology to build three-dimensional solid complex parts from the computer-aided design data without the use of tooling and human intervention. Because of this, FDM process have resulted in applications in functional prototype development, medical, automobile industries, construction industries, space applications, tool and die making [1].

A. Moonis et al. (Eds.): IEA/AIE 2014, Part I, LNAI 8481, pp. 218–226, 2014.

Literature reveals that properties of the FDM fabricated parts such as wear, tensile strength, compressive strength and surface roughness are function of various process related parameters and can be significantly improved with its proper adjustment [2]. One route is develop the new materials but this may require one to have expert knowledge about the characteristics of materials at different operating conditions [3]. Other route is to develop mathematical models that can replace the tedious and costly FDM experiments. In context of this, several physics-based models have been formulated [4-5]. The formulation of the physics-based models requires in-depth understanding of the process, and, therefore is not an easy task in presence of partial information about the process. Therefore, researchers shifted focus on developing models based on only the given data.

To develop models based on only data, several well-known artificial intelligence (AI) methods such as artificial neural networks, fuzzy logic, adaptive-network-based fuzzy inference system, genetic programming (GP) and support vector regression have been applied [6-10]. Among these methods, GP possesses the ability to evolve models structure and its coefficients automatically [11-18]. Most popular variant of GP used recently is multi-gene genetic programming (MGGP). Despite of good number applications of MGGP in manufacturing, it has limitation for producing models that *over-fit* on the testing data. This indicates that the underlying relationships of the whole data were not learned, and instead a set of relationships existing only on training cases were learned, but these have no correspondence over the whole possible set of cases. The poor performance of models on the testing data is undesirable and is likely to give falsify information about the process. *Over-fitting* is due to two reasons: inappropriate procedure of formation of MGGP model and difficulty in model selection. *Over-fitting* in MGGP is the popular problem among researchers and have been paid less attention [19-20].

Therefore, an improved MGGP (I-MGGP) approach is proposed for predicting the wear property of the FDM fabricated prototype. Unlike standard GP, each model participating in I-MGGP is made from the set of combination of genes. Proposed I-MGGP approach makes use of statistical approach of stepwise regression and classification methods such as support vector machines (SVM) and ANN to reduce *over-fitting*. The wear of the FDM fabricated prototypes is measured based on five input variables such as layer thickness, orientation, raster angle, raster width and air gap. Based on the data obtained from the experiments, the proposed I-MGGP method is applied and its performance is compared to that of standardized MGGP.

2   FDM Process

The FDM process to be used is referred from an earlier study conducted on an investigation on sliding wear of FDM built parts [8]. The input process parameters used are layer thickness (x_1), orientation (x_2), raster angle (x_3), raster width (x_4) and air gap (x_5). For each of the input process variables, values at three levels (low, centre and high) are considered as per guidelines of machine manufacturer and industrial application. For wear testing, pin on disk apparatus (Ducom, TR- 20LE-M5) is used.

Wear strength is important characteristic for the durability of part and very little work is done to understand this characteristic of RP processed part [21].

Half factorial 25 unblocked design having 16 experimental run, 10 (2K, where K=5) axial run and 6 centre run have been used to generate thirty-two set of data points. Nature of the data set collected is shown by its descriptive statistics in Table 1. In this work, Kennard-and-stone algorithm is used to select the appropriate training and testing data set. The algorithm selects the training samples in such a way, that the data is distributed uniformly throughout the domain. 25 samples are chosen as training data with the remainder used for the test samples. In the following section, we will provide an overview of two methods: MGGP and I-MGGP

Table 1. Descriptive statistics of the FDM data comprising of input and output variables

Parameter	x_1	x_2	x_3	x_4	x_5	y
Mean	0.18	15	30	0.456	0.004	0.0266
Median	0.178	15	30	0.456	0.004	0.0269
Standard deviation	0.04	11.43	22.86	0.038	0.003	0.009
Kurtosis	-1.20	-1.22	-1.22	-1.22	-1.22	-0.50
Skewness	0.35	0	0	-9.5E-1	-2.02E-1	0.10
Minimum	0.127	0	0	0.406	0	0.011
Maximum	0.254	30	60	0.506	0.008	0.048

3   Multi-Gene Genetic Programming

Principle of genetic programming (GP) is shown by flowchart in Fig. 1. The initial population of individuals is generated by combining the elements randomly from the functional and terminal set. The terminal set consists of input process parameters and random constants. The range of random constant chosen is -10 to 10. The function set can include elements such as arithmetic operators (+, -, /, ×), non-linear functions (sin, cos, tan, exp, tanh, log) or Boolean operators. The performance of the individuals in the initial population is evaluated based fitness function, namely, root mean square error (RMSE) given by:

$$RMSE = \sqrt{\frac{\sum_{i=1}^{N}\left|G_i - A_i\right|^2}{N}} \qquad (1)$$

where G_i is the valued predicted of ith data sample by the MGGP model , A_i is the actual value of the ith data sample and N is the number of training samples.

If any member of the population does not satisfy the termination criterion, the genetic operations such as selection, crossover and mutation are performed on the individuals to evolve the new population. The termination criterion is the maximum number of generations or the threshold error of the model as specified by the user. Genetic operations mainly crossover and mutation forms most of the individuals of the population. In this way, genetic operations on the initial population consequently form a new population and this iterative process of forming new populations continues until a termination criterion is met.

The key difference between GP and the MGGP is that, in the latter, the model participating in the evolutionary stage is a combination of several sets of genes/trees combined using the least squares method.

We used GPTIPS [22-23] to perform MGGP for the prediction of wear strength of FDM fabricated prototype. MGGP method is applied to the data as shown in Section 2. The best MGGP model is selected based on minimum RMSE on training data from all runs.

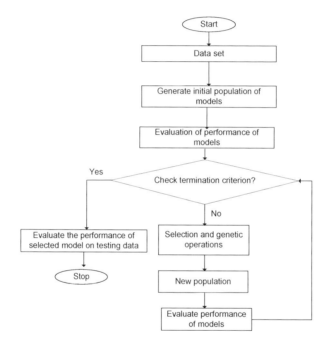

Fig. 1. The mechanism of GP

4   Improved Multi-Gene Genetic Programming

Step by step procedure of the proposed approach is shown in Fig. 2. I-MGGP approach makes use of statistical approach of stepwise regression and classification methods such as SVM and ANN to tackle two issues respectively as follows:

1) **In appropriate procedure of formation of the MGGP model**: In MGGP method, the genes are randomly chosen and regressed using least squares method. During the combination, the genes of lower performance i.e. genes having poor accuracy on training data may combine with other genes of higher performance and degrades the performance of the MGGP model on the testing data. Dotted line

represented by digit 1 shown in Fig. 2 refers to combination of genes using the stepwise regression approach. The stepwise approach selectively combines the genes of only higher performance.

2) **Difficulty in model selection**: The dotted line represented by digit 2 shown in Fig. 2 refers to the classification scheme embedded in the paradigm of MGGP. The two classifiers such as ANN and SVM algorithm make use of the valuable information such as training and number of nodes of the evolved models to form a classifier. The architecture of these two classifiers is selected based on trial-and-error approach. The classification criteria based on the training error is formulated to assign these models into the two categories of "best" or "bad". Based on the training error, first 15% of the models with least training error are classified as best models with the other categorized as "bad". The class representing best models is coded with a numeric digit 0 and the bad models are coded with digit 1. The model classified as "best" by both the classifiers is selected. If there is a tie up, the model with the lower number of nodes is chosen.

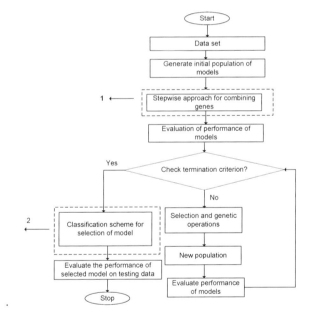

Fig. 2. Flowchart showing mechanism of I-MGGP

5   Performance Evaluation of Proposed Approach

The parameter settings of the MGGP and the I-MGGP methods are set based on trial-and-error approach (Table 2). Classification strategies used in the proposed I-MGGP classifies models into two categories. From Table 3, it is obvious that the I-MGGP model corresponding to run 34 is classified as best by both of the classifiers and therefore is selected for the analysis on the testing data.

Three metrics: RMSE, mean absolute percentage error (MAPE) and relative error (%) are used to evaluate the performance of two models. MAPE and relative error is given by:

$$MAPE\ (\%) = \frac{\sum_{i=1}^{N} \frac{|G_i - A_i|}{A_i} \times 100}{N} \tag{2}$$

$$Relative\ error\ (\%) = \frac{|G_i - A_i|}{A_i} \times 100 \tag{3}$$

where G_i is the valued predicted of ith data sample by the MGGP and I-MGGP model, A_i is the actual value of the ith data sample and N is the number of training samples

Table 4 shows the comparison of performance of the best model selected from I-MGGP and standardized MGGP approach. Fig. 3 shows the relative error (%) of two methods on the training and testing data

The results clearly show that the I-MGGP model has outperformed the standardized MGGP since it has lower value of RMSE and MAPE of 1.42 and 7.08 respectively on testing data.

Table 2. Parameter settings for I-MGGP and MGGP methods

Parameters	Values assigned
Runs	50
Population size	150
Number of generations	100
Tournament size	2
Max depth of tree	6
Max genes	8
Functional set (F)	(multiply, plus, minus, tan, tanh, square, psqroot, exp, sin)
Terminal set (T)	(x_1, x_2, x_3, x_4, x_5 [-10 10])
Crossover probability rate	0.85
Reproduction probability rate	0.10
Mutation probability rate	0.05

Table 3. Class of I-MGGP models

Run	ANN	SVM
1	Bad	Bad
2	Bad	Best
.		
.		
.		
34	**Best**	**Best**
.		
.		
.		
50	Best	Bad

Table 4. Comparison of I-MGGP and MGGP models

Data	Training		Testing	
Method	RMSE	MAPE	RMSE	MAPE
I-MGGP	0.44	4.28	1.42	7.08
MGGP	0.31	3.31	14.06	31.55

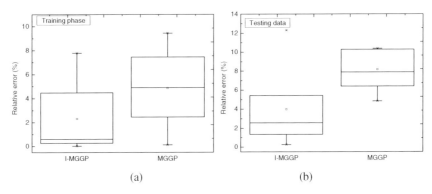

Fig. 3. Relative error of the two models on a) training data b) testing data

6   Conclusions

The paper addresses the problem of *over-fitting* in MGGP and proposes an improved MGGP approach. The results conclude that the proposed I-MGGP approach is able to evolve a model that give better performance on the testing data than that of a model selected from the standardized MGGP approach. We can conclude that the I-MGGP model can be deployed by experts in industry to predict the wear strength of the FDM process in uncertain input conditions.

The fitness function also plays important role in improving an evolutionary search in MGGP. The performance of the I-MGGP can be compared to MGGP using other fitness functions such as structural risk minimisation, Akaike information criterion, Bayesian information criterion, etc. The future work for authors includes an in-depth investigation of the performance of the I-MGGP approach and arrives at more decisive conclusions on the pros and cons of the methodology.

Acknowledgments. This work was partially supported by the Singapore Ministry of Education Academic Research Fund through research grant RG30/10, which the authors gratefully acknowledge.

References

1. Mansour, S., Hague, R.: Impact of rapid manufacturing on design for manufacture for injection moulding. Proceedings of the Institution of Mechanical Engineers, Part B: Journal of Engineering Manufacture 217, 453–461 (2003)

2. Anitha, R., Arunachalam, S., Radhakrishnan, P.: Critical Parameters Influencing the Quality of Prototypes in Fused Deposition Modelling. Journal of Materials Processing Technology 118, 385–388 (2001)
3. Stampfl, J., Liska, R.: New Materials for Rapid Prototyping Applications. Macromolecular Chemistry and Physics 206, 1253–1256 (2005)
4. Byun, H.S., Lee, K.H.: Determination of the optimal build direction for different rapid prototyping processes using multi-criterion decision making. Robotics and Computer-Integrated Manufacturing 22, 69–80 (2006)
5. Chang, D.Y., Huang, B.H.: Studies on profile error and extruding aperture for the RP parts using the fused deposition modeling process. The International Journal of Advanced Manufacturing Technology 53, 1027–1037 (2001)
6. Garg, A., Tai, K., Lee, C.H., Savalani, M.M.: A Hybrid M5 -Genetic Programming Approach For Ensuring Greater Trustworthiness of Prediction Abilit in Modelling of FDM Process. Journal of Intelligent Manufacturing (2013) (in press), doi:10.1007/s10845-013-0734-1
7. Garg, A., Savalani, M.M., Tai, K.: State-of-the-Art in Empirical Modelling of Rapid Prototyping Processes. Rapid Prototyping Journal 20(2) (2014)
8. Sood, A.K., Equbal, A., Toppo, V., Ohdar, R., Mahapatra, S.S.: An investigation on sliding wear of FDM built parts. CIRP Journal of Manufacturing Science and Technology 1, 48–54 (2011)
9. Sood, A.K., Ohdar, R.K., Mahapatra, S.S.: A hybrid ANN-BFOA approach for optimization of FDM process parameters. In: Panigrahi, B.K., Das, S., Suganthan, P.N., Dash, S.S. (eds.) SEMCCO 2010. LNCS, vol. 6466, pp. 396–403. Springer, Heidelberg (2010)
10. Sood, A., Ohdar, R., Mahapatra, S.: Parametric appraisal of fused deposition modelling process using the grey Taguchi method. Proceedings of the Institution of Mechanical Engineers, Part B: Journal of Engineering Manufacture 224, 135–145 (2010)
11. Garg, A., Tai, K.: Selection of a Robust Experimental Design for the Effective Modeling of the Nonlinear Systems using Genetic Programming. In: Proceedings of 2013 IEEE Symposium Series on Computational Intelligence and Data mining (CIDM), Singapore, April 16-19, pp. 293–298 (2013)
12. Garg, A., Bhalerao, Y., Tai, K.: Review of Empirical Modeling Techniques for Modeling of Turning Process. International Journal of Modelling, Identification and Control 20(2), 121–129 (2013)
13. Garg, A., Rachmawati, L., Tai, K.: Classification-Driven Model Selection Approach of Genetic Programming in Modelling of Turning Process. International Journal of Advanced Manufacturing Technology 69, 1137–1151 (2013)
14. Garg, A., Sriram, S., Tai, K.: Empirical Analysis of Model Selection Criteria for Genetic Programming in Modeling of Time Series System. In: Proceedings of 2013 IEEE Conference on Computational Intelligence for Financial Engineering & Economics (CIFEr), Singapore, April 16-19, pp. 84–88 (2013)
15. Garg, A., Garg, A., Tai, K.: A multi-gene genetic programming model for estimating stress-dependent soil water retention curves. Computational Geosciences 20.18(1), 45–56 (2014)
16. Vijayaraghavan, V., et al.: Estimation of mechanical properties of nanomaterials using artificial intelligence methods. Applied Physics A (2013), doi:10.1007/s00339-013-8192-3
17. Garg, A., Vijayaraghavan, V., Mahapatra, S.S., Tai, K., Wong, C.H.: Performance evaluation of microbial fuel cell by artificial intelligence methods. Expert Systems with Applications 41(4), 1389–1399 (2013)

18. Garg, A., Tai, K.: Review of genetic programming in modeling of machining processes. In: Proceedings of 2012 International Conference on Modelling, Identification and Control (ICMIC 2012), Wuhan, China, June 24-26, pp. 653–658. IEEE (2012)
19. Chan, K.Y., Kwong, C.K., Dillon, T.S., Tsim, Y.C.: Reducing Overfitting in Manufacturing Process Modeling Using a Backward Elimination Based Genetic Programming. Applied Soft Computing 11, 1648–1656 (2011)
20. Gonçalves, I., Silva, S., Melo, J.B., Carreiras, J.M.B.: Random Sampling Technique for Overfitting Control in Genetic Programming. In: Moraglio, A., Silva, S., Krawiec, K., Machado, P., Cotta, C. (eds.) EuroGP 2012. LNCS, vol. 7244, pp. 218–229. Springer, Heidelberg (2012)
21. Kumar, S., Kruth, J.-P.: Wear Performance of SLS/SLM Materials. Advanced Engineering Materials 10(8), 750–753 (2008)
22. Searson, D.P., Leahy, D.E., Willis, M.J.: GPTIPS: An Open Source Genetic Programming Toolbox for Multigene Symbolic Regression. In: International Multiconference of Engineers and Computer Scientists 2010, vol. 1, pp. 77–80 (2010)
23. Hinchliffe, M., Hiden, H., Mckay, B., Willis, M., Tham, M., Barton, G.: Modelling Chemical Process Systems Using a Multi-Gene Genetic Programming Algorithm, pp. 28–31 (1996)

Optimization of Temperature Precision for Double-Shell Oven by Lean Six Sigma Method

Kai-Hsiang Yen, Ming-Lung Lin, Hsiao-Wei Shih, and Carl Wang

Underwriters Laboratories Taiwan Co., Ltd
Jerry.Yen@ul.com

Abstract. A relatively inexpensive and highly effective design for high precision oven was developed but the control theory and optimized parameters were totally different from traditional methodologies. Among all kinds of quality programs, lean six sigma (LSS) is the most famous and widely-used method by manufacturing and service industries. The DMAIC (Define, Measure, Analyze, Improve and Control) methodology was used to figure out the key factors relative to high precision of temperature and improve the control parameters to meet the target. At the start, the baseline of Z bench was -1.99 and standard deviation was 0.255229. After lean six sigma project implemented, Z bench was up to 3.66 and standard deviation was down to 0.12822. This paper will address the critical process and analysis method for lean six sigma project, especially those for precise temperature control.

Keywords: Lean six sigma, DMAIC, Oven, Precision control, Temperature.

1   Introduction

Oven plays important roles in testing advance devices, equipment and materials [1-3]. To satisfy comprehensive test purposes, control parameters such as temperature and air velocity in ovens must be precisely controlled at required set-points. For example, polymeric and plastic materials are highly sensitive to environmental characteristics such as relative humidity and temperature [4]. This equipment for artificial aging and reliability tests also play essential character in comparing and predicting the accelerating aging performance of polymeric materials and determining the effect of different accelerating factors on the performance and safety of products. In order to obviously find out the necessary correlations between the life service and accelerating aging assessment, testing experiments accomplished in ovens must be highly reproducible and repeatable. Admittedly, a significant percentage of these reproducibility and repeatability can be attributed to strategy and methodology for precise temperature control.

Over the last two decades, a large numbers of new tools and innovative approaches have developed by industry to achieve high levels of operational efficiency. Currently, two of the most popular programs in industry are Lean Thinking and Six Sigma and both provide a well-ordered methodology to facilitate incremental process

A. Moonis et al. (Eds.): IEA/AIE 2014, Part I, LNAI 8481, pp. 227–235, 2014.

innovations [5]. Six Sigma was developed by Motorola Corporation in the 1980s but gained high impact after its adoption by General Electric in the mid-1990s and subsequently adopted by many US companies, including 3M, Agilent Technologies and Boeing. The key question driving the progression of Six Sigma was the need for quality improvement when manufacturing complex products having a large number of components, which often resulted in accordingly high probability of defective final products. Lean Thinking emerged within the Japanese automobile company such as Toyota after World War II but can be traced back to the early days of the Ford Motor Company, and has been implemented by many major US firms, including Danaher Corporation and Harley-Davidson. The driving force behind the development of Lean Thinking was the elimination of waste, especially in Japan, a country with few natural resources.

Lean Six Sigma is the synthesis of Six Sigma and Lean Thinking. Its projects are focused on improvement of some routine operations, seeking to make them more effective and more efficient. By the way, Lean Six Sigma brings understanding of the root causes of the problem, and provides a definitive and optimal solution. In this study, Lean Six Sigma's DMAIC method is used to modify the precision of temperature control for the new developed oven.

2   Define

In define phase, it is to have the team understanding the problem and reach agreement on the scope, goals and performance targets for the project. Background information is that ovens play important roles in testing material, component and system for part of UL standards. However, the precision of temperature of commercial oven is set point ± 1-3 ℃, which is not good enough for long term property evaluation and sometimes we need to conduct more tests than necessary to determine compliance. According to NIST technical roadmap, the precision of temperature, under 0.5 ℃, is necessary for future application.

After the description of problem, scope was specified clearly that how to optimize precision of temperature using new developing oven and the upper limit of temperature was 100 ℃ and the lower limit was 35 ℃. The project Y was the precision of interior chamber temperature of oven and defined that after system stability(firstly getting set-point in 3 min under control limit (± 1 ℃)), temperature of geometrical center was recorded by wireless temperature sensor(ex. iButton) for 30 minutes to compare with set point. In the meanwhile, two path Y were temperature of chamber and during time, which were used to quantify the project Y. The goal of this project was to improve the precision of temperature from set-point ± 1-3 ℃ to under ± 0.5 ℃ in three months and the benefit was to improve temperature accuracy, resulting in high consistency which will increase consumer acceptance and minimize redundant tests.   This also gave high creditability to UL in this area.

3   Measure

The significant activity in measure phase is to recognize the real situation of the project, including measurement system analysis (MSA) and Gage R&R, process capability analysis, creating a value stream map to confirm current process flow and defining potential factors (known as Xs) that affect the project index (also called Y). Four sensors were placed in the same oven at the same location at three different setting temperatures such as 40℃, 50℃ and 80℃ to record the response and perform Gage R&R analysis. Figure 1 shows the graphic results of continuous Gage R&R and it appears that differences within sensors are small and differences between sensors are also small. According to the R chart and X bar chart, samples represented a good range of temperature variations. The percentage study variance was calculated and its value was 0.66 %, which was small enough for good measurement system. The number of distinct categories was 215, greater than 5, which meant this measurement system had good resolution.

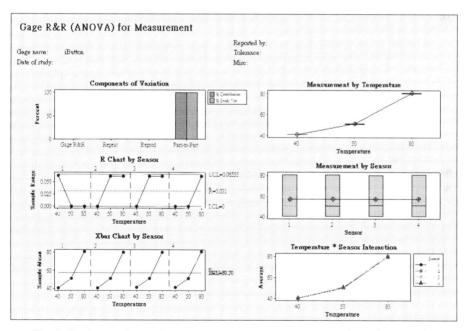

Fig. 1. Continuous Gage R&R graphic output for four sensors at 40℃, 50℃ and 80℃

There are two critical assumptions to consider when performing process capability analyses with continuous data, which are the process in statistical control and the distribution of the process considered as normal. Suppose the UCL = 86.773, the LCL = 85.241, and our target for this process is midway between the specs, i.e. at 86.007. Firstly, considering the I-MR charts in Figure 2, it seems that the distribution is stable over the period of study. From the normal probability plot graph in Figure 2, the Anderson-Darling (AD) normality test shows that it is unable to reject the null hypothesis (H0). This is due to the fact that the p-value for the A-D test is 0.164, which is greater than 0.05 - a frequently used level of significance for such a hypothesis test. The capability analysis in Figure 2 shows that with the LSL = 84.5 and USL = 85.5. Short term (and long-term) performances are also indicated, namely that approximately 976535 parts per million (ppm) would be nonconforming if only common causes of variability were present in the system, and approximately 965612 ppm in the long-term. The corresponding Z-Bench value was -1.99. It showed that this system had a serious problem about centering.

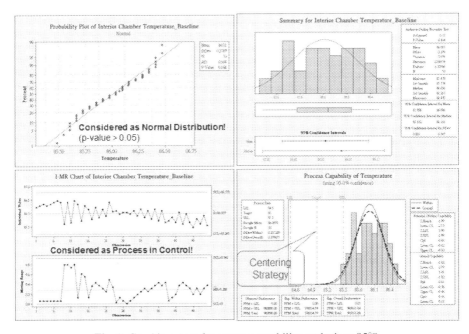

Fig. 2. Graphic output for process capability analysis at 85℃

The final step in measure phase is to investigate potential influential factors (Xs) for precision of temperature. From the process map and cause-and-effect analysis, thirty factors from related process were reviewed using cause-and-effect matrix (C&E matrix), which weighed each factor by three indexes such as accuracy, time to stability and temperature. Four potential factors were selected including, P value of exterior chamber controller (Out P), I value of exterior chamber controller (Out I), P value of Interior chamber controller (In P) and I value of Interior chamber controller (In I).

4   Analyze

In the analyze phase, two kind of analysis were performed including design of experience (DOE) for verifying key X's and hypothesis test. In this DOE, P value of exterior chamber controller, I value of exterior chamber controller, P value of Interior chamber controller and I value of Interior chamber controller were selected to proceed with a "two-level, four-factor and one center point" full factorial design. A model was built for the precision of temperature: In P and In I had a significant effect, as the Pareto chart of the standardized effects in Figure 3 and the In P*In I two-factor interactions was also significant. The p-values from factorial fit analysis are accordingly 0.002, 0.000 and 0.031. All of them are smaller than 0.05, which indicates that In P, In I and In P*In I are potential key X's. Therefore the remaining parameters which were not significant from the model were eliminated.

Hypothesis test with multiple variable X's was also used to verify key factors. A main effects plot is the tool to graph the response mean for each factor level, connected by a line. Once the connecting line is horizontal or parallel to the x-axis and this means that no main effect exists. When the line is not horizontal and this means that different levels of the factor affect the response differently. The less horizontal the line, the greater the likelihood that a main effect is statistically significant. According to matrix plot of Out P, Out I, In P, In I and Xin in Figure 4, In P and In I are positive relationship to temperature, In P* In I is negative relationship, and others are almost not important impact. After regression analysis, the model was found:

<1>Temperature of chamber increases 0.225℃ for every P value of interior controller , with I & Int value held constant.

<2> Temperature of chamber increases 0.00198℃ for every I value of interior controller , with P & Int value held constant.

<3> Temperature of chamber decreases 0.000144℃ for every Int value of interior controller , with P & I value held constant.

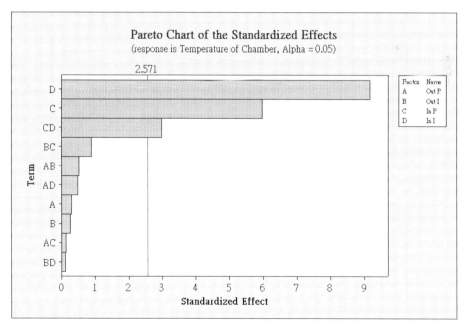

Fig. 3. Pareto chart of the standardized effects for two-level, four-factor and one center point" full factorial design

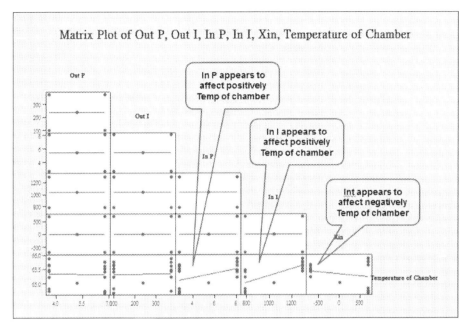

Fig. 4. Matrix plot of Out P, Out I, In P, In I and Xin for hypothesis test to verify key factors

5   Improve

The goal of the improve phase is to identify an effective and efficient solution to the problem that the LSS project focuses to address. This involves brainstorming or six thinking hats potential solutions, select and prioritize solutions, apply Lean Six Sigma best practices, Perform risk assessment etc. Often a pilot implementation is conducted prior to a full-scale rollout of improvements. After key factors with an impact on the project Y were found, the best solution for the precision of temperature was still not clear. Therefore, a series of design of experiments were implemented. In order to analyze the results, the response optimizer in Minitab was used to identify the combination of input variable settings that jointly optimize a single response or a set of responses. From the graphic results of response optimizer in Figure 5, the expected result equals to our target when In P is 5.2426 and In I is 780. This parameter is appropriate to pilot the improvements before proceeding to a full roll out. The most common piloting options include either making changes only in one group or department or making changes for a limited time period. The benefit of a pilot test is that the project team can ensure the changes result in the desired improvements before a full roll out. In addition, the team can gain insights to allow a more effective implementation during the full roll out.

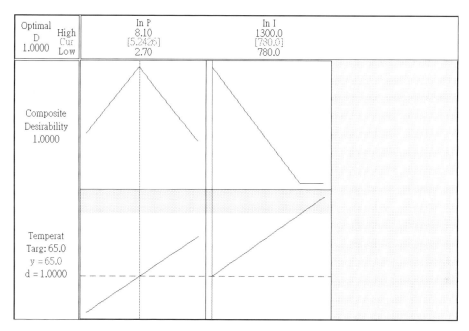

Fig. 5. Graphic results of response optimizer for In P and In I

6   Control

The principal objective of the control phase is to ensure that the results obtained from improve phase are properly maintained after the project has ended. In order to keep the achievements, it is necessary to standardize the process and document the procedures, make sure all employees are trained and communicate the project's results. Furthermore, the project team needs to create a control plan for ongoing monitoring of the process and for reacting to any problems that may happen.

After improve phase, all of the key factors causing the precision of temperature were figured out. All activities were implemented sequentially in the temperature control procedures, and the results were monitored in control phase. Compare to the results of analyze phase, the precision of temperature was more close to our target. Suppose the UCL = 85.4087, the LCL = 84.6394, and our target for this process is midway between the specs, i.e. at 85.024. Firstly, considering the I-MR charts in Figure 6, it seems that the distribution is stable over the period of study. From the normal probability plot graph in Figure 6, the Anderson-Darling (AD) normality test shows that it is unable to reject the null hypothesis (H0). This is due to the fact that the p-value for the A-D test is 0.24, which is greater than 0.05 - a frequently used level of significance for such a hypothesis test. The capability analysis in Figure 6 shows that with the LSL = 84.5 and USL = 85.5. Short term (and long-term) performances are

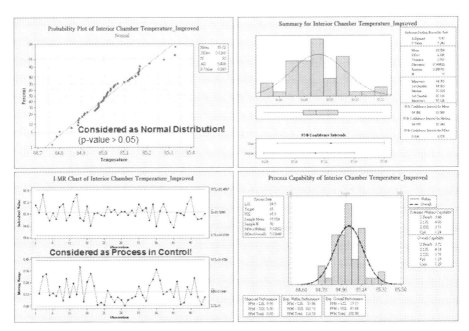

Fig. 6. Graphic output for process capability analysis at 85℃ after improvement

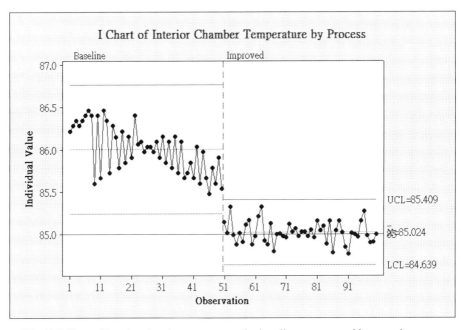

Fig. 7. I-Chart of interior chamber temperature by baseline process and improved process

also indicated, namely that approximately 125 parts per million (ppm) would be non-conforming if only common causes of variability were present in the system, and approximately 101 ppm in the long-term. The corresponding Z-Bench value was 3.66.

7   Conclusion

The capability comparison in Figure 7 shows that temperature of interior chamber is 86°C (baseline) at the beginning of this project, and closed to 85°C (set-point) after the actions were implemented. The capability analysis also shows that with the LSL = 84.5 and USL = 85.5. short term (and long-term) performances are also indicated, namely that approximately 125 parts per million (ppm) would be nonconforming if only common causes of variability were present in the system, and approximately 101 ppm in the long-term. The corresponding Z-Bench value was 3.66.

Therefore, we can conclude that key factors affecting the precision of temperature control were figured out in this LSS project, and some useful actions were proposed and efficiently implemented, and we achieved quite impressive results that went beyond the target we had set in the beginning. This study shows that Lean Six Sigma project could be used to affect factors of temperature control to improve the precision of temperature for new developing equipment.

References

1. Lee, S.C., Sanches, L., Kin, F.H.: Characterization of VOCs, ozone, and PM10 emissions from office equipment in an environmental chamber. Building and Environment 36(7), 37–42 (2001)
2. Yu, K.P., Lee, G.W., Hsieh, C.P., Lin, C.C.: Evaluation of ozone generation and indoor oganic compounds removal by air cleaners based on chamber tests. Building and Environment 45(1), 35–42 (2011)
3. Yen, K.-H., Yu, B., Chiang, H., Shih, J., Wang, C.: Strength Evaluation of Acrylic Fresnel Lens in Different Temperatures for HCPV Application. In: The 4th Asian Conference on Electrochemical Power Sources (ACEPS-4), Taipei, Taiwan, November 8-12, pp. 159–160 (2009)
4. Huanga, J., Haoa, Y., Linb, H., Zhanga, D., Songa, J., Zhou, D.: Preparation and characteristic of the thermistor materials in the thick-film integrated temperature–humidity sensor. Materials Science and Engineering 99(1-3), 523–526 (2003)
5. de Koning, H., Verver, J., van den Heuvel, J., Bisgaard, S., Does, R.: Lean Six Sigma in Healthcare. Journal of Healthcare Quality 28(2), 4–11 (2006)

A Novel Watermarked Multiple Description Scalar Quantization Coding Framework

Linlin Tang[1,*], Jeng-Shyang Pan[1], and Junbao Li[2]

[1] Harbin Institute of Technology Shenzhen Graduate School,
Shenzhen, China
[2] Harbin Institute of Technology,
Harbin, China
{linlintang2009,jengshyangpan,junbaolihit}@gmail.com

Abstract. Watermarked Multiple Description Coding techniques belong to one branch of covert communication techniques. A novel wavelet tree vector based watermarked Multiple Description Scalar Quantization coding frame has been proposed in this paper. The wavelet orientational tree is used as the tree vectors in the coding process. And the overlap of the different orientational information is used to introduce the redundancy. Good performance in the experiments has shown its efficiency.

Keywords: Multiple Description Coding (MDC), Multiple Description Scalar Quantization coding (MDSQ), Watermarking, Wireless Communication.

1   Introduction

With rapid development of the internet and wireless communication, more and more information has been digitized and transmitted through wireless channels. Therefore, two important things have been paid more and more attention: one is the transmission efficiency, the other is the transmitted information security. In fact, packet loss is unavoidable which will bring a great impact on the reconstructed quality of the multimedia signals. For example, if some of the transmitted data of ATM were lost, it would cause great loss to the bank and the customers. And if the retransmission failed again, the whole network might slip into a state of paralysis. Therefore, it is important to enhance the robustness of the transmission. In addition, the information security problems are always met. For example, how to realize the digital rights management? Actually, for the above two problems, people have done a lot on solving them respectively. How to consider the two techniques in one process? This is our topic in this paper.

On the one hand, the Multiple Description Coding (MDC) has earned widely attention for its good performance in the congested network. The earliest work in this area was invented at the Bell Labs in connection with communicating speech over the

---

* Corrresponding author.

A. Moonis et al. (Eds.): IEA/AIE 2014, Part I, LNAI 8481, pp. 236–245, 2014.

telephone network. In their work, the original sound signal is split into the odd and even parts, each being encoded and decoded separately. This MDC framework is very simple but useful. After that, various practical quantization-based MDC methods were presented, such as the MDSQ (Multiple Description Scalar Quantization) method [2, 3, 4], the MDVQ (Multiple Description Vector Quantization) method [5], and the MDLVQ (Multiple Description Lattice Vector Quantization) method [6]. Subsequently, various transform-based MDC methods were proposed, such as multiple description transform coding (MDTC) methods [7, 8, 9], pairwise correlating transform (PCT) based methods [10, 11] and frame expansion based methods [12, 13].

The main idea of MDC is to generate multiple descriptions of the source such that each description describes the source with a certain desired fidelity. If more than one description is available, then they can be combined to enhance the quality. MDC is robust due to the redundancy in multiple descriptions of the same source. Secondly, MDC may be scalable as each correctly received description improves the decoder performance. Furthermore, MDC does not require prioritized transmission, as each description is independently decodable. MDC can be viewed as a joint source and channel coding technique since it can be used to provide error resilience to media streams with a relatively small reduction in compression ratio. In addition, MDC simplifies the network design without feedback or retransmission of any lost packet, which meets the requirement in real-time interactive applications [14]. The basic block diagram of MDC is shown in Figure 1, where $\{X_K\}$ denotes the original multimedia sequence, it can be coded firstly, then divided into two descriptions. Here, $\{\hat{X}_K^{(1)}\}$ and $\{\hat{X}_K^{(2)}\}$ are the two side reconstructions, and $\{\hat{X}_K^{(0)}\}$ is the central reconstruction. To evaluate the results, we usually use D_s^1, D_s^2 and D_c^0 to represent the two side distortions and the central distortion respectively:

$$D_s^i = \sum_K \left(\hat{X}_K^{(i)} - X_K \right)^2 \qquad i = 1, 2 \tag{1}$$

$$D_c^0 = \sum_K \left(\hat{X}_K^{(0)} - X_K \right)^2 \tag{2}$$

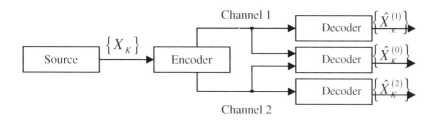

Fig. 1. Block diagram of MDC with two descriptions

Based on the different ways of introducing redundancy, MDC can probably be divided into two categories: the quantization based MDC [5, 6] and the transform based MDC [7]. In fact, more efficient MDC frame needs more complex quantizer design process. So, a lot of research has been applied on the transform based techniques [7]. But in this work, we will give a novel design for the Multiple Description Scalar Quantization (MDSQ) frame. The basic index assignment in MDSQ can be shown as in the following figure 2 which was proposed in 1993 by Vaishampayan.

(a)

	0	1	2	3	4	5	6	7
0	0	2						
1	1	3	4					
2		5	6	8				
3			7	9	10			
4				11	12	14		
5					13	15	16	
6						17	18	20
7							19	21

(b)

	0	1	2	3	4	5	6	7
0	0	1						
1	2	3	5					
2		4	6	7				
3			8	9	10			
4				11	13			
5					12	14	15	
6						16	17	19
7							18	20

Fig. 2. The index assignment matrix of MDSQ which is proposed by Vaishampayan
(a) nesting style (b) linear assignment style

As we can see from the above figure 2: the transmitted information is described as numbers in the blocks and the blank blocks are the redundancy. In the MDSQ frame, the main work is to design the filling rule. For the two channel MDC frame, the column and row indexes are usually used as the two transmission information for the two different channels.

On the other hand, watermarking can be called a classical tool which is used for copyright protecting and authorization [15-17]. It is an information hiding technique, and its basic thought is to embed secret information into the digital products, such as the digital images, audios and videos etc, in order to protect their copyrights, testify reliabilities, track piracy behaviors or supply products additional information. The secret information can be copyright symbols, consumer serials or other relevant information. Commonly they need to be embedded into digital products after proper transforms, and usually the transformed information is called Digital Watermark. Various watermarking signals are referred in many literatures, usually they can be defined as the below signal w:

$$w = \{ w_i / w_i \in \mathbf{O}, i = 0,1,2,...,M-1 \} \tag{3}$$

Where M is the length of watermarking sequences, and O represents the value range.

Our main work in this paper is to design a watermarked MDC frame to consider the transmission security and the multimedia security problems in one time. Section 2 will give an introduction of some related works about our design. Section 3 is our proposed frame. The experimental results will be shown in section 4. Conclusion and the acknowledgment will be given in section 5 and 6.

2   Related Work

2.1   Early Work for Watermarked MDC

The first MDC framework with watermark algorithm [18] was proposed by Pan in 2004. The following figure 3 shows its basic process. Many other researchers gave more research on this area and much more MDC with information hiding frames have been proposed in the following several years [19, 20].

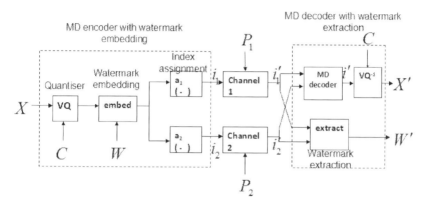

Fig. 3. The covert communication scheme based on MDSQ

In this paper, the above watermark embedding method is introduced to complete the watermarking embedding and extracting process. But different from the above frame, the orientation wavelet tree information is used for coding process. And the overlap of the orientation information is used to introduce the redundancy in the MDC frame.

2.2   Orientation Tree Vector

As we all know, the tree structure which is usually used in wavelet coding can be classified into the zerotree and the spatial tree. And the steps to achieve the tree vector are shown in the following.

Firstly, three-level wavelet transform is applied on the original image. And the tree structure can be shown as in the following figure 4.

As shown above, each orientation tree has 21 pixels. The energy value of the pixels in the highest frequency is much smaller than the other subbands'. Especially the diagonal part in this area, energy value is much smaller than others and in fact it only plays a very small role in representing the original information. So, the zerotree lies in the diagonal orientation can be chosen as set with 5 pixels. The 16 pixels lie in the highest frequency subband is cut off for little contribution for the whole image energy.

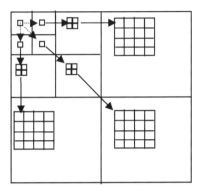

Fig. 4. Three-level wavelet ezrotree

Then after three-level wavelet decomposition, two kinds of tree vector with size 21 and one with size 5 will be achieved.

2.3   Vector Quantization

The common steps of VQ in transmission are shown in the following figure 5.

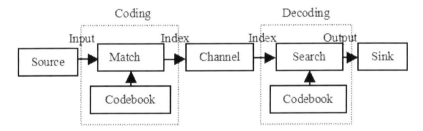

Fig. 5. VQ encoding and decoding diagram

As shown, the original information can be encoded by the index of codebook. The important design steps for this transmission process are to find efficient codebook and to improve the efficiency of codeword research.

3   Our Proposed Frame

Generally speaking, our proposed frame is a combination work of the above shown watermarking method and the coding method.

First of all, after a three-level decomposition, some tree vectors can be achieved as the former introduced. Thus, a VQ coding method can be designed for them. To show efficiency of the whole MDC framework, the classical LBG method is used to form the codebook. It is worth to say that the scalar quantization with steps 1, 2, 4, 8 is used for different wavelet subbands.

Secondly, we divide the tree information into three groups based on orientation for constructing two descriptions in the proposed MDC frame. The three orientational information groups is shown in the following figure 6.

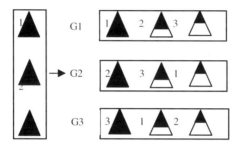

Fig. 6. Composition of the three descriptions

The number 1, 2 and 3 represent information from the three different orientations. For example, we suppose the number 1 represents the horizontal information, number 2 represent the vertical information and number 3 represent the diagonal information. $G1$, $G2$ and $G3$ represent the three different groups. Here, the three black triangles represent the refined coded information from three orientations. And the part black triangles represent the coarse encoded information.

In order to form two descriptions from the above three orientational information groups, we divide the diagonal information group into two sub groups and add them into the other two.

Thirdly, the VQ method is used onto the two descriptions to form two bit-streams for the original information.

Then, the watermarking embedding process just as mentioned in the reference [18] is applied on them.

4   Experimental Results

In our experiments, 512×512 Lena gray image has been used. The biorthogonal 9/7 three-layer wavelet transform is implemented on the original image to get better energy concentration results. And the watermark is a binary 64×64 image. Our channel is the mostly used balance channel.

The following figure 7 gives the performance under different packet losing rates.

(a) PSNR=37.44dB   (b)PSNR=36.26dB   (c) PSNR=36.20dB   (d)PSNR=27.97dB

(a1)   (b1)   (c1)   (d1)

Fig. 7. The recovered Lena image and the extracted secret information

The following table 1 gives some details results for one-channel received cases for our frame under different bit rates.

Table 1. Reconstruction results under different bit rates

PSNR(dB)	Bit rates			
	0.55bpp	1.10bpp	1.65bpp	2.20bpp
Channel 1	33.98dB	36.75dB	37.89dB	39.77dB
Channel 2	32.24dB	35.56dB	36.69dB	37.58dB

Though performance is good, it still has some space for improving. Actually, the above results are the average values in many times testing under the same channel conditions.

The following figure 8 gives some comparison between our proposed method and some other works which have some similar design. The performance is the robustness under different attacks one of the SFQ (Spatial Frequency Quantization based) is another work of authors published in 2013[19]. The definition of similarity is shown below in formula (4).

$$\rho\left(W^*,W\right) = \sum_{j=0}^{T-1}\left(x_j^* \cdot x_j\right) / \left(\sqrt{\sum_{j=0}^{T-1}\left(x_j^*\right)^2} \cdot \sqrt{\sum_{i=0}^{T-1}\left(x_j\right)^2}\right) \tag{4}$$

Here, W and W^* represents the original watermarking and extracted information respectively. The watermarking information can be represents as a sequence $\{x_j | j=0,1,\dots,T-1\}$. In one word, the all possible conditions for these two-channel frames are fixed as the same including the testing watermarking sequence. As we can check from below, our proposed method here is better than Pan's in 2004 under the same condition. The reason is the more redundancy for the two descriptions in MDC frame.

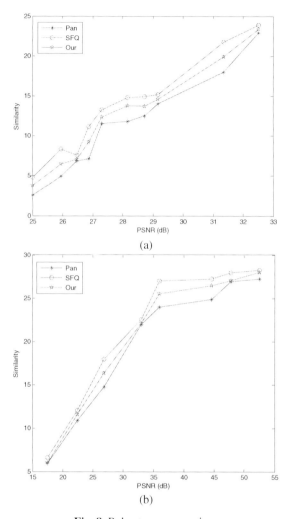

Fig. 8. Robustness comparison

(a) Robustness under JPEG compression   (b) Robustness under Gaussian noise

5   Conclusion

A novel orientation tree vector based watermarked MDC frame has been proposed in this paper. The traditional MDSQ idea is used to form the two different channels and the overlap information from all three orientations is used to introduce the redundancy. The experimental results show its efficiency. To find more efficient redundancy introducing method and better index assignment method is our future work.

Acknowledgment. The authors would like to thank for the support from the project named Research on Multiple Description Coding Frames with Watermarking Techniques in Wavelet Domain which belongs to the NSFC (National Natural Science Foundation of China) with the Grant number 61202456.

References

1. Vaishampayan, V.A.: Design of multiple description scalar quantizers. IEEE Transactions on Information Theory 39(3), 821–834 (1993)
2. Servetto, S.D., Ramchandran, K., Vaishampayan, V.A., Nahrstedt, K.: Multiple description wavelet based image coding. In: Proc. of IEEE International Conference on Image Processing, Chicago, IL, vol. 1, pp. 659–663. IEEE Computer Soc., Los Alamitos (1998)
3. Zhang, Z., Beger, T.: Multiple description source coding with no excess marginal rate. IEEE Transactions on Information Theory 41(2), 349–357 (1995)
4. Servertto, S.D., Vaishampayan, V.A., Sloane, N.J.A.: Multiple description lattice vector quantization. In: Proc. of IEEE Data Compression Conference, Snowbird, UT, pp. 13–22. IEEE Computer Soc., LA (1999)
5. Vaishampayan, V.A., Sloane, N.J.A., Servetto, S.D.: Multiple description vector quantization with lattice codebooks: design and analysis. IEEE Transactions on Information Theory 47(5), 1718–1734 (2001)
6. Wang, Y., Orchard, M., Reibman, A.R.: Multiple description images coding for noisy channels by paring transform coefficients. In: Proc. of IEEE First Workshop on Multimedia Signal Processing, pp. 419–424 (1997)
7. Goyal, V.K., Kovacevic, J.: Optimal multiple description transform coding of Gaussian vectors. In: Proc. of IEEE Data Compression Conference, Snowbird, UT, pp. 388–397. IEEE Computer Soc., LA (1998)
8. Goyal, V., Kovacevie, J., Arean, R., Vetteri, M.: Multiple description transform coding of images. In: Proc. of IEEE International Conference of Image Processing, Chicago, IL, vol. 1, pp. 674–678. IEEE Computer Soc., Los Alamitos (1998)
9. Wang, Y., Orchard, M.T., Vaishampayan, V.A., Reibman, A.R.: Multiple description coding using pairwise correlating transforms. IEEE Transactions on Image Processing 10(3), 351–366 (2001)
10. Wang, Y., Reibman, A.R., Orchard, M.T., Jafarhani, H.: An improvement to multiple description transform coding. IEEE Transactions on Signal Processing 50(11), 2843–2854 (2002)
11. Goyal, V.K., Kovacevic, J., Vetterli, M.: Quantized frame expansions as source-channel codes for erasure channels. In: Proc. of IEEE Data Compression Conference, Snowbird, UT, pp. 326–335. IEEE Computer Soc., Los Alamitos (1999)
12. Mehrotra, S., Chou, P.: Multiple description decoding of overcomplete expansions using projections onto convex sets. In: Proc. of IEEE Data Compression Conference, Snowbird, UT, pp. 72–81. IEEE Computer Soc., Los Alamitos (1999)
13. Wang, Y., Reibman, A.R., Lin, S.: Multiple description coding for video delivery. Proceedings of the IEEE 93(1), 57–70 (2005)
14. Latif, A.: An Adaptive Digital Image Watermarking Scheme using Fuzzy Logic and Tabu Search. Journal of Information Hiding and Multimedia Signal Processing (JIHMSP) 4(4), 250–271 (2013)

15. Benhocine, A., Laouamer, L., Nana, L., Pascu, A.C.: New Images Watermarking Scheme Based on Singular Value Decomposition. Journal of Information Hiding and Multimedia Signal Processing (JIHMSP) 4(1), 9–18 (2013)
16. Lin, C.-H., Yang, C.-Y.: Multipurpose Watermarking Based on Blind Vector Quantization (BVQ). Journal of Information Hiding and Multimedia Signal Processing (JIHMSP) 2(2), 239–246 (2011)
17. Pan, J.-S., Hsin, Y.-C., Huang, H.-C., Huang, K.-C.: Robust Image Watermarking Based on Multiple Description Vector Quantization. Electronics Letters 40(22), 1409–1410 (2004)
18. Tang, L.-L., Pan, J.-S., Luo, H., Li, J.-B.: Novel Watermarked MDC System Based on SFQ Algorithm. IEICE Transactions on Communication E95-B(09), 2922–2925 (2012)

Genetic Generalized Discriminant Analysis and Its Applications

Lijun Yan[1], Linlin Tang[1,*], Shu-Chuan Chu[2], Xiaorui Zhu[1],
Jun-Bao Li[3], and Xiaochuan Guo[1]

[1] Harbin Institute of Technology Shenzhen Graduate School
Xili University Town, NanShan, Shenzhen, China
linlintang2009@gmail.com
[2] School of Computer Science,Engineering and Mathematics
Flinders University of South Australia
GPO Box 2100, Adelaide, South Australia 5001, Australia
[3] Department of Automatic Test and Control,
Harbin Institute of Technology, China

Abstract. In this paper, a novel Genetic Generalized Discriminant Analysis (GGDA) is proposed. GGDA is a generalized version of Exponential Discriminant Analysis (EDA). EDA algorithm is equivalent to map the samples to a new space and then perform LDA. However, is this space is optimal for classification? The proposed GGDA uses Genetic Algorithm to search for an more discriminant diffusing map and then perform LDA in the new space. The Experimental results confirm the efficiency of the proposed algorithm.

Keywords: Feature extraction, Exponential discriminant analysis, Genetic algorithm.

1   Introduction

In the recent decades, image classification, including face recognition [1], palmprint recognition [2] and so on, becomes more and more popular. There are several stages in pattern recognition such as preprocessing, sample selection, feature extraction, classification and others. Among those stages, feature extraction is one of the most important stages for classification. A lot of different methods are proposed for feature extraction. Dimention reduction technology is one of the most popular methods. imensionality reduction technology is an essential stage in many intelligent data analysis applications and has been applied in many fields including handwriting recognition, speech recognition, face recognition, facial expression analysis, and so on. Principal Component Analysis (PCA)[3, 4], Linear Discriminant Analysis (LDA)[5, 6] are two most popular subspace-based dimensionality reduction algorithms. PCA projects the original samples to a low dimensional subspace, which is generated by the eigenvectors

---

* Corresponding author.

A. Moonis et al. (Eds.): IEA/AIE 2014, Part I, LNAI 8481, pp. 246–255, 2014.

corresponding to the largest eigenvalues of the covariance matrix of all training samples. PCA aims at minimizing the mean squared error. However, PCA is an unsupervised algorithm, which may reduce the efficiency of feature extraction. LDA is to find an optimal transformation matrix U that linearly projects the high-dimensional sample $x \in R^m$ to low-dimension space by $y = U^T x \in R^n$, where $n \ll m$. LDA can compute an optimal discriminant projection by maximizing the ratio of the trace of the between-class scatter matrix to that of the within-class scatter matrix. LDA uses the labels of the prototype samples during the training and improves the discriminant ability. However, LDA has to suffer from the famous small sample size (SSS) problem. Many effective algorithms have been introduced to solve the problem. Local Preserving Projection (LPP) [7] is to preserve the neighborhood of the samples. Some nonlinear extensions using kernel trick of these algorithms are proposed in the recent 10 years[8, 9, 10]. Besides, there are some works on feature fusion presented[11] to improve the performance of feature extraction.

In 2010, an exponential discriminant analysis (EDA) [12] was proposed by Zhang et al. to deal with the SSS problem. The EDA algorithm can extract the most discriminant information that is contained in the null space of within-class scatter matrix compared with the FIsherface algorithm. EDA is equivalent to projecting the original samples into a new space by distance diffusion mapping, and LDA is used in the new space in which the scatter between different classes is enlarged to improve the efficiency of the feature extraction.

This paper proposes a feature extraction method which generalizes the classical EDA by using Genetic Algorithms (GA) [13, 14]. In the proposed Genetic Generalized Discriminant Analysis(GGDA) algorithm, the exponential function is decomposed by Tailor expansion compared with the EDA and GA is used to choose which terms should be taken for the optimization.

The paper is organized as follows: Section II reviews the EDA method. The new subspace learning algorithm is proposed in Section III with the various experimental results in Section IV. Lastly, Section V concludes the paper.

2   Review of EDA

2.1   LDA

Suppose there are c pattern classes. N is the total number of training samples, and N_i is the number of ith class. In the class, the jth training image is denoted by $X_j^i \in R^D$. $\bar{X}^i$ is the mean matrix of training samples of the ith class. And $\bar{X}$ is the mean matrix of all training samples.

The between-class scatter matrix S_b and within-class scatter matrix S_w can be constructed by

$$S_b = \frac{1}{N} \sum_{i=1}^{c} N_i (\bar{X}^i - \bar{X})(\bar{X}^i - \bar{X})^T, \tag{1}$$

and

$$S_w = \frac{1}{N} \sum_{i=1}^{c} \sum_{j=1}^{N_i} (X_j^i - \bar{X}^i)(X_j^i - \bar{X}^i)^T. \qquad (2)$$

By LDA, sample I will be projected to the space R^d with $d \ll D$. The criterion to select the most discriminative features can be defined by

$$\max J(u) = \left(u^T S_b u\right) / \left(u^T S_w u\right). \qquad (3)$$

In fact, it is evident that the optimal projection matrix U is a set of general $\lambda_1 \geq \lambda_2 \geq \cdots \geq \lambda_d$. Let $Y = IU$. The resulting matrix Y is called the feature matrix of image I and used to represent X for classification.

2.2   EDA

Given a training sample set, each sample of X belongs to one of C class $\{L_1, L_2, \cdots, L_C\}$. Let N_i be the number of training samples in ith class. Therefore, $N = \sum_{i=1}^{C} N_i$.

Definition 1: Given an arbitrary n-order square matrix A, its exponential is defined as follows:

$$\exp(A) = I + A + \frac{A^2}{2!} + \cdots + \frac{A^m}{m!} + \cdots. \qquad (4)$$

Then the criterion of EDA is as follows:

$$J_2(w) = \arg\max_{w} \frac{\left|w^T \exp(S_b)w\right|}{\left|w^T \exp(S_w)w\right|}. \qquad (5)$$

EDA is equivalent to mapping the original samples into a new feature space by distance diffusion mapping, and LDA criterion is used in such a new space. The margin between different classes is enlarged as a result of diffusion mapping.

2.3   GA

The following method is used as GA in this paper. Let g be the number of generation, $M(g)$ be the gene population, $|M|$ be the number of individuals, m be the chromosome length, P_c be the crossover rate, P_m be the mutation rate and g_{max} be the number of maximum generations. In this paper, chromosome coding is binary coding algorithm and each chromosome constitutes of a string of "0" or "1". The procedure of GA is as follows:

Step 1, Generate the initial gene population $M(g)$ initially: $g = 0$.

Step 2, Compute the fitness function of the ith gene in $M(g)$, f_i, for all i.

Step 3, Choose the parent gene in the next generation $S(g)$, from $M(g)$ by using f_i.

Step 4, Perform crossover and mutation to $S(g)$, and construct the next generation gene population $M(g+1)$.

Step 5, Repeat Steps 2 to 4 until convergence or g equals g_{max}.

3   The Proposed Algorithm

From the definition of exponential of a matrix, EXP(A) is a Tailor extension. This means it is a sum. Therefore, a question emerges naturally. which terms of the sum is more important for feature extraction? In this section, a GGDA is proposed to choose the discriminant terms. GGDA preserves the basic procedure of EDA besides that a novel matrix function will be chose to generate the novel diffusion map.

3.1   Main Idea of GGDA

For chromosome representing the exponential function, the bit with value '0' represents the corresponding term is abandoned, and '1' represents the corresponding term is chosen. Note that the precision of representing exponential function depends on the length of the chromosome. For example, in Fig. 1, the bit strings use binary code. This chromosome means the 1st, 3rd, 4th, and 8th terms are abandoned. Given a chromosome $b = [b(0), b(1), \cdots, b(m-1)]$, a corresponding matrix function $E_b(A)$ can be generated using the following formula,

$$E_b(A) = b(0)I + b(1)A + b(2)\frac{A^2}{2!} + \cdots + b(m-1)\frac{A^m}{m!}, \tag{6}$$

where m is the length of chromosome. For example, the chromosome $b = [0, 1, 0, 0, 1, 1, 1, 0]$ in the Fig. 1 generates $E_b(A) = A + \frac{A^5}{5!} + \frac{A^6}{6!} + \frac{A^7}{7!}$.

Fig. 1. A sample of chromosome

$X = \{X_j^i, i = 1, 2, \cdots, c, j = 1, 2, \cdots, N_i,\} \in R^D$ denotes the prototype samples set, where X_j^i denote the jth sample in the ith class, c denotes the class number and N_i is the samples number in the ith class.

Then one optimization function for feature extraction can be proposed as follows:

$$J_3(w) = \arg\max_w \frac{|w^T E_b(S_b)w|}{|w^T E_b(S_w)w|}, \tag{7}$$

where S_b and S_w are defined in Eq. (1) and Eq. (2).

The projection W can be obtained by solving the following eigenvalue problem

$$E_b(S_b)w = \lambda E_b(S_w)w. \tag{8}$$

Let $w_1, w_2, \cdots, w_d$ be the eigenvectors of Eq. (8) corresponding to the d largest eigenvalues ordered according to $\lambda_1 \geq \lambda_2 \geq \cdots \geq \lambda_d$. An $D \times d$ transformation matrix $W = [w_1, w_2, \cdots, w_d]$ can be obtained to project each sample with size of $D \times 1$ x into a feature vector with size of $d \times 1$ y as follows:

$$y = W^T x. \tag{9}$$

y will be used for classification instead of x. So one chromosome leads to one transformation matrix W. Then W can be used for feature extraction.

In this paper, recognition rate is chosen as the fitness function. Thus, for the individual (chromosome) with high recognition rate leads to a high fitness value.

GGDA aims at finding out an optimal matrix function $E_{b_o}(A)$, where

$$E_{b_o}(A) = b_o(0)I + b_o(1)A + b_o(2)\frac{A^2}{2!} + \cdots + b_o(m-1)\frac{A^m}{m!}. \tag{10}$$

and $b_o(i)$ denotes the ith bit of the optimal chromosome. A higher recognition rate can be gotten using the optimal matrix function $E_{b_o}(A)$.

3.2   The Procedure of the Proposed GGDA

The procedure of the proposed GGDA is as follows:

Step 1, Partition the prototype samples to two sets: traning set X_1 and test set X_2.

Step 2, Calculate S_b and S_w using X_1 following Eq. (1) and Eq. (2).

Step 3, Generate an initial chromosome population $M(g)$, $g = 0$.

Step 4, Calculate $|M|$ matrix functions generated by chromosomes following Eq. (6) and the corresponding transformation matrix W.

Step 5, Extract the features of samples in both training set and test set.

Step 6, Classify using Nearest Neighbor Classifier and compute the recognition rate as the fitness function.

Step 7, Crossover and mutation.

Step 8, End the process when the end conditions are satisfied; Otherwise, go to Step 4.

Step 9, Output the optimal transformation matrix W.

4   Experimental Results

Since images are complicated while dealing with classification problems, face and image recognition were implemented to test the proposed GGDA. Academically, there are various open databases so that different algorithms can compare with each other on the same platform. In this experiment, Finger-Knuckle-Print database (FKP) [15],AR face database [16] are the two popular image databases on which we implemented the proposed GGDA. During the experiment, the system ran ten times in order to reduce the variation on each database. In the experiments, the performance of the proposed GGDA is compared with those of

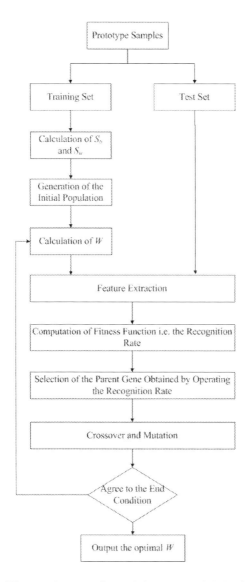

Fig. 2. The procedure of the proposed GGDA

PCA, Fisher, LPP, and EDA. For each database, half samples per subject were selected as prototype samples randomly and the others were used as query samples in each run. The system runs 20 times. The maximum average recognition rate (MARR) is used to evaluated the performance of different algorithms.

To evaluate the different feature extraction algorithms, Nearest Neighbor classifier is applied for classification. The parameters used in these experiments are as follows: $m = 8$, $|M| = 100$, $P_c = 1.0$, $P_m = 0.25$, $g_{max} = 300$.

AR face database [16] was created by Aleix Martinez and Robert Benavente in the Computer Vision Center (CVC) at the U.A.B. It contains over 4,000 color images corresponding to 126 people's faces (70 men and 56 women). Images feature frontal view faces with different illumination conditions, facial expressions, and occlusions (sun glasses and scarf). The pictures were taken at the CVC under strictly controlled conditions. Each person participated in two sessions, separated by two weeks (14 days) time. The same pictures were taken in both sessions. In the following experiments, only nonoccluded images of 120 people in AR face database are selected. Seven images per person are randomly selected for training and the other images are for testing. This system also runs 20 times. Some samples of AR face database are shown in Fig. 2. Table 1 tabulates the MARR of these algorithms on AR face database. Clearly, MARR of the proposed algorithm is higher than other approaches.

Fig. 3. Some samples of AR face database [16]

The Biometric Research Centre (UGC/CRC) at the Hong Kong Polytechnic University created FKP database [15]. There were 165 peoples fingers collected in FKP database, 125 were from men and 40 were from women. Each person provided 12 images on middle finger and index finger from both hands. First instead of treating each persons fingers as one subject, we treated each finger as one subject. Therefore, there are 660 subjects with 12 images per subjects. In the experiments, we find there are some duplicate samples in the database. To evaluate various algorithms better, the duplicate samples are removed from the database during the experiments. Besides, in order to reduce the complexity of

Table 1. MARR of different algorithms on AR face database

Algorithms	MARR	Feature dimension
Fisherface	0.9481	120
PCA	0.7604	120
LPP	0.8792	190
EDA	0.9490	150
Proposed algorithm	0.9536	120

the experiment and experimental time, PCA was taken to reduce the dimension of the samples. And 97% of energy is retained in PCA stage. Table 2 is the ARR result for FKP database under different algorithms is as follows. Compared with other learning methods, the GGDA has higher recognition rate.

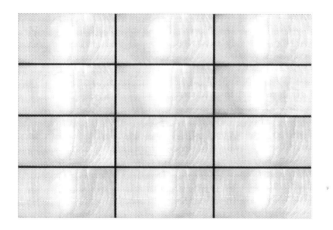

Fig. 4. Examples for FKP image database [15]

Table 2. MARR of different algorithms on FKP database

Algorithms	MARR	Feature dimension
Fisherface	0.9318	120
PCA	0.8346	120
LPP	0.8251	190
EDA	0.9432	120
Proposed algorithm	0.9521	110

5   Conclusion

This paper has proposed a new subspace learning algorithm called Genetic Generalized Discriminant Analysis (GGDA) for feature extraction. GGDA is a generalized algorithm of exponential discriminant analysis approach. Different from

EDA, GGDA can choose more discriminant terms from the Tailor extension of exponential to diffuse the scatters using genetic algorithms. In the experiments, the proposed algorithm has been compared with some popular methods. The experimental results on AR face database and FKP database show the efficiency of GGDA.

Acknowledgments. This work was supported in part by National Natural Science Foundation of China (61371178), the Peacock Project of Shenzhen, Intelligent Vehicle Cloud Client System, under Project NO. KQC201109020055A, Shenzhen Strategic Emerging Industries Program under Grants No. ZDSY20120613125016389 and China Postdoctoral International Exchanges Program.

References

[1] Li, J.B., Chu, S.C., Pan, J.S., Jain, L.C.: Multiple viewpoints based overview for face recognition. Journal of Information Hiding and Multimedia Signal Processing 3(4), 352–369 (2012)

[2] Kong, A., Zhang, D., Kamel, M.: A survey of palmprint recognition. Pattern Recognition 42(7), 1408–1418 (2009)

[3] Turk, M., Pentland, A.: Eigenfaces for recognition. J. Cogn. Neurosci. 3(1), 71–86 (1991)

[4] Roweis, S.: Em algorithms for PCA and SPCA. In: Advances in Neural Information Processing Systems, pp. 626–632. MIT Press (1998)

[5] Belhumenur, P., Hepanha, J., Kriegman, D.: Eigenfaces vs. Fisherface: Recognition using class specific linear projection. IEEE Transactions on Pattern Analysis Machine and Intelligence 19(7), 711–720 (1997)

[6] Martinez, A.M., Kak, A.: PCA versus LDA. IEEE Transactions on Pattern Analysis and Machine Intelligence 23(2), 228–233 (2001)

[7] He, X., Yan, S., Hu, Y., Niyogi, P., Zhang, H.: Face recognition using laplacianface. IEEE Transactions on Pattern Analysis Machine and Intelligence 27(3), 328–340 (2005)

[8] Xu, Y., Zhang, D., Jin, Z., Li, M., Yang, J.: A fast kernel-based nonlinear discriminant analysis for multi-class problems. Pattern Recognition 39(6), 1026–1033 (2006)

[9] Li, J., Gao, H.: Sparse data-dependent kernel principal component analysis based on least squares support vector machine for feature extraction and recognition. Neural Computing and Applications 21(8), 1971–1980 (2012)

[10] Li, J., Pan, J., Chu, S.: Kernel class-wise locality preserving projection. Information Sciences 178(7), 1825–1835 (2008)

[11] Xu, Y., Zhang, D.: Represent and fuse bimodal biometric images at the feature fevel: Complex-matrix-based fusion scheme. Optical Engineering 49(3) (March 2010) 037002–037002–6

[12] Zhang, T., Fang, B., Tang, Y.Y., Shang, Z., Xu, B.: Generalized discriminant analysis: A matrix exponential approach. IEEE Trans. Systems, Man, and Cybernetics-Part B: Cybernetics 40(1), 186–197 (2010)

[13] Goldberg, D., Holland, J.: Genetic algorithms and machine learning. Machine Learning 3(2-3), 95–99 (1988)

[14] Goldberg, D.: Genetic algorithms in search, optimization and machine learning. Addison-Wesley (1989)

[15] Zhang, L., Zhang, D., Guo, Z.: Phase congruency induced local features for finger-knuckle-print recognition. Pattern Recognition 45(7), 2522–2531 (2012)

[16] Martinez, A.M., Benavente, R.: The AR face database. In CVC Technical Report 24 (1998)

Assessing the Availability of Information Transmission in Online Social Networks

Dayong Zhang[1] and Pu Chen[2,*]

[1] Department of New Media and Arts, Harbin Institute of Technology, Harbin150001, China
No.92, West Da-zhi Street, Harbin, P.R. China 150001
yonghit@163.com
[2] Academy of Fundamental and Interdisciplinary Sciences, Harbin Institute of Technology,
Harbin 150001, China
No.92, West Da-zhi Street, Harbin, P.R. China 150001
chenpu@live.com

Abstract. This paper selects several representative social network services as samples. In order to evaluate how efficiently a system exchanges information before and after the removal of a set of nodes, we introduce the concept of the global efficiency. Especially, comparing with results of the statistic characteristics of different networks, we find that there are some significant differences between two types of structure modes. First, the individual-centered networks have the characteristics of hierarchical structure, but the group-centered networks have not the same structural feature. Second, the individual-centered networks have a higher efficiency than the group-centered networks in information diffusion. Moreover, our research sheds light on the structural properties of networks are responsible for the efficiency and reliability of networks.

Keywords: online social network, global efficiency, hierarchical structure, network reliability, network reliability.

1   Introduction

SNS (Social Network Service) is the online service platform based on the concept of "Six degree separation". Through this platform, users could share information, express individual viewpoint and enjoy interactive games, etc. As public online communities, SNS sites such as Facebook, Twitter, Myspace, blog, and renren, could help users meet their needs of information, entertainment, interest and social relations. With the ability of exchanging massive amounts of data , complex structural features as well as undeniable impact on public opinion, How efficiently the nodes of online social networks exchange information has been one of the hotspots in complex network research.

---

* Corresponding author.

A. Moonis et al. (Eds.): IEA/AIE 2014, Part I, LNAI 8481, pp. 256–262, 2014.

In recent years, the static structure and the dynamic evolutionary process of the SNS sites are the main goals in studying SNS. As to the former, it focuses on the users and their complicated relationships by analyzing the statistical data[1,2] such as the users' race, religion, gender and nationality, as well as the basic topological properties such as degree distribution, hierarchical structure, associations. And for the latter, one purpose is to provide an empirical research by analyzing the evolution of the networks' structure and users' behavior [3]; the other is to explore the role the social networks play in the dissemination of information, combining with biological dynamics and communication-related model[4,5]. In particular, the assessment of the hierarchical structure and assortativity coefficient is of great significance to revealing the structure and formation of the network. Nowadays,one research is about the cluster, which is a typical attribute of the realistic network. It is known that nodes tend to be connected closely in one cluster. While among different clusters there are only loosely connections, especially in specific networks. This kind of phenomenon can be explained with a sociological concept—clustering coefficient. There are various factions and groups in reality social networks, and the members of which are familiar with each other. With more and more people interested in developing social relationship and lifestyles in virtual networks, the structure of factions and communities has fundamentally changed. It is no longer confined to acquaintance, on the contrary, people who have common beliefs or hobby, and even the same problems or interests, form new groups. Study shows that the clustering coefficient following a power-law distribution is the most important quantitative index for hierarchical structure[6].

In this paper, we select some representative online social networks, such as weights network and non-weights network; information network and professional forum network. We hope to find some common features in different social networks to help people have a better understanding about the relation between the network topology and the information diffusion.

2   Data and Methods

At first, the online services for the users mainly focus on free or low-cost method of communication, such as E-mail and instant messaging system. With the rapid development of network technology, especially the social cooperation technology such as Web 2.0, people have much broader network of space and diverse display modes. As the diversification of technology, organization and implementation in the social network, there is a big difference in the network topology structure. Such as Facebook, which is designed as individual-centered, in order to achieve the expansion of individual external, allows users to upload pictures, add multimedia content and modify the appearance; some other online networks, such as BBS (Bulletin Board System), is assembled with the same view or the interests of users, and the exchange of information within groups is more important for it. The form of network structure of these two kinds of technology is very different.

In order to have a full understanding of the overall technical structure of social networks, the data we analyzed is from two network. The Lilac community is formerly known as the largest student forum in Northeast of China, and belongs to the online community network, and the users are students. Its purpose is to increase the exchange of information between students and help students expand their circle of friends. Sina Weibo is a broadcast social networking platform by sharing brief real-time information. Users can form their own communities to share information through the WEB, WAP and various of clients. These networks are all undirected and fully connected networks.

In order to evaluate how efficiently a system exchanges information before and after the removal of a set of nodes, we introduce the concept of the global efficiency in this section. Normally, it is easier to transfer information from one node to another if they are closer to each other. Therefore, the global efficiency in the communication between two nodes i and j is defined as the inverse of the shortest path length d_{ij} between these two nodes: $\varepsilon_{ij} = \dfrac{1}{d_{ij}} \forall i, j$, where the shortest path length d_{ij} is the smallest sum of the physical distances throughout all the possible paths in the network from node i to j. If there is no path in the network G between nodes i and j, $d_{ij} = \infty$, and $\varepsilon_{ij} = 0$, otherwise , $0 < \varepsilon_{ij} \leq 1$. Let $G = (V, E)$ be a connected graph. The global efficiency of E_{glob} can be defined as:

$$E_{glob} = \frac{\sum_{i \neq j \in G} \varepsilon_{ij}}{N(N-1)} = \frac{1}{N(N-1)} \sum_{i \neq j \in G} \frac{1}{d_{ij}} \tag{1}$$

Table 1. The effective samples of different network

Name	Lilac	Sina Weibo
nodes	3414	839
edges	10353	2112
efficiency	0.263	0.383

3   The Analysis of Hierarchical Structure

Social network analysis shows that the basic elements of social networks are persons and groups. There are a variety of different sizes and they form complex social structures[24]. In general, the lager the scale of social structure is, the more complex it is, and the more subgroups it contains. Although social groups differ in thousands of ways and change constantly, the subgroup members still maintain close connections through core nodes. However, if there is always the alternative connectivity paths between all nodes of the groups, that is, the communication and exchange of information in different sub-groups do not depend on any one node, we can say the network is resilient and shows a non-hierarchical structure. In the analysis of complex networks, Cluster coefficient is often used to reflect the closeness of nodes, the Cluster coefficient in undirected network can be defined as:

$$C_i = \frac{3 \times \text{the numbers of triangles connected with node i}}{\text{the numbers of triples connected with node i}} \qquad (2)$$

As a local parameter, the Cluster coefficient can be used to depict the extent of network groups. Empirical studies for a large number of complex networks have shown that, the Cluster coefficient is much larger compared with the equivalent size of random networks. In other words, there are many small-scale node clusters within intensive sides, and they are connected loosely. Some small clusters trend to become larger groups without changing the scale-free topology at the same time in some networks (such as metabolic network), thus forming the hierarchical network.

One of the most important quantitative signs in Hierarchical Network is that the Cluster coefficient of nodes follows a power- law distribution $C(k) \sim ck^{-\beta}$. It indicates that nodes of low degree have a high cluster coefficient and are in highly connected module. On the contrary, the hub nodes of high degree, which is to connect the different modules, have a very low cluster coefficient.

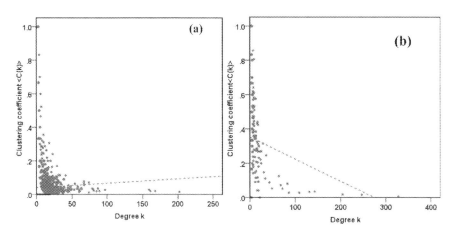

Fig. 1. The analysis of cluster coefficient, (a)~(b) is Lilac and Sina Weibo, respectively

As shown in Fig.1 (b) , the cluster coefficient $C(k)$ follows $C(k) \sim ck^{-\beta}$,and the power-law is $\beta = -0.733$ .There are distinct hierarchical structures in the kind of online social network above, and it largely due to its own technology and organization forms. It emphasizes more about the personalized design in Sina Weibo. The connections between users are point-to-point or point-to-side and the network scale can expand unlimitedly. On the other hand, in Lilac(a), the network cluster coefficient is higher than that in corresponding random network ,what is $C_{random} = 1/N$. But it is lack of obvious hierarchical structure, which means there is no clear boundaries between the clusters, that is, closely connections but no sense of community. It is mainly because sharing information is more important for online community network, and there are stringent restrictions in users' identity and networks' size, in which the connections ways among users are face-to-face or point-to-point.

4   The Analysis of Centrality

An online social network is a set of users and relationships of different kinds among them, where some users often play a vital , or at least important role, e.g., in forming and directing public opinion. Centrality is an important concept in studying nodes' prominent positions. Centrality of a node is a structural index, and seeks to quantify an individual node's prominence within a network by summarizing structural relations among the nodes. Generally, a more "central" node has a stronger influence on other nodes in the same network. Typical measures of the centrality include degree and betweeness centralities.

Degree centrality: One of the most used centrality measures is degree centrality, which can be interpreted as a measure of immediate influence, as opposed to long-term effect in the network. The degree centrality of a node i refers to the number of edges attached to the node, and is defined as

$$C_D(N_i) = \sum_{i=1}^{n} x_{ij} (i \neq j)$$
(3)

Each of the links is defined by a couple of nodes i and j , and is denoted as x_{ij} . $\sum_{i=1}^{n} x_{ij}$ stands for the degree of a node i , and is defined in terms of the adjacency matrix. The standardized degree centrality is

$$C'_D(N_i) = \frac{C_D(N_i)}{n-1}$$
(4)

Betweenness centrality: In the context of social network theory, the importance of a node for spreading is often associated with the betweenness centrality, which is believed to determine who has more 'interpersonal influence' on others. Betweenness centrality characterizes how influential a node is in communicating between node pairs, and quantifies how many times a node can interrupt the shortest paths between the two nodes of the pair. The standardized betweenness centrality is

$$C'_B(N_i) = \frac{C_B(N_i) \times 2}{(n-1)(n-2)}$$
(5)

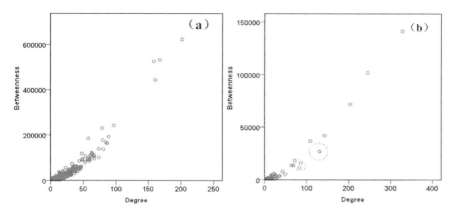

Fig. 2. The analysis of centralies, (a)~(b) is Lilac and Sina Weibo, respectively

In this section, we have examined empirically the relation between the different criteria to determine the importance of a node, and find that there is a high correlation the degree centrality and betweeness centrality except for a few nodes in the two representative social networks, which means the criterion of centrality can measure the reliability of network, and find out the critical nodes in holding the network connectivity and facilitating information flow.

5   Conclusions

In this paper, several representative online social networks are selected to carry on the analysis, including personal communication networks and professional BBS. The experimental results show that: 1) the cluster coefficient follows a power-law distribution in the individual user-centric social network with obvious hierarchical structure. While in no-center online networks there is no clear hierarchical structure and members tend to be diversified, what's more, there are no clear boundaries and the members have no sense of community. 2) In all the four representative social networks, there is a high correlation between the degree centrality and betweeness centrality, indicating the indices play as an important role in holding the network connectivity and facilitating information flow.

Acknowledgments. This work was partly supported by the NSF of PRC under Grant No. 70903016 and the national science & technology support program under Grant No. 2012BAH81F03.

References

1. Eom, Y.-H., Jeon, C., Jeong, H., Kahng, B.: Evolution of weighted scale-free networks in empirical data. Physical Review E 77, 056105 (2008)
2. Panzarasa, P., Opsahl, T., Carley, K.M.: Patterns and Dynamics of Users' Behavior and Interaction: Network Analysis of an Online Community. Journal of the American Society for Information Science and Technology 60, 911–932 (2009)

3. González, M.C., Herrmann, H.J., Kertész, J., Vicsek, T.: Statistical Mechanics and its Applications. Physica A 379, 307–316 (2007)
4. Grabowski, A., Kruszewska, N., Kosiński, R.A.: Dynamic phenomena and human activity in an artificial society. Physical Review E 78, 066110 (2008)
5. Zhao, P., Zhang, C.-Q.: A new clustering method and its application in social networks. Pattern Recognition Letters 32, 2109–2118 (2011)
6. Costa, L.F., Silva, F.N.: Hierarchical characterization of complex networks. Journal of Statistical Physics 125, 845–876 (2006)

Feature Curve Metric for Image Classification

Qingxiang Feng[1], Jeng-Shyang Pan[1], and Tien-Szu Pan[2]

[1] Innovative Information Industry Research Center
Harbin Institute of Technology Shenzhen Graduate School
Shenzhen, China
{fengqx1988,jengshyangpan}@gmail.com
[2] Research Institute of Applied Engineering Science
Kaohsiung University of Applied Sciences
Taiwan
tien.pan@msa.hinet.net

Abstract. In this paper, an improved classifier based on nearest feature line (NFL), shortest feature line segment (SFLS) and nearest feature center (NFC), called the feature cure metric (FCM), is proposed for hand gesture recognition and face recognition. Borrowing the concept from the NFC and SFLS classifiers, the proposed classifier uses the novel distance metric between the test sample and the pair of prototype samples. A large number of experiments on Jochen Triesch Static Hand Posture (JTSHP) Database, Yale face database and JAFFE face database are used to evaluate the proposed algorithm. The experimental results demonstrate that the proposed approach achieves better recognition rate than the other well-known classifiers, such as nearest neighbor (NN) classifier, NFL classifier, nearest neighbor line (NNL) classifier, center-based nearest neighbor (CNN) classifier, extended nearest feature line (ENFL) classifier, SFLS classifier and NFC classifier.

Keywords: Nearest Feature Line, Nearest Feature Center, Face Recognition, Hand Gesture Recognition.

1 Introduction

Nearest neighbor (NN) [1] is the one of well-known approaches in pattern recognition. However, the number of prototype samples, which is usually very small, makes the classification be very difficult. So the nearest feature line (NFL) [2] [3] was proposed for face recognition by Stan Z. Li et al. in 1999. The NFL attempts to heighten the representational capacity of a sample set with limited size by using the line passing through each pair of the samples belonging to the same class. All the lines constituted by samples of the same class are called the feature lines (FL) of the corresponding class. Li and Lu in [2] explained that a feature line provides information about the possible linear variants of two sample points.

A. Moonis et al. (Eds.): IEA/AIE 2014, Part I, LNAI 8481, pp. 263–272, 2014.
© Springer International Publishing Switzerland 2014

The NFL improves the classification accuracy successfully contrasted to the nearest neighbor (NN) classifier. However, the NFL also has some shortcomings which limit their further applications in practice. Firstly, the NFL classifier will have the large computational complexity problem when there are a lot of samples in each class. Secondly, there is also an extrapolation inaccuracy problem that is shown in Fig. 2.

After the NFL being proposed, Zhou et al proposed the nearest feature midpoint (NFM) [4] in 2000. Zheng et al. proposed the nearest neighbor line (NNL) [5] in 2004. GAO et al. suggested the center-based nearest neighbor (CNN) [6] in 2007. Zhou et al. recommended the extended nearest feature line (ENFL) [7] classifier in 2004. Han et al. proposed the shortest feature line segment classifier (SFLS) [8] in 2011. Feng et al. suggested the nearest feature center (NFC) [9] classifier in 2012 and some other improved classifiers [10]-[12].

Motivated by concepts of the NFL, SFLS and NFC classifiers, the feature curve metric (FCM) is proposed for hand gesture recognition and face recognition. Borrowing the concept from the NFC classifier and shortest feature line segment (SFLS) classifier, the proposed classifier uses the novel distance metric between the test sample and the pair of prototype samples, which is called feature ellipse metric. The new distance metric is superior to some existing metric. A large number of experiments on Jochen Triesch Static Hand Posture (JTSHP) Database, Yale face database and JAFFE face database are used to evaluate the proposed classifier. The experimental results show that the proposed approach achieves better recognition rate than some other classifiers, such as nearest neighbor (NN) classifier, NFL classifier, nearest neighbor line (NNL) classifier, center-based nearest neighbor (CNN) classifier, extended nearest feature line (ENFL) classifier, SFLS classifier and NFC classifier.

2   Review

Let $Y = \{ y_i^c, c = 1, 2, \cdots, M, i = 1, 2, \cdots, N_c \} \subset R^D$ be the prototype space, where y_i^c is the ith prototype sample belonging to the cth class, M is the number of classes, and N_c is the number of prototype samples belonging to the cth class.

2.1   NFL Classifier

The core of the NFL is the feature line metric. Instead of computing the distance between query sample y and prototype sample y_i^c and calculating the distance between query sample y and prototype sample y_j^c, the NFL classifier calculates the feature line distance between query sample y and the feature line $\overline{y_i^c \, y_j^c}$. The feature line distance between query sample y and feature line $\overline{y_i^c \, y_j^c}$ is defined as

$$d(y, \overline{y_i^c \, y_j^c}) = \| y - y_p^{ij,c} \| \tag{1}$$

where $y_p^{ij,c}$ is the projection point of y on the feature line $\overline{y_i^c \, y_j^c}$, $\|.\|$ means the L_2-norm.

2.2 ENFL Classifier

The ENFL classifier does not calculate the distance between the query sample and the feature line. Instead, the ENFL calculates the product of the distances between query sample and two prototype samples. Then the result is divided by the distance between the two prototype samples. The new distance metric of the ENFL is described as

$$d_{ENFL}(y, \overline{y_i^c \, y_j^c}) = \frac{\| y - y_i^c \| \times \| y - y_j^c \|}{\| y_i^c - y_j^c \|} \tag{2}$$

2.3 CNN Classifier

Be different with NFL classifier, CNN classifier consider another kind line for classification, which is formed by the mean sample o^c of the cth class and one sample chosen randomly in the corresponding class. The metric of CNN is defined as

$$d(y, \overline{y_i^c \, o^c}) = \| y - y_p^{i,c} \| \tag{3}$$

where $y_p^{i,c}$ is the projection of y on the novel feature line.

2.4 SFLS Classifier

The SFLS classifier attempts to find the shortest feature line segment which satisfies the given geometric relation constraints together with the query sample. As shown in Fig. 1 (a), the pair of samples of the sample class forms a feature line segment. If the query sample is inside or on the hyper sphere centered at the midpoint of the feature line segment, the corresponding feature line segment will be tagged and the distance metric of SFLS can be calculated as

$$d_{SFLS}(y, \overline{y_i^c \, y_j^c}) = \| y_i^c - y_j^c \| \tag{4}$$

In the worst case, there is no tagged feature line for a query sample y, and then the SFLS utilizes the rule of the NN to make the classification decision for the query y.

2.5   NFC Classifier

Shown in the Fig. 1 (b), NFC uses the feature center metric, which is defined as the Euclidean distance between query sample y and the feature center $y_o^{ij,c}$, which is $d_{NFC}(y, \overline{y_i^c y_j^c}) = \| y - y_o^{ij,c} \|$, where $y_o^{ij,c}$ is the center of inscribed circle of the triangle $\Delta y y_i^c y_j^c$.

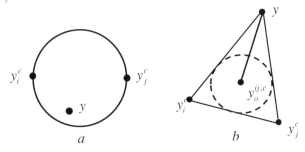

Fig. 1. The metric of SFLS for (a) and the metric of NFC for (b)

3   Proposed Classifier

In this section, the motivation is given firstly in part One. What is more, the proposed classifier, called feature curve metric (FCM) classifier is proposed in part Two. At last, the advantage of FCM is described in part Three.

3.1   Extrapolation Inaccuracy of NFL

In Fig. 2, the query sample y is surrounded by the samples of the sth class, but it is classified to c^{th} class with the decision rule of NFL classifier. The misclassification is called the extrapolation inaccuracy of NFL.

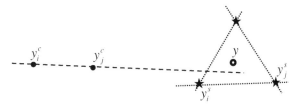

Fig. 2. Extrapolation inaccuracy of NFL

3.2   Novel Classifier

In order to improve the extrapolation inaccuracy of the NFL, feature curve metric (FCM) based on the SFLS and NFC is proposed in this part. The feature curve metric

(FCM) classifier supposes that at least two prototype samples are available for each class, which is analogous to the NFL, ENFL, NNL, CNN, SFLS and NFC. However, the metric is different. A better distance metric, called as feature ellipse metric, is proposed in this subsection. The basic idea is shown in Fig.3.

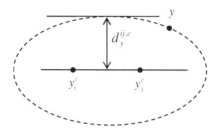

Fig. 3. The metric of FCM

As shown in the Fig. 3, the feature ellipse metric is defined as the length of the minor axis of ellipse. The two focuse points of the ellipse are y_i^c and y_j^c. The query sample point y is on the ellipse. The feature ellipse metric can be computed by formula (5).

$$\mathrm{d}_{NFE}(y, \overline{y_i^c y_j^c}) = \frac{\sqrt{(b_{yi}^c + b_{yj}^c) \times (b_{yi}^c + b_{yj}^c) - b_{ij}^c \times b_{ij}^c}}{2} \tag{5}$$

where $b_{yi}^c = \| y - y_i^c \|$, $b_{yj}^c = \| y - y_j^c \|$ and $b_{ij}^c = \| y_i^c - y_j^c \|$.

The detailed classification process of the FCM is depicted as follows. Firstly, a number of the feature ellipse distances between the test sample y and each pair prototype samples y_i^c and y_j^c are calculated. Secondly, the distances will be sorted with the ascending order and each distance is given with a class label and two prototypes. Then, the FCM distance can be determined as the first rank distance shown in the following formula (6).

$$d(y, \overline{y_{i^*}^{c^*} y_{j^*}^{c^*}}) = \min_{1 \leq c \leq L, 1 \leq i < j \leq N_c} d(y, \overline{y_i^c y_j^c}) \tag{6}$$

The first rank associates the best matched c^*-class and the two best matched prototypes i^* and j^* of the class. The query sample y will be classified into the c^*-class.

3.3   Improve the Exploration Inaccuracy

The FCM can improve the extrapolation inaccuracy in some cases which is explained as follows. From the Fig. 4, the distance between query sample y and the cth class is d_c . The distance between query sample y and the sth class is d_s . Since the distance d_s is shorter than the distance d_c , the query sample y is classified into s^{th} class. So FCM classifier can improve the inaccuracy of NFL in some situation.

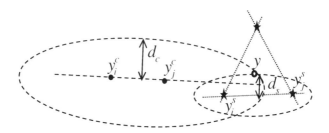

Fig. 4. Improve the exploration inaccuracy of NFL

4   Experimental Result

The classification accuracy of the FCM is contrasted to that of the NN classifier, NFL classifier, ENFL classifier, NNL classifier, CNN classifier, SFLS classifier and NFC classifier. "First N" scheme is taken for comparison: N samples per person are chosen as prototype samples from the database. The rest samples of the database are used for testing. The recognition rate (RR) is used to evaluate the effectiveness of new classifier.

4.1   Hand Gesture Recognition on JTSHP Database

Jochen Triesch Static Hand Posture (JTSHP) Database [13] consists of 10 hand signs performed by 24 persons against three backgrounds. For each person the ten postures were recorded in front of uniform light, uniform dark and complex background. In the experiments, the subset of JTSHP database includes 440 images of 10 hand postures in front of uniform light and uniform dark background. All images are recorded 8-bit grey-scale images of 64×64 pixels.

In the first experiment, the "First N" scheme is used on JTSHP database. The results are shown as Fig. 5. From the Fig. 5, we can know that the recognition rate (RR) of the FCM classifier is superior to the ARR of the other several classifiers, when first 4, 5 and 6 samples are used as prototype set. The average recognition rate (ARR) of the several classifiers with first 4, 5 and 6 samples are shown in Table 1. Compared with the NN classifier, NFL classifier, ENFL classifier, NNL classifier, CNN classifier, SFLS classifier and NFC classifier, the average recognition rate (ARR) of the FCM classifier outperforms the ARRs of them with 15.22%, 4.70%, 5.29%, 7.08%, 3.94%, 12.94%, and 2.32%, respectively.

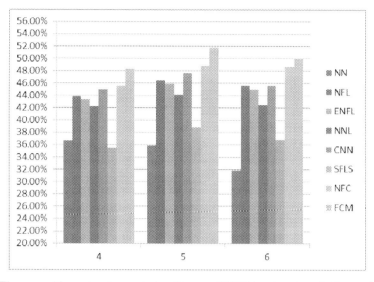

Fig. 5. The recognition rate of several classifiers on JTSHP hand gesture database with 'First N' scheme

Table 1. The ARR of Several Classifiers on JTSHP Hand Gesture Database

Classifie	NN	NFL	ENFL	NNL	CNN	SFLS	NFC	FCM
ARR	34.81	45.33	44.74	42.95	46.09	37.09	47.71	50.03

4.2   Face Recognition on Yale Database

The Yale Face Database [14] contains 165 grayscale images in GIF format of 15 individuals. There are 11 images per subject, one per different facial expression or configuration: center-light, w/glasses, happy, left-light, w/no glasses, normal, right-light, sad, sleepy, surprised, and wink. All images in Yale face database were manually cropped into 25×25 pixels.

In the second experiment, the "First N" scheme is adopted on Yale face database. The results are shown as Fig. 6. From the Fig. 6, we can learn that the recognition rate (RR) of FCM classifier is superior to the RRs of the other several classifiers, when first 4, 5 and 6 samples are used as prototype set. The average recognition rate (ARR) of several classifiers with first 4, 5 and 6 are shown in Table 2. Contrasted to NN classifier, NFL classifier, ENFL classifier, NNL classifier, CNN classifier, SFLS classifier and NFC classifier, the average recognition rate (ARR) of the FCM classifier outperforms the ARRs of them with 2.96%, 1.13%, 2.88%, 0.69%, 1.87%, 3.52%, and 2.64%, respectively.

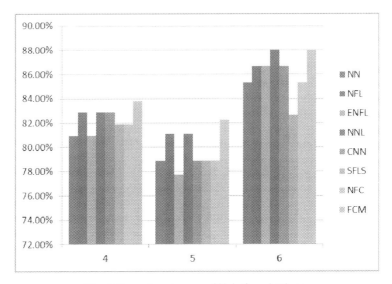

Fig. 6. Some face images of Yale face database.

Table 2. The ARR of Several Classifiers on Yale Face Database

Classifier	NN	NFL	ENFL	NNL	CNN	SFLS	NFC	FCM
ARR (%)	81.72	83.55	81.80	83.99	82.81	81.16	82.04	84.68

4.3   Face Recognition on JAFFE Database

Japanese female facial expression (JAFFE) database [15] contains 213 images of 7 facial expressions (6 basic facial expressions + 1 neutral) posed by 10 Japanese female models. Each image has been rated on 6 emotion adjectives by 60 Japanese subjects. In this experiment, each person chooses 19 images, in total, 190 images are used and are cropped into 32×32 pixels.

In the third experiment, the "First N" scheme is utilized on JAFFE face database. The results are shown as Fig. 7. From the Fig. 7, we can be aware of that the recognition rate (RR) of FCM classifier is superior to the RR of the other several classifiers, when first 4, 5 and 6 samples are used as prototype set. The average recognition rate (ARR) of the several classifiers with the first 4, 5 and 6 are shown in Table 3. Compared with NN classifier, NFL classifier, ENFL classifier, NNL classifier, CNN classifier, SFLS classifier and NFC classifier, the average recognition rate (ARR) of the FCM classifier outperforms the ARRs of them with 4.49%, 6.67%, 7.85%, 9.71%, 4.64%, 7.58%, 4.01%, respectively.

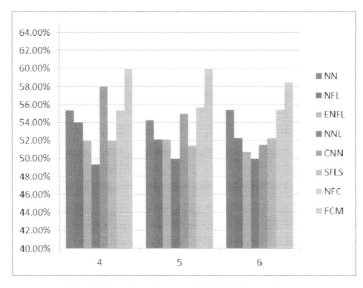

Fig. 7. Some face images of JAFFE face database

Table 3. The ARR of Several Classifiers on JAFFE Face Database

Classifier	NN	NFL	ENFL	NNL	CNN	SFLS	NFC	FCM
ARR (%)	55.00	52.82	51.64	49.78	54.85	51.91	55.48	59.49

5   Conclusion

In this paper, we propose the nearest feature ellipse classifier for hand posture recognition and face recognition based on the nearest feature center classifier and nearest feature line classifier. The proposed classifier takes the advantages of the nearest feature line and suppresses the exploration inaccuracy of the nearest feature line. The novel classifier achieves the better recognition rate than those of the NN, NFL, ENFL, NNL, CNN, SFLS and NFC classifiers. The experimental results confirm the efficiency of the proposed algorithm.

References

1. Cover, T.M., Hart, P.E.: Nearest neighbor pattern classification. IEEE Trans. Inform. Theory 13(1), 21–27 (1967)
2. Li, S.Z., Lu, J.: Face recognition using the nearest feature line method. IEEE Trans. Neural Networks 10(2), 439–443 (1999)
3. Li, S.Z., Chan, K.L., Wang, C.L.: Performance evaluation of the nearest feature line method in image classification and retrieval. IEEE Trans. Pattern Anal. Machine Intell. 22(11), 1335–1339 (2000)
4. Zhou, Z., Kwoh, C.K.: The pattern classification based on the nearest feature midpoints. In: International Conference on Pattern Recognition, vol. 3, pp. 446–449 (2000)

5. Zheng, W., Zhao, L., Zou, C.: Locally nearest neighbor classifiers for pattern classification. Pattern Recognition 37(6), 1307–1309 (2004)
6. Gao, Q.B., Wang, Z.Z.: Center-based nearest neighbor classifier. Pattern Recognition 40(1), 346–349 (2007)
7. Zhou, Y.L., Zhang, C.S., Wang, J.C.: Extended nearest feature line classifier. In: Proc. of 8th Pacific Rim Int. Conf. Artificial Intelligence, Auckland, New Zealand, pp. 183–190 (August 2004)
8. Han, D.Q., Han, C.Z., Yang, Y.: A novel classifier based on shortest feature line segment. Pattern Recognition Letters 32(3), 485–493 (2011)
9. Feng, Q., Pan, J.S., Yan, L.: Nearest feature centre classifier for face recognition. Electronics letters 48(18), 1120–1122 (2012)
10. Feng, Q., Huang, C.-T., Yan, L.: Repsentation-based Nearest Feature Plane for Pattern Recognition. Journal of Information Hiding and Multimedia Signal Processing 4(3), 178–191 (2013)
11. Feng, Q., Pan, J.S., Yan, L.: Restricted Nearest Feature Line with Ellipse for Face Recognition. Journal of Information Hiding and Multimedia Signal Processing 3(3), 297–305 (2012)
12. Feng, Q., Pan, J.S., Yan, L.: Two classifiers based on nearest feature plane for recognition. In: Proc. of IEEE International Conference on Image Processing, Melbourne, Australia, pp. 3316–3319 (September 2013)
13. Triesch, J., Malsburg, C.: Jochen Triesch Static Hand Posture Database (1996), http://www.idiap.ch/resource/gestures/
14. The Yale faces database (2001), http://cvc.yale.edu/projects/yalefaces/yalefaces.html
15. Lyons, M.J., Akamatsu, S., Kamachi, M., Gyoba, J.: Coding Facial Expressions with Gabor Wavelets. In: IEEE International Conference on Automatic Face and Gesture Recognition, pp. 200–205 (April 1998)

Multiscale Slant Discriminant Analysis for Classification

QingLiang Jiao

Heihe University,
Heihe, Heilongjiang, China
hhxy2000@163.com

Abstract. In this paper, a novel feature extraction algorithm in Curvelet domain is proposed. Slant Discriminant Analysis (SDA) is a powerful matrix-based approach for feature extraction. The proposed algorithm aims to extract the most discriminant features of the samples in Curvelet Domain. Compared with several classical algorithms, the efficiency of the proposed algorithm is confirmed by the experimental results.

Keywords: Feature Extraction, Discriminant Analysis, Image Classification.

1 Introduction

Image classification has recently attracted wide attention of the researchers. In face recognition [1], two-dimensional face images samples are usually projected into vectors through row by row or column by column concatenation. The resulting vectors of face samples usually result in a high dimensional space, where it is difficult to evaluate the scatter accurately because of the large size and the small number of the prototype image samples.

Recently, more and more dimensionality reduction algorithms are proposed to deal with this problem. Two of classical dimensionality reduction approaches are principal component analysis (PCA) [2, 3] and linear discriminant analysis (LDA) [4, 5]. PCA is a fundamental feature extraction technique widely applied in computer vision and pattern recognition. PCA is to find a linear projection, which uses maximizing the trace of feature variance to preserve the total variance. The optimal transformation matrix of PCA is corresponding to the first k-largest eigenvalues of the image samples total variance matrix. Thus, PCA can preserve the total variance through maximizing the trace of feature variance. However, PCA cannot preserve the discriminant and local feature due to pursuing maximal variance. LDA aims to find the optimal set of transformation vectors that maximize the determinant of the between-class scatter and at the same time minimize the determinant of the within-class scatter. But, the dimension of sample vectors is high and the number of classes is small, usually tens or hundreds of classes. An intrinsic limitation of classical LDA is that it fails to work

A. Moonis et al. (Eds.): IEA/AIE 2014, Part I, LNAI 8481, pp. 273–281, 2014.
© Springer International Publishing Switzerland 2014

because the within-class scatter matrix will become singular when the classes is less than the dimension of the samples, which is the well-known small sample size (SSS) problems. Many effective approaches have been presented to solve it [6, 7]. What is more, LDA is an interclass and intraclass scatters based algorithm, which is optimal only in cases that the samples of each class is Gaussian distributed approximately, which cannot always be satisfied in image classification tasks.

Compared with traditional PCA, Two dimensional principal component analysis (2DPCA) [8] extracts the features of the image samples directly from image matrices rather than one-dimensional vectors. Therefore, the image matrices do not need to be projected into vectors. 2DPCA extracts the feature through the total image covariance matrix from the original image matrices. The optimal transformation matrix consists of its eigenvectors corresponding to its biggest eigenvalues. As the smaller size of image variance matrix than that of the original variance matrix, 2DPCA needs less time to extract features of the image samples and gets a better recognition rate. The idea of 2DPCA is extended directly, applied to LDA and leads to two-dimensional linear discriminant analysis (2DLDA) [9]. With-class variance matrix and between class variance matrix were constructed for image samples in 2DLDA. Some improved algorithms were also introduced [10].

Besides, many dimensionality reduction approaches based on Nearest Feature Line (NFL) [11] were presented in recent years. Uncorrelated discriminant nearest feature line analysis [12], Neighborhood Discriminant Nearest Feature Line Analysis (NDNFLA) [13], and some other discriminant analysis based on NFL [14] were some popular approaches. These algorithms use feature line metric to evaluate the scatters of the image samples rather than Euclidean distance.

In the last decade, wavelet transformation has been widely applied for classification purposes [15]. The success of wavelets is mainly because of its good performance for piecewise smooth functions in one dimension. However, image classification is not the case. Usually, wavelets are good at extracting zero-dimensional or point singularities. Wavelets in two dimensions are calculated by a tensor-product of one dimensional wavelet and isolating the discontinuity across an edge is its advantage, but the smoothness along the edge will not be perfect. To overcome the weakness of wavelets in two and higher dimensions, Donoho et. al. [16] designed another multiscale transform entitled Curvelet transform which is presented to handle curve discontinuities well.

Dimensionality reduction in transformation domain also becomes a hot research topic. In this paper, a novel image feature extraction method in curvelet domain is proposed.

The paper is organized as follows: Section 2 gives some preliminaries. The new feature extraction algorithm is proposed in Section 3. The various experimental results is shown in Section 4. Lastly, Section 5 concludes the paper.

2   Preliminaries

2.1   Curvelet

Donoho et. al. developed a novel multiscale transform named curvelet transform that was presented to detect edges and other singularities along curves much more efficiently than traditional multiscale transforms, i.e., requiring fewer coefficients for a given accuracy of reconstruction. Calculation of curvelet transform involves the following 4 steps: (a) Sub-band decomposition, (b) Smooth partitioning, (c) Renormalization, (d) Ridgelet transform.

(a) Sub-band Decomposition: The image is first decomposed into log_2^D sub-bands (D is the size of the image) using wavelet and Curvelet Sub-bands are abtained through partial reconstruction from these wavelet sub-bands at levels. So the Curvelet Sub-band, $s = 1$ corresponds to wavelet sub-bands $j = 0, 1, 2, 3$, Curvelet Sub-band, $s = 2$ corresponds to wavelet sub-bands $j = 4, 5$, etc.

(b) Smooth Partitioning: The sub-band $s = 2$ is divided into an serials of 64×64, 50% overlapping sub-blocks, sub-band $s = 3$ is divided into an serials of 32×32, 50% overlapping sub-blocks, etc. Given an image of size of 256×256, there will be 64 64×64 sub-blocks for $s = 2$ and 256 32×32 sub-blocks for $s = 3$.

(c) Renormalization: The partitioning leads to redundancy because a pixel belongs to 4 neighboring sub-blocks. So, each square which is generated in the previous stage should be renormalized to unit scale.

(d) Ridgelet transform [17] is performed on each square resulting from the previous stage.

2.2   Digital line

Consider the digitization of rays [18, 19]

$$D_{r,b} = \{(x, rx + b) : 0 \le x < \infty\} \tag{1}$$

in the set $N^2 = \{(i, j) : i, j \in N\}$ of all points with non-negative integer coordinates in the plane. Assume $0 \le r \le 1$; This is reasonable to the symmetry of the grid. Given a r, such a ray generates a sequence of intersection points corresponding to $D_{r,b}$ with the vertical grid lines. Let $(n_x, n_y) \in N^2$ be the grid point. Define

$$I_{r,b} = \{n\_x, n\_y : n\_x \ge 0 \wedge n\_y = \lfloor rn\_x + b \rfloor\} \tag{2}$$

2.3   2DLDA [9]

Suppose there are c pattern classes. N is the total number of training samples, and N_i is the number of ith class. In the ith class, the jth training image is denoted by X_i^j. $\bar{X}_i$ is the mean matrix of training samples of the ith class. And $\bar{X}$ is the mean matrix of all training samples.

Fig. 1. An example of the slant image [20]

The between-class scatter matrix G_b and within-class scatter matrix G_w can be constructed by

$$G_b = \frac{1}{N} \sum_{i=1}^{c} N_i (\bar{X}_i - \bar{X})^T (\bar{X}_i - \bar{X}) \tag{3}$$

$$G_w = \frac{1}{N} \sum_{i=1}^{c} \sum_{j=1}^{N_i} (X_i^j - \bar{X}_i)^T (X_i^j - \bar{X}_i) \tag{4}$$

By 2DLDA, an image matrix $T \in R^{m \times n}$ will be projected to the space $T \in R^{m \times p}$ with $p \ll n$. The criterion to select the most discriminant features can be defined by

$$\max J(u) = (u^T G_b u)/(u^T G_w u) \tag{5}$$

In fact, the optimal discriminant feature vectors $U = (u_1, u_2, \cdots, u_p)$ is the set of generalized eigenvectors of G_b and G_w corresponding to the p largest eigenvalues, and $G_b u_i = \lambda_i G_w u_i$, where $\lambda_1 \geq \lambda_2 \geq \cdots \geq \lambda_p$. U is called the optimal image projection matrix. Let

$$F = TU \tag{6}$$

The resulting matrix F is called the feature matrix of image T and used to represent T for classification.

2.4   SDA [20]

Given a r_0, all points in $I_{r_0,b}$ constitute a digital line with slope r. Then an image can be decomposed by such parallel digital lines. The main idea of this paper is performing 2DLDA in these digital line to extract the slant feature of images. So the proposed SDA aims to preserve the directional feature of the samples. Given a r_0, $I_{r_0,0}$ is the digital line through $[0,0]$ of the original image. Then choose the available b, we can get a set of digital lines, which is a partition of the plain. Then an image can be reshaped by the similar method in [?]. Fig. 1 show an example of the slant image. The left is the original image. The middle is the slant image with slope 0.3 and the right is the slant image with slope 0.8.

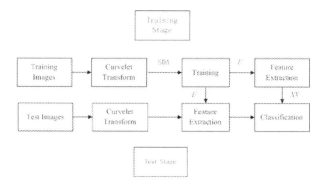

Fig. 2. The Procedure of the Proposed Algorithm

The procedure of SDA

Training stage:

Step 1, reshape the prototype images to slant images;

Step 2, apply 2DLDA to find the optimal transformation matrix W on the new prototype samples set;

Step 3, extract the features of reduced prototype samples using Eq. (6).

Classification stage:

Step 1, reshape the prototype images to slant images;

Step 2, extract the features of query following Eq. (6);

Step 3, classify.

3   The Proposed Algorithm

In this section, we will introduce a novel feature extraction algorithm in Curvelet Domain. The detailed procedure of the proposed algorithm is given by Fig. 2 as follows:

Step 1, Perform Curvelet transform on prototype samples.

Step 2, Transform SDA on the resulting data in the last step to find the optimal transformation matrix W.

Step 4, Extract the feature of the prototype samples using W.

Given a query sample, the same Curvelet transform is performed and then we use the optimal transformation matric W to extract its feature. At last, the feature will be used for classification instead of the original query sample.

4   Experimental Results

4.1   Experimental Results on COIL20 Database

COIL20 database [21] is the one of most popular image databases. There are 1400 image samples of 20 individuals in the COIL20 database. Each individual has 72 image samples with the size of 200×200. To reduce the computation complexity, all the image samples in COIL database are cropped to 32×32. During this experiment, 10 images per class are selected randomly for training and the others are for test. Nearest Neighbor classifier is applied for classification. The system ran ten times in order to reduce the variation on each database. Therefore, the average recognition rate (ARR) is used to evaluate the performance of different algorithms. In the experiments, the performance of the proposed algorithm is compared with those of PCA, LDA, 2DPCA, 2DLDA and SDA. Label 1 gives the ARR of different algorithm on COIL. From the label, we can get the proposed algorithm has the higher ARR than the other classical algorithms.

Fig. 3. Some image samples in COIL database [21]

Table 1. ARR of different algorithms on COIL20 database

Approaches	ARR	Feature dimension
PCA	0.6521	75
LDA	0.6972	19
2DPCA	0.6742	30*32
2DLDA	0.7568	19*32
SDA	0.7714	19*32
Proposed algorithm	0.7926	19*32

4.2   Experimental Results on Stanford Gender Database

The Stanford Gender Database [22] consists of 200 male and 200 female images. The images were all 200×200 pixels, black and white photos, that were taken using a frontal-view. Some examples of images in the database are shown in Fig. 4. To reduce the computation complexity, all the image samples in COIL database are cropped to 64×64. During this experiment, 100 images per class are selected randomly for training and the others are for test. Nearest Neighbor classifier is applied for classification. The system ran ten times in order to reduce the variation on each database. Therefore, the average recognition rate (ARR) is used to evaluate the performance of different algorithms. In the experiments, the performance of the proposed algorithm is compared with those of PCA, LDA, 2DPCA, 2DLDA and SDA. Label 2 gives the ARR of different algorithm on COIL. From the label, we can get the proposed algorithm has the higher ARR than the other classical algorithms.

Fig. 4. Some image samples in Gender database [22]

Table 2. ARR of different algorithms on Gender database

Approaches	ARR	Feature dimension
PCA	0.7016	120
LDA	0.7262	1
2DPCA	0.7135	25*64
2DLDA	0.7328	1*64
SDA	0.7594	1*64
Proposed algorithm	0.7741	1*64

5   Conclusion

This paper has proposed a new feature extraction algorithm in Curvelets domain. The proposed algorithm performs Slant Discriminant Analysis in the Curvelet domain. Different SDA, the proposed algorithm extracts the multiscale and multidirection features of the samples. The experimental results on COIL20 database and Gender database confirm the efficiency of the proposed algorithm.

References

[1] Li, J.B., Chu, S.C., Pan, J.S., Jain, L.C.: Multiple viewpoints based overview for face recognition. Journal of Information Hiding and Multimedia Signal Processing 3(4), 352–369 (2012)

[2] Abdi, H., Williams, L.J.: Principal component analysis. Wiley Interdisciplinary Reviews: Computational Statistics 2(4), 433–459 (2010)

[3] Jolliffe, I.: Principal component analysis. In: Lovric, M. (ed.) International Encyclopedia of Statistical Science, pp. 1094–1096. Springer, Heidelberg (2011)

[4] Zhao, H., Yuen, P.C.: Incremental linear discriminant analysis for face recognition. IEEE Transactions on Systems, Man, and Cybernetics, Part B: Cybernetics 38(1), 210–221 (2008)

[5] Erenguc, S.S., Koehler, G.J.: Survey of mathematical programming models and experimental results for linear discriminant analysis. Managerial and Decision Economics 11(4), 215–225 (1990)

[6] Bandos, T., Bruzzone, L., Camps-Valls, G.: Classification of hyperspectral images with regularized linear discriminant analysis. IEEE Transactions on Geoscience and Remote Sensing 47(3), 862–873 (2009)

[7] Regularization studies of linear discriminant analysis in small sample size scenarios with application to face recognition. Pattern Recognition Letters 26(2), 181–191 (2005)

[8] Yang, J., Zhang, D., Frangi, A., Yang, J.: Two-dimensional pca: a new approach to appearance-based face representation and recognition. IEEE Trans. Pattern Analysis and Machine Intelligence 26(1), 131–137 (2004)

[9] Yang, J., Zhang, D., Yong, X., Yang, J.: Two-dimensional discriminant transform for face recognition. Pattern Recognition 37(7), 1125–1129 (2005)

[10] Yan, L., Pan, J.S., Chu, S.C., Muhammad, K.K.: Adaptively weighted subdirectional two-dimensional linear discriminant analysis for face recognition. Future Generation Computer Systems 28(1), 232–235 (2012)

[11] Li, S., Lu, J.: Face recognition using the nearest feature line method. IEEE Transactions on Neural Networks 10(2), 439–443 (1999)

[12] Lu, J., Tan, Y.P.: Uncorrelated discriminant nearest feature line analysis for face recognition. IEEE Signal Processing Letters 17(2), 185–188 (2010)

[13] Yan, L., Zheng, W., Chu, S., Roddick, J.: Neighborhood discriminant nearest feature line analysis for face recognition. Journal of Internet Technology 14(1), 344–347 (2013)

[14] Yan, L., Wang, C., Chu, S.-C., Pan, J.-S.: Discriminant analysis based on nearest feature line. In: Yang, J., Fang, F., Sun, C. (eds.) IScIDE 2012. LNCS, vol. 7751, pp. 356–363. Springer, Heidelberg (2013)

[15] Arivazhagan, S., Ganesan, L.: Texture classification using wavelet transform. Pattern Recognition Letters 24(9-10), 1513–1521 (2003)

[16] Donoho, D.L., Duncan, M.R.: Digital curvelet transform: strategy, implementation, and experiments (2000)

[17] Ridgelets, J.C.: theory and applications. Ph. d, Stanford University (1998)

[18] Rosenfeld, A., Klette, R.: Digital straightness. Electronic Notes in Theoretical Computer Science 46, 1–32 (2001)

[19] Chattopadhyay, S., Das, P.: A new method of analysis for discrete straight lines. Pattern Recognition Letters 12(12), 747–755 (1991)

[20] Zhao, L.Y., Zou, D., Gao, G.: Slant discriminant analysis for image feature extraction. In: Proceedings of IIHMSP 2013. IEEE (2013)

[21] Nene, S.A., Nayar, S.K., Murase, H.: Columbia object image library (coil-20). In Technical Report CUCS-005-96 (1996)

[22] Valentin, D., Abdi, H., Edelman, B., O'Toole, A.J.: Principal component and neural network analyses of face images: What can be generalized in gender classification? Journal of Mathematical Psychology 41(4), 398–413 (1997)

A Novel Feature Extraction Algorithm in Shearlets Domain

Dan Su

Heihe University,
Heihe, Heilongjiang, China
sudan1108@163.com

Abstract. In this paper, a novel feature extraction algorithm in Shearlet domain is proposed. Neighborhood Discriminant Nearest Feature Line Analysis (NDNFLA) is a powerful tool for feature extraction. To enrich the research of NDNFLA, we design a approach to perform NDNFLA in Shearlet domain. The experimental results demonstrates the efficiency of the proposed algorithm.

Keywords: Subspace learning, Nearest Feature Line, Image Classification.

1 Introduction

In the recent years, image classification [1], including palmprint recognition [2], face recognition [3], and so on, becomes more and more hot for researchers. Many statistical machine learning problems include some issues of dimensionality reduction either implicitly or explicitly. Feature selection (to select a subset of discriminant features) and feature transformation (to find the linearnonlinear combination of the original sample set of features and select a few transformed features) are two classical ways for dimensionality reduction. The dimensionality reduction is to map the samples in a high dimensional space to a low dimensional space while preserving some intrinsic or discriminant features of the original samples. Principal Component Analysis (PCA)[4, 5], Linear Discriminant Analysis (LDA)[6, 7], Local Preserving Projection, (LPP) [8, 9] are several most popular subspace-based dimensionality reduction approaches. PCA is a well-known dimensionality reduction algorithm widely applied in the signal processing, computer vision, and pattern recognition and so on. In PCA, the original samples will be mapped to a low dimensional subspace which restores the largest possible variance of the original samples. PCA is order to minimize the mean squared error. However, PCA is an unsupervised algorithm, which may reduce the efficiency of feature extraction. LDA can calculate an optimal discriminant projection by maximizing the ratio of the trace of the between-class scatter matrix to that of the within-class scatter matrix. LDA uses the labels of the prototype samples during the training and improves the discriminant ability. However, LDA has to suffer from the well-known small sample size (SSS) problem. Many effective

A. Moonis et al. (Eds.): IEA/AIE 2014, Part I, LNAI 8481, pp. 282–289, 2014.

algorithms have been introduced to solve the problem [10, 11]. LPP, also known as Laplacianfaces, aims to search for an embedding project that preserves the local features and get a face subspace that best preserves the essential structure of the face manifold. If the prototype samples are not sufficient enough and data dimension is too high, especially for image samples, LPP will face the singularity of matrices and can not be applied directly. Some nonlinear extensions using kernel trick of these algorithms are proposed in the recent 10 years[12, 13, 14]. Besides, there are some works on feature fusion presented[15] to improve the performance of feature extraction.

The algorithm mentioned above need perform on the vector sample data. It means one has to use these algorithms after transforming the image samples to vectors. This results in a big computation complexity and needs big storage space. To avoid this problem, some matrix-based approaches were proposed. Two-dimensional Principal Component Analysis (2DPCA) [16] and two-dimensional linear discriminant analysis (2DLDA) are two most popular algorithms which can perform on the image matrix data directly. Some modified algorithms were also presented [17].

Besides, some subspace learning algorithms based on Nearest Feature Line (NFL) [18] were proposed in recent years. Yan201302Nearest feature line non-parametric discriminant analysis [19], Neighborhood Discriminant Nearest Feature Line Analysis (NDNFLA) [20], and some other discriminant analysis based on NFL [21] were some popular approaches. In these algorithms, feature line metric is used to evaluate the scatter of samples instead of Euclidean distance. Authors claimed the efficiency of these algorithms.

Dimensionality reduction in transformation domain is also a hot research topic. Shearlets transformation [22, 23] was proposed by K. Guo et. al. and is a powerful multi-dimensional signal processing tool. This paper proposes a feature extraction method in Shearlets domain.

The paper is organized as follows: Section 2 reviews the some preliminaries. The new feature extraction algorithm is proposed in Section 3 with the various experimental results in Section 4. Lastly, Section 5 concludes the paper.

2   Preliminaries

2.1   Continuous Shearlet

The continuous shearlet is a parabolic scaling matrices,shear matrices and translations based systems. The parabolic scaling matrices A_a is the means the deal with the solution, where

$$A_a = \begin{bmatrix} a & 0 \\ 0 & a^{1/2} \end{bmatrix}, a > 0. \tag{1}$$

Shear matrices is the means to handle the orientation, where

$$S_s = \begin{bmatrix} 1 & s \\ 0 & 1 \end{bmatrix}, s \in R. \tag{2}$$

The translations is to change the positioning. Compared to Curvelets, Shearlets apply shearings rather than rotations. The shear operator S_s results in the integer lattice invariant if $s \in Z$. It leads to a unified treatment of digital realm and continuum, so giving a powerful digital implementation.

For $\psi \in L^2(R^2)$, the continuous shearlet system gotten by ψ is as follows:

$$SH_{cont}(\psi) = \{\psi_{a,s,t} = a^{3/4}\psi(S_s A_a(\cdot - t))|a > 0, s \in R, t \in R^2\}, \qquad (3)$$

then, the continuous shearlet transformation is as follows:

$$f \to SH_\psi f(a, s, t) = \langle f, \psi_{a,s,t}\rangle \, f \in L^2(R^2), (a, s, t) \in R^+ \times R \times R^2\}, \qquad (4)$$

where $R^+ = \{a > 0, \text{ and } a \in R\}$.

2.2   Discrete Shearlet

Discretizing the parameters a, s, t in $SH_{cont}(\psi)$, the discrete shearlets can be obtained. One of most famous approaches for discretization of parameters is as follows:

$$\{(2^j, k, A_{2j}^{-1} S_k^{-1} m)|j \in Z, k \in Z, m \in Z^2\} \subseteq R^+ \times R \times R^2. \qquad (5)$$

Then, the discrete shearlet system associated with ψ is as follows:

$$SH(\psi) = \{\psi_{j,k,m} = 2^{3j/4}\psi(S_k A_{2j} \cdot -m)|j \in Z, k \in Z, m \in Z^2\}, \qquad (6)$$

and the corresponding discrete shearlet transformation is as follows:

$$f \to SH_\psi f(a, s, t) = \langle f, \psi_{a,s,t}\rangle, f \in L^2(R^2), j \in Z, k \in Z, m \in Z^2. \qquad (7)$$

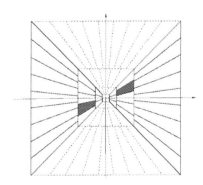

Fig. 1. A example of frequency partition of Shearlet

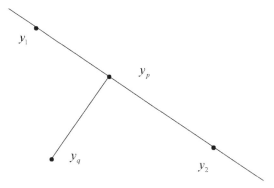

Fig. 2. NFL metric

2.3 Nearest Feature Line

Nearest feature line is a classifier. It is first presented by Stan Z. Li and Juwei Lu. Given a training samples set, $X = \{x_n \in R^M : n = 1, 2, \cdots, N\}$, denote the class label of x_i by $l(x_i)$, the training samples sharing the same class label with x_i by $P(i)$, and the training samples with different label with x_i by $R(i)$. NFL generalizes each pair of prototype feature points belonging to the same class: $\{x_m, x_n\}$ by a linear function $L_{m,n}$, which is called the feature line. The line $L_{m,n}$ is expressed by the span $L_{m,n} = sp(x_m, x_n)$. The query x_i is projected onto $L_{m,n}$ as a point $x^i_{m.n}$. This projection can be computed as

$$x^i_{m,n} = x_m + t(x_n - x_m) \tag{8}$$

where $t = [(x_i - x_n)(x_m - x_n)]/[(x_m - x_n)^T(x_m - x_n)]$.

The Euclidean distance of x_i and $x^i_{m,n}$ is termed as FL distance. The less the FL distance is, the more probability that x_i belongs to the same class as x_m and x_n. Fig. 2 shows a sample of FL distance. In Fig. 2, the distance between y_p and the feature line $L_{1,2}$ equals to the distance between y_q and y_p, where y_p is the projection point of y_q to the feature line $L_{1,2}$.

2.4 NDNFLA [?]

Let's introduce two definitions firstly.

Definition 1 Homogeneous neighborhoods: For a sample x_i, its k nearest homogeneous neighborhood N^o_i is the set of k most similar data which are in the same class with x_i.

Definition 2 Heterogeneous neighborhoods: For a sample x_i, its k nearest Heterogeneous neighborhoods N^e_i is the set of k most similar data which are not in the same class with x_i.

In NDNFLA approach, the optimization problem is as follows:

$$\max J(W) = (\sum_{i=1}^{N} \frac{1}{NC_i^2 |N_i^c|} \sum_{x_m,x_n \in N_i^c} \left\| W^T x_i - W^T x_{m,n}^i \right\|^2$$
$$- \sum_{i=1}^{N} \frac{1}{NC_i^2 |N_i^o|} \sum_{x_m,x_n \in N_i^o} \left\| W^T x_i - W^T x_{m,n}^i \right\|^2) \tag{9}$$

Using matrix computation, the problem becomes

$$\max J(W) = \text{tr}[W^T(A-B)W] \tag{10}$$

where

$$A = \sum_{i=1}^{N} \frac{1}{NC_i^2 |N_i^c|} \sum_{x_m,x_n \in N_i^c} [(x_i - x_{m,n}^i)(x_i - x_{m,n}^i)^T] \tag{11}$$

$$B = \sum_{i=1}^{N} \frac{1}{NC_i^2 |N_i^o|} \sum_{x_m,x_n \in N_i^o} [(x_i - x_{m,n}^i)(x_i - x_{m,n}^i)^T] \tag{12}$$

A length constraint $w^T w = 1$ is imposed on the proposed NDNFLA. Then, the optimal projection W of NDNFLA can be obtained by solving the following eigenvalue problem.

$$(A-B)w = \lambda w \tag{13}$$

Let $w_1, w_2, \cdots, w_q$ be the eigenvectors of formula(13) corresponding to the q largest eigenvalues ordered according to $\lambda_1 \geq \lambda_2 \geq \cdots \geq \lambda_q$. An $M \times q$ transformation matrix $W = [w_1, w_2, \cdots, w_q]$ can be obtained to project each sample $M \times 1$ x_i into a feature vector $q \times 1$ y_i as follows:

$$y_i = W^T x_i, \qquad i = 1, 2, \cdots, N \tag{14}$$

3   The Proposed Algorithm

In this paper, a novel feature extraction algorithm is proposed. The Fig. 3 gives the detailed procedure of the novel algorithm as follows:

Step 1, Perform Shearlets transform on prototype samples.
Step 2, Transform the samples to vectors.
Step 3, Find the optimal transformation matrix W using the NDNFLA.
Step 4, Extract the feature using W.

For query samples, the same Shearlets transform are used and its feature can be extracted by the optimal transformation matrix W. Then, the features will be used for classification.

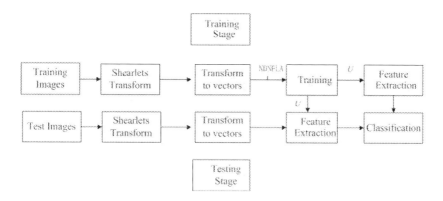

Fig. 3. The Procedure of the Proposed Algorithm

4   Experimental Results

There are various open databases so that different algorithms can compare with each other on the same platform. In this experiment, COIL20 database [24] is the one of most popular image databases. During the experiment, the system ran ten times in order to reduce the variation on each database. Therefore, the average recognition rate (ARR) is used to evaluate the different algorithms. In the experiments, the performance of the proposed algorithm is compared with those of PCA, LDA, NDNFLA, and UDNFLA. For each database, half samples per subject were selected as prototype samples randomly and the others were used as query samples in each run. To evaluate the different feature extraction algorithms, Nearest Neighbor classifier is applied for classification.

There are 1400 image samples of 20 individuals in the COIL20 database [24]. Each individual has 72 image samples with the size of 200×200. To reduce the computation complexity, all the image samples in COIL database are cropped to 48×48.

Table 1. ARR of different algorithms on COIL20 database

Approaches	ARR	Feature dimension
PCA	0.681	80
LDA	0.773	19
NDNFLA	0.792	70
UDNFLA	0.785	90
Proposed algorithm	0.824	95

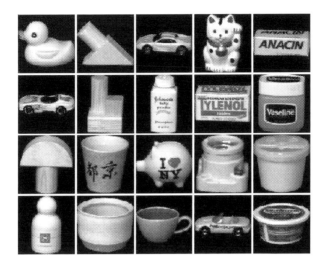

Fig. 4. Some image samples in COIL database

5   Conclusion

This paper has proposed a new feature extraction algorithm in Shearlets domain. The proposed algorithm is an application of Neighborhood Discriminant Nearest Feature Line Analysis (NDNFLA). Different NDNFLA, the proposed algorithm extracts the features of the samples in the Shearlets Domain rather than time domain. The experiments on COIL20 database confirms the efficiency of the proposed algorithm.

References

[1] Lu, D., Weng, Q.: A survey of image classification methods and techniques for improving classification performance. International Journal of Remote Sensing 28(5), 823–870 (2007)

[2] Lu, G., Zhang, D., Wang, K.: Palmprint recognition using eigenpalms features. Pattern Recognition Letters 24(9-10), 1463–1467 (2003)

[3] Li, J.B., Chu, S.C., Pan, J.S., Jain, L.C.: Multiple viewpoints based overview for face recognition. Journal of Information Hiding and Multimedia Signal Processing 3(4), 352–369 (2012)

[4] Abdi, H., Williams, L.J.: Principal component analysis. Wiley Interdisciplinary Reviews: Computational Statistics 2(4), 433–459 (2010)

[5] Ding, C., He, X.: K-means clustering via principal component analysis. In: Proceedings of the Twenty-first International Conference on Machine Learning, pp. 29–36. ACM, New York (2004)

[6] Haeb-Umbach, R., Ney, H.: Linear discriminant analysis for improved large vocabulary continuous speech recognition. In: 1992 IEEE International Conference on Acoustics, Speech, and Signal Processing, ICASSP 1992, vol. 1, pp. 13–16 (March 1992)

[7] In:

[8] He, X., Yan, S., Hu, Y., Niyogi, P., Zhang, H.: Face recognition using laplacianface. IEEE Transactions on Pattern Analysis Machine and Intelligence 27(3), 328–340 (2005)

[9] Xu, Y., Song, F., Feng, G., Zhao, Y.: A novel local preserving projection scheme for use with face recognition. Expert Systems with Applications 37(9), 6718–6721 (2010)

[10] Bandos, T., Bruzzone, L., Camps-Valls, G.: Classification of hyperspectral images with regularized linear discriminant analysis. IEEE Transactions on Geoscience and Remote Sensing 47(3), 862–873 (2009)

[11] Regularization studies of linear discriminant analysis in small sample size scenarios with application to face recognition. Pattern Recognition Letters 26(2), 181–191 (2005)

[12] Xu, Y., Zhang, D., Jin, Z., Li, M., Yang, J.: A fast kernel-based nonlinear discriminant analysis for multi-class problems. Pattern Recognition 39(6), 1026–1033 (2006)

[13] Li, J., Gao, H.: Sparse data-dependent kernel principal component analysis based on least squares support vector machine for feature extraction and recognition. Neural Computing and Applications 21(8), 1971–1980 (2012)

[14] Li, J., Pan, J., Chu, S.: Kernel class-wise locality preserving projection. Information Sciences 178(7), 1825–1835 (2008)

[15] Xu, Y., Zhang, D.: Represent and fuse bimodal biometric images at the feature fevel: Complex-matrix-based fusion scheme. Optical Engineering 49(3) (March 2010) 037002–037002-6

[16] Yang, J., Zhang, D., Frangi, A., Yang, J.: Two-dimensional pca: a new approach to appearance-based face representation and recognition. IEEE Trans. Pattern Analysis and Machine Intelligence 26(1), 131–137 (2004)

[17] Yan, L., Pan, J.S., Chu, S.C., Muhammad, K.K.: Adaptively weighted subdirectional two-dimensional linear discriminant analysis for face recognition. Future Generation Computer Systems 28(1), 232–235 (2012)

[18] Li, S., Lu, J.: Face recognition using the nearest feature line method. IEEE Transactions on Neural Networks 10(2), 439–443 (1999)

[19] Zheng, Y., Yang, J., Yang, J., Wu, X., Jin, Z.: Nearest neighbour line nonparametric discriminant analysis for feature extraction. Electron. Lett. 42(12), 679–680 (2006)

[20] Yan, L., Zheng, W., Chu, S., Roddick, J.: Neighborhood discriminant nearest feature line analysis for face recognition. Journal of Internet Technology 14(1), 344–347 (2013)

[21] Yan, L., Wang, C., Chu, S.-C., Pan, J.-S.: Discriminant analysis based on nearest feature line. In: Yang, J., Fang, F., Sun, C. (eds.) IScIDE 2012. LNCS, vol. 7751, pp. 356–363. Springer, Heidelberg (2013)

[22] Guo, K., Kutyniok, G., Labate, D.: Sparse multidimensional representations using anisotropic dilation and shear operators. In: Proceedings of the International Conference on the Interactions between Wavelets and Splines, Athens, GA, vol. 8055, pp. 16–19 (2005)

[23] Guo, K., Labate, D.: Optimally sparse multidimensional representation using shearlets. SIAM J. Math. Anal. 39, 298–318 (2007)

[24] Nene, S.A., Nayar, S.K., Murase, H.: Columbia object image library (coil-20). In Technical Report CUCS-005-96 (1996)

Prediction of Team Performance and Members' Interaction: A Study Using Neural Network

Feng-Sueng Yang[1] and Chen-Huei Chou[2]

[1] Aletheia University, New Taipei City, Taiwan, R.O.C.
`fsyang@mail.au.edu.tw`
[2] College of Charleston, Charleston, SC, USA
`chouc@cofc.edu`

Abstract. Team based organization has become more widespread because of rapidly changing environment. The way to form a team and therefore improve team performance has become an important topic. Limited amount of studies have focused on forecasting team performance based on individual behaviors. If one's future team behavior and performance can be identified, it is beneficial for the process of human resource planning, recruitment, and training. This study aims to use data mining approach on the prediction of team performance and members' interaction based on individual personality traits. Using undergraduates' Big Five personality traits, we established neural network models in order to predict their team performance and members' interaction. We found that neural network can effectively predict team performance with the accuracy rate over 99%. It also correctly predicts members' interaction in more than 72% of cases. Our findings potentially contribute to the development of human resource management academically and practically.

Keywords: personality traits, team performance, team members' interaction, neural networks, forecasting.

1 Introduction

Due to rapid changes and complexity in the environment as well as the advent of the knowledge economy, the formation of a team and the ways to improve team performance have become important and popular topics for recent studies [10], [11]. It is found that not all teams perform well and achieve high performance [40], [43]. Previous studies found that team performance is affected by many factors such as leadership [27], team structure [49], characteristics of team members [7], and team composition [18]. Many of these studies also found that the team performance is influenced by the progress of teamwork and interaction of team members [14], [49]. Particularly, cooperation, communication, trust, and leadership are major factors affecting team performance [1], [5], [31].

Team performance has been well studied through observation, survey, or explanation in prior empirical studies [7], [18], [49]. However, there are limited amount of studies forecasting team performance based on individual behaviors. The

A. Moonis et al. (Eds.): IEA/AIE 2014, Part I, LNAI 8481, pp. 290–300, 2014.
© Springer International Publishing Switzerland 2014

findings of such type of studies would benefit human resource management area by predicting team performance based on one's individual behaviors. It is also beneficial for the process of human resource planning, recruitment, hiring, performance evaluation, training, etc. Therefore, this study attempts to fill in the gap in examining a variety of individual characteristics, which may influence team performance, for the prediction of team performance using data mining approach.

Student team project is one of the most common pedagogy used in higher education. It utilizes the learning by doing methodology which allows students to apply the learned knowledge practically into the project with the collaboration with other teammates. However, some common complaints about unsuccessful student team projects include poor interactions, poor communications, free riding, unfair loading, etc. These issues negatively affect team performance and neglect the synergy of individual behaviors. Student team project is one example of real world teamwork. It is also analogous to other team collaborations in various settings. Although the rigor, complexity, and richness of student team projects are far less than those in enterprise team projects, it truly reflects the details and phenomena of team collaboration. Therefore, this study sets the domain of team collaboration on university team projects and utilizes Artificial Neural Network to predict team performance based on individual member's personality traits.

The rest of the paper is organized as follows. First, we review the personality traits, team performance and members' interaction, and the use of personality traits in human resource management. Next, we provide the methodology and experimental design of the use of neural network for the prediction of team performance and team members' interaction using personality traits. Finally, experimental findings are reported and discussed.

2 Background

2.1 Personality Traits

"Personality" is originated from the Latin word "persona" which means "mask" wore in ancient Greek drama. Different mask represents its drama role and also implicitly represents its personality. Therefore, personality also represents individual's true self, which includes his or her intrinsic motivation, emotions, habits, thoughts, etc. Personality has been well studied in psychology, behavioral science, cognitive science and many other fields.

However, psychologists do not have consensus for the definition of personality. The most commonly used definition is that "the dynamic organization within the individual of those psychophysical systems that determine his unique adjustment to his environment" [4] . In 1961, Allport [3] redefined that "personality is the dynamic organization within the individual of those psychophysical systems that determine his characteristic behavior and thought" (p.28). In addition, Guilford [24] stated that personality is unique pattern of traits, where a trait is "any distinguishable, relatively enduring way in which one individual differs from others."   David and Stanley [17] argued that personality is a set of traits that can be used to determine the differences and similarities from one individual and another.

The personality was studied and explained by many different theories including psychoanalysis theory, phenomenology, cognitive theory, trait theory, behavioral theory, and social cognitive theory, etc. Although the emphasis of these theories is different, they deal with different aspects of personality. Trait theory is one of the theories focusing on the composition and structure of personality. It is argued that personality is composed of a series of personality traits which are building blocks to construct one's personality and form one's behavior [42].

Also, it is believed that personality is the overall image provided by personality traits. It cannot be described by a single trait. The traits are the unique characteristics which influence one's interaction with the environment. One's characteristic behavior and thought are determined by individual's psychophysical systems [3]. A trait is one of the behavioral facets such as sociality and independency, used to represent consistent behavior in a certain phenomenon. Traits reflect one's unique personality characteristics such as shyness, aggressiveness, submissiveness, laziness, loyalty, fear, and so on. If the characteristics continue to appear in different phenomena, these are one's personality traits. Therefore, personality traits are stable and important components in one's life [15].

Cattell [13] identified that personality can be determined by 16 personality traits. Norman [37], [38] further validated and grouped the traits into five factors. Such Big Five Personality Traits method was proved to be stable in evaluating one's personality [22], [41]. The five personality traits first appeared in the Lexical Hypothesis [20] which used more than a thousand personality characteristic words to interpret personality. With the reference of Webster New International Dictionary, Allport and Odbert [2] extended the set to include more than seventeen thousand words which were related to one's personality and behavior. Cattell [12] further condensed the set into 171 words. In order to measure personality, Cattell also prepared a corresponding questionnaire called 16PF with a focus on 16 factors including warmth, reasoning, emotional stability, dominance, liveliness, rule-consciousness, social boldness, sensitivity, vigilance, abstractedness, privateness, apprehension, openness to change, self-reliance, perfectionism, and tension. Moreover, five factors were formed in the studies [19], [51], derived from Cattell [12]. The five factors identified by Tupes and Christal [51] were surgency, agreeableness, dependability, emotional stability, and culture. Goldberg [22] suggested the use of five-factor approach for studies investigating personality. Finally, the Five Factor Model theory proposed by Costa and McCrae [16] is commonly used. The constituent traits in the five factors include openness to experience, conscientiousness, extraversion, agreeableness, and neuroticism. The Big Five Personality Traits method was successfully used to investigate work performance [6], [47], [50] and team performance [29], [30].

2.2   Team Performance and Members' Interaction

Knowledge acquisition has been far more complex due to the growth of available knowledge. It becomes harder for one individual to independently and effectively gain knowledge as well. Group learning, with group members' individual abilities and characteristics, may benefit the knowledge acquisition process. A team consists of a group of interdependent individuals who have complementary skills for a common goal [28]. These individuals can bring in different skills and knowledge, and share the loading of projects. A team is a temporary organization in which members, with the responsibility for the success or failure, accomplish tasks and solve problems together [33]. The definition of "team" varies due to researchers' different aspects. Most of them agreed that a team needs to meet the following conditions. First, a team includes at least two members. Second, a team relies on members' interdependence and coordination to complete the work. Third, members work together in order to accomplish the common goal.

In general, teamwork can generate a positive synergy to outperform a task completed by one individual [44]. However, there is no guarantee of high performance due to the teamwork. There is a need for good cooperation and interaction among the members. The team performance may be influenced by team members' ability to work, group size, level of conflict with others, group norms, organizational strategies, reward system, role perception, status, resources, team cohesion, and so on [23], [26], [31].

Team performance can be measured at group level or individual level. At group level, it is common to measure the performance output [32], team growth [25], team satisfaction [21], quality, time, and error rate [46], task performance, and customer satisfaction. In addition, Brown and Eisenhardt [9] measured the product development team performance by process performance (e.g. development speed, productivity) as well as financial performance (e.g. profits, sales, market share). At individual level, it is usual to measure the individual performance [36], member's growth [25], team commitment, satisfaction on cooperation, and satisfaction on member's contribution.

Prior studies also investigated the relationship between the combination of team members' personality traits and team performance [6], [7], [29] [35]. Barrick and Mount [6], [50] investigated the connection between individual personality traits and job performance. Kichuk and Wiesner [29] used Big Five personality traits to examine the influence to team performance of product design.

2.3   The Use of Personality Traits in Human Resource Management

Human resource management normally deals with individual's current and future performance through the process of selection, promotion, training, and evaluation. In terms of selection, capability and personality traits are the indicators for future job performance [39]. In order to understand the individual personality traits, interviews or personality tests are common approaches used. Cattell [12] suggested that personality can be understood by observing.

Therefore, employers often use interviews to observe applicant's personality. However, the validity for the use of interviews to predict job performance is between 0.3 and 0.6. Also, sometimes interviews may cause subjective bias issues such as interviewer's expectancy confirmation behavior [8] and therefore affect the fairness of the selection. As a result, employers may rather use a systematic approach—

personality test—to understand applicant's personality traits. The popular personality scales include NEO Personality Inventory [15], California Psychological Inventory (CPI), Personality Characteristics Inventory (PCE), Myers-Briggs Type Indicator (MBTI), and so on. Nevertheless, it was found that the average validity of personality tests was only 0.206, based on a meta-analysis of eight studies between 1962 and 1984 by Neal Schmitt et. al [34]. Thus, the use of these personality tests is controversial.

Although personality test is not the most suitable tool for selection, it was found that appropriate collection of personality data would aid the decision making on selection [45]. Due to the popularity of Big Five Model, these personality tests have been gradually adopted by recent researchers. Because of the advances of information systems and intelligent computing, personality analysis can be carried out quickly by computers and data can be visualized graphically. The validity and reliability of these tests have been further enhanced. In the next section, the methodology and experimental design of the study is discussed.

3   Methodology and Experimental Design

Neural network has been successfully applied for predictions. We propose the use of neural network along with personality traits to predict team performance and team members' interaction. Neural network is a computer-based algorithm to mimic biological brain functioning. Through non-linear computations, it carries the ability to learn and find relationships and patterns in the data. It examines the relationship between the output variable and the whole set of input variables together, rather than just one input variable at one time. It is questionable to present one's personality based on one personality trait. To describe one's personality, neural network can provide an overall description of a set of personality traits. Therefore, we argue that neural network can be used to better manage personality traits to predict team performance and team members' interaction.

3.1   Data Collection

We set the domain of team collaboration on university team projects. The team observations were collected in an upper-level required course for the major in Management Information Systems. A total of 83 students participated in the study. The participants were randomly assigned into 20 teams. Throughout the semester, the students were asked to provide a team presentation and a paper on each of eight different technical topics including Wiki, SOA, Web 2.0, XML, 3G, RFID, Blog, and P2P in eight weeks. Before each topical presentation, the members in each team were asked to discuss using various methods such as e-mail, sharing on learning management system to the team or to the class, instant messenger, telephone, or face-to-face. The presentation was limited to 10 minutes long. For each topic, each team was required to use PowerPoint to present its definition, history, associated technologies, applications, and references. Within two days of each presentation, participants were asked to provide evaluation for each team, cross evaluation of team members, and self-evaluation.

In order to get a large amount of team performance data, which may be associated with different combinations of members' personality traits, the 83 students were randomly assigned into 20 teams for the first presentation topic. In the next seven topics, for the teams in which its performance was lower than average, members in these teams were dismissed to unassigned pool and then the ones in the unassigned pool were then randomly assigned to form new teams. The teams with performance above average would keep the same members for the next topic. The Big Five Personality Traits data were also collected from each of participants before the first presentation.

3.2   Variables of Neural Network

In this section, we discuss the input and output variables used in the neural network. The Big Five Personality Traits were used as input variables in this study. Followed by Costa and McCrae [16], they were neuroticism, extraversion, openness to experience, agreeableness, and conscientiousness. The traits were measured by the Mini-Marker proposed by Saucier [48] in a nine-point scale, where 1 means that the trait was extremely inaccurate and 9 means that the trait was extremely accurate.

Due to the lack of guidance for team project performance indicators in the literature, we delivered unstructured questionnaire to professors in Management Information Systems and concluded five indicators: richness, completeness, correctness, innovativeness, and understandability. Each indicator was rated between 0 (worst) and 10 (best). The team performance for each topic was the average of the five indicators computed as follows:

$$[\sum_{i=1}^{p}\left(\frac{\text{Richness}_i + \text{Completeness}_i + \text{Correctness}_i + \text{Innovativeness}_i + \text{Understandability}_i}{5}\right)]1/p$$

, where p is the number of raters. As a result, we used this team performance measure as one of the output variables.

We also concluded nine team member interaction indicators from the participants through an open-ended questionnaire. The indicators were interactivity, leadership, documentary, activeness, cooperation, contribution, integration, professionalism, and understandability. Each of one's indicators was rated by other members in a nine-point scale, where 1 was the worst and 9 was the best. The overall interaction indicator was also prepared by taking the average of the nine indicators. Each of team member's interaction indicators, the overall indicator and nine separate indicators, was then assigned as output variable for neural network modeling.

We adopted one of most popular neural network—back propagation network. Using the classic one hidden layer back propagation neural network, we established one predictive model for team performance and another for team member's interaction. Different number of hidden nodes, between one and 20, were experimented. Overfitting issue was minimized when nine hidden nodes were used;

therefore, we set this value for all experiments. The neural network was coded by MATLAB 6.5 and functions provided by Neural-Network Toolbox.

In this study, we built two sets of neural network models. The first model was trained to predict team performance whereas the second model was to predict team members' interaction. The two sets of models shared the same inputs which were Richness, Completeness, Correctness, Innovativeness, and Understandability. The output variable of the first model was team performance of assigned projects. When predicting team members' interaction, we built 10 different models using different output variables. We captured nine different aspects such as interactivity and leadership in evaluating members' interaction and thus used each aspect as one output variable to build nine different models. Finally, we used the average of the nine aspects as an overall interaction output variable to establish the last model.

4   Results

4.1   Prediction of Team Performance and Team Member's Interaction

After removing incomplete data entries, we had 626 performance evaluations for eight project topics. We randomly selected 426 entries (around 70%) for training and used the rest of 200 entries for testing in neural network. Each setting was repeated 30 times. The evaluation measure—accuracy rate—was the average of the 30 runs.

When the full training set was used for performance evaluation (see the second row in Table 1), the accuracy rate was above 99.5% when the estimated team performance was within 2 points difference (written as Error in ± 2 in Table 1), compared to the actual team performance. When the difference was restricted to 1 point, the accuracy rate was 97.86%. The chosen configurations for the neural network demonstrate the effectiveness of prediction. When the test set was evaluated by the trained neural network model (see the third row in Table 1), high accuracy rate 99.9% was still reached if the estimated team performance was within 2 points away from the actual team performance. The accuracy rate decreased to 74.37% when a high precision case was considered (difference within 1 point of actual value).

Table 1. Accuracy Rate in Predicting Team Performance Using Training Set

Evaluation Method	Predictive Variable	Mean Squared Error	Accuracy Rate Using Training Set		
			Error in ± 2	Error in ± 1.5	Error in ± 1
Training Set	Team Performance	10217.73	99.50%	98.83%	97.86%
Test Set	Team Performance	10115.39	99.90%	98.83%	74.37%

Instead of predicting team performance using five personality traits, we built up other neural network models to test the efficacy of predicting team members' interactions. The results are reported in Table 2. The team members' interactions were measured in nine different ways including interactivity, leadership, and so on. The overall interaction is an average measure of the nine measures. Based on our experimental results using five personality traits as predictors, the ability to estimate the nine members' interactions as well as overall interaction was similar.

Table 2. Accuracy Rate in Predicting Team Member's Interaction

Predictive Variable	Mean Squared Error	Accuracy Rate Using Test Set		
		Error in ± 2	Error in ± 1.5	Error in ± 1
Overall Interaction	1118.40	83.04%	72.02%	52.60%
Interactivity	1060.24	80.50%	68.80%	52.10%
Leadership	1188.92	81.97%	68.08%	50.98%
Documentary	1053.42	83.57%	70.24%	50.35%
Activeness	1028.65	82.98%	70.54%	53.07%
Cooperation	1072.08	84.50%	71.52%	51.75%
Contribution	1165.43	84.18%	70.57%	52.77%
Integration	1093.68	79.82%	69.18%	49.38%
Professionalism	1132.18	82.43%	70.98%	53.45%
Understandability	1124.36	81.68%	69.73%	49.87%

4.2   Validation of Established Neural Network

In order to perform external validity of established neural network model, 56 data observations were collected in the same way from a final project of a system analysis and design course. Table 3 shows the validation results. Although the team project was different from the ones used to train the model, the trained neural network model demonstrated the effectiveness for predicting team performance based on members' personality traits.

Table 3. Validation Results in Predicting Team Member's Interaction

Predictive Variable	Accuracy Rate Using Test Set		
	Error in ± 2	Error in ± 1.5	Error in ± 1
Overall Interaction	75.57%	65.52%	44.47%
Interactivity	60.82%	49.06%	36.32%
Leadership	69.71%	57.89%	42.63%
Documentary	69.34%	57.59%	40.96%
Activeness	70.40%	56.37%	40.66%
Cooperation	64.62%	52.57%	38.95%
Contribution	64.44%	50.58%	34.97%
Integration	68.13%	58.19%	41.93%
Professionalism	70.53%	57.25%	40.12%
Understandability	66.78%	55.44%	40.53%

5   Conclusions and Discussions

In this study, we empirically investigated the relationships between team member's personality traits and two important team-based aspects: team performance and team members' interactions. Using eight team projects done by each of 20 teams formed by

83 undergraduate students, we established neural network models to predict team performance and team members' interactions based on students' individual personality traits. We found that if the tolerance of precision was set to two points away from the actual team performance, the neural network model reached 99.5% accuracy rate. The accuracy rate decreased a little to 98.83%, when the tolerance of precision was 1.5 points away. The promising results open a new venue of research in using data mining approach for predicting team performance and members' interactions. Future studies may use team projects in varies industries to generalize the feasibility of a data mining method for predicting team performance and members' interactions.

References

1. Adair, J.: Effective Teambuilding. Gower Pub., Aldershot (1986)
2. Allport, G.W., Odbert, H.S.: Trait Names: A Psycho-lexical Study. Psychological Monographs 47(211) (1936)
3. Allport, G.W.: Pattern and Growth in Personality (1961)
4. Allport, G.W.: Personality: A Psychological Interpretation (1937)
5. Barczak, G., Lassk, F., Mulki, J.: Antecedents of team creativity: an examination of team emotional intelligence, team trust and collaborative culture. Creativity and Innov. Manage. 19(4), 332–345 (2010)
6. Barrick, M.R., Mount, M.K.: The big five personality dimensions and job performance: A meta-analysis. Person. Psychol. 44, 1–26 (1991)
7. Barrick, M.R., Stewart, G.L., Neubert, M.J., Mount, M.K.: Relating member ability and personality to work-team processes and team effectiveness. J. Appl. Psychol. 83(3), 377–391 (1998)
8. Hall, B.J.: Confirming first impressions in the employment interview: A field study of interview behavior. J. Appl. Psychol. 79, 659–663 (1994)
9. Brown, S.L., Eisenhardt, K.M.: Product development: past research, present findings, and future direction. Acad. Manage. Rev. 20(2), 343–378 (1995)
10. Buchholz, S., Roth, T.: Creating the high-performance team. John Wiley, New York (1987)
11. Carton, A.M., Cummings, J.N.: A theory of subgroups in work teams. Acad. Manage. Rev. 37(3), 441–470 (2012)
12. Cattell, R.B.: The description of personality: Basic trait resolved into clusters. J. Abnormal and Soc. Psychol. 38, 476–506 (1943)
13. Cattell, R.B.: Description and Measurement of Personality. World Book, New York (1946)
14. Chang, H.H., Chuang, S.S., Chao, S.H.: Determinants of cultural adaptation, communication quality, and trust in virtual teams' performance. Total Qual Manage & Bus Excellence 22(3), 205–329 (2011)
15. Costa, P.T., McCrae, R.R.: Revised NEO Personality Inventory and New Five-Factor Inventory. Psychological Assessment Resources, FL (1992)
16. Costa, P.T., McCrae, R.R.: The NEO Personality Inventory Manual. Psychological Assessment Resources, FL (1985)
17. David, V.D., Stanley, B.S.: Personality and Job Performance: Evidence of Incremental Validity. Pers. Psychol. 42, 25–36 (1989)

18. Earley, P.C., Mosakowski, E.: Creating hybrid team cultures: An empirical test of transnational team functioning. Acad. Manage J. 43(1), 26–49 (2000)
19. Fiske, D.W.: Consistency of the factoral structure of personality ratings from different sources. J. Pers. and Abnormal Psychol. 44, 329–344 (1949)
20. Galton, F.: Measurement of character. Fortnightly Rev. 36, 179–185 (1884)
21. Gladstein, D.L.: Groups in context: A model of task group effectiveness. Admin. Sci. Quart. 29, 499–517 (1984)
22. Goldberg, L.R.: Language and individual differences: The search for universals in personality lexicons. In: Rev. of Pers and Soc. Psychol., Beverly Hill, CA, pp. 141–165 (1981)
23. Goodman, P.S., Ravlin, E., Schminke, M.: Understanding Groups in Organizations. Res. in Org. Behav. 9, 121–174 (1987)
24. Guilford, T.P.: Personality. Megraw-Hill, New York (1959)
25. Hackman, J.R.: A normative model of work team effectiveness. Yale University, CT (1983)
26. Huang, J.C.: The relationship between conflict and team performance in Taiwan: the moderating effect of goal orientation. Int. J. Hum. Resour. Manage. 23(10), 2126–2143 (2012)
27. Ishikawa, J.: Leadership and performance in Japanese R&D teams. Asia Pac. Bus. Rev. 18(2), 241–258 (2012)
28. Katzenbach, J., Smith, D.: The Discipline of teams. Harvard Bus. Rev. 71(2), 111–120 (1993)
29. Kichuk, S.L., Wiesner, W.H.: The big five personality factors and team performance: Implications for selecting successful product design teams. J. Eng. and Tech. Manage. 14, 195–221 (1997)
30. LePine, J.A., Buckman, B.R., Crawford, E.R., Methot, J.R.: A review of research on personality in teams: Accounting for pathways spanning levels of theory and analysis. Hum. Resour. Manage. Rev. 21(4), 311–330 (2011)
31. Mach, M., Dolan, S., Tzafrir, S.: The differential effect of team members' trust on team performance: The mediation role of team cohesion. J. Occup. & Organ. Psychol. 83(3), 771–794 (2010)
32. McGrath, J.E.: Social psychology: A brief introduction. Rinehart & Winston, Holt (1964)
33. Mohrman, S.A., Cohen, S.G., Mohrman, A.J.: Designing team-based organization-new forms for knowledge work. Jossey-Bass Press, San Francisco (1995)
34. Schmitt, N., Gooding, R., Noe, R., Kirsch, M.: Metaanalyses of validity studies published between 1964 and 1982 and the investigation of study characteristics. Pers. Psychol. 37, 407–422 (1984)
35. Neuman, G.A., Wagner, S.H., Christiansen, N.D.: The relationship between work-team personality composition and the job performance of teams. Group & Org. Manage. 24(1), 28–46 (1999)
36. Nieva, V.F., Fleishman, E.A., Rieck, A.: Team dimensions: their identity, their measurement, and their relationship. Adv. Res. Resour. Org., Washington, DC (1978)
37. Norman, W.T.: 2800 personality trait descriptors: Normative operating characteristics for a university population. University of Michigan, Ann Arbor (1967)
38. Norman, W.T.: toward and adequate taxonomy of personality attributes: Replicated factor structure. J. Abnormal and Soc. Psychol. 66, 574–583 (1963)
39. Osborne, R.E.: Personality Traits. Choice, 36 (1998)
40. Paul, R.J.: Why users cannot 'get what they want'. Int. J. Manuf. Sys. Design. 1(4), 389–394 (1994)

41. Peabody, D.: Selecting representative trait adjectives. Journal of Personality and Social Psychology 52, 59–77 (1987)
42. Pervin, L.A.: Personality: Theory and Research, 6th edn. John Wily & Sons, New York (1993)
43. Poulymenakou, A., Holmes, A.: A contingency framework for the investigation of information systems failure. Eur. J. Inform. Sys. 5(1), 34–56 (1996)
44. Robbins, S.P.: Organizational Behavior, 9th edn. Prentice-Hall Press, New York (2001)
45. Hogan, R.T.: Personality and Personality Measurement. In: Dunnette, M., Hough, L. (eds.) The Handbook of Industrial and Organizational Psychology, 2nd edn. (1991)
46. Salas, E., Dickinson, T.L., Converse, S.A., Tannenbaum, S.I.: Toward an understanding of team performance and training, In Swezey R, pp. 3–29. Ablex Publishing, Greenwich (1992)
47. Salgado, J.F.: The 5-factor model of personality and job-performance in the European-community. J. Appl. Psychol. 82(1), 30–43 (1997)
48. Saucier, G.: Mini-Markers: A Brief Version of Goldberg's Unipolar Big-Five Markers. J. Personality Assess. 63(3), 506–516 (1994)
49. Stewart, G.L., Barrick, M.R.: Team structure and performance: Assessing the mediating role of intrateam process and the moderating role of task type. Acad. Manage J. 43(2), 135–148 (2000)
50. Tett, R.P., Jackson, D.N., Rothstein, M.: Personality measures as predictors of job performance: A Meta-analytic Review. Personality Psychol. 77, 703–742 (1991)
51. Tupes, E.C., Christal, R.E.: Recurrent personality factors based on trait ratings. J. Personality 60, 225–251 (1961)

A Modified Maintenance Algorithm for Updating FUSP Tree in Dynamic Database

Ci-Rong Li[1], Chun-Wei Lin[2,3,*], Wensheng Gan[2], and Tzung-Pei Hong[4,5]

[1] Faculty of Management,
Fuqing Branch of Fujian Normal University, Fujian, China
[2] Innovative Information Industry Research Center (IIIRC)
[3] Shenzhen Key Laboratory of Internet Information Collaboration
School of Computer Science and Technology,
Harbin Institute of Technology Shenzhen Graduate School
HIT Campus Shenzhen University Town, Xili, Shenzhen, China
[4] Department of Computer Science and Information Engineering
National University of Kaohsiung, Kaohsiung, Taiwan, R.O.C.
[5] Department of Computer Science and Engineering
National Sun Yat-sen University, Kaohsiung, Taiwan, R.O.C.
{cirongli,wsgan001}@gmail.com, jerrylin@ieee.org,
tphong@nuk.edu.tw

Abstract. In the past, we proposed a pre-large FUSP tree to preserve and maintain both large and pre-large sequences in the built tree structure. In this paper, the pre-large concept is also adopted for maintaining and updating the FUSP tree. Only large sequences are kept in the built tree structure for reducing computations. The PreFUSP-TREE-MOD maintenance algorithm is proposed to reduce the rescans of the original database due to the pruning properties of pre-large concept. When the number of modified sequences is smaller than the safety bound of the pre-large concept, better results can be obtained by the proposed PreFUSP-TREE-MOD maintenance algorithm for sequence modification in the dynamic database.

Keywords: Data mining, pre-large concept, dynamic database, FUSP tree, sequence modification.

1 Introduction

It is a critical issue to efficiently mine the desired knowledge or information to aid managers in decision-making from a very large database. The mostly common knowledge can be classified as association-rule mining [1, 2, 6], classification [12], clustering [5], and sequential pattern mining [4, 11, 21], among others [16-18]. Finding sequential patterns in temporal transaction database has become an important issue since it allows the modeling of customer behaviors.

---

[*] Corresponding author.

A. Moonis et al. (Eds.): IEA/AIE 2014, Part I, LNAI 8481, pp. 301–310, 2014.
© Springer International Publishing Switzerland 2014

Agrawal et al. then proposed AprioriAll algorithm [4] for mining sequential patters in a level-wise way. Although customer behaviors can be efficiently extracted by several sequential-pattern-mining algorithms [4, 11, 21] to assist managers in making decisions, the discovered sequential patterns may become invalid since sequences are inserted [13, 20], deleted [15] or modified [14] in real-world applications. Developing an efficient approach to maintain and update sequential patterns is thus a critical issue in real-world applications.

Few studies [13-15, 20] are, however, designed to handle the sequential patterns in the dynamic database compared to those on maintaining association rules. In the past, a pre-large concept [9] was adopted to maintain the build pre-large FUSP tree. Since the pre-large sequences are kept in the tree structure, more computations are required to maintain both the large and pre-large sequences for finding the corresponding branches [19]. In this paper, the pre-large concept is also adopted in the FUSP tree but only large sequences are kept in the built tree structure. A pre-large fast updated sequential pattern tree for sequence modification (PreFUSP-TREE-MOD) maintenance algorithm is designed to easier facilitate the updating process of the built FUSP tree. A FUSP tree [13] is initially built to completely preserve customer sequences with only large items in the given databases. When some sequences are modified from the original database, the proposed PreFUSP-TREE-MOD maintenance algorithm is then processed to maintain the built FUSP tree and the Header_Table. Experimental results show that the proposed PreFUSP-TREE-MOD maintenance algorithm balances the trade-off between execution time and tree complexity and has the better performance than the batch methods.

2   Review of Related Works

In this section, works related to mining sequential patterns, FUSP-tree structure, and maintenance approach of pre-large concept are briefly reviewed.

2.1   Mining Sequential Patterns

Agrawal et al. first proposed the AprioriAll algorithm [4] for level-wisely mining sequential patterns in a static database. Conventional approaches may re-mine the entire database to update the sequential patterns in the dynamic database. Lin et al. thus proposed the FASTUP algorithm [20] to maintain sequential patterns. Hong and Wang et al. then extended the pre-large concept of association-rule mining [9] to handle the sequential patterns [10, 22]. Cheng et al. proposed the IncSpan (incremental mining of sequential patterns) algorithm for efficiently maintaining sequential patterns in a tree structure [7]. Lin et al. designed a fast updated sequential pattern (FUSP)-tree and developed the algorithms for efficiently handling sequence insertion [13], sequence deletion [15], and sequence modification [14] to maintain and update the built FUSP tree for efficiently discovering sequential patterns in the dynamic database. Other algorithms for mining various sequential patterns are still developed in progress [11, 20].

2.2   FUSP-Tree Structure

The FUSP tree [13] is used to store customer sequences with only large 1-sequences in the original database. An example is given to briefly show the FUSP tree. Assume a database shown in Table 1 is used to build the FUSP tree.

Table 1. Original customer sequences

Customer ID	Customer sequence
1	(AC)(E)(I)
2	(A)(I)(B)
3	(BE)(CD)
4	(AC)(DF)
5	(A)(B)(F)
6	(A)(B)(D)(EF)
7	(AC)(B)(E)
8	(AC)(E)(F)(G)
9	(BG)(D)
10	(DE)(GH)

Also assume that the minimum support threshold is set at 60%. For the given database, the large 1-sequnces are (A), (B), and (E), from which Header_Table can be constructed. The results are shown in Figure 1.

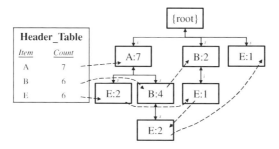

Fig. 1. Initial constructed FUSP tree with its Header_Table.

In Figure 1, only the customer sequences with large items (1-sequences) are stored in the FUSP tree. The link between two connected nodes is marked by the symbol *s* (representing the sequence relation) if the sequence is within the sequence relation in a sequence; otherwise, the link is marked by the symbol *i*, which indicates the sequence is within the itemset relation in a sequence [7]. A FP-growth-like algorithm can be used to mine the sequential patterns [8].

2.3   Maintenance Approach of Pre-large Concept

A pre-large sequence is not truly large, but has highly probability to be large when the database is updated [9-10]. A lower support threshold and an upper support threshold are used to define pre-large concept. Pre-large sequences act like buffers and are used to reduce the movement of sequences directly from large to small and vice-versa in the maintenance process. Therefore, when few sequences are modified, the originally small sequences will at most become pre-large and cannot become large, thus reducing the amount of rescanning necessary. Considering an original database and some customer sequences to be modified, the following nine cases in Figure 2 may arise.

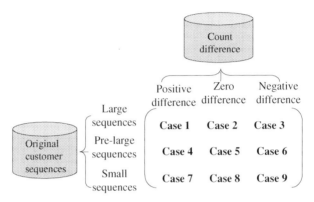

Fig. 2. Nine cases arising from the original database and the modified sequences

Cases 1, 2, 5, 6, 8 and 9 above will not affect the final sequential patterns. Case 3 may remove some existing sequential patterns, and cases 4 and 7 may add some new sequential patterns. It has been formally shown that a sequence in case 7 cannot possibly be large for the entire updated database when the number of modified sequences is smaller than the number f shown below [22]:

$$f = \lfloor (S_u - S_l)d \rfloor,$$

where f is the safety number of modified sequences, S_u is the upper threshold, S_l is the lower threshold, and d is the number of customer sequences in the original database.

3   Proposed PreFUSP-TREE-MOD Algorithm

In this paper, the pre-large concept is adopted to maintain and update the built FUSP tree [13] for sequence modification [14] in dynamic database. A FUSP tree must be built in advance from the initial original database. When sequences are modified from the original database, the FUSP tree and its corresponding Header_Table are required to be modified and updated by the proposed PreFUSP-TREE-MOD maintenance algorithm. The details of the proposed algorithm are stated below.

***Proposed PreFUSP-TREE-MOD maintenance algorithm*:**
INPUT: An old database consisting of d sequences, its corresponding Header_Table, and the built FUSP tree, a lower support threshold S_l, an upper support threshold S_u, a set of pre-large 1-sequences *Prelarge_Seqs* from the original database, and a set of t modified sequences.

OUTPUT: An updated FUSP tree.

STEP 1: Calculate the safety number f for modified sequences as [22]:
$$f = \lfloor (S_u - S_l)d \rfloor.$$

STEP 2: Find all 1-sequences in the t modified sequences before and after modification. Denote them as a set of modified 1-sequences, M.

STEP 3: Find the count difference (including zero) of each 1-sequence in M for the modified sequences.

STEP 4: Divide the 1-sequences in M into three parts according to whether they are large, pre-large or small in the original database.

STEP 5: For each 1-sequence s in M which is large in the original database (appearing in the Header_Table), do the following substeps (for **cases 1, 2** and **3**):

    Substep 5-1: Set the new count $S^U(s)$ of s in the entire updated database as:
$$S^U(s) = S^D(s) + S^M(s),$$
        where $S^D(s)$ is the count of s in the Header_Table (from the original database) and $S^M(s)$ is the count difference of s from sequence modification.

    Substep 5-2: If $S^U(s) \geq (S_u \times d)$, update the count of s in the Header_Table as $S^U(s)$, and put s in both the sets of *Increase_Seqs* and *Decrease_Seqs*, which will be further processed to update the FUSP tree in STEP 9;

        Otherwise, If $(S_l \times d) \leq S^U(s) \leq (S_u \times d)$, connect each parent node of s directly to its corresponding child nodes; remove s from the FUSP tree and the Header_Table; put s in the set of *Prelarge_Seqs* with its updated count $S^U(s)$;

        Otherwise, s is small after the database is updated; connect each parent node of s directly to its corresponding child nodes and remove s from the FUSP tree and the Header_Table;

STEP 6: For each 1-sequence s in M which is pre-large in the original database (in the set of pre-large sequences), do the following substeps (for cases 4, 5 and 6):

    Substep 6-1: Set the new count $S^U(s)$ of s in the entire updated database as:
$$S^U(s) = S^D(s) + S^M(s).$$

    Substep 6-2: If $S^U(s) \geq (S_u \times d)$, 1-sequence s will become large after the database is updated; remove s from the set of *Prelarge_Seqs*, put s in the set of *Branch_Seqs* with its new count $S^U(s)$, and put s in the set of *Increase_Seqs*;

        Otherwise, if $(S_l \times d) \leq S^U(s) \leq (S_u \times d)$, 1-sequence s is still pre-large after the database is updated; update s with its new count $S^U(s)$ in the set of *Prelarge_Seqs*;

        Otherwise, 1-sequence s is small after the database is updated; remove s from the set of *Prelarge_Seqs*.

STEP 7: For each 1-sequence s which is neither large nor pre-large in the original database but has positive count difference in M (for **case 7**), put s in the set of *Rescan_Seqs*, which is used when rescanning the database in STEP 8 is necessary.

STEP 8: If $(t + c) \leq f$ or the set of *Rescan_Seqs* is ***null***, do nothing; Otherwise, do the following substeps for each 1-sequence s in the set of *Rescan_Seqs*:

Substep 8-1: Rescan the original database to determine the original count $S^D(s)$ of s (before modification).

Substep 8-2: Set the new count $S^U(s)$ of s in the entire updated database as:
$$S^U(s) = S^D(s) + S^M(s).$$

Substep 8-3: If $S^U(s) \geq (S_u \times d)$, 1-sequence s will become large after the database is updated; put s in both the sets of *Increase_Seqs* and *Branch_Seqs*;

Otherwise, if $(S_l \times d) \leq S^U(s) \leq (S_u \times d)$, 1-sequence s will become pre-large after the database is updated; put s in the set of *Prelarge_Seqs* with its updated count $S^U(s)$;

Otherwise, neglect s.

STEP 9: For each updated sequence before modification (T) and with a 1-sequence J existing in the *Decrease_Seqs*, find the corresponding branch of J in the FUSP tree and subtract 1 from the count of the J node in the branch; if the count of the J node becomes zero after subtraction, remove node J from its corresponding branch and connect the parent node of J directly to the child node of J.

STEP 10: Insert the 1-sequences in the *Branch_Seqs* to the end of the Header_Table according to the descending order of their counts.

STEP 11: If the set of *Branch_Seqs* is ***null***, nothing is done in this step;

Otherwise, for each unmodified sequence (D^-) with a 1-sequence J in *Branch_Seqs*, if J has not been at the corresponding branch of the FUSP tree, then insert J at the end of the branch and set its count as 1; Otherwise, add 1 to the count of the node J.

STEP 12: For each updated sequence after modification (T') with a 1-sequence J existing in *Increase_Seqs*, if J has not been at the corresponding branch of the FUSP tree, insert J at the end of the branch and set its count as 1; Otherwise, add 1 to the count of the J node.

STEP 13: If $(t + c) > f$, set $c = 0$; otherwise, set $c = t + c$.

After STEP 13, the final updated FUSP tree is thus maintained by the proposed PreFUSP-TREE-MOD maintenance algorithm for sequence modification. Based on the FUSP tree, the desired large sequences can then be found by the FP-growth-like mining approach [8].

4   An Illustrated Example

A FUSP tree [13] was firstly built from the original database shown in Figure 2. An upper support threshold was set at 60% and the lower support threshold is set at 30%.

The pre-large 1-sequences with their counts are then found and then put in the set of *Prelarge_Seqs*. The results are shown in Table 2, and the modified customer sequences are shown in Table 3.

Table 2. Pre-large 1-sequences

1-sequence	Count
(C)	5
(D)	5
(F)	4
(G)	3

Table 3. Modified customer sequences

Cust_ID	Before modification	After modification
2	(A)(I)(B)	(A)(BC)(FH)
10	(DE)(GH)	(BD)(H)

The safety bound for the modified sequences is calculated as $f = \lfloor (0.6 - 0.3) \times 10 \rfloor$ (= 3). The proposed PreFUSP-TREE-MOD maintenance algorithm is the performed to maintain and update the FUSP tree by the following steps. The count differences of the sequences before and after modification are then calculated. After that, the results of count difference are then divided into three parts according to whether they are large (appearing in Header_Table), pre-large (appearing in the set of *Prelarge_Seqs*) or small in the original customer sequences.

For each 1-sequence from the divided part, which is large in the original database (appearing in Header_Table shown in Figure 1), is then processed. For each 1-sequence from the divided part, which is pre-large in the original database (appearing in *Prelarge_Seqs* shown in Table 2), is then processed. For each 1-sequence from the divided part, which is small in the original database (not appearing either in the Header_Table or the set of *Prelarge_Seqs*) but has positive count difference is then processed. Since only two sequences are modified from the original database, which is smaller than the safety bound (2 < 3); nothing has to be processed for the set of *Rescan_Seqs*.

The 1-sequences in the *Decrease_Seqs* are then processed to subtract their counts before sequence modification in the built FUSP tree. The 1-sequenecs in the *Branch_Seqs* are then sorted in descending order of their counts. The 1-sequences from *Branch_Seqs* are then inserted into the end of the Header_Table. The corresponding branches of 1-sequences in *Branch_Seqs* are then found from the unmodified sequences in the original database. The 1-sequences in the *Increase_Seqs* are then processed to find the corresponding branches after sequence modification. After that, the final result of the FUSP tree is shown in Figure 3. Since the number of the modified sequences is 2 in this example, variable c is then accumulated as $(0 + 2)$ (= 2), which indicates that one more sequence can be modified without rescanning the original database for case 7.

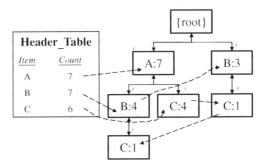

Fig. 3. Final updated FUSP tree

5   Experimental Results

Experiments were made to compare the performance of the GSP algorithm [21], the FUSP-TREE-BATCH [13] algorithm, the pre-large maintenance algorithm for sequence modification (defined as PRE-APRIORI-MOD) [22], and the proposed PreFUSP-TREE-MOD maintenance algorithm. The S10I4N1KD10K is a simulated database generated from IBM Quest Dataset Generator [3]. The percentage of the modified sequences is set at 1%. The S_l values are respectively set as 50% of S_u values for the PRE-APRIORI-MOD algorithm and the proposed PreFUSP-TREE-MOD maintenance algorithm. The performance of execution time is shown in Figure 4.

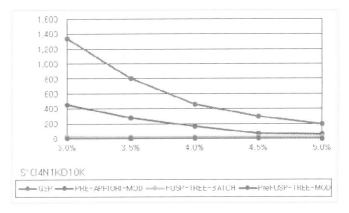

Fig. 4. The comparisons of the execution time

From Figure 4, the proposed PreFUSP-TREE-MOD maintenance algorithm has the better performance than the other algorithms. The number of tree nodes are also compared to show the performance. Since only the tree-based approaches involved tree nodes for the comparisons, the FUSP-TREE-BATCH algorithm and the proposed PreFUSP-TREE-MOD maintenance algorithm are then compared and shown in Figure 5.

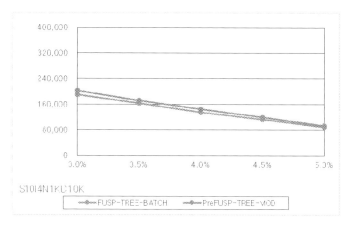

Fig. 5. The comparisons of the tree nodes

From Figure 5, it can be seen that the proposed PreFUSP-TREE-MOD maintenance algorithm generates nearly the same number of tree nodes.

6   Conclusion

In this paper, the pre-large fast updated sequential pattern tree for sequence modification (PreFUSP-TREE-MOD) maintenance algorithm is proposed to efficiently and effectively maintain the FUSP tree based on pre-large concept for deriving sequential patterns. From the experiments, the proposed PreFUSP-TREE-MOD maintenance algorithm can thus achieve a good trade-off between execution time and tree complexity.

Acknowledgement. This research was partially supported by the Shenzhen Peacock Project, China, under grant KQC201109020055A, by the Natural Scientific Research Innovation Foundation in Harbin Institute of Technology under grant BD29100003, and by the Shenzhen Strategic Emerging Industries Program under grant ZDSY20120613125016389.

References

1. Agrawal, R., Imielinski, T., Swami, A.: Database mining: A performance perspective. IEEE Transactions on Knowledge and Data Engineering 5, 914–925 (1993)
2. Agrawal, R., Srikant, R.: Fast algorithms for mining association rules in large databases. In: The International Conference on Very Large Data Bases, pp. 487–499 (1994)
3. Agrawal, R., Srikant, R.: Quest synthetic data generator (1994), http://www.Almaden.ibm.com/cs/quest/syndata.html
4. Agrawal, R., Srikant, R.: Mining sequential patterns. In: The International Conference on Data Engineering, pp. 3–14 (1995)
5. Berkhin, P.: A survey of clustering data mining techniques. In: Grouping Multidimensional Data, pp. 25–71 (2006)

6. Chen, M.S., Han, J., Yu, P.S.: Data mining: An overview from a database perspective. IEEE Transactions on Knowledge and Data Engineering 8, 866–883 (1996)
7. Cheng, H., Yan, X., Han, J.: Incspan: Incremental mining of sequential patterns in large database. In: ACM SIGKDD International Conference on Knowledge Discovery and Data Mining, pp. 527–532 (2004)
8. Han, J., Pei, J., Yin, Y., Mao, R.: Mining frequent patterns without candidate generation: A frequent-pattern tree approach. Data Mining and Knowledge Discovery 8, 53–87 (2004)
9. Hong, T.P., Wang, C.Y., Tao, Y.H.: A new incremental data mining algorithm using pre-large itemsets. Intelligent Data Analysis 5, 111–129 (2001)
10. Hong, T.P., Wang, C.Y., Tseng, S.S.: An incremental mining algorithm for maintaining sequential patterns using pre-large sequences. Expert Systems with Applications 38, 7051–7058 (2011)
11. Huang, Z., Shyu, M.L., Tien, J.M., Vigoda, M.M., Birnbach, D.J.: Prediction of uterine contractions using knowledge-assisted sequential pattern analysis. IEEE Transactions on Biomedical Engineering 60, 1290–1297 (2013)
12. Kotsiantis, S.B.: Supervised machine learning: A review of classification techniques. In: The Conference on Emerging Artificial Intelligence Applications in Computer Engineering: Real Word AI Systems with Applications in eHealth, HCI, Information Retrieval and Pervasive Technologies, pp. 3–24 (2007)
13. Lin, C.W., Hong, T.P., Lu, W.H., Lin, W.Y.: An incremental fusp-tree maintenance algorithm. In: The International Conference on Intelligent Systems Design and Applications, pp. 445–449 (2008)
14. Lin, C.W., Hong, T.P., Lu, W.H., Chen, H.Y.: An fusp-tree maintenance algorithm for record modification. In: IEEE International Conference on Data Mining Workshops, pp. 649–653 (2008)
15. Lin, C.W., Hong, T.P., Lu, W.H.: An efficient fusp-tree update algorithm for deleted data in customer sequences. In: International Conference on Innovative Computing, Information and Control, pp. 1491–1494 (2009)
16. Lin, C.W., Hong, T.P., Lu, W.H.: An effective tree structure for mining high utility itemsets. Expert Systems with Applications 38, 7419–7424 (2011)
17. Lin, C.W., Hong, T.P.: A new mining approach for uncertain databases using cufp trees. Expert Systems with Applications 39, 4084–4093 (2012)
18. Lin, C.W., Lan, G.C., Hong, T.P.: An incremental mining algorithm for high utility itemsets. Expert Systems with Applications 39, 7173–7180 (2012)
19. Lin, C.W., Hong, T.P., Lee, H.Y., Wang, S.L.: Maintenance of pre-large FUSP trees in dynamic databases. In: International Conference on Innovations in Bio-inspired Computing and Applications, pp. 199–202 (2011)
20. Lin, M.Y., Lee, S.Y.: Incremental update on sequential patterns in large databases. In: IEEE International Conference on Tools with Artificial Intelligence, pp. 24–31 (1998)
21. Srikant, R., Agrawal, R.: Mining sequential patterns: Generalizations and performance improvements. In: The International Conference on Extending Database Technology: Advances in Database Technology, pp. 3–17 (1996)
22. Wang, C.Y., Hong, T.P., Tseng, S.S.: Maintenance of sequential patterns for record modification using pre-large sequences. In: IEEE International Conference on Data Mining, pp. 693–696 (2002)

A New Method for Group Decision Making Using Group Recommendations Based on Interval Fuzzy Preference Relations and Consistency Matrices

Shyi-Ming Chen and Tsung-En Lin

Department of Computer Science and Information Engineering,
National Taiwan University of Science and Technology,
Taipei, Taiwan, R. O. C.

Abstract. This paper presents a new method for group decision making using group recommendations based on interval fuzzy preference relations and consistency matrices. First, it constructs consistency matrices from interval fuzzy preference relations. Then, it constructs a collective consistency matrix, constructs a weighted collective preference relation, and constructs a group collective preference relation. Then, it constructs a consensus relation for each expert and calculates the group consensus degree for the experts based on the constructed consensus relations. If the group consensus degree is smaller than a predefined threshold value, then it modifies the interval fuzzy preference values in the interval fuzzy preference relations. The above process is performed repeatedly, until the group consensus degree is larger than or equal to the predefined threshold value. Finally, based on the group collective preference relation, it calculates the score of each alternative. The larger the score of the alternative, the better the preference order of the alternative. The proposed method can overcome the drawbacks of the existing methods for group decision making using group recommendations.

Keywords: Consistency Matrices, Group Consensus Degree, Group Decision Making, Group Recommendations, Interval Fuzzy Preference Relations.

1   Introduction

Some group decision making methods have been presented [2]-[19]. In [18], Xu presented a method for group decision making based on the consistency of interval fuzzy preference relations. In [19], Xu and Liu presented a group decision making method based on interval multiplicative preference relations and interval fuzzy preference relations by using the projection with a consensus process. However, in [19], Xu and Liu pointed out that Xu's method [18] has the drawbacks that the weights of experts are not considered, which is not reasonable. Furthermore, it does not consider the consensus level which is necessary in group decision making. Moreover, in this paper, we also found that Xu and Liu's method [19] has the following drawbacks: 1) It has the "divided by zero" problem when the interval preference

A. Moonis et al. (Eds.): IEA/AIE 2014, Part I, LNAI 8481, pp. 311–320, 2014.

relation of an expert and the group collective preference relation of all experts are the same and 2) It is unreasonable that their method which calculates the consensus degree in the consensus relation for each expert does not hold the commutative law. Therefore, we must develop a new method for group decision making using group recommendations based on interval fuzzy preference relations and consistency matrices to overcome the drawbacks of Xu's method [18] and Xu and Liu's method [19].

In this paper, we present a new method for group decision making using group recommendations based on interval fuzzy preference relations and consistency matrices. The proposed method can overcome the drawbacks of Xu's method [18] and Xu and Liu's method [19] for group decision making using group recommendations.

2   Preliminaries

In this section, we briefly review the concept of interval fuzzy preference relations from [18], briefly review the concept of consistency matrices from [11], and briefly review the concept of consistency degrees from [6].

Definition 2.1 **[18]:** Let P be an interval fuzzy preference relation for the set X of alternatives, where $X = \{x_1, x_2, \ldots, x_n\}$, shown as follows:

$$P = (p_{ij})_{n \times n} = \begin{bmatrix} p_{11} & p_{12} & \cdots & p_{1n} \\ p_{21} & p_{22} & \cdots & p_{2n} \\ \vdots & \vdots & \ddots & \vdots \\ p_{n1} & p_{n2} & \cdots & p_{nn} \end{bmatrix}, \tag{1}$$

where $p_{ij} = [p_{ij}^-, p_{ij}^+]$ denotes an interval preference value for alternative x_i over x_j. Then, $0 \le p_{ij}^- \le p_{ij}^+ \le 1$, $p_{ji} = 1 - p_{ij} = [1 - p_{ij}^+, 1 - p_{ij}^-]$, $p_{ii}^+ = p_{ii}^- = 0.5$, $1 \le i \le n$, and $1 \le j \le n$.

Definition 2.2 **[11]:** Given a complete fuzzy preference relation $P = (p_{ij})_{n \times n}$, where p_{ij} denotes preference value for alternative x_i over alternative x_j, $p_{ij} + p_{ji} = 1$, $p_{ii} = 0.5$, $1 \le i \le n$, and $1 \le j \le n$. The consistency matrix $\overline{P} = (\overline{p}_{ik})_{n \times n}$ is constructed based on the complete fuzzy preference relation P, shown as follows:

$$\overline{p}_{ik} = \frac{1}{n} \sum_{j=1}^{n} (p_{ij} + p_{jk}) - 0.5. \tag{2}$$

The consistency matrix $\overline{P} = (\overline{p}_{ik})_{n \times n}$ has the following properties:

(1) $\overline{p}_{ik} + \overline{p}_{ki} = 1$,

(2) $\overline{p}_{ii} = 0.5$,

(3) $\overline{p}_{ik} = \overline{p}_{ij} + \overline{p}_{jk} - 0.5$,

(4) $p_{ik} \le p_{is}$ for all $i \in \{1, 2, \ldots, n\}$, where $k \in \{1, 2, \ldots, n\}$ and $s \in \{1, 2, \ldots, n\}$.

Definition 2.3 **[6]:** Let $\overline{P} = (\overline{p}_{ik})_{n \times n}$ be a consistency matrix constructed by a fuzzy preference relation $P = (p_{ij})_{n \times n}$ given by an expert. The consistency degree d between P and $\overline{P}$ is defined as follows:

$$d = 1 - \frac{2}{n(n-1)} \sum_{i=1}^{n} \sum_{\substack{j=1 \\ j \neq i}}^{n} \left| p_{ij} - \overline{p}_{ij} \right|, \tag{3}$$

where $d \in [0, 1]$, p_{ij} denotes the preference value in the fuzzy preference relation P for alternative x_i over alternative x_j, $\overline{p}_{ij}$ denotes a preference value in the consistency matrix $\overline{P}$ for alternative x_i over alternative x_j, $1 \leq i \leq n$, and $1 \leq j \leq n$. The larger the value of d, the more consistent the fuzzy preference relation given by the expert. If the value of d is close to one, then the information of the fuzzy preference relation given by the expert is more consistent.

3   A New Method for Group Decision Making Using Group Recommendations Based on Interval Fuzzy Preference Relations and Consistency Matrices

In this section, we present a new method for group decision making using group recommendations based on interval fuzzy preference relations and consistency matrices. Assume that there are m interval fuzzy preference relations $P^1, P^2, \ldots,$ and P^m given by m experts $E_1, E_2, \ldots,$ and E_m, respectively, and assume that there are n alternatives $x_1, x_2, \ldots,$ and x_n. Assume that the interval fuzzy preference relation P^k given by expert E_k for alternative x_i over x_j is shown as follows:

$$P^k = (p_{ij}^k)_{n \times n} = \begin{bmatrix} p_{11}^k & p_{12}^k & \cdots & p_{1n}^k \\ p_{21}^k & p_{22}^k & \cdots & p_{2n}^k \\ \vdots & \vdots & \ddots & \vdots \\ p_{n1}^k & p_{n2}^k & \cdots & p_{nn}^k \end{bmatrix}, \tag{4}$$

where p_{ij}^k is an interval-valued preference value, $p_{ij}^k = [p_{ij}^{-k}, p_{ij}^{+k}]$, $0 \leq p_{ij}^{-k} \leq p_{ij}^{+k} \leq 1$, $p_{ji}^k = 1 - p_{ij}^k = [1 - p_{ij}^{+k}, 1 - p_{ij}^{-k}]$, $p_{ii}^{+k} = p_{ii}^{-k} = 0.5$, $1 \leq i \leq n$, $1 \leq j \leq n$, and $1 \leq k \leq m$. The proposed method is now presented as follows:

Step 1: Initially, let $r = 0$. Construct the fuzzy preference relation $B^k = (b_{ij}^k)_{n \times n}$ for expert E_k, construct the consistency matrix $\overline{B}^k = (\overline{b}_{ij}^k)_{n \times n}$ for expert E_k, construct the collective consistency matrix $\overline{B}^* = (\overline{b}_{ij}^*)_{n \times n}$ for all experts, and calculate the consistency degree d_k of expert E_k, shown as follows:

$$b_{ij}^k = \frac{1}{2}(p_{ij}^{-k} + p_{ij}^{+k}), \tag{5}$$

$$\bar{b}_{ij}^{k} = \frac{1}{n}\sum_{t=1}^{n}(b_{it}^{k} + b_{tj}^{k}) - 0.5, \tag{6}$$

$$d_{k} = 1 - \frac{2}{n(n-1)}\sum_{i=1}^{n}\sum_{j=1}^{n}\left|b_{ij}^{k} - \bar{b}_{ij}^{k}\right|, \tag{7}$$

$$\bar{b}_{ij}^{*} = \frac{1}{m}\sum_{k=1}^{m}\bar{b}_{ij}^{k}, \tag{8}$$

where $1 \leq i \leq n$, $1 \leq j \leq n$ and $1 \leq k \leq m$.

Step 2: Calculate the weight λ_{k} of expert E_{k} using the consistency degree d_{k}, shown as follows:

$$\lambda_{k} = \frac{d_{k}}{\sum_{t=1}^{m}d_{t}}, \tag{9}$$

where $1 \leq k \leq m$. Construct the weighted collective preference relation $P^{*} = \left(p_{ij}^{*}\right)_{n \times n}$ for all experts and construct the group collective preference relation $U = \left(u_{ij}^{*}\right)_{n \times n}$ for all experts, shown as follows:

$$p_{ij}^{*} = \sum_{k=1}^{m}\lambda_{k}(p_{ij}^{k}) = \left[\sum_{k=1}^{m}\lambda_{k}\left(p_{ij}^{-k}\right), \sum_{k=1}^{m}\lambda_{k}\left(p_{ij}^{+k}\right)\right] = \left[p_{ij}^{-*}, p_{ij}^{+*}\right], \tag{10}$$

$$u_{ij}^{*} = \left[\frac{p_{ij}^{-*} + \bar{b}_{ij}^{*}}{2}, \frac{p_{ij}^{+*} + \bar{b}_{ij}^{*}}{2}\right] = \left[u_{ij}^{-*}, u_{ij}^{+*}\right], \tag{11}$$

where $1 \leq i \leq n$, $1 \leq j \leq n$, and $1 \leq k \leq m$.

Step 3: Construct the consensus relation $C^{k} = \left(c_{ij}^{k}\right)_{n \times n}$ for expert E_{k} and calculate the group consensus degree CD for all experts, shown as follows:

$$c_{ij}^{k} = 1 - \frac{1}{2}\left|u_{ij}^{*} - p_{ij}^{k}\right| = 1 - \frac{1}{2}\left(\left|u_{ij}^{-*} - p_{ij}^{-k}\right| + \left|u_{ij}^{+*} - p_{ij}^{+k}\right|\right), \tag{12}$$

$$CD = \frac{\sum_{i=1}^{n}\sum_{j=1, j \neq i}^{n}\sum_{k=1}^{m}c_{ij}^{k}}{m \times (n^{2} - n)}, \tag{13}$$

where $c_{ii}^{k} = 1$, $1 \leq i \leq n$, $1 \leq j \leq n$, and $1 \leq k \leq m$. If the group consensus degree CD is smaller than the predefined threshold value γ, where $\gamma \in [0, 1]$, then let $r = r + 1$ and go to Step 4. Otherwise, Step 9.

Step 4: Construct the proximity relation $F^{k} = \left(f_{ij}^{k}\right)_{n \times n}$ for expert E_{k}, shown as follows:

$$f_{ij}^{k} = \left[u_{ij}^{-} - p_{ij}^{-k}, u_{ij}^{+} - p_{ij}^{+k} \right] = \left[f_{ij}^{+k}, f_{ij}^{-k} \right], \tag{14}$$

where $\gamma \in [0, 1]$, $1 \le i \le n$, $1 \le j \le n$, and $1 \le k \le m$. If the consensus value c_{ab}^{k} in the consensus relation C^{k} is smaller than the group consensus degree CD, then get the set H^{k} of pairs (a, b) of alternatives x_{a} and x_{b} which satisfy "$c_{ab}^{k} < CD$", shown as follows:

$$H^{k} = \left\{ (a,b) \middle| c_{ab}^{k} < CD \right\}, \tag{15}$$

where the corresponding preference value of c_{ab}^{k} in the interval fuzzy preference relation P^{k} given by expert E^{k} is p_{ab}^{k}, $1 \le a \le n$, $1 \le b \le n$, and $1 \le k \le m$. Construct the modified interval fuzzy preference relation $P^{k(r)} = (p_{ij}^{k(r)})_{n \times n}$ for expert E_{k} using the proximity relations F^{k} and the modified constant δ, where $\delta \in (0, 1]$ and $1 \le k \le m$, shown as follows:

$$p_{ij}^{k(r)} = \begin{cases} p_{ij}^{k(r-1)} - \delta \times f_{ij}^{k}, & \text{if } (i, j) \in H^{k} \\ p_{ij}^{k(r-1)}, & \text{otherwise} \end{cases} \tag{16}$$

where $\delta \in (0, 1]$, (r) denotes the rth round, $1 \le i \le n$, $1 \le j \le n$, and $1 \le k \le m$. Go to Step 5.

Step 5: Based on Eqs. (5)-(8), update the fuzzy preference relation $B^{k} = (b_{ij}^{k})_{n \times n}$ for expert E_{k}, update the consistency matrix $\bar{B}^{k} = (\bar{b}_{ij}^{k})_{n \times n}$ for expert E_{k}, calculate the consistency degree d_{k} of expert E_{k}, and update the collective consistency matrix $\bar{B}^{*} = (\bar{b}_{ij}^{*})_{n \times n}$ for all experts, respectively.

Step 6: Based on Eq. (9), calculate the weight λ_{k} of expert E_{k} using the updated consistency degree d_{k}, where $1 \le k \le m$. Based on Eqs. (10) and (28), update the weighted collective preference relation $P^{*} = (p_{ij}^{*})_{n \times n}$ for all experts, and update the group collective preference relation $U = (u_{ij}^{*})_{n \times n}$ for all experts, respectively.

Step 7: Based on Eqs. (12) and (13), update the consensus relation $C^{k} = (c_{ij}^{k})_{n \times n}$ for expert E_{k} and calculate the group consensus degree CD for all experts, respectively, where $1 \le k \le m$. If the consensus degree CD is smaller than the predefined threshold value γ, where $\gamma \in [0, 1]$, then let $r = r + 1$ and go to Step 8. Otherwise, go to Step 9.

Step 8: Based on Eq. (14)-(16), update the proximity relations $F^{k} = (f_{ij}^{k})_{n \times n}$ for expert E_{k}, where $1 \le k \le m$, get the set H^{k} of pair (a, b) of alternatives x_{a} and x_{b}, and update the modified interval fuzzy preference relation $P^{k(r)} = (p_{ij}^{k(r)})_{n \times n}$ for expert E_{k} using the proximity relations F^{k} and the modified constant δ, respectively, where $\delta \in (0, 1]$ and $1 \le k \le m$. Go to Step 5.

Step 9: Based on the group collective preference relation U for all experts, calculate the score $R(x_i)$ of each alternative x_i, shown as follows:

$$R(x_i) = \frac{1}{n^2} \sum_{j=1}^{n} (u_{ij}^+ + u_{ij}^-), \tag{17}$$

where $1 \leq i \leq n$ and $1 \leq j \leq n$. The larger the value of $R(x_i)$, the better the preference order of alternative x_i, where $1 \leq i \leq n$.

In the following, we use an example to illustrate the group decision making process of the proposed method.

Example 3.1 **[18]:** Assume that there are five alternatives x_1, x_2, x_3, x_4 and x_5 and assume that the interval fuzzy preference relations P^1, P^2 and P^3 given by the experts E_1, E_2 and E_3, respectively, are shown as follows:

$$P^1 = \begin{pmatrix} [0.5, 0.5] & [0.6, 0.8] & [0.7, 1] & [0.2, 0.3] & [0.4, 0.5] \\ [0.2, 0.4] & [0.5, 0.5] & [0.4, 0.6] & [0.7, 0.8] & [0.3, 0.5] \\ [0, 0.3] & [0.4, 0.6] & [0.5, 0.5] & [0.6, 0.9] & [0.4, 0.7] \\ [0.7, 0.8] & [0.2, 0.3] & [0.1, 0.4] & [0.5, 0.5] & [0.3, 0.4] \\ [0.5, 0.6] & [0.5, 0.7] & [0.3, 0.6] & [0.6, 0.7] & [0.5, 0.5] \end{pmatrix},$$

$$P^2 = \begin{pmatrix} [0.5, 0.5] & [0.5, 0.7] & [0.8, 0.9] & [0.3, 0.5] & [0.3, 0.6] \\ [0.3, 0.5] & [0.5, 0.5] & [0.6, 0.7] & [0.5, 0.6] & [0.4, 0.5] \\ [0.1, 0.2] & [0.3, 0.4] & [0.5, 0.5] & [0.7, 0.9] & [0.6, 0.7] \\ [0.5, 0.7] & [0.4, 0.5] & [0.1, 0.3] & [0.5, 0.5] & [0.5, 0.6] \\ [0.4, 0.7] & [0.5, 0.6] & [0.3, 0.4] & [0.4, 0.5] & [0.5, 0.5] \end{pmatrix},$$

$$P^3 = \begin{pmatrix} [0.5, 0.5] & [0.7, 0.9] & [0.8, 1] & [0.4, 0.5] & [0.3, 0.4] \\ [0.1, 0.3] & [0.5, 0.5] & [0.6, 0.7] & [0.4, 0.7] & [0.4, 0.6] \\ [0, 0.2] & [0.3, 0.4] & [0.5, 0.5] & [0.7, 0.8] & [0.5, 0.8] \\ [0.5, 0.6] & [0.3, 0.6] & [0.2, 0.3] & [0.5, 0.5] & [0.4, 0.7] \\ [0.6, 0.7] & [0.4, 0.6] & [0.2, 0.5] & [0.3, 0.6] & [0.5, 0.5] \end{pmatrix},$$

Assume that the predefined threshold value $\gamma = 0.94$ and assume that the modified constant $\delta = 2/3$. Table 1 shows the scores of the alternatives and the group consensus degree for each round by applying the proposed method. Fig. 1 shows the scores of the alternatives for different rounds when the predefined threshold value $\gamma = 1$ by applying the proposed method. Fig. 2 shows the group consensus degrees for different rounds when the predefined threshold value $\gamma = 1$ by applying the proposed method.

Table 2 makes a comparison of the experimental results of the proposed method with Xu and Liu's method [19] and Xu's method [18]. From the Table 2, we can see that the preference order of the alternatives x_1, x_2, x_3, x_4 and x_5 obtained by the proposed method and Xu and Liu's method [19] are the same, i.e., $x_1 > x_5 > x_3 > x_2 > x_4$. However, the preference order of the alternatives x_1, x_2, x_3, x_4 and x_5 obtained by Xu's method [18] is: $x_1 > x_2 > x_5 > x_3 > x_4$. In [19], Xu and Liu [19] have pointed out that Xu's method [18] has the drawbacks that 1) the weights of experts are not considered, which is not reasonable and 2) it does not consider the consensus level which is necessary in group decision making. Therefore, Xu's method [18] gets an unreasonable result of the preference order of the alternatives in this situation.

Table 1. The scores of the alternatives and the group consensus degree at the rth round by the proposed method for Example 3.1

Round Number r	Scores of the Alternatives					Consensus Degrees CD
	$R(x_1)$	$R(x_2)$	$R(x_3)$	$R(x_4)$	$R(x_5)$	
0	0.2279	0.1974	0.1934	0.1786	0.2027	0.8989
1	0.2252	0.1978	0.1945	0.1798	0.2027	0.9268
2	0.2211	0.1972	0.1986	0.1784	0.2047	0.9436
3	0.2196	0.1950	0.2008	0.1806	0.2041	0.9571
⋮	⋮	⋮	⋮	⋮	⋮	⋮
33	0.2169	0.1958	0.2026	0.1794	0.2053	0.9999
34	0.2169	0.1958	0.2026	0.1794	0.2053	1.0000

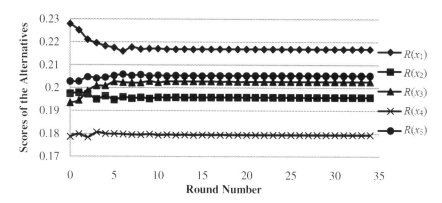

Fig. 1. The scores of the alternatives for different rounds when the predefined threshold value $\gamma = 1$ by the proposed method.

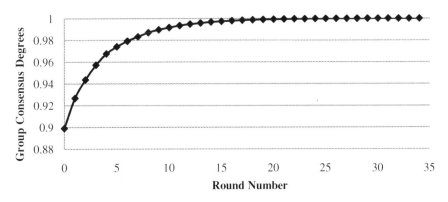

Fig. 2. The group consensus degrees for different rounds when the predefined threshold value $\gamma = 1$ by the proposed method

Table 2. A comparison of the experimental results for different methods for Example 3.1

Methods	Preference Order
Xu's Method [18]	$x_1 > x_2 > x_5 > x_3 > x_4$
Xu and Liu's Method [19]	$x_1 > x_5 > x_3 > x_2 > x_4$
The Proposed Method	$x_1 > x_5 > x_3 > x_2 > x_4$

4   Conclusions

We have presented a new method for group decision making using group recommendations based on interval fuzzy preference relations and consistency matrices. It can overcome the drawbacks of Xu's method [18] and Xu and Liu's method [19], where Xu's method [18] has the drawbacks that the weights of experts are not considered, which is not reasonable, and it does not consider the consensus level which is necessary in group decision making; Xu and Liu's method [19] has the following drawbacks: 1) It has the "divided by zero" problem when the interval preference relation of an expert and the group collective preference relation of all experts are the same and 2) It is unreasonable that their method which calculates the consensus degree in the consensus relation for each expert does not hold the commutative law. The proposed method provides us with a useful way for group decision making using group recommendations based on interval fuzzy preference relations and consistency matrices.

Acknowledgements. This work was supported in part by the National Science Council, under Grant NSC 101-2221-E-011-171-MY2.

References

1. Alonso, S., Chiclana, F., Herrera, F., Herrera-Viedma, E., Alcala-Fdez, J., Porcel, C.: A consistency-based procedure to estimate missing pairwise preference values. International Journal of Intelligent Systems 23(1), 155–175 (2008)
2. Alonso, S., Herrera-Viedma, E., Chiclana, F., Herrera, F.: A web based consensus support system for group decision making problems and incomplete preferences. Information Sciences 180(23), 4477–4495 (2010)
3. Ben-Arieh, D., Chen, Z.: Linguistic-labels aggregation and consensus measure for autocratic decision making using group recommendations. IEEE Transactions on Systems, Man, and Cybernetics – Part A: Systems and Humans 36(3), 558–568 (2006)
4. Chen, S.M., Lee, L.W.: Autocratic decision making using group recommendations based on the ILLOWA operators and likelihood-based comparison relations. IEEE Transactions on Systems, Man, and Cybernetics-Part A: Systems and Humans 42(1), 115–129 (2012)
5. Chen, S.M., Lee, L.W., Yang, S.W., Sheu, T.W.: Adaptive consensus support model for group decision making systems. Expert Systems with Applications 39(16), 12580–12588 (2012)
6. Chen, S.M., Lin, T.E., Lee, L.W.: Group decision making using incomplete fuzzy preference relations based on the additive consistency and the order consistency. Information Sciences 259, 1–15 (2014)
7. Herrera-Viedma, E., Alonso, S., Chiclana, F., Herrera, F.: A consensus model for group decision making with incomplete fuzzy preference relations. IEEE Transactions on Fuzzy Systems 15(5), 863–877 (2007)
8. Herrera-Viedma, E., Chiclana, F., Herrera, F., Alonso, S.: Group decision-making model with incomplete fuzzy preference relations based on additive consistency. IEEE Transactions on Systems, Man and Cybernetics-Part B: Cybernetics 37(1), 176–189 (2007)
9. Herrera-Viedma, E., Herrera, F., Chiclana, F.: A consensus model for multiperson decision making with different preference structures. IEEE Transactions on Systems, Man and Cybernetics-Part A: Systems and Humans 32(3), 394–402 (2002)
10. Herrera-Viedma, E., Martínez, L., Mata, F., Chiclana, F.: A consensus support system model for group decision-making problems with multigranular linguistic preference relations. IEEE Transactions on Fuzzy Systems 13(5), 644–658 (2005)
11. Lee, L.W.: Group decision making with incomplete fuzzy preference relations based on the additive consistency and the order consistency. Expert Systems with Applications 39(14), 11666–11676 (2012)
12. Liu, F., Zhang, W.G., Wang, Z.X.: A goal programming model for incomplete interval multiplicative preference relations and its application in group decision-making. European Journal of Operational Research 218(3), 747–754 (2012)
13. Mata, F., Martinez, L., Herrera-Viedma, E.: An adaptive consensus support model for group decision-making problems in a multigranular fuzzy linguistic context. IEEE Transactions on Fuzzy Systems 17(2), 279–290 (2009)
14. Perez, I.J., Cabrerizo, F.J., Herrera-Viedma, E.: A mobile decision support system for dynamic group decision-making problems. IEEE Transactions on Systems, Man and Cybernetics-Part A: Systems and Humans 40(6), 1244–1256 (2010)
15. Tapia-García, J.M., del Moral, M.J., Martínez, M.A., Herrera-Viedma, E.: A consensus model for group decision making problems with linguistic interval fuzzy preference relations. Expert Systems with Applications 39(11), 10022–10030 (2012)
16. Wu, J., Li, J.C., Li, H., Duan, W.Q.: The induced continuous ordered weighted geometric operators and their application in group decision making. Computers and Industrial Engineering 56(4), 1545–1552 (2009)

17. Wu, Z., Xu, J.: A concise consensus support model for group decision making with reciprocal preference relations based on deviation measures. Fuzzy Sets and Systems 206, 58–73 (2012)
18. Xu, Z.: Consistency of interval fuzzy preference relations in group decision making. Applied Soft Computing 11(5), 3898–3909 (2011)
19. Xu, G.L., Liu, F.: An approach group decision making based on interval multiplicative and fuzzy preference relations by using projection. Applied Mathematical Modelling 37(6), 3929–3943 (2013)

Using Red-Otsu Thresholding to Detect the Bus Routes Number for Helping Blinds to Take Bus

Ching-Ching Cheng[1] and Chun-Ming Tsai[2,*]

[1] Graduate Institute of Education, Providence University,
No. 200, Sec. 7, Taiwan Boulevard, Shalu Dist., Taichung City 43301, Taiwan
puchild@gmail.com
[2] Department of Computer Science, University of Taipei,
No. 1, Ai-Kuo W. Road, Taipei 100, Taiwan
cmtsai2009@gmail.com

Abstract. Current methods to help blind persons read text like menus, business cards, and book covers are inadequate because they assumed both the user and the captured scene are stationary. Furthermore, these methods cannot overcome the illumination problem. This paper presents an intelligent system to threshold the bus route number on a moving bus and to identify the bus route number from the binary images. Experimental results show that the proposed method uses less time complexity and achieves a higher detection rate than the gray level Otsu thresholding method.

Keywords: red-Otsu thresholding, bus route number detection, bus route number identification, blind persons, illumination noise.

1 Introduction

Most blind people are not able to drive a motor vehicle because of their visual impairment. Thus, public transportation is a major key to independence, productivity, and community participation for blind people [1]. Fortunately, every metropolitan area in the world has a public transit bus system. In Taiwan, an e-bus system [2] has been built by the Taipei and New Taipei city governments. This system displays information about the bus route number and time of next bus at each e-bus station. However, the e-bus system is affected by the climate, especially rain, and the system only displayed information visibly on a lighted panel. So the information cannot be seen by the blind.

In Taiwan, there are now some 56,953 blind persons [3] and the number is increasing annually. For blind persons, inability to use the bus has a huge impact on interpersonal relations. Their scope of action is restricted. When want to take the bus, they need to be helped by the other people; but if no one can or will help them, they must

---

* This paper is supported by the National Science Council, R.O.C., under Grants NSC 101-2221-E-133-004-.

A. Moonis et al. (Eds.): IEA/AIE 2014, Part I, LNAI 8481, pp. 321–330, 2014.

try to help themselves, perhaps by writing the desired bus number on a card, which they show the bus driver to see it. However, if the bus driver does not see it, the blind person cannot take their bus.

Computer applications that provide support to blind persons have become an important topic [4]. Several devices have been designed to help them "read" text using an alternative sense such as sound or touch, but these developments are still at an early stage. Ezaki et al. [4] proposed a two-mode text-reading system to help blind persons read a menu or book cover. Tsai [5] proposed an intelligent system to estimate the text in moving business cards. Zandifar et al. [6] proposed a prototype device to read scene text.

Several devices have been proposed using a glasses-mounted camera [7], RFID [8], or RFID combined with GIS [9] and sound to help blind people to "see" bus route numbers. However, the development of such devices is still at an early research stage.

Pan et al. [10] proposed a primary image-based detection system to obtain the route information at a bus station. Their system achieved high accuracy of bus region detection and the text information of bus route number can be successfully retrieved by their scene text extraction algorithm. However, their bus dataset is captured by cell phone, which requires that the blind person be able to use a cell phone. Also, their system is an image-based system. For real-time application, a video-based system should be implemented.

Moreover, most of the above-mentioned methods cannot overcome the illumination problem.

A text detection method proposed by Tsai and Yeh [7] detected the bus route number in the text region on the bus façade panel, achieving moving bus detection, bus panel detection, and text detection. However, their method as described in [7] only detected the text region on the panel; the bus route number itself was not extracted. In this paper, we will propose an efficient and effective detection method to detect the bus route number in the bus façade panel of the moving bus. The proposed method, called Red-Otsu thesholding method, is based on red component in the RGB color space [11] and Otsu thresholding [12] method.

2   Red-Otsu Thresholding

In the system proposed here a "glasses" video camera is used to "see" the bus route number in the moving bus as it approached a bus station, where a blind person is assumed to be standing.

2.1   Motivation for Using Red Component

The reason for using red component is explained as follows. Simulating a blind trying to "see" a moving bus, videos containing 91 frames are captured by a glasses video camera for use as training examples. An initial coarse segmentation image is created by selecting a rectangle for each training example. This rectangle image includes foreground pixels (bus route number) and background pixels. The flood fill method [13] is used to mark the foreground pixels as "Red" pixels. These "Red" pixels are the bus route number in the selected rectangle. The total luminance (Y), red (R), green (G), and blue (B) histograms of all marked bus route numbers as well as the mean and the standard deviation for luminance, red, green, and blue are computed from the luminance, red, green, and blue component, respectively, in all the training examples.

Figure 1 shows an example. The source frame (#31), the selected red rectangle of a bus route number, the coarse segmentation image, the marked "Red" pixels of the bus route number, the total red, green, and blue histograms, and the total luminance histogram are shown in Figs. 1(a)-(h), respectively. The mean and the standard deviation for luminance, red, green, and, blue are 248.95 and 6.4, 243.95 and 11.04, 243.15 and 12.16, and 244.94 and 8.7, respectively. In Figure 1 example, the standard deviation of the red component is the smallest, while the mean of the red component is the biggest. Furthermore, the red histogram is more centralized and the color transformation is omitted. Thus, the red component is used to be processed to obtain the threshold value.

2.2   Modified Otsu Thresholding Method

The Otsu thresholding method [12] is famous in the field of binarization. This method assumes the presence of two distributions, one for the foreground and another for background. A threshold value is selected by minimizing the within-class variance or by maximizing the between-class variance of the two groups of pixels. This statement is equivalent to solving the following optimization problem:

$$\sigma_B^2(t^*) = \max_{0 \leq t \leq 255} \sigma_B^2(t), \tag{1}$$

where σ_B^2 and t^* are the between-class variance and the optimal threshold value, respectively. The between-class variance is defined as follows:

$$\sigma_B^2(t) = \frac{[\mu_T \varpi(t) - \mu(t)]^2}{\varpi(t)[1 - \varpi(t)]}, \tag{2}$$

where

$$\varpi(t) = \sum_{i=0}^{t} p_i, \tag{3}$$

and

$$\mu(t) = \sum_{i=0}^{t} i p_i, \tag{4}$$

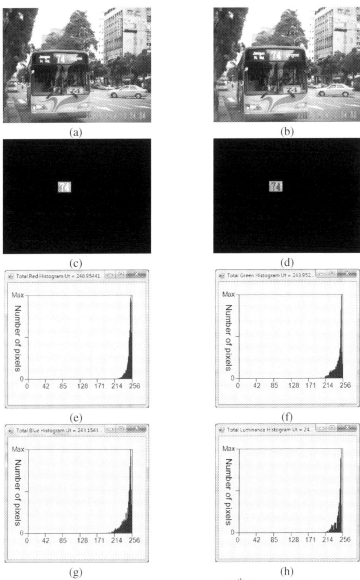

Fig. 1. Example for why to use the red component. (a) 31th frame (b) A bus route number is selected by a red rectangle (c) An initial coarse segmentation image (d) The pixels of the bus route number are marked by "Red" (e) The total red histogram (f) The total green histogram (g) The total blue histogram (h) The total luminance histogram.

are the zeroth- and the first-order cumulative moments of the histogram up to the tth level, respectively, and

$$\mu_T = \sum_{i=0}^{255} i p_i,$$ (5)

(a)   (b)

Fig. 2. Example of MOtsu and Otsu threshoding methods are applying for Fig. 1(a). (a) Binary image by using the proposed method (b) Binary image by using the Otsu thresholding method.

is the total mean level of the original image. Furthermore, p_i is the ith probability distribution and is defined as follows:

$$p_i = {n_i}/{N},\tag{6}$$

where n_i is the number of pixels at level i and N is the total number of pixels in the original image.

The proposed modified Otsu (MOtsu) thresholding method is defined as follows:

$$T = t^* + k\sigma_T,\tag{7}$$

where T and k are the threshold value and the noise removal constant, respectively, and σ_T is the total standard deviation of the original image. Here, k is used to adjust the threshold value to obtain a binary image with less noise (herein, k is set as 1.5 by using a pre-learning method).

Figure 2 shows an example of MOtsu and Otsu threshoding methods are used to threshold Fig. 1(a) on red and gray components, respectively. The thresholding results of the proposed MOtsu and the Otsu thresholding methods are shown in Fig. 2(a) and Fig. 2(b), respectively. The threshold values of MOtsu and Otsu are 226 and 131, respectively. As Fig. 2(b), shows the bus route number "74" cannot be detected, but in Fig. 2(a), the bus route number "74" can be detected. The noise removal constant (k) and the standard deviation (σ_T) can help to remove the illumination noise. However, many noises still exist.

3   Bus Route Number Identification

After thresholding the original image with the threshold value which obtained by the proposed MOtsu method, connected component labeling [14] is applied to detect the candidates for the bus route number. These candidates are represented by the bounding-boxes (BBs), which may be gray sky, advertising, taxi, bus façade body, start text, end text, traffic marking, or bus route number. Figure 3(a) shows an example of applying the connected component labeling method to label the bus route number

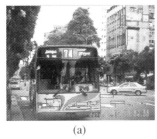   

(a)   (b)

Fig. 3. Example of bus route number localization. (a) The labeled result via connected component labeling. (b) The localization result via rule-based identification method.

candidates in the image in Fig. 2(a). From Fig. 3(a), 265 bounding-box candidates have been labeled, but most are noises and should be removed.

A rules-based identification method is proposed to remove the non-bus route number candidate noises, merge concentrated broken BBs, and thereby identify the bus route number. The proposed identification method includes eleven cascaded rules and is described as follows.

(1) Removing large width BBs rule: The bus is moving from the far to the near. When the bus is closest, the width of the bus route number is the largest. The *largest width* of the bus route number is used to remove as noise bounding-boxes with larger width than that value. This *removing large width BBs rule* is defined as: If the width in a bounding-box is greater than a predefined value (T_{LWI}, set as 62 for frame image with width is 640), this bounding-box is removed.

(2) Removing large height BBs rule: When the bus is closest, the height of the bus route number is the largest. This *largest height* of the bus route number is used to remove as noise bounding-boxes with larger height. This *removing large height BBs rule* is defined as: If the height in a bounding-box is greater than a predefined value (T_{LHI}, set as 62 for frame image with height 480), this bounding-box is removed.

(3) Removing small height BBs rule: After applying above-mentioned rules, many small noise BBs still remains. The *small height* of the bus route number is used to remove the noise bounding-box with small height. This *removing small height BBs rule* is defined as: If the height in a bounding-box is smaller than a predefined value (T_{SHI}, set as 2 for frame image with height 480), this bounding-box is removed.

(4) Merging horizontal closed BBs rule: After applying the above-mentioned rules, sometimes the bus route number is broken. In order to obtain the complete bus route number, a BB-based closing [15] is used. In this merging-closed-BBs rule, to horizontally merge adjacent broken BBs, the horizontal dilation and erosion constants are considered. This *merging horizontal closed BBs rule* is defined as: The horizontal dilation constants are set as T_{MHD}; a BB-based dilation operation uses this dilation constant to dilate the broken BBs, a geometry-based BB merging operation is used to merge the dilated broken BBs, and a BB-based erosion

operation uses the erosion constant $(=T_{MHD})$ to erode the merged BBs. The merged result depends on the size of the dilation and erosion constants. In order to obtain an optimal result, a supervised learning method is applied to set dilation and erosion constants as $T_{MVD} = 2$.

(5) Removing large width BBs rule: After applying above-mentioned rules, some additional BBs with large width are produced. These large width BBs are removed by the removing-large-width-BBs rule as defined as rule1. In addition, the predefined value, T_{LW2}, is set as 51 for frame image with width 640.

(6) Removing large height BBs rule: Sometimes, some additional BBs with large height are also produced in the merging and must be removed, using the removing-large-height-BBs rule as defined as rule2. In addition, the predefined value, T_{LH2}, is set as 54 for frame image with height 480.

(7) Removing small height BBs rule: When the bus is moving at a distance, the height of the bus route number is small. The *smallest height* of the bus route number is used to remove the noise bounding-boxes with smaller height. This removing small height BBs rule is defined as rule3. In addition, the predefined value, T_{SH2}, is set as 9 for frame image with height 480.

(8) Removing small width BBs rule: When the bus is moving at a distance, the width of the bus route number is small. The *smallest width* of the bus route number is used to remove the noise bounding-boxes with smaller width. This *removing small width BBs rule* is defined as: If the width in a bounding-box is smaller than a predefined value (T_{SW}, is set as 10 for frame image with width 640), this bounding-box is removed.

(9) Removing illegal ratio BBs rule: When a bus is approaching, the ratio of bus route numbers is approximately constant. Thus, the *ratio* of the bus route number is used to remove the illegal ratio BBs. This *removing illegal ratio BBs rule* is defined as: If the ratio in a bounding-box is smaller than a pre-learning value (T_{rL}) or greater than a pre-learning value (T_{rH}), this BB is removed. The T_{rL} and T_{rH} are set as 0.85 and 1.24, respectively, for two digits in frame image with 640*480.

(10) Removing boundary BBs rule: The bus route number is usually located at the center of the bus panel. If the detected BBs are close to the image boundary, they are noise. To remove these noise BBs near the image boundary, the *distance* between a BB and the image boundary is used. This *removing boundary BBs rule* is defined as: If the distance between a BB and the image boundary is less than a pre-learning value (T_d), this BB is removed.

(11) Identifying bus route number BBs rule: After applying all above-mentioned rules, the remaining BBs are the bus route number, start text, end text, and traffic marking. In order to identify the bus route number, start place and end place in the bus panel are used. Usually, the start place, the bus route number, and the end place are shown in the bus panel from left to right. In Fig. 1(a), the text of the start place is "Jianguo N. Rd.", the bus route number is "74", and the text of the end place is "Jingmei." Thus, the *left distance* between start place and bus route number and the *right distance* between bus route number and end place are used to identify the bus route number. This *identifying bus route number*

rule is defined as: If a BB has left start text and right end text and the left and the right distances are less than T_{Ld} and T_{Rd}, respectively, this BB is the bus route number.

After applying all above-mentioned rules for Fig. 3(a), the image of the bus route number localization is shown in Fig. 3(b). From this figure, the bus route number "74" has been located.

4   Experimental Results

The Red-Otsu Thresholding method proposed was implemented as a Microsoft Visual C# 2012 Windows-based application on an Intel(R) Core(TM) i7-3667U CPU @ 2.00GHz Notebook, carried out on two video clips. These two video clips are captured by a "glasses" camera with 640 x 480 pixel resolution, simulating a visually impaired person "seeing" the moving bus at the station. The video clips 1 and 2 are used as the training set and the testing set of 92 and 161 frame images, respectively. The training set is used to obtain the pre-learning parameters in Section 3, which the proposed method used to detect the bus route number in the training and testing sets, respectively.

To demonstrate the performance of the proposed method, it is compare to the gray-Otsu thresholding [12]. The comparisons include the bus route number detection rate and the execution complexity.

Table 1. Detection performance of the proposed red-Otsu thresholding method

Video clip	Original frames	Original bus route number	Detected bus route number	Detected rate
Training set	92	92	92	100%
Testing set	161	161	161	100%

Table 1 shows the detection performance of the proposed method. The actual bus route number in the training and testing sets is 92 and 161, respectively, which is the route number detected by the proposed method in the two sets, i.e., the proposed method detected correctly in both sets. By contrast, as Table 2 shows, the detection performance of the gray-Otsu thresholding method was less good: The number of correct bus route number detections by the gray-Otsu thresholding method in the two sets are only 59 and 121, respectively, i.e., detected rates for the gray-Otsu thresholding method in two sets are 64.1% and 75.2%. The reason is that the gray-Otsu thresholding method is affected by the illumination, while the proposed red-Otsu thresholding method is not affected by the illumination.

Table 2. Detection performance of the gray-Otsu thresholding method

Video clip	Original frames	Original bus route number	Detected bus route number	Detected rate
Training set	92	92	59	64.1%
Testing set	161	161	121	75.2%

Table 3. Execution times performance of the proposed red-Otsu thresholding method

Video clip	frames	binary	CCL	identify	whole	FPS
Training set	92	0.440(s)	0.617(s)	0.162	1.219(s)	75.47
Testing set	161	0.756(s)	1.039(s)	0.238	2.033(s)	79.19

Table 3 shows the execution time performance of the proposed red-Otsu thresholding method. The size of both training and testing sets is 640 x 480. The numbers of the frame for training and testing are 92 and 161, respectively. The total execution times for the training and the testing videos are 1.219 and 2.033 seconds, and the FPSs are 75.47 and 79.19, respectively. These times are fast enough to make the system useful to a person waiting for a bus.

Table 4. Execution times performance of the gray-Otsu thresholding method

Video clip	frames	binary	CCL	identify	whole	FPS
Training set	92	0.916(s)	0.598(s)	0.187(s)	1.701(s)	54.08
Testing set	161	1.611(s)	1.015(s)	0.340(s)	2.966(s)	54.28

Table 4 shows, for comparison, that the execution time performance of the gray-Otsu thresholding method for the training and the testing videos are 1.701 and 2.966 seconds, and the FPSs are 54.08 and 54.28, respectively. Thus, the binary execution time of the proposed method is less than the gray-Otsu thresholding method. Thus, the proposed method is faster than gray-Otsu thresholding method. The reason is that the proposed method is directly implemented at red component, while the gray-Otsu thresholding method needs convert the RGB color model into gray level and find the threshold value.

5   Conclusions

A method is proposed to detect the bus route number of a moving bus. The proposed method modified the Otsu threshoding method and applied on red component in the RGB color model to identify the bus route number. Experiment showed that the proposed method is more accurate and faster than the gray-Otsu thresholding method. In the future, the actual text of the bus route number will be recognized and translated into voice to form the full system to "see" the bus route number.

Acknowledgements. The author would like to express his gratitude to Dr. Jeffrey Lee and Walter Slocombe, who assisted editing the English language for this article.

References

1. American Foundation for the Blind, http://www.afb.org/section.aspx?SectionID=40&TopicID=168&DocumentID=907
2. Taipei e-bus System, http://www.e-bus.taipei.gov.tw/new/english/en_index_6_1.aspx

3. Department of Statistics, Ministry of the Interior, Taiwan,
   http://statis.moi.gov.tw/micst/stmain.jsp?sys=100
4. Ezaki, N., Bulacu, M., Schomaker, L.: Text detection from natural scene images towards a system for visually impaired persons. In: Proceeding of ICPR 2004, Tampa, FL, vol. 2, pp. 683–686 (2004)
5. Tsai, C.-M.: Local Skew Estimation in Moving Business Cards. In: Ali, M., Bosse, T., Hindriks, K.V., Hoogendoorn, M., Jonker, C.M., Treur, J. (eds.) IEA/AIE 2013. LNCS, vol. 7906, pp. 644–653. Springer, Heidelberg (2013)
6. Zandifar, A., Duraiswami, R., Chahine, A., Davis, L.: A video based interface to information for the visually impaired. In: Proceeding of IEEE 4th International Conference on Multimodal Interfaces, pp. 325–330 (2002)
7. Tsai, C.M., Yeh, Z.M.: Text detection in bus panel for visually impaired people 'seeing' bus route number. In: ICMLC 2013, pp. 1234–1239 (2013)
8. Noor, M.Z.H., Ismail, I., Saaid, M.F.: Bus detection device for the blind using RFID application. In: International Colloquium on Signal Processing & Its Applications, pp. 247–249 (2009)
9. Mustapha, A.M., Hannan, M.A., Hussain, A., Basri, H.: UKM campus bus identification and monitoring using RFID and GIS. In: IEEE Student Conference on Research and Development, pp. 101–104 (2009)
10. Pan, H., Yi, C., Tian, Y.: A Primary Travelling Assistant System of Bus Detection and Recognition for Visually Impaired People. In: IEEE Workshop on Multimodal and Alternative Perception for Visually Impaired People (MAP4VIP), in conjunction with ICME 2013 (2013)
11. Gonzalez, R.C., Woods, R.E.: Digital Image Processing, 3rd edn. Prentice Hall (2008)
12. Otsu, N.: A threshold selection method from gray-level histogram. IEEE Trans. SMC 9, 62–66 (1979)
13. Dunlap, J.: Queue-linear flood fill: A fast flood fill algorithm, CodeProject (2006),
    http://www.codeproject.com/Articles/16405/
    Queue-Linear-Flood-Fill-A-Fast-Flood-Fill-Algorith
14. Chang, F., Chen, C.J., Lu, C.J.: A linear-time component-labeling algorithm using contour tracing technique. CVIU 93(3), 206–220 (2004)
15. Tsai, C.-M.: Intelligent post-processing via bounding-box-based morphological operations for moving objects detection. In: Jiang, H., Ding, W., Ali, M., Wu, X. (eds.) IEA/AIE 2012. LNCS, vol. 7345, pp. 647–657. Springer, Heidelberg (2012)

Stock Portfolio Construction Using Evolved Bat Algorithm

Jui-Fang Chang[1], Tsai-Wei Yang[1], and Pei-Wei Tsai[2,*]

[1] National Kaohsiung University of Applied Science, Kaohsiung, Taiwan
rose@cc.kuas.edu.tw, 1101346112@kuas.edu.tw
[2] National Kaohsiung Marine University, Kaohsiung, Taiwan
peri.tsai@gmail.com

Abstract. Investment portfolio construction is always a popular issue for the investors. In this paper, we utilize Investment Satisfied Capability Index (ISCI) to sieve out the potential candidate stocks and then use Evolved Bat Algorithm (EBA) to construct the portfolio for stock investment. Three years historical daily Return on Investment (ROI) data from 2008 to 2010 is included in the experiment in order to test and verify the performance of the constructed stock portfolio by EBA. The experimental result is compared with the investment results from Taiwan broader market and Taiwan Top 50 Tracker Fund of Taiwan Yuanta Securities. The experimental result indicates that our proposed approach provides an efficient portfolio for the stock investment.

Keywords: Stock portfolio, investment portfolio, evolved bat algorithm, swarm intelligence.

1 Introduction

Investment is one way to expend the value of the property. According to the published information from Taiwan Directorate-General of Budget, Accounting and Statistics (DGBAS), the National Income (NI) of Taiwanese is increasing every year. It implies that people have more chance to put their funding in the investment. The stock investment is one of the big market for investment in Taiwan. In many cases, the investment portfolios are made according to the decisions from the investment managers no matter bases on their previous experiences or their hunch. To construct the investment portfolio, many people choose stocks to invest in accordance with the company's fundamental and technical information. Some investors use Time-series models such as ARCH, GARCH, GJRGARCH, etc. to assist the capital allocation; some of them utilize methods from artificial intelligence as the tools to deal with the same task. However, researchers have noticed that the ROI is not the only evaluation criterion for the investors to measure the investment performance. Chang et al. proposed ISCI [3], [5] in 2009 providing the investors to measure their own risk acceptance.

In this paper, we propose an approach to construct the stock portfolio. Our method can be split into two phases: the candidate stock selection and the capital allocation.

A. Moonis et al. (Eds.): IEA/AIE 2014, Part I, LNAI 8481, pp. 331–338, 2014.
© Springer International Publishing Switzerland 2014

In the first phase, five major conditions and necessary requirements [7] proposed by Chou in 2009 are employed to sieve out the stocks with better potential to be the candidate stocks. Furthermore, we use ISCI to select 15 investment targets from the candidates for later usage in the second phase. In the second phase, EBA (Tsai et al., 2011) [11] is exploited to optimize the capital allocation for investing on the selected targets. The purpose of this work is to provide the investors an investment portfolio with optimum reward. The sliding window strategy is involved in the experiment to test the performance of our design. The experimental result is compared with the broader market and Yuanta Securities portfolio. The rest of the article is composed as follows: The literature review on portfolio theory and EBA are given in section 2; our proposed approach is introduced in section 3; the experiment design and the experimental result are given in section 4; and finally, the conclusion is made in the last.

2   Literatures Review

2.1   Portfolio Theory

The concern on the risk diversification of stocks are led in to the stock portfolio analysis by two American economists, Markowitz (1952) and Sharpe (1964). The portfolio theory is first proposed by Markowitz in 1952. This theory contains two major elements: the "Mean-variance analysis method" and the "efficient frontier model portfolio." The Mean-variance analysis method makes the portfolio providing two useful information to the investors: the first one is the maximum returns under the fixed risk; and the second one is the minimum risk under the fixed return. [10]

Besides the issue on the investment return and the risk level, Evans and Archer (1968) bring the idea of diversification of the portfolio into the discussion. [1] Based on their finding, the conclusions are summarized as follows:

1. The investment risk would be fully diversified under the condition that the stock portfolio is composed by 10 to 15 stocks.
2. The diversification is helpful for reducing the investment risk, however, the investment risk would not drop when the portfolio is composed by more than 15 stocks.

Forthergill and Coke (2001) also claim similar conclusion that the risk diversification can be achieved only if the portfolio contains 15 to 20 stocks. [1]

Nowadays, many researchers propose new stock portfolio construction methods with the assistance from the Artificial Intelligence (AI) algorithms. For instance, Lin (1998) utilize Genetic Algorithm (GA) to give assistant on composing the stock portfolio for the investors. [9] To further consider the satisfactory level of the investors received from their ROI, Chang et al. propose ISCI [3], [5] in 2008 to quantitate the satisfactory level of the investors. In 2009, Chang and Chen (2009) utilize PSO with ISCIFCM [4], which combines ISCI and Fuzzy C-Means clustering, to construct the stock portfolio. Chang et al. (2011) use ISCIFCM with DEA Investment Portfolio Efficiency Index (DPEI) to select the potential stocks for investing. [2] They use

Genetic Algorithm (GA) and PSO to decide the capital allocation for the investment portfolio. Based on their findings, the potential of using AI methods to assist the construction of the stock portfolios is clear to see.

2.2   Evolved Bat Algorithm (EBA)

Swarm intelligence brings up many researchers' interest in this field in recent years. Most of the algorithms in the field of swarm intelligence are developed based on the evolutionary computing. They have also been successfully used to solve optimization problems in the engineering, the financial, and the management fields. In 2010, Tsai et al. propose Evolved Bat Algorithm (EBA) [11] based on utilizing the basic structure of Bat Algorithm (Yang, 2010) [12], [13] and re-estimating the characters used in BA. The details of EBA are given as follows.

In EBA, the medium for spreading the sonar wave should be decided at the first beginning. We follow Tsai et al.'s experience and define the medium to be the air. It results in the distance between a virtual bat to a target coordinate can be calculated by Eq. (1):

$$D = 170 \cdot \Delta T \left({m}/{sec.} \right) = 0.17 \cdot \Delta T \left({km}/{sec.} \right) \tag{1}$$

where D denotes the distance and ΔT means the time difference between sending the sound wave and receiving the echo. The value of ΔT is assigned to be a random number in the range of [-1 , 1]. The negative part of ΔT comes from the moving direction in the coordinate. ΔT is given with a negative value when the transmission direction of the sound wave is opposite to the axis of the coordinate.

The movement of the bat in EBA is defined by Eq. (2):

$$x_i^t = x_i^{t-1} + D \tag{2}$$

where x_i denotes the location of the ith bat in the solution space and t indicates the current iteration.

Moreover, if a virtual bat moves into the random walk process, its location will be updated by Eq. (3):

$$x_i^{t_R} = \beta \cdot (x_{best} - x_i^t) \tag{3}$$

where $x_i^{t_R}$ indicates the new location of the bat after the random walk process, β is a random number in the range of [0 , 1], x_{best} denotes the coordinate of the near best solution found so far, and x_i^t is the present location of the virtual bat.

The processes of EBA can be depicted in 4 steps:

Step 1. Initialization: Randomly spread the bats into the solution space.
Step 2. Move the bats by Eq. (1)-(2). Generate a random number. If it is greater than the pulse emission rate, move the bat by the random walk process, which is defined by Eq. (3).
Step 3. Evaluate the fitness of the bats and update the global near best solution.
Step 4. Check the termination condition to decide whether go back to step 2 or terminate the program and output the near best result.

3   Methodology and Our Proposed Method

Our proposed method can be split into two phases: the candidate stock selection and the capital allocation. The details are given as follows.

3.1   The Candidate Stock Selection

When selecting the candidate stock with ISCI, an index value called C_{SL} is calculated. C_{SL} is a quantification measurement, which reflects the investor's ROI. We adopt Pearn et al.'s (2001) design [6], which utilizes the Probability Density Function (PDF) of C_{SL} with the minimum unbiased estimator to provide a complete statistical test procedure and the evaluation criteria for assisting the investors on analyzing the performance of individual stocks. In general, every investor has different feelings of satisfaction about the ROI. It implies that the Lower Reward Limit (LRL) for every investor should also be different. Thus, we set the *LRL* as the risk-free rate plus the annual rate of inflation as the minimum criterion in our design. For example, the one-year Treasury bill rate is 0.5957% and the inflation rate is 0.1687% in 2011; then the daily *LRL*, which is calculated by Eq. (4), is equal to 0.0031%.

$$LRL = [(0.5957\% + 0.1687\%)/247] \qquad (4)$$

where 247 is the count of the trading days in 2011.

To sieve out the candidate stocks with higher rate of return and better performance from the large amount of listed stocks, we use *LRL* as a threshold. The market price can be treated as a random variable (X) because it is changing at any time. When $X < LRL$, it implies that the performance of the individual stock is lower than the investor's expectation. Therefore, the daily returns of each stock's ISCI can be calculated by Eq. (5):

$$C_{SL} = \frac{\mu - LRL}{3\sigma} \qquad (5)$$

where μ is the average value of the stock's daily return and σ is the standard deviation. The higher the C_{SL} value is, the better performance of the stock is.

The statistical sampling method is used to estimate the number of the nonparametric (μ, σ). Let x_i be the ith sample data of the daily return of the individual stock. Under the normality assumption $X_i \sim N(\mu_i, \sigma_i)$, we calculate the sample mean $(\bar{X})$ and the standard deviation (s) to estimate the mean of the population (μ) and the standard deviation (σ). The natural estimator of ISCI can is given in Eq. (6)-(8):

$$\hat{C}_{SL} = \frac{\bar{X} - (R_f + CPI_a)}{3S} \qquad (6)$$

$$\bar{X} = \sum_{f=1}^{n} \frac{x_i}{n_i} \qquad (7)$$

$$S^2 = (n-1)^{n-1} \sum_{i=1}^{n} (X_i - \bar{X})^2 \qquad (8)$$

where $\bar{X}$ and S^2 denote the average and the variance of the daily returns of individual stocks, R_f is the one-year Treasury bills for each year, and CPI_a is the annual growth rate of the inflation of each year.

Chou and Owen (1989) claim that the distribution estimation formula $\hat{C}_{SL}$ can be written as $C_n t_{n-1}(\delta)$ under the normal distribution assumption, where $C_n = (3\sqrt{n})^{-1}$ and $t_{n-1}(\delta)$ have the non-central t distribution, the degree of freedom equals to $n - 1$, the non-central parameter is $\delta = 3\sqrt{n}C_{SL}$ [8].

The investors' satisfaction under the assumption of normality can be calculated by Eq. (9):

$$PI = P(X > LRL) = P\left(\frac{\mu-X}{3\sigma} < \frac{\mu-LRL}{3\sigma}\right) = 1 - \Phi(-3C_{SL}) = \Phi(-3C_{SL}) \qquad (9)$$

where LRL is the lower limit of the daily return, X is the daily return of the stock, μ is the average daily return, σ is the standard deviation of the daily return, and Φ is the standard normal cumulative distribution function.

Table 1 shows the C_{SL} value and the corresponding Investment Satisfaction Degree (ISD). For example, when the value of C_{SL} is 0.0844, the ISD is equal to 60%; when the value of C_{SL} is equal to 1, the ISD is 99.87 %.

Table 1. Satisfaction Capability Index of the Investment Performance

C_{SL}	Investment Satisfied Degree(ISD)	C_{SL}	Investment Satisfied Degree(ISD)
0.0844	0.6	0.3455	0.85
0.1284	0.65	0.4272	0.9
0.1748	0.7	0.5483	0.95
0.2248	0.75	1	0.9987
0.2805	0.8		

3.2 The Capital Allocation

We first select candidate stocks from "Money Line Tycoon." The candidates all satisfy five conditions listed as follows:

1. The listed scale is above 20 million.
2. The company is established over 5 years.
3. The gross margin must be greater than 10% annually in the past five years.
4. Operating profit rate must be greater than 7% annually in the past five years.
5. The stock price is higher than $5 and the trading volume of 5-day average must not than 500.

The reason we compose the candidate stocks with conditions listed above is because the company has a complete data indicating the fund manager's operating performance, and they may be better to fight against the impact of market risk if their capital is above 20 million. After excluding stocks without complete information, 92 candidates are remained from 9728 Taiwan listed and OTC stocks. Based on the candidate stocks, we use ISCI to further select top 15 stocks to be the potential investment targets. The weighting for the capital allocation will be optimized by EBA only with the selected investment targets. The top 15 Taiwan stocks selected by ISCI is listed as follows: Hiwin (2049), Quaker (1227), Polylite (1813), Flexium (6269), Yeashin (5213), Tong Yang Group (4105), Yulon-motor (9941), Lotes (3533), Cheng-Shin Tire (2105), Highwealth (2542), Honchuan (9939), Ememory (3529), TSRC (2103), Elaser (3450), and Grape king (1707).

The historical data from 2008 to 2010 is included in our experiment. We use the past two continuous days' daily ROIs of the target stocks to be the input for EBA for optimizing the weightings for the capital allocation. The weightings are used to invest in the next day and the ROIs of the investment targets are multiplied by their weightings, accordingly, and summed up for calculating the ROI. To test the performance of our proposed method, the sliding window strategy is employed to run over from the first trading day in 2008 to the last trading day of 2010. The parameter setting for EBA is listed in Table 2.

Table 2. Parameter Settings for EBA

Pulse Emission Rate	0.5	Population Size	32
ΔT	[-1, 1]	Iteration	800
β	[0, 1]	Runs	30

4   Experimental Result

We sample part of the data points and list them in Table 3.

Table 3. Accumulative ROIs of "Yuanta Top50 Tracker Fund," "Broader Market," and "EBA Portfolio."

Date	Yuanta (%)	Broader Market (%)	EBA (%)
2008/01/15	2.303	2.664	4.121
2008/03/03	1.138	1.333	11.422
2008/06/19	0.802	-0.560	46.852
2008/09/23	-14.942	-25.141	35.151
2008/12/31	-50.324	-52.594	10.776
2009/01/15	-6.178	-5.765	-1.333
2009/03/03	-2.606	-2.842	22.444
2009/06/19	28.802	32.693	114.888
2009/09/23	49.624	50.114	241.697
2009/12/31	56.865	60.878	304.879
2010/01/15	2.352	2.078	8.655
2010/03/03	-6.698	-6.700	-8.360
2010/06/19	-7.440	-6.053	11.323
2010/09/23	2.681	1.363	58.333
2010/12/31	9.601	10.499	75.786

The experimental result indicates that the investment performance with our proposed method presents better ROIs than Yuanta and Broader market from 2008 to 2010. The results are drawn year-by-year in Fig. 1 to Fig. 3.

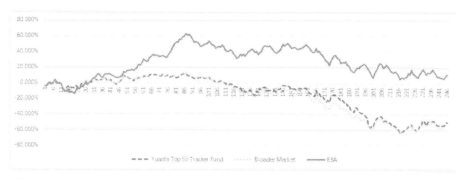

Fig. 1. Accumulated ROIs of Yuanta Top 50 Tracker Fund, Broader Market, and EBA in 2008

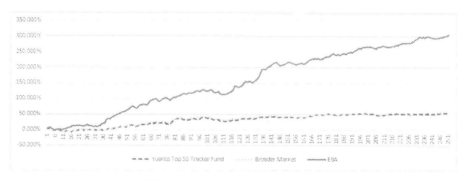

Fig. 2. Accumulated ROIs of Yuanta Top 50 Tracker Fund, Broader Market, and EBA in 2009

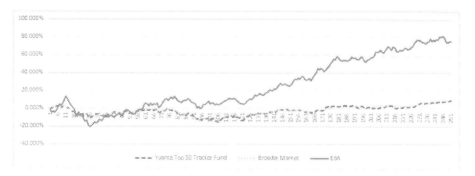

Fig. 3. Accumulated ROIs of Yuanta Top 50 Tracker Fund, Broader Market, and EBA in 2010

Although our method doesn't get good investment result when looking at some specific days, the final results of investing over all year are still positive. When the investment result is worse than Yuanta and Broader market, it means that EBA failed to locate the optimum weighting distribution in our designed iterations. This may be overcome by extending the iteration or increase the population size.

5   Conclusions

In this paper, we use ISCI to select the investment targets from totally 9728 Taiwan listed and OTC stocks. Furthermore, EBA is employed to optimize the weight distribution of the funds to construct the portfolio, which may be more close to the investor's expectancy. The results show that using ISCI and EBA to construct portfolio outperforms the investment result than Yuanta Securities portfolio and Broader market. In the future, we aim at increasing the experiment scale by considering the qualitative data and the information from the market in our design to further improve the performance.

References

1. Benjelloun, H.: Evans and Archer – forty years later. Investment Management and Financial Innovations 7(1), 98–104 (2010)
2. Chang, J.-F., Chang, C.-W.: Comparison of DPEI and ISCIFCM Indexes Based on PSO Capital Weighting. ICIC Express Letters 5(4(B)), 1353–1358 (2011)
3. Chang, J.-F., Chen, J.-F., Lin, S.-H.: Applying Investment Satisfied Capability Index and Particle Swarm Optimization to Construct the Stocks Portfolio. In: 3rd International Conference on Innovative Computing, Information and Control. IEEE Press, Dalian (2008)
4. Chang, J.-F., Chen, K.-L.: Applying New Investment Satisfied Capability Index and Particle Swarm Optimization to Construct Stock Portfolio. ICIC Express Letters 3(3), 349–355 (2009)
5. Chang, J.-F., Shi, P.: Using Investment Satisfaction Capability Index Based Particle Swarm Optimization to Construct a Stock Portfolio. Information Sciences 181(14), 2989–2999 (2011)
6. Chen, K.S., Pearn, W.L.: Capability indices for processes with asymmetric tolerances. Journal of the Chinese Institute of Engineers 24(5), 559–568 (2001)
7. Chou, M.-H.: Chou: Analysis the Timing Stock Picking Ability of China's Open-end Funds. Web Journal of Chinese Management Review 12(1) (2009)
8. Chou, Y.-M., Owen, D.-B.: On the distributions of the estimated process capability indices. Communications in Statistics: Theory and Methods 18(2), 4549–4560 (1989)
9. Lin, P.-C.: An Application of Genetic Algorithms on User-Oriented Portfolio Selection. Journal of Information Management 7(1) (2000)
10. Markowitz, H.: Portfolio Selection. The Journal of Fiance 7(1), 77–91 (1952)
11. Tsai, P.-W., Pan, J.-S., Liao, B.-Y., Tsai, M.-J., Vaci, I.: Bat Algorithm Inspired Algorithm for Solving Numerical Optimization Problems. Applied Mechanics and Materials 148-149, 134–137 (2012)
12. Yang, X.-S.: A New Metaheuristic Bat-Inspired Algorithm. In: González, J.R., Pelta, D.A., Cruz, C., Terrazas, G., Krasnogor, N. (eds.) NICSO 2010. SCI, vol. 284, pp. 65–74. Springer, Heidelberg (2010)
13. Yang, X.-S.: Nature-Inspired Metaheuristic Algorithms. Luniver Press, United Kindom (2010)

Semantic Frame-Based Natural Language Understanding for Intelligent Topic Detection Agent

Yung-Chun Chang[1,2], Yu-Lun Hsieh[1], Cen-Chieh Chen[3], and Wen-Lian Hsu[1]

[1] Institute of Information Science,
Academia Sinica, Taipei, Taiwan,
[2] Department of Information Management,
National Taiwan University, Taipei, Taiwan
[3] Department of Computer Science,
National Chengchi University, Taipei, Taiwan
{changyc,morphe,can,hsu}@iis.sinica.edu.tw

Abstract. Detecting the topic of documents can help readers construct the background of the topic and facilitate document comprehension. In this paper, we proposed a semantic frame-based method for topic detection that simulates such process in human perception. We took advantage of multiple knowledge sources and identified discriminative patterns from documents through frame generation and matching mechanisms. Results demonstrated that our novel approach can effectively detect the topic of a document by exploiting the syntactic structures, semantic association, and the context within the text. Moreover, it also outperforms well-known topic detection methods.

Keywords: Topic Detection, Semantic Frame, Semantic Class, Partial Matching, Sequence Alignment.

1 Introduction

Due to recent technological advances, we are overwhelmed by the sheer number of documents and have difficulty assimilating knowledge of interest from them. To promote research on automatically detect topics and track related documents, the Defense Advanced Research Projects Agency initiated the Topic Detection and Tracking project. In essence, a topic is associated with specific times, places, and persons [1]. To manifest topic associated features, one often needs to annotate them, which is rarely done in most machine learning models [2]. Nevertheless, current machine learning models have encountered bottlenecks due to knowledge shortage. The purpose of machine learning is to learn patterns that are general enough to be applied to unseen texts. However, they can only achieve a mediocre score. This fact is especially obvious when comparing the similarity of two sentences [3]. One can easily find two literally different sentences with similar semantics, which confuse most machine learning models.

We model topic detection as a classification problem. Our method is unique in that we took advantage of knowledge sources, and implemented a random walk algorithm to generate semantic frames that represent discriminative patterns in documents. Furthermore, we developed an alignment-based frame matching algorithm. Results demonstrated that it is effective in detecting topics of online news articles. In addition, the

A. Moonis et al. (Eds.): IEA/AIE 2014, Part I, LNAI 8481, pp. 339–348, 2014.

proposed method successfully exploits the syntactic structures, semantic association, and the content within the text. Consequently, the method outperforms the word vector model-based method [4] and the widely-used latent dirichlet allocation (LDA) method [5].

2 Related Work

Several machine learning-based approaches were used to recognize discriminative features for topic mining. For instance, Nallapati and Feng [1] attempted to find characteristics of topics by clustering keywords using a statistical similarity measure for grouping documents into clusters, each of which represents a topic. Furthermore, some researches treated topic detection as a supervised classification problem [5,6]. Given a training corpus containing manually-tagged examples of predefined topics, a supervised classifier is employed to assign (i.e. classify) topics. On the other hand, ontology is a conceptualization of a domain into a human understandable, machine-readable format consisting of entities, attributes, relationships, and axioms [7]. It is also re-usable, making it very powerful for representing knowledge. Some document detection methods made use of ontology and Wikipedia to enhance the performance [8].

Our method differs from existing approaches in a number of aspects. First, we proposed a semantic frame-based approach that mimics the perceptual behavior of humans. Second, the generated frames can be considered as the domain knowledge required for detecting topics. In addition, we considered syntactic features, context and semantic associations in articles. Finally, unlike other Chinese researches that rely on word segmentation for preprocessing, we utilize ontology for semantic class labeling.

The remainder of this paper is structured as follows: Section 3 describes the architecture of our topic detection system. Section 4 presents the experimental results and discussions. Finally, Section 5 concludes this work.

3 System Architecture

Our system consists of two mechanisms, the Semantic Frame Generation Mechanism (SFGM) and the Topic Detection Mechanism (TDM), as shown in Fig. 1. The SFGM first uses prior knowledge of each topic to mark the semantic classes of words. Then it collects frequently co-occurring tuples, and generates frames for each topic by a Probability Graphical Model. They are stored in the Topic-dependent Knowledge Base as our domain-specific knowledge. In the TDM, an article is first labeled using the same knowledge as mentioned above. Then an alignment-like algorithm obtains the similarity between topics and articles to determine the topic of an article. Details are explained in the following sections.

3.1 Semantic Frame Generation Mechanism, SFGM

As depicted in Fig. 1, the SFGM first labels words with their semantic classes in order to generalize and extract distinctive semantic features. Then a graph-based frequent

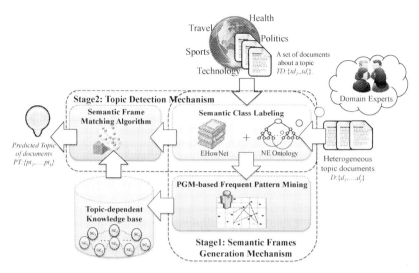

Fig. 1. Architecture of our semantic frame-based topic detection system

pattern mining algorithm is used to generate semantic frames, which form the basis of the topic detection mechanism that follows. In contrast, most Chinese topic detection researches rely on word segmentation, which is error-prone and can undermine the accuracy of a system. In light of this, we use the following two knowledge bases instead:

- **Named Entity Ontology, NEO**

We adopt a novel structure to construct the NE ontology which represents critical components in human knowledge. The architecture of the NEO includes a *topic layer*, a *semantic layer*, and an *instance layer*. There are five topics in the *topic layer*, namely "Sports", "Politics", "Travel", "Health", and "Technology". Moreover, there are 40 semantic classes in the *semantic layer*, including "醫生 (doctor)", "景點 (hot spot)", etc. Each class denotes a general semantic meaning of NEs that can be aggregated from many topics. The *instance layer* contains 6,323 NEs extracted from five topics by the Stanford NER[1]. Domain experts further annotated each NE with their corresponding semantic classes. Each instance can have multiple semantic classes. For example, in Fig. 2, the NE "蘋果 (apple)" can be generalized to both "水果 (fruit)" and "3C產品 (3C product)". Thus, the semantic class "水果 (fruit)" is mentioned in both "Health" and "Technology" topics.

- **Extended HowNet, EHowNet**

Extended HowNet, or EHowNet, an extension of HowNet [9] constructed by the CKIP of Academia Sinica, is a structural representation of knowledge and semantics. It linked approx. 90,000 words in the CKIP Chinese Lexical Knowledge Base and HowNet, and included additional frequent words that are specific in Traditional Chinese. It also adopted a different formulation of words to better fit its semantic representation, as well as distinct definitions of function and content words. A total of four basic semantic classes were applied, namely, *object*, *act*, *attribute*, and *value*. Furthermore, compared

---
[1] http://nlp.stanford.edu/ner

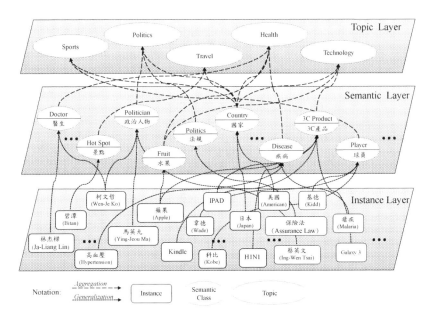

Fig. 2. Architecture of named entity ontology

to HowNet, EHowNet possesses a layered definition scheme and complex relationship formulation, and uses simpler concepts to replace *sememes* as the basic element when defining a complex concept or relationship. To illustrate, EHowNet defines "狗食 (dog food)" as:

Simple Definition:
{food|食物 : $telic$ = {feed|餵 : $target$ = {dog|狗}, $patient$ = {~}}}
Expanded Definition:
{food|食物 : $telic$ = {feed|餵 : $target$ = {livestock|牲畜 : $telic$ = {TakeCare|照料 : $patient$ = {family|家庭}, $agent$ = {~}}}, $patient$ = {~}}}

We can see that EHowNet not only contains semantics of a word, but also its relations to other words or entities. This enables us to combine or dissect the meaning of words by using its components.

Our research assigns clause as the unit for semantic labeling. As shown in Fig. 3, the input clause C_n = "國民隊的王建民又投出勝投 (Chien-Ming Wang, the pitcher of the Washington Nationals, throws a winning pitch again)" is first labeled by the NEO to tag all the named entities. Subsequently, the clause becomes "[球隊$_{team}$]的[球員$_{player}$]又投出勝投 ([player] of [team] throws a winning pitch again)". Then it is further labeled by EHowNet to tag the main definition of all remaining words, as shown in the following example, "[球隊$_{team}$] 的 [球員$_{player}$] 又 [投出$_{throw}$] [贏$_{victory}$] 投 ([player] of [team] [pitch] a [victory] pitch again)". Lastly, all the non-labeled words are removed to form the semantic class sequence, "[球隊$_{team}$] [球員$_{player}$] [投出$_{throw}$] [贏$_{victory}$]". This process can eliminate the errors caused by Chinese word segmentation as well as group synonyms of a word together by the same label, in order to find distinctive and prominent semantic classes for each topic.

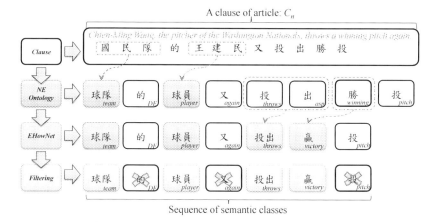

Fig. 3. Semantic class labeling process

We formulate semantic frame generation as a frequent pattern mining problem. Based on the co-occurrence of semantic classes, we can construct a graph to describe the strength of relations between them. Since semantic classes are of an ordered nature, the graph is directed and can be made with association rules. To avoid generating frames with insufficient length, we empirically set $support_{min}=50$, $confidence_{min}=0.3$ in our association rules. Thus, an association rule can be represented as (1).

$$confidence(SC_i \Rightarrow SC_j) = P(SC_j|SC_i) = \frac{support(SC_i \cup SC_j)}{support(SC_i)} \qquad (1)$$

Fig. 4 is an illustration of such a graph, in which vertices (SC_x) represent semantic classes, and edges represent the co-occurrence of two classes, SC_i and SC_j, where SC_i precedes SC_j. The number on the edge denotes the confidence of two connecting vertices. After constructing all semantic graphs, we then generate semantic frames by applying the random walk theory [10]. It consists of a series of random selections on the graph, in search of frequent and representative classes for each topic. Let a semantic graph G be defined as $G = (V, E)$, where $|V| = p, |E| = k$. Every edge (SC_n, SC_m) has its own weight M_{nm}, which denotes the probability of a semantic class SC_n, followed by another class SC_m. For each class, the sum of weight to all neighboring classes $N(SC_n)$ is defined as (2), and the whole graph's probability matrix is defined as (3).

$$\forall SC_n \sum_{m \in N(SC_n)} M_{nm} = 1 \qquad (2)$$

$$Pr = [X_{t+1} = SC_m | X_t = SC_n, X_{t-1} = SC_k, ..., X_0 = SC_i]$$
$$= Pr[X_{t+1} = SC_m | X_t = SC_n] = M_{nm} \qquad (3)$$

As a result, a series of a random walk process becomes a Markov Chain. According to Li et al. [4], the cover time of such a process on a normal graph is $\forall SC_n, C_{SC_n} \leq 4k^2$.

We can conclude that using random walk to find frequent patterns on semantic graphs would help us capture even the low probability combinations and shorten the processing time. Although this process can help us generate frames from frequent patterns, it can also create some redundancy. Hence, we reduce the redundancy by only retaining the longest and highest-coverage frames, and remove those that are completely covered by another frame. For example, we will remove the frame "[Country]-[Team]-[Player]" for it is completely covered by "[Country]-[League]-[Team]-[Player]-[Match]-[Lost]".

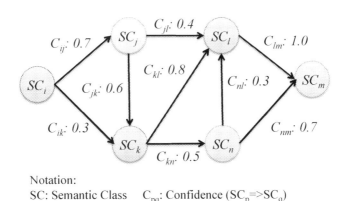

Notation:
SC: Semantic Class C_{pq}: Confidence ($SC_p \Rightarrow SC_q$)

Fig. 4. A semantic graph for frame generation

3.2 Topic Detection Mechanism, TDM

Most of the previous machine-learning based topic detection studies focus on feature engineering to enhance the performance. However, once the range of topics is changed, the effort must be repeated in order to find another optimal set of features. Unlike human knowledge, it is difficult to accumulate and share the knowledge collected from different topics. Moreover, the human perception of a topic is obtained through the recognition of important events or semantic contents to rapidly pinpoint possible candidates. For example, when strongly correlated terms like "Kobe Bryant (basketball player)", "LA Lakers (basketball team)" and "NBA (basketball league)" occur simultaneously, it is natural to conclude that this is a sport-related article, with a less likelihood of a technology-related one. This phenomenon can explain why humans can skim through an article to quickly capture the topic. In light of this rationale, we proposed a novel approach for topic detection that simulates such process in human perception.

We use semantic frames derived from the frame generation mechanism as basic knowledge for topic detection. First, a new article is first labeled with semantic classes, and a matching algorithm is applied to determine the topic. The matching algorithm is a modified version of sequence alignment [11] that enables a single frame to accurately match multiple semantically similar expressions. It compares all sequences of semantic classes $S = \{s_1, ..., s_n\}$ in an article to all the frames $F = \{f_1, ..., f_m\}$ in each topic, and calculates the sum of scores for each topic. Unlike normal templates that involve mostly rigid left-right relation of slots, we utilize these relations as scoring criteria during frame alignment. An illustration of the alignment process of a sequence of semantic classes to a semantic frame is shown in Fig. 5. The matched and unmatched contents between the two sequences were given different scores according to their type. An insertion is defined as a label that is present in the article but not in the frame, and is given a negative score computed from the entropy of this label, which can be thought of as the uniqueness or generality of this label. On the other hand, a deletion is defined as a label that is in the frame but not in the article, and a negative score is computed from the count of this semantic class divided by the total number of classes in this topic. Lastly, a match between the two sequences is given a positive score obtained from the frequency of the semantic class in a topic times 100. The topic with the highest sum of scores is considered as the winner.

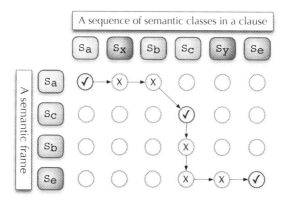

Fig. 5. Illustration of a semantic frame alignment process

4   Performance Evaluation

4.1   Dataset and Experimental Settings

To the best of our knowledge, there is no official corpus for Chinese topic detection. Therefore, we compiled a corpus for the evaluations. It contains five topics, i.e., "Sports", "Health", "Politics", "Travel", and "Tech", from a news agency database between the year 2000 and 2013. Each topic consists of 10,000 Chinese news documents. To derive credible evaluation results, we applied a 10-fold cross validation. The evaluation metrics are the precision, recall, and F1-measure, presented by macro-average to ensure the fairness of our evaluations. Three widely used methods were also implemented and evaluated [12]. The first is a word vector model which chooses keywords based on the TF-IDF score and uses the cosine distance to measure the similarity (denoted as TF-IDF). Another is a cluster-based method that integrates K-means with vector space model [4] (denoted as K-means). Lastly is a probabilistic graphical model which uses the LDA model as document representation to train an SVM to classify the documents as either topic relevant or irrelevant [5] (denoted as LDA-SVM).

4.2   Results and Discussion

Fig. 6 depicts the performance of our system on five topics. Our system performs the best on the topic "Sports", with precision, recall, and F1-measure scores of 76.22%, 49.7%, and 60.17%, respectively. This is because there are plenty of specific nouns in the articles within this topic. In addition, unique sports terms like "先發 (Starter)" and "晉級 (Advance)" are also common. These are the reasons why semantic frames for the topic "Sports" are very stable and distinct. On the other hand, high precision and low recall were found in topics "Health" and "Politics". In particular, the precision of "Politics" is the highest among all topics, i.e. 93.23%. We speculate that it is because named entities of politicians are common among articles about "Politics". Thus, the semantic class "Politicians" is useful in identifying this topic. However, considering the fact that building named entity ontology is time-consuming, we only included famous politicians in the ontology. The same is for the topic "Health". Only a limited number of medical terms were included in the ontology of "Health". The shortage of knowledge may be the cause of a restricted coverage of the semantic frames in these two topics. Nevertheless, the topic "Tech" has a low precision and high recall. We found that they frequently contain country names like "Taiwan", "China", or "USA". Consequently, the semantic class label "國家 (Countries)" exists in many frames of this topic as well as in another topic, "Travel". It is possible that the confusion of these topics caused the misjudgment of the topic "Tech".

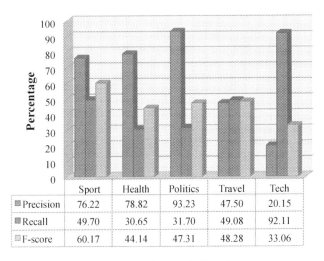

	Sport	Health	Politics	Travel	Tech
▨ Precision	76.22	78.82	93.23	47.50	20.15
▨ Recall	49.70	30.65	31.70	49.08	92.11
▨ F-score	60.17	44.14	47.31	48.28	33.06

Fig. 6. Performance of our topic detection system

As shown in Table 1, the random method was the worst among all methods, with PRF values of around 20%, as expected. TF-IDF method outperforms the random method, since it can weigh words in different topics by vectors in order to find unique words in each topic. For example: "NBA" and "林書豪 (Jeremy Lin)" are common in sports articles, while "治療 (Treatment)" and "患者 (Patient)" are frequent in health-related articles. TF-IDF method can assign different weights to these words according to their distribution, and discover the most distinctive ones. K-means outperforms TF-IDF and random method as it applied vector space model to calculate similarity between topic documents for K-means algorithm. However, K-means method ignores the context information within the text, which contains semantic content related to the topic. It is therefore inferior to our method.

Since TF-IDF and K-means both consider similarity between two words as a cosine function with independent dimensions, there is no way to represent relations among many words. By contrast, our method can extract semantics in addition to words, and generate frames that can capture the relations between them. As a result, we can achieve a better performance, with overall precision, recall, and F1-score of 63.18%, 50.65%, and 56.23%, respectively. However, when compared to LDA-SVM, our system has a higher precision and lower recall, which resulted in a lower F1-measure. It may be attributed to the use of Chinese word segmentation tool in LDA-SVM for constructing a word dictionary as background knowledge, in addition to a probabilistic graph with weighted edge representing between-word relations. In contrast, our system relies on topic-specific NE ontology for semantic class labeling and frame generation, which is constrained by the scope of the ontology. Despite the lower recall, our system is unique in that it can generate and accumulate knowledge during the process. In contrast to other researches, we can capture more crucial information for a topic over time. The semantic frames generated by our system can describe the semantic relations within a document and assist in detecting the topic. We consider them as the foundation for a deeper understanding of topics that extends beyond the surface words.

Table 1. Performance of topic detection systems

Method	Precision	Recall	F1-measure
Random	16.70%	20.03%	18.22%
TF-IDF	43.92%	39.15%	41.40%
K-means	55.62%	38.91%	45.80%
LDA-SVM	61.84%	**59.95%**	**60.88%**
Our Method	**63.18%**	50.65%	56.23%

5   Concluding Remarks

This research proposes a novel approach based on knowledge sources and semantic frame generation to the topic detection task. It differs from popular machine learning methods as it can create a topic-dependent knowledge base that is flexible and extendable. Results showed that this framework can effectively detect the topic of articles, as well as assist the user in constructing background knowledge of each topic in order to better understand the essence of them. In the future, we will expand the ontology to improve the effect of semantic class labeling and frame generation. Moreover, we will reduce the human effort and rapidly broaden the coverage of the knowledge ontology through automatic construction. Furthermore, we will also modularize the semantic labeling mechanism for the ease of use in other researches.

Acknowledgments. This research was supported by the National Science Council of Taiwan under grant NSC102-3111-Y-001-012, NSC102-3113-P-001-006 and NSC 102-3114-Y-307-026.

References

1. Nallapati, R., Feng, A., Peng, F., Allan, J.: Event threading within news topics. In: Proceedings of the Thirteenth ACM International Conference on Information and Knowledge Management, pp. 446–453. ACM (2004)
2. Scott, S., Matwin, S.: Feature engineering for text classification. In: ICML, vol. 99, pp. 379–388. Citeseer (1999)
3. Hsu, W., Chen, Y., Wang, Y.: A context sensitive model for concept understanding. In: Proceeding of 3rd International Conference on Information Theoretic Approaches to Logic, Language, and Computation (1998)
4. Li, S., Lv, X., Wang, T., Shi, S.: The key technology of topic detection based on k-means. In: 2010 International Conference on Future Information Technology and Management Engineering (FITME), vol. 2, pp. 387–390. IEEE (2010)
5. Blei, D.M., Ng, A.Y., Jordan, M.I.: Latent dirichlet allocation. Journal of Machine Learning Research 3, 993–1022 (2003)
6. Zhang, X., Wang, T.: Topic tracking with dynamic topic model and topic-based weighting method. Journal of Software 5(5), 482–489 (2010)
7. Tho, Q.T., Hui, S.C., Fong, A.C.M., Cao, T.H.: Automatic fuzzy ontology generation for semantic web. IEEE Transactions on Knowledge and Data Engineering 18(6), 842–856 (2006)
8. Grineva, M., Grinev, M., Lizorkin, D.: Extracting key terms from noisy and multitheme documents. In: Proceedings of the 18th International Conference on World Wide Web, pp. 661–670. ACM (2009)
9. Dong, Z., Dong, Q., Hao, C.: Hownet and its computation of meaning. In: Proceedings of the 23rd International Conference on Computational Linguistics: Demonstrations, pp. 53–56. Association for Computational Linguistics (2010)
10. Lovász, L.: Random walks on graphs: A survey. Combinatorics, Paul Erdos is Eighty 2(1), 1–46 (1993)
11. Needleman, S.B., Wunsch, C.D.: A general method applicable to the search for similarities in the amino acid sequence of two proteins. Journal of Molecular Biology 48(3), 443–453 (1970)
12. Manning, C.D., Schütze, H.: Foundations of statistical natural language processing, vol. 999. MIT Press (1999)

Turbo Machinery Failure Prognostics

Inhaúma Neves Ferraz[1] and Ana Cristina Bicharra Garcia[2,*]

[1] ADDLabs – Active Documentation and Design Laboratory
Instituto de Computação - Universidade Federal Fluminense
Av. Gen. Milton Tavares de Souza, s/n - 24210-340 Niterói, RJ, Brazil
{ferraz@addlabs.uff.br}
[2] Instituto de Computação -Universidade Federal Fluminense
Av. Gen. Milton Tavares de Souza, s/n - 24210-340 Niterói, RJ, Brazil
{bicharra@ic.uff.br}

Abstract. The turbomachinery systems on oil platforms are extremely sensitive, as their failure can cause total shutdown of the platform's production activity. Scheduled stoppages for preventive maintenance are effective, since qualified personnel and replacement parts are available and ready. On the other hand, shutting down a turbo machine ceases petroleum production. Not rarely, maintenance people conclude the machine could have worked much longer. Monitoring the machines' behaviors to predict the best time to stop has been an industry trend. Failure prognosis has been performed using statistical methods and Artificial Intelligence techniques, such as Neural Networks, Fuzzy Systems and Expert Systems. This study presents the results obtained by developing an expert system based on case-based reasoning for turbomachinery failure prognostics.

Keywords: Turbo machines, vibration, failure prognosis, Artificial Intelligence, Case-Based reasoning.

1   Introduction

The equipment used for oil extraction and exploration operates under severe conditions. High pressure, high temperatures, aggressive working conditions, high throughput and long shifts can have a critical effect on any component.

The turbomachinery systems are the most sensitive equipment on a platform, since any interruption causes total shutdown of platform activity, resulting in high financial cost. To avoid interruption to turbomachinery operation it is very important to carry out preventive maintenance that is scheduled to occur during circumstances and at times when there is a high degree of control. The choice of when to carry out this preventive maintenance is based on the equipment's failure prognostics. Failure models that predict time or cycles to failure have been available for materials and simple structures for years and they are generally based on vibration analysis (Pusey & Roemer 1999). Using regular measuring, it is possible to draw a trend graph that will

---

* Corresponding author.

A. Moonis et al. (Eds.): IEA/AIE 2014, Part I, LNAI 8481, pp. 349–358, 2014.

make advance scheduling possible, by identifying the best window of opportunity for equipment maintenance. Experts examine the data and use the information to determine potential trends and fault models. These professionals are expensive, and hard to find. This gives rise to a window of opportunity for providing intelligent systems that perform at least as well as human specialists in failure prognostics.

This work describes the results obtained from developing an intelligent system for failure prognostics on turbomachinery installed on oil platforms. A mandatory specification was to consider the context so that the observations stored about the machinery, when obtained from different equipment for example, would have a weighting to adjust the similarity. These weights are historic heuristic values and are stored by the maintenance team. In the case of turbomachinery, there is little available data and it is extremely uneven. When training set is unbalanced, the conventional least square error (LSE) training strategy is less efficient to train neural network (NN) for classification because it often lead the NN to overcompensate for the dominant group (Li et al. 2006). Then the use of Neural Networks and Fuzzy Systems has ceased to be a preferred option when solutions are being sought. When using specialist rule-based systems, the conclusion, as per Bergmann (2000), is that there is too much interdependency between the rules, that the rules obtained were difficult for non-specialists in Artificial Intelligence to understand and that the effects of changes (updates) to the rules were difficult to predict. As such, the most appropriate way to address the task was to use Case-Based Reasoning.

This paper presents CBRTurbo, a case-based reasoning system that uses a set of cases and a set of heuristic rules to reach a diagnosis for a machine behavior. A case is composed of: (1) machine behavior data: a set of 16 interpreted sensors' data and the trend for each sensor, (2) a set of parameters describing the environment, such as the environment temperature and the petroleum flow, and (3) the set of diagnosis related to the machine in the conditions described by (1) and (2). We compared our results with SVN and pure neural networks.

2   Overview of Prognostics Technology

The experienced based prognostics are used in absence of a model of the equipment and lack of sufficient sensors and require the failure history or recommendations of the designer of the component under similar operation. This kind of prognostics can be updated frequently for better results. Evolutionary prognostics rely on gauging the proximity and rate of change of the current component condition to measure the degradation of the component (Roemer et al. 2006). Usually this approach considers the mechanical system as a whole and requires multiple sensors information in order to assess the current condition of the system or subsystem and relative level of uncertainty in this measurement. The state estimator prognostics are a technique that uses filters, such as the Kalmann filters or various other tracking filters to obtain the desired prognostics. In this case, the minimization of error between a model and measurement is used to predict future feature behavior. At the end, the tracking filter approach is used to control and smooth out the features related to the prediction of progression of a given failure mode and thus is used in diagnosis and prognosis. A potential failure mode is the manner in which a failure can occur—that is, the ways in which the item fails to perform its intended design function, or performs the function but fails to meet its objectives (Lipol & Haq 2011). Failure modes are closely related to the functional and performance requirements of the product. Physics of failure is an approach that utilizes knowledge of a product's life cyclic loading and failure mechanisms to design for and assess reliability. This approach is based on the identification of potential failure modes, failure mechanisms and failure sites of the product as a function of the product's life cycle loading conditions (Pusey & Roemer 1999). A physics-based stochastic model can be used to evaluate the distribution of remaining useful component life as a function of uncertainties in component strength/stress or condition for a particular fault. Often the results from such a model can then be used to create a neural network or probabilistic-based autonomous system for real-time failure prognostic predictions.

Feature progression and AI based prognostics tracks the degradation paths of selected features that progress through a failure. For this approach is necessary to know the values that characterize the component failure being studied. These real world values are used for training in a machine learning procedure. Using the features of input and the expected output method AI will learn and provide as output the desired outcome. The difference between the prediction and the value obtained from the real world is the error that should be minimized in the learning system such as neural network, SVM, expert system, etc. At present, it is crucial for predictive maintenance to perfect and automate the ability to interpret vibration data, which represents a wide scope for applying Artificial Intelligence, particularly Neural Networks and Fuzzy Systems (Lopes, 1997). Whenever enough data is available these techniques present significant results.

Behavior models that should be monitored can be identified by statistical methods, such as Robust Estimation and Systems Identification, as well as techniques based on Artificial Intelligence, such as Neural Networks, Fuzzy Systems and Expert Systems. Expert Systems can adopt Rule-Based Reasoning and Case-Based Reasoning. Fault identification algorithms widely use pattern recognition techniques, mostly different Artificial Neural Networks (Loboda et. al 2012). In order to avoid direct use of complex statistical recognition methods, Pipe (1987) used a simplified linear and non-linear discriminant analysis technique. Neural networks or other AI techniques are trained on features that progress through a failure (Byington et. al 2002).

3   Case-Based Reasoning

Case-Based Reasoning (CBR) a technique which coordinates past events with current events to enable generalization and prediction (Schank 1982). It is a methodology for solving problems and not a specific technology (Watson 1999). Methodology is an organized set of principles which guide action in trying to "manage" (in the broad sense) real-world problem situations (Checkland & Scholes 1990).

CBR solves new problems by adapting similar solutions that were used to solve old problems to new problems (Riesbeck & Schank 1989). A case describes one particular diagnostic situation and records several features and their specific values occurred in that situation (Bergmann 2000). Conceptually CBR is commonly described by the CBR-cycle made up of four actions: retrieve, reuse, revise and retain (Watson 1995). Retrieval is the action of finding and returning cases similar to the one under analysis. Reusing is the action of adapting the solution retrieved so that it adapts to the new problem. Revision is the action of assessing the solution in terms of the current case, evaluating its effectiveness, and possibly reformulating it based on knowledge of the domain. This happens when the user does not agree with the solution and asks for a new case, defining which input features are relevant and what the correct fault and time to fault should be for this new case. Retention is the action of storing a newly recognized case in memory, for future use.

There are two key aspects of CBR technology: defining what "similar" means and defining the adaptation function.

Prognostics on a machine means detecting a fault that will lead to failure and predicting how much time will elapse before this probable failure occurs.

The CBR algorithm is characterized by its use of two functions: the similarity function and the adaptation function. The similarity function, usually implemented using the k-Nearest Neighbor algorithm, calculates the distance between the current case and the cases stored in the case base. This function should select cases that can be adapted easily to the current problem or select cases that have (nearly) the same solution for the current problem. The basic assumption is that similar problems have similar solutions (Bergmann 2000). The adaptation function starts with the current case and the case in the case base that is most similar to the current case. It returns the set of transformations / adaptations to be applied to the case pulled from the case base, for solutions to the current problem.

4   CBRTurbo Model

The model proposed uses CBR technology. As described in figure 1, the user inputs the data from the machine he needs a prognosis. The data input are temporal series of vibration from the sensors and the petroleum flow data. After pre-processing the data to segment the time series and to reach trend values for each series, the Controller calculates the distance between the current case data and all cases in the database. The three best fits are retrieved. The Controller triggers the Diagnosis module to reconcile the results and reach a probable set of diagnosis. After obtaining the probable diagnosis, the Controller triggers the Prognosis module to get the most probable date when the machine will break considering the trends. The user follows all the process and adjusts it by ratifying or rectifying the results. Based on this interaction, the weights for calculating distances might be modified by the Learning Module.

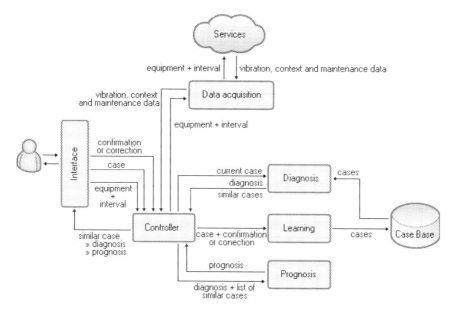

Fig. 1. System Model

The user interacts with the system via an interface module. The confirmed cases on the system are stored in a case base and the turbomachinery information is obtained using a data acquisition module. A diagnostic module identifies the most probable future fault for a case under analysis. When the user disagrees with the system's diagnosis and identifies a new one, a new case is generated and recorded on the case base by the learning module. A prognosis module generates the most probable time to future failure as a result of the fault diagnosed. A controller module oversees the activity of the other modules.

Each case was configured containing the equipment, a list of faults found, the user who confirmed the case, the platform where the equipment is installed and a list of vibration readings (for one or more sensors). The similarity of contexts compares equipment, sensor, rotation, total RMS (root mean square), number of peaks and greatest peak for the case in the case base, with corresponding values for the current case using heuristic weighting supplied by the platform's engineering team. Table 1 shows these weightings.

Table 1. Context weight values

Equipment/ Rotation / Total RMS / #peak / Greatest peak	
equal	1.00
same type	0.90
different	0.70
Sensor	
equal	1.00
same type, component and position	0.95
same component and position	0.90
different	0.70

The search for the most similar case combines classic vibration analysis (Mitchel 1993) with the equipment's environment (context). Mitchell makes the fault diagnosis via vibration by discrete analysis of the vibration readings' frequency spectrum. Discretization is done in multiple/submultiple frequency bands of the equipment's fundamental frequency, for which the total RMS is calculated. If the fundamental frequency is called N, the submultiple bands correspond to 0-40, 40-55 and 55-90 of N. The multiple bands correspond to 1N, 2N, 345N, with 345N being the band comprising the frequencies from 3N to 5N. In most cases the fundamental frequency was 182.5 Hz.

The cases in the base are stored with only one spectrum, the one with the total RMS closest to the corresponding shutdown value. The influence of the frequency spectrum gives origin to what is called internal weight (intWeight), which is calculated as the Euclidean distance between the total RMS vectors of the database case's spectrum and the closest failure spectrum of the current case. The influence of context gives origin to the external weight (extWeight). The similarity weight (w) will be expressed as the sum of the internal weight and the external weight.

Using the diagnostics selected by similarity, failure prognostics by regression of the machinery history are carried out. This is possible because the base cases store vibration readings over time. Starting from the point where the RMS of the current case reading coincides with the chosen regression curve ordinate (taken from the most similar case in the base), this curve is applied to the readings of the case under analysis, and, when the curve attains the shutdown RMS corresponding to the sensor of the reading under analysis, the time to failure prognosis will have been found.

CBRTurbo model was implemented using Visual C + + with MFC (Microsoft Foundation Classes) and was integrated to an environment accessing a Plant Information Database (PI).

The algorithm time complexity is $O(n^3)$ manly due to the prognosis part for which n is equal the number of sensor readings stored in the case base. So far there is no need for a parallel solution since the response time has been admissible in the range of few seconds on dual core computers with 4 GB memory. However, as the number of cases increase, a parallel solution will be called for. Our algorithm is easily paralleled since it can send each similar case to a different processor for adjustments.

5   Data Set

The maintenance team on the actual platform studied provided a set of tests showing 475 cases. There were 10 diagnosable faults (imbalance, rotor rub, angular misalignment, parallel misalignment, defective bearing, loose bearing or support, defective coupling, friction instability and oil film instability, as well as no fault detected). The initial number of confirmed cases was 43. There were 475 cases available, with 34 classified as no fault detected.

The overall hit rate achieved was **87.13%**, as described below. The hit rate by fault is shown in table 2.

Table 2. Fault diagnostic hit rate

Fault	Number of test runs with the fault	Hit rate
Imbalance	264	93.82%
Angular misalignment	63	83.60%
Parallel misalignment	79	80.26%
Defective bearing	26	84.61%
Loose bearing or support	2	100.00%
Defective coupling	0	ND
Friction instability	1	42.85%
Oil Film Instability	6	ND

The relative influence of the frequency spectra and context were analyzed using tests to compute the quantity of hits regardless of faults, annulling some of the singularity weight components (frequency spectrum and context factors). In this way the results shown in table 3 were obtained:

Table 3. Fault diagnostic hit rate X similarity weight component

Context attribute inconsideration	Overall Hit rate
All attributes	86.31%
Equipment attribute	89.50%
Sensor attribute	87.40%
Rotation attribute	89.50%
Total RMS attribute	87.40%
Peak number attribute	88.40%
Greatest peak attribute	87.40%
None (using all attributes)	89.50%

By calculating similarity only via the frequency spectrum, the worst result is obtained (86.31%). Of the context components in the sample of available cases, it was clear that the Total RMS and Greatest Peak attributes have the strongest influence on the hit rate, since not considering them caused the largest drop in the rate. Similarly, the Equipment attribute has the lowest influence on the hit rate since not considering it caused an insignificant drop in the hit rate.

In another test of the results obtained did the clustering of the data available in four models using the k-Means algorithm obtaining clusters sets KM1, KM2, KM4 and Km3. We compared the sizes of the clusters obtained with the cardinalities of fault types obtained with CBRTurbo getting the results shown in Figure 2.

One can see that the results are quite consistent presenting distortions only in fault types that are rare in the sample available such as Loose bearing or support, Defective coupling, instability Friction and Oil Film Instability. This means that the CBRTurbo acquires knowledge as more data are analyzed.

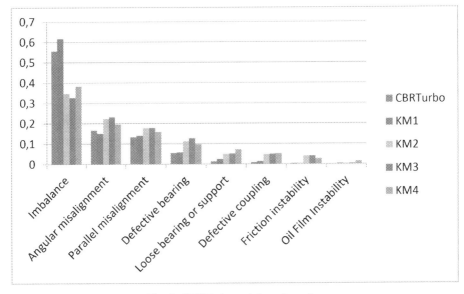

Fig. 2. CBRTurbo and data clustering

6   Conclusion

The analysis of vibrations in rotating machines has been addressed by several techniques (Yang et al. al 2003). The use of Statistical Methods such as PCA analysis and Autocorrelation Matrix was made in (Thiry et . al 2004). Multi -agent decision fusion was reported by (Niu et al. 2007). Fuzzy measure and fuzzy integral data fusion techniques can be found in (Liu et . al 2009) .

These studies are predominantly theoretical without displaying hit rates for comparison with the treatment presented in this article. In a paper on Ensemble Classifiers applied to pumps platforms (Wandekokem et . al 2010) we found some results that can be used for comparison. In this work, the hit rate for misalignment varies between 76.5 % and 77.8 % (SVM with or without ensemble classifiers) . CBRTurbo model provided us with results of 80.3% for angular misalignment and 93.8 % for parallel misalignment.

To unbalance (Wandekokem et . al 2010) present values between 82.2% and 83.3% while CBRTurbo model presents the 93.8% value.

The CBRTurbo results were very good for the prognosis using synthetic data, but were inconclusive using real data since the information available gives fault dates (confirming the diagnostics) but does not show any failure dates (to confirm the failures). Platform maintenance does not allow the vibration peaks to get high enough, so that when regression is applied to the prognosis the failure dates obtained are for periods of 10 or more years, which is not satisfactory.

Comparing the results obtained, we can conclude that the maintenance team overvalued the context components. It seems that the good results obtained by the context components are due to the fact that in the case base the frequency spectra of cases that are most similar to the current cases are chosen because of being similar equipment and therefore, likely to show similar faults.

Considering only the frequency spectra is less contradictory, and results of 81.10% can be considered quite satisfactory.

References

1. Bergmann, R.: Introduction to Case-Based Reasoning – Lecture Notes Centre for Learning Systems and Applications - Department of Computer Science - University of Kaiserslautern - Kaiserslautern, Germany (2000), `http://www.dfki.uni-kl.de/~aabecker/Mosbach/Bergmann-CBR-Survey.pdf`
2. Byington, C., Roemer, M., Galie, T.: Prognostic Enhancements to Diagnostic Systems for Improved Condition-Based Maintenance. In: Aerospace Conference Proceedings, vol. 6, pp. 6-2815–6-2824 (2002)
3. Checkland, P., Scholes, J.: Soft Systems Methodology in Action. Wiley, NY (1990)
4. Li, B.-Y., Peng, J., Chen, Y.-Q., Jin, Y.-Q.: Classifying unbalanced pattern groups by training neural network. In: Wang, J., Yi, Z., Żurada, J.M., Lu, B.-L., Yin, H. (eds.) ISNN 2006. LNCS, vol. 3972, pp. 8–13. Springer, Heidelberg (2006)
5. Lipol, L., Haq, J.: Risk analysis method: FMEA/FMECA in the organizations. International Journal of Basic & Applied Sciences IJBAS-IJENS 11(05) (2011)
6. Liu, X., Ma, L., Mathew, J.: Machinery fault diagnosis based on fuzzy measure and fuzzy integral data fusion techniques. Mechanical Systems and Signal Processing 23(3), 690–700 (2009)
7. Loboda, I., Feldshteyn, Y., Ponomaryov, V.: Neural Networks for Gas Turbine Fault Identification: Multilayer Perceptron or Radial Basis Network? Int. J. Turbo Jet-Engines 29(1), 37–48 (2012)
8. Lopes, T.A.P., Troyman, A.C.R.: Neural Networks on the Predictive Maintenance of Turbomachinery. In: Proceedings IFAC, pp. 988–993 (1997)
9. Mitchel, J.: Turbomachinery Analysis and Monitoring. PennWell Books (1993)
10. Niu, G., Han, T., Yang, B., Tan, A.: Multi-agent decision fusion for motor fault diagnosis. Mechanical Systems and Signal Processing 21, 1285–1299 (2007)
11. Pipe, K.: Application of advanced pattern recognition techniques in machinery failure prognosis for Turbomachinery. In: Proc. Condition Monitoring 1987 Int. Conf., British Hydraulic Research Association, UK, pp. 73–89 (1987)
12. Pusey, H., Roemer, M.: An Assessment of Turbomachinery Condition Monitoring and Failure Prognosis Technology. The Shock and Vibration Digest 31(5), 365–371 (1999)
13. Riesbeck, C., Schank, R.: Inside case-based reasoning. Lawrence Erlbaum Associates, Pubs., Hillsdale (1989)
14. Roemer, M., Byington, C., Kacprzynski, G., Vachtsevanos, G.: An Overview of Selected Prognostic Technologies With Application to Engine Health Management. In: ASME Turbo Expo 2006: Power for Land, Sea, and Air, Barcelona, Spain, May 8-11. Aircraft Engine; Ceramics; Coal, Biomass and Alternative Fuels; Controls, Diagnostics and Instrumentation; Environmental and Regulatory Affairs, vol. 2 (2006)
15. Schank, R.: Dynamic Memory: A Theory of Learning in Computers and People. Cambridge University Press, New York (1982)
16. Thiry, C., Yan, A.-M., Golinval, J.: Damage Detection in Rotating Machinery Using Statistical Methods: PCA analysis and Autocorrelation Matrix. Surveillance 5 CETIM Senlis, October 11-13 (2004)

17. Wandekokem, E., Mendel, E., Fabris, F., Varejão, F., Rauber, T., Batista, R.J.: Constructing feature-based ensemble classifiers for real-world machines fault diagnosis. In: IECON 2010 - 36th Annual Conference on IEEE Industrial Electronics Society, Glendale, AZ, November 7-10, pp. 1099–1104 (2010)
18. Watson, I.: An Introduction to Case-Based Reasoning. In: Watson, I.D. (ed.) UK CBR 1995. LNCS, vol. 1020, pp. 3–16. Springer, Heidelberg (1995)
19. Watson, I.: CBR is a methodology not a technology. Knowledge Based Systems Journal 12(5-6), 303–308 (1999)
20. Yang, H., Mathew, J., Ma, L.: Vibration Feature Extraction Techniques for Fault Diagnosis of Rotating Machinery - A Literature Survey. In: Asia Pacific Vibration Conference, Gold Coast, Australia, November 12-14 (2003)

Extremely Fast Unsupervised Codebook Learning for Landmark Recognition

Yilin Guo and Wanming Lu

HTC Beijing Advanced Technology and Research Center
Beijing, P.R. China
{guoyilin1987,louiselu0214}@gmail.com

Abstract. Traditional landmark recognition methods work by using local image features, k-means vector quantization and classifiers like SVM to recognize landmarks. However, the inefficient codebook learning by k-means constraints the possibility of using high-dimensional feature spaces, large numbers of image descriptors and large codebooks which are needed for good results. In this paper we introduce a fast unsupervised codebook learning - Extremely Random Projection Forest (ERPF), which is an ensemble of random projection tree with randomly splitting direction. We evaluate our approach on two public datasets and ERPF significantly outperforms other spatial tree methods and k-means.

Keywords: Landmark Recognition, Random Projection Tree, Codebook Learning.

1 Introduction

Recently, the massiveness of sharing photos on the Internet and the prevalence of mobile camera harness the research in landmark recognition. It is practically useful to recognize an unknown image when a user captures a landmark picture. Landmark recognition has brought great commercial potential in geo-location [1,2] and tourist guide[3].

In this paper, we treat landmark recognition as a classification task. The typical image classification pipeline is composed of the following three steps: (1) Image feature Extraction (e.g., SIFT[4], SURF[5]). (2) Codebook learning and vector quantization of each image (e.g., histogram of image descriptors). (3) Classification (e.g., SVM, Logistic Regression and Boosting). Given an image classification model on a pre-defined landmark list, which has low memory and storage costs, the image query time is very short.

Due to the image background clutter, object occlusion, changes of illumination, various scales and orientations, it is a great challenge to extract image feature efficiently while keeping the discrimination between categories and invariance inside each class. Broadly speaking, previous studies on landmark recognition ignore the impact of different local features especially the binary features[6]. Recently, the limited computational power and storage space in mobile devices

A. Moonis et al. (Eds.): IEA/AIE 2014, Part I, LNAI 8481, pp. 359–368, 2014.
© Springer International Publishing Switzerland 2014

have driven the researchers to propose binary feature which has impact representation. In this paper, we firstly provide a comprehensive empirical analysis on different local features involving the recent wave of binary feature descriptors.

Conventional image classification usually uses k-means to learn visual codebook and bag-of-words model for image representation[7]. However, k-means method is computationally expensive owing to its high $O(tNK)$ processing time (t is the number of iterations, N is the number of data points, K is the number of clusters) and the linear cost of assigning local descriptors to each visual word to find the nearest neighbor during training and testing.

Other unsupervised codebook learning methods, such as Gaussian Mixture Models[8] and mean-shift[9]have been generally used in codebook learning. But the cost of finding nearest visual words for each visual descriptor is expensive. Instead of these flatten codebook learning methods, spatial trees that involve hierarchical k-means tree[10], PCA-tree[11] and Random Projection Tree (RP tree)[11], not only efficiently encode visual descriptors into discrete codes, but also take logarithmic time to traverse the tree.

Recently, we have witnessed spatial trees applied in face recognition[12] and music similarity search[13]. To our knowledge, we have not seen a thorough analysis of its impact on landmark recognition. In this paper, we not only comprehensively evaluate these methods, but also firstly introduce the idea of random forest[14] by extending RP tree as an ensemble of trees, which we call Random Projection Forest (RP Forest). Our experiments show that RP Forest is more effective than other methods in codebook construction. Furthermore, in order to choose a projection direction promptly, we adopt the method of [13], which selects a direction maximizing the projected diameter. However, the complexity of $O(mN)$ (m is number of directions) in each node is still costly when applying to large-scale data. Consequently, we randomly generate a direction without sacrificing accuracy in our experiment, which we call Extremely Random Projection Forest(ERPF). Our ERPF only needs $O(N)$ time for splitting node. Then the total forest construction time is $O(KN)$.

We evaluate our approach on two public datasets: PKUBench[15] and Landmark-3D benchmark[1]. Experiments show that ERPF outperforms k-means, ensembles of PCA-tree, RP tree and 2-means tree in time and classification performance.

2   Related Work

Most previous works consider landmark recognition as image retrieval [3,16,17] or classification problems [1,2,18,19]. Although image retrieval method is very fast to return similar landmark images when users input a query, it is costly to store the inverted index like vocabulary tree[10] in memory. In this paper, we treat our problem as a classification task.

The bag-of-words model[7] is proved to be efficient in image classification. Most of previous works have focused on the building of visual vocabulary [9,20,21], which is called codebook learning. The existing methods could be divided into two categories: generative and discriminative. The derived vocabularies by discriminative method are not universal[21]. When the category of landmark increases, the codebook has to be learned again. Generative codebook learning methods include k-means[7], Gaussian Mixture Models[8] and mean-shift[9]. These flatten methods give impressive results but they are computationally expensive owing to the cost of assigning visual words to visual descriptors.

Recently, spatial trees method demonstrate its efficiency in face recognition[12] and music similarity search[13]. Spatial trees are tree structures which recursively partition data according to some splitting rules. Due to different rules, spatial trees involve hierarchical k-means[10], PCA-tree[11], random projection tree (RP tree)[11] and so on. RP tree is a simple variant of k-d tree[24] which automatically adapts to intrinsic low dimensional structure. Due to the high computation of PCA-tree, RP tree approximately picks a random direction for node splitting. Practical experience indicates that RP tree generally outperforms PCA-tree in high dimension space which makes possible for the large dimensionality of data to encode abundant informative information[11]. Freund [11] mathematically proves that the quantization error depends only on the low intrinsic dimension rather than the high apparent dimension. If the data has intrinsic dimension d, then each split pares off about a $\frac{1}{d}$ fraction of the quantization error. Furthermore, the hierarchical spatial partition of RP tree benefits us the rapid search of nearest visual words in the leaf node.

Our paper firstly comprehensively evaluates spatial trees on landmark recognition and extends RP tree as an ensemble of RP tree using the idea from random forest. Although Bergamo[18] also adopts RP partition rule for codebook construction, it is used for supervised codebook learning, which is different from our way. The complexity of RP tree lies in splitting direction choose in each node and distributing data into descendant nodes. McFee [13] practically selects a direction maximizing the projected diameter that only needs $O(mN)$ time for direction selection. Experimental results on landmark datasets demonstrate that there is no necessity to select a good direction. We randomly choose a direction without sacrificing accuracy, which we call Extremely Random Projection Forest (ERPF).

3   Extremely Random Projection Forest vs. Other Methods

In this section, we give an overview of Extremely Random Projection Forest (ERPF) and compare with k-means, an ensemble of spatial trees (2-means tree, PCA-tree and RP tree). Firstly, we briefly describe classical k-means algorithm. K-means is an heuristic algorithm to find a suboptimal solution of minimizing formula 1:

Given N number of data $x_1, x_2, ..., x_n \in \mathbb{R}^D$, the number of cluster k, we get a partition of k clusters $\chi_1, \chi_2, ..., \chi_k$ and cluster centers $u_1, u_2, .., u_k$. We want to minimize:

$$\sum_{i=1}^{k} \sum_{j \in \chi_i} \|x_j - u_i\|^2 \qquad (1)$$

The first weakness of k-means algorithm is the costly $O(tkN)$ time(t is the number of iterations) which it needs to compare each point with each cluster at each iteration. Secondly, finding the global optimal solution of formula 1 is an NP-hard optimization problem.

Then, we introduce spatial trees. Spatial trees recursively bisect data $\chi \subset \mathbb{R}^D$ by projecting onto a direction $w \subset \mathbb{R}^D$ and splitting at the median, forming two fine subsets, which are efficient for nearest neighbor retrieval[13] and vector quantization[22]. 2-means recursively obtains k-clusters node by using k-means algorithm at each tree level. PCA-tree is a variant of the k-d tree which uses principal components analysis (PCA) rule to choose a splitting direction which can adapt to the low dimensional manifold. RP tree is also a variant of the k-d tree which randomly selects a direction for projection. Next up, we briefly introduce RP tree.

3.1   Random Projection Tree

Random projection tree is a simple variant of the k-d tree[24] which automatically adapts to intrinsic low dimensional structure in data. In the original paper[11], there are two splitting rules depending on the relative size of squared average diameter $\Delta_A^2(\chi)$ and the squared diameter of cell χ $\Delta^2(\chi)$ (the distance between the two furthest points in the set).The first type of rule splits data along a random direction and selects an appropriate splitting position which maximally decreases vector quantization (VQ) error. VQ error is quantified by the average squared Euclidean distance between a vector in the set and the representative vector to which it is mapped. On the other hand, the second type of rule splits data based on the distance from the mean of the cell. Practically, this rule is not efficient for large data, so we adopt a more simple splitting rule. We randomly sample $\log(k)$(k is the clusters) projection directions, select the direction which maximizes the projected diameter and split along the projected median:

$$\underset{w_i}{argmax} \ \underset{x_1,x_2 \in \chi}{max} \ w_i^T x_1 - w_i^T x_2 \qquad (2)$$

Since the time to choose the direction of RP tree only needs $O(\log(k)Dn)$, where $\log(k)$ is much less than D, splitting rule of RP tree is faster than PCA-tree.

In order to overcome the drawback of points near the splitting boundary become isolated from their neighbors across the partition, we employ the overlapping between the left and right subtrees by allowing the points near the median have the opportunity to enter the left and right subtrees. The overlapping stands for visual word ambiguity avoiding the mismatch of hard assignment with the nature of continuous image features. Section 4.3 suspects this effect on the overall classification performance.

Table 1. Algorithm of Extremely Random Projection Forest

Algorithm of Extremely Random Projection Forest
Input: data $\mathcal{X} \subset \mathbb{R}^D$, depth h, number of tree t
Output: t numbers of RP tree over $\mathcal{X}$
ERPF($\mathcal{X}, h, t$)
1:    **for** $i = 1$ to t
2:        RPtree $(\mathcal{X}, h)$
RPtree $(\mathcal{X}, h)$
1:    **if** $h = 0$ **then**
2:        **return** leaf index i
3:    **else**
4:        $w \leftarrow$ randomly select a direction from $\mathcal{N}(\mu, I_d)$
5:        $m \leftarrow$ median($\{w^T x \| x \in \mathcal{X}\}$)
6:        $\mathcal{X}_l \leftarrow \{x \| w^T x \le m, x \in \mathcal{X}\}$
7:        $\mathcal{X}_r \leftarrow \{x \| w^T x > m, x \in \mathcal{X}\}$
8:        leaf indices $I \leftarrow I \cup$ RPtree($\mathcal{X}_l, h - 1$)
9:        leaf indices $I \leftarrow I \cup$ RPtree($\mathcal{X}_r, h - 1$)

3.2   Extremely Random Projection Forest

Inspired by random forest[14], we learn an ensemble of RP tree which stacks the leaf indices from each tree into an codebook vector. Although the time to choose the direction of RP tree only needs $O(\log(k)Dn)$, we further reduce the time by randomly choose a direction, which we call Extremely Random Projection Forest (ERPF). Table 1 lists ERPF algorithm.

Finally, given the projection direction of each non-leaf node in each tree and the leaf indices from each tree, each image is characterized with the occurrences of each visual word by traversing down each tree to search the nearest leaf indices for each descriptor. Then the image histogram formed by the occurrences of each visual word is normalized to total sum 1.

4   Experiments

In this section, we present the detailed landmark recognition result of EPRF on PKUBench[15] and Landmark-3D[1] datasets. the original PKUBench dataset contains 13179 images of 198 landmarks in Peking University. Each landmark is captured by digital cameras and mobilephone cameras from various shot sizes and viewing angles. Landmark-3D includes 25 landmarks, 45180 training images from Flickr website and 10000 positive test images.

Before our experiments, we resize all images to 320*240 pixels or 240*320 pixels. Unless otherwise stated, 60% images of each landmark category in PKUBench are used for model training, the left are for testing. After extracting visual descriptors for total images, we randomly sample 10% descriptors from each image

Table 2. Comparison of the classification accuracy (%) of various feature extraction methods on PKUBench

SIFT	SURF	BRIEF	ORB	BRISK	ROOTSIFT[23]
72.17±0.27	80.14±0.33	52.01±0.43	33.27±0.48	16.59±0.47	71.91±0.31

Table 3. Comparison of the classification accuracy (%) of various feature extraction methods on Landmark-3D

SIFT	SURF	BRIEF	ORB	BRISK	ROOTSIFT
69.85	80.08	52.62	33.59	16.85	68.89

for codebook learning since 10% is enough for achieving the same accuracy with 100% in our experiment. We set each RP tree's height to 9, use bag-of-words model for image vectorization and LIBLINEAR[25] for classification. We employ dual L2-regularized L2-loss SVM method and try 5 cross validation on the training data and find the optimal parameter C as 32768. So, we use this value in the following experiments.

We measure classification performance with accuracy rate. We report our means and variances on 10 runs for RP Forest and find the variance is very small. Therefore, all other experiments use only one time running. The test machine runs on Linux platform, has Intel Core i7-3370, CPU 3.40GHz and 8G RAM. All our source codes are implemented in C++.

4.1   Comparing Various Features

We test various visual descriptors with 10 time running on PKUBench. Due to large time consumption of feature extraction, we only test one time on Landmark-3D. The best choice depends on the particular database. Table 2 and 3 both show that SURF is the best choice for PKUBench and Landmark-3D. Although binary features (BRIEF, ORB, BRISK) are computed quickly and compact, they are less discriminant than SIFT and SURF in image classification. Due to integral images for image convolutions and Fast-Hessian detector, SURF is about 10 times faster and works well comparatively with SIFT. We adopt SURF as feature extraction.

4.2   Comparing Codebook Learning Methods

Firstly, we compare accuracy rates of ERPF with k-means, ensembles of RP tree, 2-means tree, PCA-tree on PKUBench. Fig. 1(a) shows some quantitative differences when the codebook size is increasing with times of 512. It is obvious that k-means is superior when the codebook size is less than 1536. However, k-means cannot withstand the prohibitive learning time when the size beyond 6512.

On the other hand, we observe that ensembles of PCA-tree and 2-means tree are less discriminative than k-means and RP Forest. At first, the performance of RP Forest grows quickly with the increase of codebook size even it is eclipsed by k-means. when the codebook size surpasses about 2000, the accuracy of RP Forest exceeds other methods. When the number of RP tree arrives to 10, the accuracy grows slowly. Consider the slow rising accuracy and the costly codebook learning time, we decide to use 10 RP trees in our RP Forest, which achieves high accuracy rate. Furthermore, we compare accuracy rates of ERPF and RP Forest with the rising of tree number in Fig. 1(b) and find the accuracy is almost the same even we randomly choose a direction at each node for ERPF.

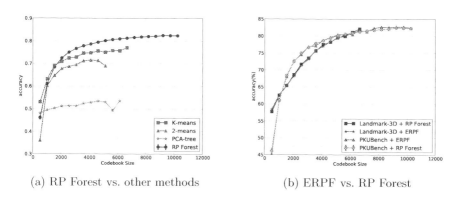

(a) RP Forest vs. other methods   (b) ERPF vs. RP Forest

Fig. 1. Comparison of the classification accuracy of ERPF and other methods

Secondly, we compare codebook learning time of ERPF, RP Forest and k-means. There are three main elements which influence codebook learning time: visual descriptor dimension, the number of training descriptors and codebook size. Since the dimension of SURF is fixed to 64, we observe the other two elements. Fig. 2(a) lists codebook learning time when the number of training features ascends in PKUBench (training images increase from 20% to 80%, codebook size is set to 512). Fig. 2(b) shows the learning time with the increase of codebook size in PKUBench (training images are 60% of total data). Obviously, RP Forest consumes much less time when compare with k-means. On the other hand, Fig. 2(c) shows the learning time with the increase of codebook size between ERPF and RP Forest. We find ERPF is extremely fast learning.

At the same time, after the codebook learning phase, we adopt bag-of-words model to gain each image histogram. Fig. 3 presents the image histogram time of ERPF and k-means with the increase of codebook size for PKUBench data. ERPF only needs the logarithmic time to traverse each tree. If the codebook size is $t * k$, we construct t trees with each tree learning k words. Image histogram time of ERPF is proportional to $O(t*\log(k))$, whereas k-means needs to compare with each visual word for each descriptor, which is linear growth with data size.

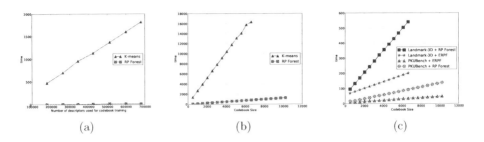

(a)   (b)   (c)

Fig. 2. Comparison of codebook learning time between ERPF, RP Forest and k-means

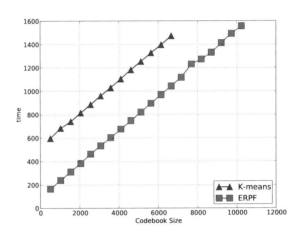

Fig. 3. Comparison of the image histogram time of ERPF and k-means for PKUBench data

4.3   Spill Threshold for ERPF

In this section, we suspect the spill thresholds impact on ERPF. All standard deviation is 0.003 in Table 4. We observe that the larger spill threshold (0.1) has subtle accuracy increase because there exists some data concentrating near the median value after projection, the hard partitioning will have detrimental effects on finding the nearest visual word for each descriptor. The larger spill threshold allows us to increase the number of nearest visual words for each descriptor. In some sense, the overlapping stands for visual word ambiguity avoiding the mismatch of hard assignment with the nature of continuous image features.

Table 4. Comparison of different spill thresholds for ERPF

	0.0	0.01	0.05	0.10
ACCURACY	0.786±0.003	0.788±0.003	0.796±0.003	0.801±0.003

5   Conclusion

Our paper proposes an efficient unsupervised codebook learning method for landmark recognition, which we call Extremely Random Projection Forest (ERPF) - a randomly splitting direction choose for an ensemble of random projection tree. Our ERPF significantly outperforms classical k-means, ensembles of PCA-tree, 2-means tree in classification performance while achieves faster codebook construction, which enables the possibility of using high-dimensional feature spaces, large numbers of image descriptors and large codebooks for large-scale landmark classification.

References

1. Hao, Q., Cai, R., Li, Z., Zhang, L., Pang, Y., Wu, F.: 3d visual phrases for landmark recognition. In: IEEE International Conference on Computer Vision and Pattern Recognition, pp. 3594–3601 (2012)
2. Li, Y., Crandall, D.J., Huttenlocher, D.P.: Landmark classification in large-scale image collections. In: IEEE International Conference on Computer Vision and Pattern Recognition, pp. 1957–1964 (2009)
3. Zheng, Y.-T., Zhao, M., Song, Y., Adam, H., Buddemeier, U., Bissacco, A., Brucher, F., Chua, T.-S., Neven, H.: Tour the world: building a web-scale landmark recognition engine. In: IEEE International Conference on Computer Vision and Pattern Recognition, pp. 1085–1092 (2009)
4. Lowe, D.G.: Distinctive image features from scale invariant keypoints. International Journal of Computer Vision 60(2), 91–110 (2004)
5. Bay, H., Tuytelaars, T., Van Gool, L.: SURF: Speeded up robust features. In: Leonardis, A., Bischof, H., Pinz, A. (eds.) ECCV 2006, Part I. LNCS, vol. 3951, pp. 404–417. Springer, Heidelberg (2006)
6. Heinly, J., Dunn, E., Frahm, J.-M.: Comparative evaluation of binary features. In: Fitzgibbon, A., Lazebnik, S., Perona, P., Sato, Y., Schmid, C. (eds.) ECCV 2012, Part II. LNCS, vol. 7573, pp. 759–773. Springer, Heidelberg (2012)
7. Csurka, G., Dance, C., Fan, L., Willamowski, J., Bray, C.: Visual categorization with bags of keypoints. In: Int. Workshop on Statistical Learning in Computer Vision, ECCV, vol. 1, p. 22 (2004)
8. Perronnin, F., Dance, C., Csurka, G., Bressan, M.: Adapted vocabularies for generic visual categorization. In: Leonardis, A., Bischof, H., Pinz, A. (eds.) ECCV 2006. LNCS, vol. 3954, pp. 464–475. Springer, Heidelberg (2006)
9. Jurie, F., Triggs, B.: Creating efficient codebooks for visual recognition. In: International Conference on Computer Vision, vol. 1, pp. 604–610 (2005)
10. Nister, D., Stewenius, H.: Scalable recognition with a vocabulary tree. In: IEEE International Conference on Computer Vision and Pattern Recognition, vol. 2, pp. 2161–2168 (2006)
11. Freund, Y., Dasgupta, S., Kabra, M., Verma, N.: Learning the structure of manifolds using random projections. In: Advances in Neural Information Processing Systems, pp. 473–480 (2007)
12. Cao, Z., Yin, Q., Tang, X., Sun, J.: Face recognition with learning-based descriptor. In: IEEE International Conference on Computer Vision and Pattern Recognition, pp. 2707–2714 (2010)

13. McFee, B., Lanckriet, G.R.G.: Large-scale music similarity search with spatial trees. In: International Society for Music Information Retrieval, pp. 55–60 (2011)
14. Breiman, L.: Random forests. Machine Learning 45(1), 5–32 (2001)
15. Ji, R., Duan, L.-Y., Chen, J., Yang, S., Huang, T., Yao, H., Gao, W.: Pkubench: a context rich mobile visual search benchmark. In: International Conference on Image Processing, pp. 2545–2548 (2011)
16. Chen, D.M., Baatz, G., Koser, K., Tsai, S.S., Vedantham, R., Pylvanainen, T., Roimela, K., Chen, X., Bach, J., Pollefeys, M.: City-scale landmark identification on mobile devices. In: IEEE International Conference on Computer Vision and Pattern Recognition, pp. 737–744 (2011)
17. Li, Y., Snavely, N., Huttenlocher, D.P.: Location recognition using prioritized feature matching. In: IEEE International Conference on Computer Vision and Pattern Recognition, pp. 791–804 (2010)
18. Bergamo, A., Sinha, S.N., Torresani, L.: Leveraging structure from motion to learn discriminative codebooks for scalable landmark classification. In: IEEE International Conference on Computer Vision and Pattern Recognition, pp. 763–770 (2013)
19. Fritz, G., Seifert, C., Paletta, L.: A mobile vision system for urban detection with informative local descriptors. In: International Conference on Computer Vision Systems, p. 30 (2006)
20. Cai, H., Yan, F., Mikolajczyk, K.: Learning weights for codebook in image classification and retrieval. In: IEEE International Conference on Computer Vision and Pattern Recognition, pp. 2320–2327 (2010)
21. Perronnin, F., Dance, C.: Fisher kernels on visual vocabularies for image categorization. In: IEEE International Conference on Computer Vision and Pattern Recognition, pp. 1–8 (2007)
22. Dasgupta, S., Freund, Y.: Random projection trees for vector quantization. IEEE Trans. Information Theory 55(7), 3229–3242 (2009)
23. Arandjelovic, R., Zisserman, A.: Three things everyone should know to improve object retrieval. In: IEEE International Conference on Computer Vision and Pattern Recognition, pp. 2911–2918 (2012)
24. Bentley, J.L.: Multidimensional binary search trees used for associative searching. Communications of the ACM 18(9), 509–517 (1975)
25. Fan, R.-E., Chang, K.-W., Hsieh, C.-J., Wang, X.R., Lin, C.-J.: Liblinear: A library for large linear classification. The Journal of Machine Learning Research 9, 1871–1874 (2008)

A New Rough Set Based Classification Rule Generation Algorithm(RGA)

Honghai Feng[1,2], Yanyan Chen[3], Kaiwei Zou[1], Lijuan Liu[1], Qiannan Zhu[1], Zhuo Ran[1], Li Yao[1], Lijin Ji[1], and Sai Liu[1]

[1] School of Computer and Information Engineering
[2] Institute of Data and Knowledge Engineering
[3] Library
Henan University, Kaifeng, Henan, 475001, China
honghf@126.com

Abstract. Rough sets theory has taken an important role in data mining. This paper introduces a new rough set based classification rule generation algorithm. It has three features: the first is that the new algorithm can be used in inconsistent systems. The second is its ability to calculate the core value without attributes reduction before. The third is that every example gives a rule and the core values are added first in rule generation process. Experimental results indicate that the classification performanceismuch better than the standard rough set, its variants andJRIPPER, a little better thanCBA and KNN,andcompetive to C4.5in terms of 8 measures. The higher performance of the new algorithm may get benefit from its enough higher accuracy rules and having some properties like KNN.

Keywords: Rough sets, classification rule, C4.5, JRIPPER, CBA.

1 Introduction

In artificial intelligenceone of the main predicting tasks is classification.Accuracy is always a concern. Even though many machine learning theories can, the rule based methods are indispensable and have their own merits. Firstly, the classification accuracy of the rule based methods is comparable to the top classification algorithms. As a well-known fact, C4.5 [1] and JRIPPER [2] are good classifiers in classification accuracy. Secondly, a rule based method is a white box, which can induce understandable knowledge that may be essential in special area such as medical fields. Thirdly, rule based classification methods are validated andeffective in practical applications.

The main rule based techniques are decision trees, sequential covering, associative rules, and rough set [3] based methods, representative algorithms or softwares are C4.5, RIPPER, CBA [4], and ROSETTA [5].

By running the free software WEKA, we can find that C4.5 and RIPPER are two excellent classifiers in classification performance; meanwhile there are some other rule based algorithms comparable to them in classification accuracy [6-10]including the rough sets based classifiers.

A. Moonis et al. (Eds.): IEA/AIE 2014, Part I, LNAI 8481, pp. 369–378, 2014.

In the standard rough set theory, attribute reduction only preserves the dependencies of decision attributes on condition attributes in consistent examples, i.e., new inconsistencies will not be allowed to be brought about after attribute reduction only in the consistent examples. This paper introduces a new rough set based classification rule generation algorithm that is an extension of the standard rough set method to inconsistent systems. The new algorithm can preserve the dependencies of decision attributes on condition attributes in all examples including inconsistent examples, i.e., new inconsistencies will not be allowed to be brought about after attribute reduction in all examples. In addition, in standard rough set method an attribute reduction step should be carried out before rule generation, which may remove some important attributes and some attribute values may become indispensable that may not in original attribute set. In order to get better predictive performance the new algorithm generates the core value for every example first and then generates rule on basis of it directly. In the course of rule generation ofthe new algorithm the core values are added first and every example generates one rule.Meanwhile, the new algorithm does not reduce attributes before rule generation and the sub-optimum attribute reducts can being gotten by the way after rule generation.

2   New Rough Set Based Classification Rule Induction Algorithm

Table 1. Algorithm RGA

Algorithm RGA
Input: Data set U, condition attribute set C, decision attribute set D

Output: Labels of every test examples

begin

  getIND(A, I){

$IND(A,I) = \phi$

          for each attribute $a \in A$

            for each example $x \in I$

$IND(A,I) = \{x\}$

for each example y ($y \neq x$);

  if $a(x) = a(y)$

add y to $IND(A,I)$

return $IND(A,I)$

    }

  getInconsistentExamples(){

For every x

Table 1. (*continued*)

Calculate $\bigcap_i [x]_{C_i}$ and $[x]_D$

If $\bigcap_i [x]_{C_i} \not\subset [x]_D$

x is an inconsistent example
   }

getCoreValue1(){

Get all the consistent examples
For every consistent example
For every condition attribute C_i

  Calculate $\bigcap_{i=1,\ i\neq j}^{n} [x]_{C_i}$ and $[x]_D$

  If $\bigcap_{i=1,\ i\neq j}^{n} [x]_{C_i} \not\subset [x]_D$

$C_j(x)$ is a core value

   }

getCoreValue2(){

Get all the inconsistent examples;
 For every inconsistent example
For every condition attribute C_i,

Calculate $\bigcap_{i=1,\ i\neq j}^{n} [x]_{C_i}, [x]_D$ and get $\{x^{'},\cdots,x^{(n)}\}$, $\{y,y^{'},\cdots,y^{(n)}\}$

If $\bigcap_{i=1,\ i\neq j}^{n} [x]_{C_i} \not\subset [x]_D \cup \{x^{'},\cdots,x^{(n)}\} \cup \{y,y^{'},\cdots,y^{(n)}\}$

$C_j(x)$ is a core value

   }

getRules(){

rule set $RULES = \varphi$

 rule $RULE = \varphi$

condition attribute set A

for every example x

$RULE = \wedge(C_i = C_i(x))$ where $C_i(x) \in CORE(x)$

   }

classifyInstance(){

     matched rules $MRS = \phi$

Table 1. (*continued*)

for every test example x

for every rule r in $RULES$

if r matches x

add r to MRS

if $T = |MRS| \neq \phi$

$$H(x) = \text{sign} \sum_{t=1}^{T} \alpha_t r_t(x) \text{, where } \alpha = \text{coverage of } r_t(x)$$

else

$H(x)$ assigns the class label that the majority examples hold originally

}

end

For every example x, rule $r_t(x)$ outputs only one class label which the example x belongs to with a biggest confidence value. The total time complexity of the new algorithm is $O(mn^2)$ where m is the number of the condition attributes, and n is the number of the examples.

3   Data Sets

In order to get faithful results 77 data sets are used in this experiment, and all the data sets are obtained from the repository of Machine Learning databases at UCI [11], see their characteristics in Table 2. Some data sets are discretized by supervised discretization methods with WEKA and denoted as like australian_dis, and some data sets are discretized by unsupervised discretization methods with WEKA and denoted as like autos_undis. The java class weka.filters.supervised.attribute. AttributeSelection in WEKA is used for supervised discretization and discretization is by Fayyad &Irani's MDL method (the default) [12]. weka.filters.unsupervised.attribute.Discretize is used for unsupervised discretization, and discretization is by simple binning. The default value of bins is 10.

Table 2. Data sets

Data sets	features	classes	cases
adult-stretch	4	2	20
audiology	69	224	24
australian_dis	14	2	690
autos_undis	24	7	205
balance-scale_sup	4	3	625
blood_tranfusion	3	2	748

Table 2. (*continued*)

breastCancer	9	2	286
b-c-w(Prognostic)	8	2	699
b-c-w-image	32	2	196
bridges_dis	9	4	105
bridges_version2	11	7	107
car	6	4	1728
cleve_dis	11	2	303
cmc_dis	9	3	1473
colic_sup_missing	16	2	368
cpu	8	8	209
crx_dis	15	2	690
cylinder-bands	31	2	540
Dermatology_dis	34	6	366
diabetes_sup	6	2	768
echocardiogram	11	3	132
ecoli_sup	6	8	336
flag_dis	26	8	194
flare_data1	12	6	323
flare_data2	12	6	1066
german_dis	19	2	1000
glass_undis	7	6	214
haberman_unsup	3	2	306
hayes-roth_dis	4	3	132
heart-c_sup	11	2	303
heart-h_unsup	12	2	294
heart-statlog_sup	9	2	270
hepatitis_unsup	19	2	155
ionosphere_sup	33	2	351
iris_dis	4	3	150
labor_dis	16	2	57
led7	7	10	3200
led_24	24	10	1000
lenses_dis	4	3	24
liverdisorders_unsup	6	2	345
lung-cancer	56	3	32
Lymphography	18	4	148
mammo _dis	5	2	961
molecular-biology	57	2	106
monks1	6	2	432
monks2	6	2	432
monks3	6	2	432
new_thyroid_dis	5	3	215
post-operative	8	3	90
primary-tumor	17	21	339
promoter_gene	57	2	106

Table 2. (*continued*)

Robot_FailureLP4_dis	90	3	118
Robot_FailureLP5_dis	90	5	165
shuttle-landing	6	2	15
solar-flare_1	11	6	323
solar-flare_2	11	6	1066
sonar_unsup	60	2	208
soybean_unsupmissing	35	19	683
space_shuttle_disun	2	3	23
spect_train	22	2	80
sponge	44	3	76
tae	2	3	151
Teaching Assistant	4	3	151
tic tac toe	9	2	958
trains	32	2	10
urinary	6	4	120
Vehicle_dis	18	4	846
vote_unsup_missing	16	2	435
vowel	11	11	900
wine	13	3	178
yeast_dis	8	10	1484
yellow-small	4	2	20
zoo	16	7	101
Arrhythmia_supdis	133	13	452
b-c-w-cell	272	2	569
libras_movement_dis	74	15	360
Mammals_unsup	464	4	1000
Spectrometer	93	48	531

4   Experimental Results

The new algorithm (RGA), CBA, Explore, LEM2, the standard rough set methods with genetic selecting attribute before rule generation(SRGeS) and the variable precision rough set(VPR) are programmed with JAVA and embedded into WEKA 3.6.5. The C4.5 and Jrip are transformed from J48 and Jrip in WEKA 3.6.5. The KNN(IB1, K=1) is from WEKA and with no modification and transformation.

The experiment uses a ten-fold cross validation procedure that performs 10 randomized train and test runs on the dataset.

The experimental results in term of mean absolute error are listed in Table 3. VPR and SRGeS represent variable precision rough set algorithm and standard rough set method with genetic selecting attribute respectively. The detailed experimental results about every algorithm on every data set in terms of other measures like percent correct, weighted average area under ROC, weighted average F-measure, weighted average IR precision, weighted average IR recall, weighted average true negative rate, and weighted average true positive rate have not been offered, but the comprehensive results are provided in Table 4. The first line of Table 4 lists the 8 algorithms except for RGA, the first column represents the 8 performance measures and other 6 metrics for analysis and the others stand for comparison of performances in terms of 8 measures across 9 algorithms. The count (xx/ yy/ zz) of the number of times represents that the other listed schemes are bigger than (xx), the same as (yy), or smaller than (zz) the baseline scheme (the new algorithm, RGA).

Table 3. Mean_absolute_error results

Data sets	RGA	CBA	Explore	C4.5	Jrip	LEM2	VPR	SRGe	KNN
adult-stretch	0.01 |	0.00	0.02	0.00	0.24 v	0.00	0.04	0.53 v	0.10
audiology	0.02 |	0.03 v	0.06 v	0.02	0.05 v	0.02	0.02	0.05 v	0.02
australian	0.16 |	0.14	0.19	0.20 v	0.36 v	0.23 v	0.25 v	0.24 v	0.20 v
autos	0.06 |	0.05	0.07	0.07	0.14 v	0.07	0.08	0.18 v	0.04 *
balance	0.19 |	0.21 v	0.34 v	0.27 v	0.28 v	0.19	0.31 v	0.19	0.20
transfusion	0.31 |	0.27 *	0.34 v	0.36 v	0.37 v	0.31	0.33	0.31	0.34
b-c-w-w	0.07 |	0.05 *	0.05	0.08	0.27 v	0.08	0.07	0.09	0.04
b-c-w-d	0.38 |	0.37	0.24 *	0.36	0.36	0.49 v	0.32	0.34	0.36
breast-cancer	0.32 |	0.34	0.32	0.36	0.41 v	0.30	0.37	0.36	0.34
bridges	0.11 |	0.10	0.15	0.13	0.20 v	0.14	0.13	0.21 v	0.12
bridges2	0.14 |	0.10 *	0.12	0.14	0.16	0.14	0.14	0.19 v	0.13
car	0.05 |	0.05	0.05	0.04	0.19 v	0.05	0.08 v	0.05	0.11 v
cleve	0.19 |	0.20	0.21	0.31 v	0.37 v	0.24	0.35 v	0.21	0.21
cmc	0.34 |	0.34	0.36 v	0.36 v	0.41 v	0.34	0.37 v	0.34	0.35
horse-colic	0.17 |	0.19	0.21	0.24 v	0.36 v	0.24	0.27 v	0.27 v	0.26 v
cpu	0.08 |	0.07	0.09	0.09	0.13 v	0.08	0.09	0.10	0.08
crx	0.17 |	0.14 *	0.18	0.19	0.36 v	0.23 v	0.31 v	0.35 v	0.20
cylinder	0.19 |	0.23	0.29 v	0.41 v	0.41 v	0.25	0.21	0.37 v	0.21
Dermatology	0.04 |	0.03	0.08 v	0.03	0.16 v	0.03	0.04	0.22 v	0.02 *
pima_diabetes	0.28 |	0.25 *	0.34 v	0.31 v	0.40 v	0.29	0.38 v	0.28	0.29
echocard	0.34 |	0.28	0.31	0.27	0.36	0.32	0.35	0.33	0.29
ecoli	0.05 |	0.05	0.08 v	0.06	0.13 v	0.05	0.10 v	0.05	0.06
flags	0.11 |	0.11	0.13	0.12	0.14 v	0.12	0.11	0.17 v	0.11
flare_data1	0.10 |	0.10	0.11	0.11	0.14 v	0.12 v	0.12	0.14 v	0.11
flare_data2	0.10 |	0.10	0.11 v	0.10 v	0.19 v	0.11 v	0.11 v	0.15 v	0.11
german	0.29 |	0.26	0.35 v	0.34 v	0.42 v	0.30	0.31	0.35 v	0.31
Glass	0.10 |	0.08 *	0.12 v	0.10	0.15 v	0.10	0.14 v	0.10	0.09
haberman	0.33 |	0.31	0.39	0.38 v	0.39 v	0.32	0.35	0.33	0.32
hayes-roth	0.10 |	0.09	0.18	0.13	0.18 v	0.15 v	0.20 v	0.10	0.15
heart-c	0.08 |	0.08	0.08	0.11 v	0.15 v	0.10	0.15 v	0.08	0.09
heart-h	0.09 |	0.08	0.09	0.12	0.15 v	0.10	0.10	0.16 v	0.08
heart-s	0.19 |	0.17	0.20	0.24	0.36 v	0.19	0.38 v	0.19	0.19
hepatitis	0.15 |	0.19	0.20	0.27 v	0.34 v	0.26 v	0.16	0.30 v	0.23 v

Table 3. (*continued*)

ionosphere	0.10 |	0.09	0.08	0.13	0.29 v	0.11	0.09	0.20 v	0.07
iris	0.05 |	0.05	0.06	0.06	0.20 v	0.05	0.06	0.05	0.04
labor	0.13 |	0.14	0.24	0.17	0.34 v	0.23	0.11	0.39 v	0.08
led7	0.08 |	0.07 *	0.13 v	0.08	0.10 v	0.07 *	0.08	0.08	0.08
LED_24	0.07 |	0.12 v	0.18 v	0.07	0.12 v	0.10 v	0.07	0.11 v	0.11 v
lenses	0.25 |	0.20	0.21	0.15 *	0.31	0.31	0.20	0.25	0.24
liver-dis	0.39 |	0.31 *	0.37	0.45	0.48 v	0.40	0.47 v	0.39	0.37
lung-cancer	0.41 |	0.43	0.39	0.34	0.39	0.48	0.44	0.39	0.41
lymphography	0.11 |	0.10	0.14	0.13	0.21 v	0.12	0.09	0.19 v	0.10
mammographic	0.23 |	0.21	0.24	0.25 v	0.37 v	0.24	0.30 v	0.23	0.24
promoters	0.33 |	0.14 *	0.38	0.20	0.35	0.39	0.41	0.49	0.16
monks1	0.00 |	0.00	0.21 v	0.04	0.29 v	0.00	0.15 v	0.37 v	0.28 v
monks2	0.47 |	0.32 *	0.43 *	0.44	0.44	0.44	0.49	0.47	0.45
monks3-weka.	0.00 |	0.00	0.02	0.00	0.26 v	0.00	0.14 v	0.48 v	0.20 v
new_thyroid	0.03 |	0.04	0.04	0.06	0.18 v	0.03	0.07	0.04	0.02
postoperative	0.28 |	0.27	0.24 *	0.28	0.28	0.25	0.27	0.27	0.28
primary-tumor	0.06 |	0.06	0.07 v	0.06	0.07 v	0.06	0.06 v	0.06	0.06
promoter_gene	0.37 |	0.13 *	0.42	0.24	0.36	0.26	0.41	0.51	0.16 *
Robot_F_LP4	0.11 |	0.10	0.15	0.08	0.21 v	0.12	0.11	0.24 v	0.05 *
Robot_F_LP5	0.15 |	0.18	0.15	0.12	0.19 v	0.17	0.15	0.25 v	0.11 *
Shuttle	0.30 |	0.55	0.25	0.41	0.42	0.45	0.37	0.46	0.30
solar-flare	0.10 |	0.10	0.11	0.11	0.17 v	0.11 v	0.11 v	0.15 v	0.11
sonar	0.41 |	0.37	0.47	0.36	0.44	0.39	0.44	0.45	0.21 *
soybean	0.01 |	0.02	0.07 v	0.01 *	0.03 v	0.01 *	0.01	0.08 v	0.01 *
space_shuttle	0.18 |	0.17	0.19	0.27	0.27	0.17	0.26	0.18	0.17
spect	0.38 |	0.34	0.32	0.37	0.44	0.43	0.45	0.42	0.46
sponge	0.07 |	0.07	0.05	0.08	0.10	0.09	0.04	0.06	0.04
tae	0.39 |	0.38	0.44 v	0.41 v	0.44 v	0.41 v	0.42 v	0.39	0.39
Teaching	0.33 |	0.35	0.36	0.38	0.42 v	0.33	0.37	0.33	0.34
tic-tac-toe	0.08 |	0.00 *	0.19 v	0.17 v	0.24 v	0.02 *	0.08	0.14 v	0.18 v
trains	0.51 |	0.30	0.66	0.20	0.25	0.20	0.36	0.71	0.40
urinary	0.00 |	0.00	0.15 v	0.00	0.19 v	0.00	0.00	0.12 v	0.00
vehicle	0.14 |	0.15	0.31 v	0.16	0.26 v	0.16	0.21 v	0.23 v	0.15
vote	0.06 |	0.05	0.08	0.06	0.28 v	0.06	0.13 v	0.12 v	0.08
vowel	0.04 |	0.06 v	0.16 v	0.05 v	0.09 v	0.03	0.05 v	0.04	0.03 *
wine	0.01 |	0.02	0.04	0.05 v	0.22 v	0.03	0.01	0.22 v	0.01
yeast	0.10 |	0.10	0.13 v	0.11	0.13 v	0.10	0.12 v	0.15 v	0.10
yellow-small	0.00 |	0.00	0.02	0.00	0.24 v	0.00	0.09	0.00	0.15
zoo	0.01 |	0.02	0.06 v	0.02	0.12 v	0.03	0.02	0.12 v	0.01
Arrhythmia	0.05 |	0.07 v	0.05	0.04	0.07 v	0.04	0.07 v	0.08 v	0.04
breast-c-w-c	0.04 |	0.04	0.04	0.06	0.26 v	0.07	0.04	0.11 v	0.04
libras_m	0.04 |	0.12 v	0.13 v	0.05	0.08 v	0.04	0.04	0.11 v	0.03 *
Mammals	0.00 |	0.00	0.00	0.00	0.10 v	0.00	0.00	0.02 v	0.00
spectrometer	0.02 |	0.03 v	0.03 v	0.02	0.03 v	0.02	0.02	0.04 v	0.02 *

```
(v/ /*) (7/58/12 (25/49/3(18/57/2(62/15/0)(10/64/3(25/52/0(38/39/0)(8/59/10)
```

Table 4. Comparison of performances in term of 8 measuresacross 9 algorithms

	CBA	Explore	C4.5	Jrip	LEM2	VPRS	SRGeS	KNN
1.	(7/58/12)	(25/49/3)	(18/57/2)	(62/15/0)	(10/64/3)	(25/52/0)	(38/39/0)	(8/59/10)
2.	(4/67/6)	(1/55/21)	(8/64/5)	(3/65/9)	(3/62/12)	(0/63/14)	(1/49/27)	(6/55/16)
3.	(8/61/8)	(7/47/23)	(16/56/5)	(8/63/6)	(5/56/16)	(5/65/7)	(2/58/17)	(3/49/25)
4.	(5/64/8)	(1/49/27)	(7/65/5)	(3/64/10)	(3/65/9)	(0/63/14)	(0/47/30)	(7/56/14)
5.	(5/63/9)	(1/52/24)	(3/67/7)	(1/66/10)	(1/67/9)	(0/63/14)	(0/48/29)	(5/59/13)
6.	(4/67/6)	(1/55/21)	(8/64/5)	(3/65/9)	(3/62/12)	(0/63/14)	(1/49/27)	(6/55/16)
7.	(5/64/8)	(3/51/23)	(11/60/6)	(4/57/16)	(4/63/10)	(0/53/24)	(0/44/33)	(7/61/9)
8.	(4/67/6)	(1/55/21)	(8/64/5)	(3/65/9)	(3/62/12)	(0/63/14)	(1/49/27)	(6/55/16)
9.	(0/1/76)	(46/6/25)	(5/9/63)	(0/0/77)	(23/8/46)	(10/30/37)	(12/48/17)	
10.	(3/48/26)	(16/44/17)	(0/15/62)	(1/12/64)	(7/59/11)	(0/76/1)	(0/27/50)	
11.	(2/9/66)	(41/8/2)	(23/17/37)	(0/1/76)	(54/9/14)	(1/28/48)	(15/42/20)	
12.	(73/4/0)	(50/9/18)	(50/14/13)	(77/0/0)	(42/23/12)	(27/30/10)	(4/46/27)	
13.	(17/30/30)	(10/45/22)	(0/6/71)	(0/1/76)	(0/37/40)	(1/30/46)	(2/73/2)	
14.	(0/2/75)	(47/6/24)	(4/8/65)	(0/0/77)	(16/6/55)	(14/38/25)	(11/49/17)	

1.Mean_absolute_error.   2. Percent_correct.

3. Weighted_avg_area_under_ROC.   4. Weighted_avg_F_measure.

5. Weighted_avg_IR_precision6. Weighted_avg_IR_recall

7. Weighted_avg_true_negative_rate 8. Weighted_avg_true_positive_rate

9. Total_Length_of_All_Rules in the rule set 10. amount of attributes in the rule set

11. Mean length of the rules in the rule set 12. Mean coverage of the rules in the rule set

13. Mean accuracy of the rules in the rule set 14. Amount of rules in the rule set.

5   Conclusions, Discussions and Future Works

5.1   Conclusions

It can be seen from Table 4 that in term of Weighted_avg_IR_precision RGA ranks first. In term of the Mean_absolute_error CBA ranks first, RGA second and C4.5 fifth. C4.5 worsen classification performances on lots of data sets slightly, but improve the classification performances on some data sets significantly. In terms of Percent_correct, weighted average F_measure, weighted average IR recall, weighted average true negative rate, and weighted average true positive rate C4.5 ranks first and RGA second. In term of Weighted_avg_area_under_ROC C4.5 ranks first, Jrip second and RGA third.

5.2   Discussions

(1) CBA and RGA have the lowest measure of mean absolute error. It can be found in Table 4 that the only consistent factor between CBA and RGA is that in term of rules' mean accuracy CBA ranks first and RGA the second. So we can guess that the measure of mean absolute error relates most to the metric of mean accuracy of rules in rule set.

(2)The rules in LEM2 have bigger mean coverage, but longer mean length and lower mean accuracy than RGA. The bigger mean coverage is dueto that LEM2 select the attribute values with biggest coverage to construct a rule. The longer mean length and lower mean accuracy is due to that in LEM2 the equivalence class of the condition attributes should be included in the equivalence class of the decision class, this does not be satisfied for inconsistent examples and as a result very long rules will be generated for inconsistent examples, and the final classification performance is impacted.

(3) The two differences of schema between RGA and SRGeS are that SRGeS has the attribute reduction step before rule generation and does not handle the inconsistent examples. So the metric of amount of attribute in rule set in SRGeS is very small (see Table 4) and may remove some significant attributes.

(4)RGA generates a rule for one example, whereas KNN treat every original example as a rule. Obviously, the length of a generated rule is shorter and more abstract than an original example in KNN. So RGA has higher performance than KNN.

References

1. Quinlan, R.: C4.5: Programs for Machine Learning. Morgan Kaufmann Publishers, San Mateo (1993)
2. Cohen, W.W.: Fast Effective Rule Induction. In: Twelfth International Conference on Machine Learning, pp. 115–123 (1995)
3. Pawlak, Z.: Rough sets. International Journal of Computer and Information Sciences 11, 341–356 (1982)
4. Liu, B., Hsu, W., Ma, Y.: Integrating Classification and Association Rule Mining. In: Fourth International Conference on Knowledge Discovery and Data Mining, pp. 80–86 (1998)
5. http://www.lcb.uu.se/tools/rosetta/
6. Thabtah, F.A., Cowling, P.I.: A greedy classification algorithm based on association rule. Applied Soft Computing 7, 1102–1111 (2007)
7. Yin, X., Han, J.: CPAR: classification based on predictive association rule. In: Proceedings of the SDM, San Francisco, CA, pp. 369–376 (2003)
8. Lim, T.-S., Loh, W.-Y.: A Comparison of Prediction Accuracy, Complexity, and Training Time of Thirty-Three Old and New Classification Algorithms. Machine Learning 40, 203–228 (2000)
9. Thabtah, F., Cowling, P., Hammoud, S.: Improving rule sorting, predictive accuracy and training time in associative classification. Expert Systems with Applications 31, 414–426 (2006)
10. Li, R., Wang, Z.-O.: Mining classification rules using rough sets and neural networks. European Journal of Operational Research 157, 439–448 (2004)
11. Murphy, P.M., Aha, D.W.: UCI repository of machine learning databases, machine-readable data repository, Irvine, CA, University of California, Department of Information and Computer Science (1992)
12. Fayyad, U.M., Irani, K.B.: Multi-interval discretization of continuousvalued attributes for classification learning. In: Thirteenth International Joint Conference on Articial Intelligence, pp. 1022–1027 (1993)

Using the Theory of Regular Functions to Formally Prove the ε-Optimality of Discretized Pursuit Learning Algorithms

Xuan Zhang[1], B. John Oommen[2,1], Ole-Christoffer Granmo[1], and Lei Jiao[1]

[1] Dept. of ICT, University of Agder, Grimstad, Norway
[2] School of Computer Science, Carleton University, Ottawa, Canada[*]

Abstract. Learning Automata (LA) can be reckoned to be the founding algorithms on which the field of Reinforcement Learning has been built. Among the families of LA, Estimator Algorithms (EAs) are certainly the fastest, and of these, the family of Pursuit Algorithms (PAs) are the pioneering work. It has recently been reported that the previous proofs for ε-optimality for *all* the reported algorithms in the family of PAs have been flawed[1]. We applaud the researchers who discovered this flaw, and who further proceeded to rectify the proof for the Continuous Pursuit Algorithm (CPA). The latter proof, though requires the learning parameter to be continuously changing, is, to the best of our knowledge, the current best and only way to prove CPA's ε-optimality. However, for all the algorithms with absorbing states, for example, the Absorbing Continuous Pursuit Algorithm (ACPA) and the Discretized Pursuit Algorithm (DPA), the constrain of a continuously changing learning parameter can be removed. In this paper, we provide a new method to prove the ε-optimality of the Discretized Pursuit Algorithm which does not require this constraint. We believe that our proof is both unique and pioneering. It can also form the basis for formally showing the ε-optimality of the other EAs with absorbing states.

Keywords: Pursuit Algorithms, Discretized Pursuit Algorithm, ε-optimality.

1 Introduction

Learning automata (LA) have been studied as a typical model of reinforcement learning for decades. An LA is an adaptive decision-making unit that learns the optimal action from among a set of actions offered by the Environment it operates in. At each iteration, the LA selects one action, which triggers either a *stochastic* reward or a penalty as a response from the Environment. Based on the response and the knowledge acquired in the past iterations, the LA adjusts its action selection strategy in order to make a "wiser" decision in the next iteration. In such a way, the LA, even though it

---

[*] *Chancellor's Professor*; *Fellow: IEEE* and *Fellow: IAPR*. The Author also holds an *Adjunct Professorship* with the Dept. of ICT, University of Agder, Norway.

[1] This flaw also renders the finite time analysis of these algorithms [12] to be incorrect. This is because the latter analysis relied on the same condition used in the flawed proofs, i.e., they considered the *monotonicity* property of the probability of selecting the optimal action.

A. Moonis et al. (Eds.): IEA/AIE 2014, Part I, LNAI 8481, pp. 379–388, 2014.
ⓒ Springer International Publishing Switzerland 2014

lacks a complete knowledge about the Environment, is able to learn through repeated interactions with the Environment, and adapts itself to the optimal decision. Hence, LA have been applied to a variety of fields where complete knowledge of the Environment can not be obtained. These applications include game playing [1], parameter optimization [2], solving knapsack-like problems and utilizing the solution in web polling and sampling [3], vehicle path control [4], resource allocation [5], service selection in stochastic environments [6], and numerical optimization [7].

One of the most important part in the design and analysis of LA consists of the formal proofs of their convergence accuracies. Among all the different types of LA (FSSA, VSSA, Discretized etc.), the most difficult proofs involve the family of EAs. This is because the convergence involves two intertwined phenomena, i.e., the convergence of the reward estimates *and* the convergence of the action probabilities themselves. Ironically, the *combination* of these in the updating rule is what renders the EA fast.

Prior Proofs: As the pioneering work of the study of EAs, the ε-optimality of the families of PAs have been studied and presented in [9], [10], [11], [12] and [13]. The basic result stated in these papers is that by utilizing a sufficiently small value for the learning parameter (or resolution), PAs will converge to the optimal action with an arbitrarily large probability.

Flaws in the Existing Proofs: The premise for this paper is that the proofs reported *for almost three decades* for PAs have a common flaw, which involves a very fine argument. In fact, the proofs reported in these papers "deduced" the ε-optimality based on the conclusion that after a sufficiently large but finite time instant, t_0, the probability of selecting the optimal action is monotonically increasing, which, in turn, is based on the condition that the reward probability estimates are ordered properly *forever* after t_0. This ordering is, indeed, true by the law of large numbers if all the actions are chosen infinitely often and if the time instant, t_0, is defined to be infinite. But if such an "infinite" selection does not occur, the proper ordering happens only in probability. In other words, the authors of these papers misinterpreted the concept ordering "forever" with the ordering "most of the time" after t_0. As a consequence of this misinterpretation, the condition supporting the monotonicity property is false, which leads to an incorrect "proof" for the PAs being ε-optimal.

Discovery of the Flaw: Even though this has been the accepted argument for almost *three decades*, we credit the authors of [14] for discovering this flaw, and further rectifying the proof for the CPA. The latter proof is also based on the monotonicity property of the probability of selecting the optimal action, and requires that the learning parameter be decreasing over time. This methodology, to the best of our knowledge, is the current best way in proving the ε-optimality of the CPA.

Problem Statement: This paper aims at correcting the above-mentioned flaw by presenting a new proof for the DPA, which is a type of EAs with absorbing states. As opposed to all the previous EAs' proofs, we will show that while the monotonicity property is sufficient for convergence, it is not really *necessary* for proving that the DPA is ε-optimal. Rather, we will present a completely new proof methodology which is based on the convergence theory of submartingales and the theory of Regular functions [15]. The new proof is distinct in principle and argument from the proof reported in [14], and does not require the learning parameter to be continuously decreasing. Besides, the new proof can be extended to formally demonstrate the ε-optimality of other EAs with absorbing states.

2   Overview of the DPA

We first present the notations used for the DPA:

r: The number of actions.

α_j: The jth action that can be selected by the LA.

P: The action probability vector. $P = [p_1, p_2, \ldots, p_r]$ and $\sum_{j=1\ldots r} p_j = 1$. An action is selected by randomly sampling from P.

D: The reward probability vector. $D = [d_1, d_2, \ldots, d_r]$, with each d_j being the probability that the corresponding action α_j will be rewarded.

v_j: The number of times α_j has been selected.

u_j: The number of times α_j has been rewarded.

$\hat{d}_j$: The jth element of the reward probability estimates vector $\hat{D}$, $\hat{d}_j = \frac{u_j}{v_j}$.

m: The index of the optimal action.

h: The index of the largest element of $\hat{D}$.

R: The response from the Environment, where $R = 0$ corresponds to a Reward, and $R = 1$ to a Penalty.

Δ: The discretized step size, where $\Delta = \frac{1}{rN}$, with N being a positive integer.

The algorithm of DPA can be described as a sequence of iterations. In iteration t:

1. The LA selects an action by sampling from $P(t)$, suppose the selected action is α_i.
2. The selected action triggers either a reward or a penalty from the Environment. The LA updates $\hat{D} = [\hat{d}_1(t), \hat{d}_2(t), \ldots, \hat{d}_r(t)]$ based on the Environment's response as:
   $u_i(t) = u_i(t-1) + (1 - R(t)); v_i(t) = v_i(t-1) + 1; \hat{d}_i(t) = \frac{u_i(t)}{v_i(t)}$.
3. Suppose $\hat{d}_h(t)$ is the largest element in $\hat{D}(t)$, then α_h is considered as the current best action. The LA increases its action probability as:
   If $R(t) = 0$ **Then**
   $p_j(t+1) = max\{p_j(t) - \Delta, 0\}, j \neq h$
   $p_h(t+1) = 1 - \sum_{j \neq h} p_j(t+1)$.

   Else
   $P(t+1) = P(t)$.
   EndIf

We now visit the proofs of the DPA's convergence.

3   Prior Proof for DPA's ε-Optimality

The formal assertion of the ε-optimality of the DPA [9] is stated in Theorem 1.

Theorem 1. *Given any small ε, δ > 0, there exist a $N_0 > 0$ and a $t_0 < \infty$ such that for all time $t \geq t_0$ and for any positive learning parameter $N > N_0$, $Pr\{p_m(t) > 1 - \varepsilon\} > 1 - \delta$.*

The earlier reported proofs for the ε-optimality of the DPA follow the strategy that consists of four steps. Firstly, given a sufficiently large value for the learning parameter N, all actions will be selected enough number of times before a finite time instant, t_0. Secondly, for all $t > t_0$, $\hat{d}_m$ will remain as the maximum element of the reward proba-bility estimates vector, $\hat{D}$. Thirdly, suppose $\hat{d}_m$ has been ranked as the largest element in $\hat{D}$ since t_0, the action probability sequence of $\{p_m(t)\}$, with $t > t_0$, will be mono-tonically increasing, whence one concludes that $p_m(t)$ converges to 1 with probability 1. Finally, given that the probability of $\hat{d}_m$ being the largest element in $\hat{D}$ is arbitrarily close to unity, and that $p_m(t) \to 1$ w.p. 1, ε-optimality follows from the axiom of total probability.

The formal assertions of these steps are listed below.

1. The first step of the proof can be described mathematically by Theorem 2.

   Theorem 2. *For any given constants $\hat{\delta} > 0$ and $M < \infty$, there exist an $N_0 > 0$ and $t_0 < \infty$ such that under the DPA algorithm, for all positive $N > N_0$,*

   $$Pr\{All\ actions\ are\ selected\ at\ least\ M\ times\ each\ before\ time\ t_0\} > 1 - \hat{\delta}.$$

   The detailed proof for this result can be found in [9].

2. The sequence of probabilities, $\{p_m(t)_{(t>t_0)}\}$, is stated to be *monotonically* increas-ing. The previous proofs attempted to do this by showing that:

   $$|p_m(t)| \leq 1, \text{ and}$$
   $$\Delta p_m(t) = E[p_m(t+1) - p_m(t)|\bar{K}(t_0)] = p_m(t) + d_m c_t \Delta \geq 0,\ t > t_0, \tag{1}$$

   where $c_t = 1, 2, ..., r - 1$, and $\bar{K}(t_0)$ is the condition that $\hat{d}_m$ remains the largest element in $\hat{D}$ after time t_0. If this step of the "proof" was flawless[2], $p_m(t)$ can be shown to converge to 1 w.p. 1.

3. Since $p_m(t) \to 1$ w.p. 1, if it can, indeed, be proven that $Pr\{\bar{K}(t_0)\} > 1 - \delta$, by the axiom of total probability, one can then see that:

   $$Pr\{p_m(t) > 1 - \varepsilon\} \geq Pr\{p_m(t) \to 1\} \cdot Pr\{\bar{K}(t_0)\} > 1 \cdot (1 - \delta) = 1 - \delta,$$

   and ε-optimality is proved.

According to the sketch of the proof above, the key is to prove $Pr\{\bar{K}(t_0)\} > 1 - \delta$, i.e.,

$$Pr\{\hat{d}_m(t) > \hat{d}_j(t)_{j \neq m}, \forall t > t_0\} > 1 - \delta. \tag{2}$$

In the reported proofs for the EAs, Eq. (2) is reckoned true if the following assumption is true.

Let w be the difference between the two *highest* reward probabilities, if all actions are selected a large number of times, each of the $\hat{d}_i$ will be in a $\frac{w}{2}$ neighborhood of d_i with an arbitrarily large probability. In other words, the probability of $\hat{d}_m(t)$ being greater than $\hat{d}_j(t)_{j \neq m}$ will be arbitrarily close to unity. This assumption can be easily "proven" by the weak law of large numbers as per [9].

---

[2] The error in the proofs lies precisely at this juncture, as we shall show presently.

Flaw in the Argument: There is a flaw in the above argument. In fact, the above assumption does not guarantee $Pr\{\bar{K}(t_0)\} > 1 - \delta$. To be specific, let us define $K(t) = \{\hat{d}_m(t)$ is the largest element in $\hat{D}(t)\}$. Then the result that can be deduced from the assumption when $t > t_0$ is that $Pr\{K(t)\} > 1 - \delta$. But, indeed, the condition that Eq. (1) is based on is: $\bar{K}(t_0) = \bigcap_{t>t_0} K(t)$, which means that for every single time instant in the future, i.e., $t > t_0$, $\hat{d}_m(t)$ needs to be the largest element in $\hat{D}(t)$. The previous flawed proofs have mistakenly reckoned that $K(t)$ is equivalent to $\bar{K}(t_0)$.

The flaw is documented in [14], which focused on the CPA, and further provided a way of correcting the flaw, i.e., by proving $Pr\{\bar{K}(t_0)\} > 1 - \delta$ instead of proving $Pr\{K(t)\} > 1 - \delta$. However, the proof requires a sequence of *decreasing* values of the learning rate λ (for CPA), which is not necessary for proving the ε-optimality of the DPA. We thus present a new proof for the DPA that is quite distinct (and uses completely different techniques) than that reported in [14]. The new proof also follows a four-step sketch but instead of examining the monotonicity property, is rather based on the convergence theory of submartingales, and on the theory of Regular functions.

4 DPA's ε-Optimality: A New Proof

4.1 The Moderation Property of DPA

The property of moderation can be described precisely by Theorem 2, which has been proven in [9]. This implies that under the DPA, by utilizing a sufficiently large value for the learning parameter, N, each action will be selected an arbitrarily large number of times.

4.2 The Key Condition $\bar{G}(t_0)$ for $\{p_m(t)_{t>t_0}\}$ being a Submartingale

In our proof strategy, instead of examining the condition for $\{p_m(t)_{t>t_0}\}$ being *monotonically increasing*, we will investigate the condition for $\{p_m(t)_{t>t_0}\}$ being a *submartingale*. This is based on $\bar{G}(t_0)$, defined as: :

$$q_j(t) = Pr\{\hat{d}_m(t) > \hat{d}_j(t), j \neq m\},$$
$$q(t) = Pr\{\hat{d}_m(t) > \hat{d}_j(t), \forall j \neq m\} = \prod_{j \neq m} q_j(t), \tag{3}$$
$$G(t) = \{q(t) > 1 - \bar{\delta}\}, \bar{\delta} \in (0, 1),$$
$$\bar{G}(t_0) = \{\bigcap_{t>t_0} \{q(t) > 1 - \bar{\delta}\}\}, \bar{\delta} \in (0, 1). \tag{4}$$

Our goal is to prove the result given in Theorem 3.

Theorem 3. *Given a* $\bar{\delta} \in (0, 1)$, *there exists a time instant* $t_0 < \infty$, *such that* $\bar{G}(t_0)$*holds. In other words, for this given* $\bar{\delta}$, *there exists a* $t_0 < \infty$, *such that* $\forall t > t_0$: $q(t) > 1 - \bar{\delta}$.

Proof: To prove Theorem 3, we are to prove Eq. 2. As mentioned in Section 3, this can be proven by showing that each of the $\hat{d}_j$ is within a $\frac{w}{2}$ neighborhood of d_j with a large enough probability. The latter, in turn, can be directly proven by invoking the weak law of large numbers.

4.3 $\{p_m(t)_{t>t_0}\}$ is a Submartingale under the DPA

We now prove the submartingale properties of $\{p_m(t)_{t>t_0}\}$ for the DPA.

Theorem 4. *Under the DPA, the quantity $\{p_m(t)_{t>t_0}\}$ is a submartingale.*

Proof: Firstly, as $p_m(t)$ is a probability, we have

$$E[p_m(t)] \leq 1 < \infty.$$

Secondly, we explicitly calculate $E[p_m(t)]$. Using the DPA's updating rule, we describe the update of $p_m(t)$ as per Table 1. Thus, we have:

$$
\begin{aligned}
&E[p_m(t+1)|P(t)]\\
&= \sum_{j=1...r} p_j\left(d_j(q(p_m+c_t\Delta)+(1-q)(p_m-\Delta))+(1-d_j)p_m\right)\\
&= \sum_{j=1...r}(p_jd_jqc_t\Delta) - \sum_{j=1...r}p_jd_j\Delta + \sum_{j=1...r}p_jd_jq\Delta + \sum_{j=1...r}p_jp_m\\
&= p_m + \sum_{j=1...r}p_jd_j(q(c_t\Delta+\Delta)-\Delta).
\end{aligned}
$$

In the above, $p_m(t)$ and $q(t)$ are respectively written as p_m and q in the interest of conciseness. The difference between $E[p_m(t+1)]$ and $p_m(t)$ can be expressed as:

$$Diff_{p_m(t)} = E[p_m(t+1)|P(t)] - p_m(t) = \sum_{j=1...r}p_j(t)d_j(q(t)(c_t\Delta+\Delta)-\Delta).$$

Given that $p_j(t) > 0$ and $d_j > 0$, if we denote

$$Z_t = \frac{\Delta}{c_t\Delta+\Delta} = \frac{1}{c_t+1},$$

we see that if $\forall t > t_0$, $q(t) > Z_t$, then, $Diff_{p_m(t)} > 0$, and so $\{p_m(t)_{t>t_0}\}$ is a submartingale.

Table 1. The various possibilities for updating p_m for the next iteration under the DPA

	Responses	The greatest element in $\hat{D}$	Updating p_m
$p_m(t+1)$	Reward, (w.p. d_j)	$\hat{d}_m$, (w.p. $q(t)$)	$p_m(t)+c_t\Delta$
		$\hat{d}_j, j \neq m$, (w.p. $1-q(t)$)	$p_m(t)-\Delta$
	Penalty, (w.p. $1-d_j$)	$\hat{d}_j, j=1...r$, (1)	$p_m(t)$

As per the action probability updating rules of the DPA, $c_t = 1,2,...,r-1$, implying that $Z_t \in [\frac{1}{r},\frac{1}{2}]$. Let the quantity $1 - \bar{\bar{\delta}} = max\{Z_t\} = \frac{1}{2}$, Then, according to Theorem 3, there exists a time instant t_0 such that $\forall t > t_0$, $q(t) > \frac{1}{2} = max\{Z_t\}$. Consequently, $\{p_m(t)_{t>t_0}\}$ is a submartingale, and the theorem is proven.

4.4 $Pr\{p_m(\infty) = 1\} \to 1$ **under the DPA**

We now prove the ε-optimality of the DPA.

Theorem 5. *The DPA is ε-optimal in all random Environments. More formally, given any* $1 - \bar{\delta} \geq \frac{1}{2}$, *there exists a positive integer* $N_0 < \infty$ *and a time instant* $t_0 < \infty$, *such that for all resolution parameters* $N > N_0$ *and for all* $t > t_0$, *the quantities* $q(t) > 1 - \bar{\delta}$, *and* $Pr\{p_m(\infty) = 1\} \to 1$.

Proof: According to the submartingale convergence theory [15],

$$p_m(\infty) = 0 \text{ or } 1.$$

If we denote e_j as the unit vector with the jth element being 1, then our task is to prove:

$$
\begin{aligned}
&\Gamma_m(P) \\
&\equiv Pr\{p_m(\infty) = 1 | P(0) = P\} \\
&= Pr\{P(\infty) = e_m | P(0) = P\} \\
&\to 1.
\end{aligned}
\tag{5}
$$

To prove Eq. (5), we shall use the theory of Regular functions [15]. Let $\Phi(P)$ as a function of P. If we define an operator U as

$$U\Phi(P) = E[\Phi(P(n+1))|P(n) = P],$$

then the result of the n-step invocation of U is:

$$U^n\Phi(P) = E[\Phi(P(n))|P(0) = P].$$

We refer to the function $\Phi(P)$ as being:

- Superregular: If $U\Phi(P) \leq \Phi(P)$. Then applying U repeatedly yields:

$$\Phi(P) \geq U\Phi(P) \geq U^2\Phi(P) \geq \dots \geq U^\infty\Phi(P).
\tag{6}$$

- Subregular: If $U\Phi(P) \geq \Phi(P)$. In this case, if we apply U repeatedly, we have

$$\Phi(P) \leq U\Phi(P) \leq U^2\Phi(P) \leq \dots \leq U^\infty\Phi(P).
\tag{7}$$

- Regular: If $U\Phi(P) = \Phi(P)$. In such a case, it follows that:

$$\Phi(P) = U\Phi(P) = U^2\Phi(P) = \dots = U^\infty\Phi(P).
\tag{8}$$

Moreover, if $\Phi(P)$ satisfies the boundary conditions

$$\Phi(e_m) = 1 \text{ and } \Phi(e_j) = 0, (\text{for } j \neq m),
\tag{9}$$

then, as per the definition of Regular functions and the submartingale convergence theory, we have

$$U^\infty\Phi(P) = E[\Phi(P(\infty))|P(0) = P]$$

$$= \sum_{j=1}^{r} \Phi(e_m)Pr\{P(\infty) = e_j|P(0) = P\}$$

$$= Pr\{P(\infty) = e_m|P(0) = P\}$$

$$= \Gamma_m(P). \tag{10}$$

Comparing Eq. (10) with Eq. (8), we see that $\Gamma_m(P)$ is exactly the function $\Phi(P)$ upon which if U is applied an infinite number of times, the sequence of operations will lead to a function that equals the *Regular* function $\Phi(P)$. Consequently, $\Gamma_m(P)$ can be indirectly obtained by investigating a Regular function of P. However, such a Regular function is not easily found, although its *existence* is guaranteed. Fortunately, Eq. (7) tell us that $\Gamma_m(P)$, i.e., the Regular function of P, can be bounded from below by the subregular function of P. As we are most interested in the lower bound of $\Gamma_m(P)$, our goal is to find such a *Subregular* function of P, which also satisfies the boundary conditions given by Eq. (9), which then will guarantee to bound $\Gamma_m(P)$ from below.

To find a proper subregular function of P, we need to firstly find the corresponding superregular function. Consider $\Phi_m(P) = e^{-x_m P_m}$ as a specific instantiation of Φ, where x_m is a positive constant. Then, under the DPA,

$$U(\Phi_m(P)) - \Phi_m(P) \tag{11}$$

$$=E[\Phi_m(P(n+1))|P(n) = P] - \Phi_m(P)$$

$$=E[e^{-x_m P_m(n+1)}|P(n) = P] - e^{-x_m P_m}$$

$$= \sum_{j=1...r} e^{-x_m(p_m+c_t\Delta)}p_j d_j q + \sum_{j=1...r} e^{-x_m(p_m-\Delta)}p_j d_j(1-q)$$

$$+ \sum_{j=1...r} e^{-x_m P_m}p_j(1-d_j) - e^{-x_m P_m}$$

$$= \sum_{j=1...r} p_j d_j e^{-x_m P_m}\left(q(e^{-x_m c_t\Delta} - e^{x_m\Delta}) + (e^{x_m\Delta} - 1)\right).$$

We must determine a proper value for x_m such that $\Phi_m(P)$ is superregular, i.e., $U(\Phi_m(P)) - \Phi_m(P) \leq 0$. This is equivalent to solving the following inequality:

$$q(e^{-x_m c_t\Delta} - e^{x_m\Delta}) + (e^{x_m\Delta} - 1) \leq 0. \tag{12}$$

We know that when $b > 0$ and $x \to 0$, $b^x \dot{=} 1 + (\ln b)x + \frac{(\ln b)^2}{2}x^2$. If we set $b = e^{-x_m}$, when $\Delta \to 0$, Eq. (12) can be re-written as

$$q\left((\ln b)(c_t + 1)\Delta + \frac{(\ln b)^2}{2}(c_t^2 - 1)^2\Delta^2\right) - (\ln b)\Delta + \frac{\ln b^2}{2}\Delta^2 \leq 0.$$

Substitute b with e^{-x_m}, we have $x_m(x_m - \frac{2(q(c_t+1)-1)}{\Delta(q(c_t^2-1)+1)}) \leq 0$. As x_m is defined as a positive constant, we have

$$0 < x_m \leq \frac{2(q(c_t+1)-1)}{\Delta(q(c_t^2-1)+1)}. \tag{13}$$

If we denote $x_{m_0} = \frac{2(q(c_t+1)-1)}{\Delta(q(c_t^2-1)+1)}$, we see that when $\Delta \to 0$, $x_{m_0} \to \infty$ as $c_t = 1, 2, ..., r-1$ and $q(t)_{(t>t_0)} > \frac{1}{2}$.

We now introduce another function $\phi_m(P) = \frac{1-e^{-x_m p_m}}{1-e^{-x_m}}$, where x_m is the same as defined in $\Phi_m(P)$. Moreover, we observe the property that if $\Phi_m(P) = e^{-x_m p_m}$ is a super-regular (subregular), then $\phi_m(P) = \frac{1-e^{-x_m p_m}}{1-e^{-x_m}}$ is a subregular (superregular) [15]. There-fore, the x_m, as defined in Eq. (13), which renders $\Phi_m(P)$ to be superregular, makes the $\phi_m(P)$ be subregular.

Obviously, $\phi_m(P)$ meets the boundary conditions, i.e.,

$$\phi_m(P) = \frac{1-e^{-x_m p_m}}{1-e^{-x_m}} = \begin{cases} 1, & \text{when } P = e_m, \\ 0, & \text{when } P = e_j. \end{cases}$$

Therefore, according to Eq. (7),

$$\Gamma_m(P) \geq \phi_m(P) = \frac{1-e^{-x_m p_m}}{1-e^{-x_m}}. \tag{14}$$

As Eq. (14) holds for every x_m bounded by Eq. (13), we can choose the largest value x_{m_0}, and when $x_{m_0} \to \infty$, $\Gamma_m(P) \to 1$. We have thus proved that for the DPA, $Pr\{p_m(\infty) = 1\} \to 1$, implying its ε-optimality.

5   Conclusions

Estimator algorithms are acclaimed to be the fastest Learning Automata (LA), and within this family, the set of *Pursuit* algorithms have been considered to be the pio-neering schemes. The ε-optimality of Pursuit Algorithms (PAs) are of great importance and has been studied for decades. The convergence proofs for the PAs in all reported papers have a common flaw which was discovered by the authors of [14]. This paper corrects the flaw and provides a new proof for the CPA.

Rather than examining the monotonicity property of the $\{p_m(t)_{(t>t_0)}\}$ sequence as done in the previous papers and in [14], the current new proof in this paper studies the *submartingale* property of $\{p_m(t)_{(t>t_0)}\}$ in the DPA. Thereafter, by virtue of the submartingale property and the weaker condition, the new proof invokes the theory of Regular functions, and does not require the learning parameter to decrease/increase gradually in proving the ε-optimality of the DPA.

We believe that our submitted proof is both novel and pioneering. Further, as opposed to the proof in [14], the new method does not require the parameter to change continu-ously. Also, the new proof can be extended to formally demonstrate the ε-optimality of other EAs with absorbing states.

References

1. Oommen, B.J., Granmo, O.C., Pedersen, A.: Using stochastic ai techniques to achieve un-bounded resolution in finite player goore games and its applications. In: IEEE Symposium on Computational Intelligence and Games, Honolulu, HI (2007)
2. Beigy, H., Meybodi, M.R.: Adaptation of parameters of bp algorithm using learning automata. In: Sixth Brazilian Symposium on Neural Networks, JR, Brazil (2000)
3. Granmo, O.C., Oommen, B.J., Myrer, S.A., Olsen, M.G.: Learning automata-based solutions to the nonlinear fractional knapsack problem with applications to optimal resource allocation. IEEE Trans. Syst., Man, Cybern. B 37(1), 166–175 (2007)
4. Unsal, C., Kachroo, P., Bay, J.S.: Multiple stochastic learning automata for vehicle path control in an automated highway system. IEEE Trans. Syst., Man, Cybern. A 29, 120–128 (1999)
5. Granmo, O.C.: Solving stochastic nonlinear resource allocation problems using a hierarchy of twofold resource allocation automata. IEEE Trans. Computers 59(4), 545–560 (2010)
6. Yazidi, A., Granmo, O.C., Oommen, B.J.: Service selection in stochastic environments: A learning-automaton based solution. Applied Intelligence 36, 617–637 (2012)
7. Vafashoar, R., Meybodi, M.R., Momeni, A.A.H.: Cla-de: a hybrid model based on cellular learning automata for numerical optimization. Applied Intelligence 36, 735–748 (2012)
8. Thathachar, M.A.L., Sastry, P.S.: Estimator algorithms for learning automata. In: The Platinum Jubilee Conference on Systems and Signal Processing, Bangalore, India, pp. 29–32 (1986)
9. Oommen, B.J., Lanctot, J.K.: Discretized pursuit learning automata. IEEE Trans. Syst., Man, Cybern. 20, 931–938 (1990)
10. Lanctot, J.K., Oommen, B.J.: On discretizing estimator-based learning algorithms. IEEE Trans. Syst., Man, Cybern. B 2, 1417–1422 (1991)
11. Lanctot, J.K., Oommen, B.J.: Discretized estimator learning automata. IEEE Trans. Syst., Man, Cybern. B 22(6), 1473–1483 (1992)
12. Rajaraman, K., Sastry, P.S.: Finite time analysis of the pursuit algorithm for learning automata. IEEE Trans. Syst., Man, Cybern. B 26, 590–598 (1996)
13. Oommen, B.J., Agache, M.: Continuous and discretized pursuit learning schemes: various algorithms and their comparison. IEEE Trans. Syst., Man, Cybern. B 31(3), 277–287 (2001)
14. Ryan, M., Omkar, T.: On ε-optimality of the pursuit learning algorithm. Journal of Applied Probability 49(3), 795–805 (2012)
15. Narendra, K.S., Thathachar, M.A.L.: Learning Automata: An Introduction. Prentice Hall (1989)

Q-Learning Algorithm Used by Secondary Users for QoS Support in Cognitive Radio Networks

Jerzy Martyna

Institute of Computer Science, Jagiellonian University,
ul. Prof. S. Lojasiewicza 6, 30-348 Cracow

Abstract. In this paper, we propose a stochastic game that allows multiple secondary users for Quality of Service (QoS) support in cognitive radio networks. A state-transition model is built for the QoS provisioning transmission in these networks. At each stage of the game, secondary users observe the spectrum availability, channel quality and the strategy of the QoS provisioning transmission for all players. According to this observation, they will then decide how many channels they should be imposed to take into the QoS constraints. By using Q-learning, each group of secondary users can learn the optimal policy that maximises the expected sum of discounted payoff sum, defined as the spectrum-efficient throughput. The proposed stationary policy is shown to achieve much better performance than the policy obtained by ordinary stochastic games, which only maximise each stage's payoff. Our results suggest that the proposed method can be used to obtain performance gains through better adaptation to the QoS limitations.

1   Introduction

A key technology for dynamic spectrum access (DSA) is cognitive radio (CR). CR networks are new type of wireless network, that allow the secondary user (SUs) to utilise the spectrum usage more efficiently in an opportunistic fashion, without interfering with primary users. As shown in Fig. 1, the secondary users belonging to three groups are able to make opportunistic use of transmission channels without disturbing the primary users PU_1, PU_2, PU_3. One of the most challenging problems is how to assign the available channels providing quality of service (QoS) guarantees. It is obvious that providing certain QoS assurance is crucial in order to provide acceptable performance and quality.

Spectrum sensing and resource allocation (channel assignment, rate, allocating transmit power, etc.) schemes have been the main focus of research efforts on CR networks. The optimal power control policies to maximise the ergodic capacity of an SU assuming a CR network model with one PU and one SU coexisting in the same channel without a QoS requirement has been derived from the paper by Kang et al. [1]. The same model as in [1], but with a QoS defined as minimum rate and acceptable BER was considered in the paper by Attar [2]. Unfortunately, the solution presented in [2] is formulated as non-convex optimisation problem.

Various AI methods have been proposed in the literature, such as a genetic algorithm approach [3] to determine transmit power, modulation type and rate for secondary users

A. Moonis et al. (Eds.): IEA/AIE 2014, Part I, LNAI 8481, pp. 389–398, 2014.

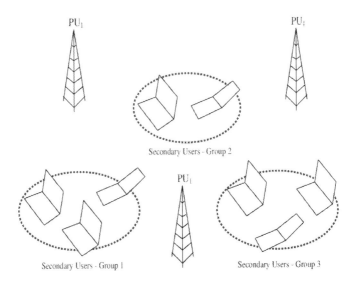

Fig. 1. An example of CR network

in both single and multi-carrier based transmission in CR networks. Due to slow convergence and the high complexity of genetic algorithm, the presented solution is not suitable to implementation. A neural network-based learning scheme for cognitive radio systems to predict the data rate has been proposed by Tsagkaris [4]. An approach using the Fuzzy Logic System (FLS) to control spectrum access and the spectrum for secondary users was proposed by Kaniezhil et al. [5]. However, both of the above-mentioned papers did not take the QoS constraints into consideration.

Stochastic game theory is an essential tool for cognitive radio networks. Among the game-theoretic approaches to addressing resource management in these networks, stochastic game is able to exploit the correlated channels in analysing decentralised behaviours of cognitive radios. Tembino et al. [6] demonstrate the existence of equilibriums and conditions for evolutionarily stable strategies under good and bad weather conditions based on a stochastic game for modeling the remaining energy of the battery for each radio device. The secondary use rate adaptation problem in cognitive radio networks with a constrained general-sum switching control Markovian dynamic game as the original problem, as considered by J.W. Huang and V. Krishnamurthy [7], has been transformed and solved using the Nash equilibrium policy.

In this study, we propose a stochastic game framework for the QoS of service provisioning over cognitive radio channels. Secondary users perform channel sensing to detect the activity of primary users. Depending on the presence or absence of active primary users, the secondary users transmit the data at two average power levels. The contributions of this paper are the following:

1) Formulating a state-transition model for QoS service provisioning over cognitive radio channels.
2) Providing a stochastic game for QoS service provisioning.

3) Secondary users can learn the optimal policy maximising the expected sum of discounted payoffs (defined as spectrum-efficient throughput) by using the Q-learning algorithm.
4) Incorporating the power and rate adaptation into the model by considering different assumptions on the availability of channel side information (CSI) at the transmitter.

The remaining of the paper is organised as follows. In Section 2, we describe the state transition model for cognitive transmission and provide the channel detection threshold, false alarm and detection probabilities for the fixed-power and fixed-rate transmission schemes with the QoS constraints. In Section 3, a stochastic game is formulated by defining the states, actions and objective functions as well as the state transition rules. In Section 4, the optimal policy of the secondary user using the Q-learning algorithm is given. In section 5, the simulation results are presented. Finally, the conclusion is given in Section 6.

2   System Model

In this section, we present the model of the secondary user network, the cognitive transmission under QoS constraints and the effective capacity concept.

2.1   Secondary User Network

We assume that there is a secondary base station in the CR network that coordinates the spectrum usage for all secondary users and secondary users of second type (class). In our approach, the secondary network is a time-slotted system. To avoid interference with primary users, the secondary users of both types need to listen to the spectrum before their transmissions. Thus, the secondary users and secondary user of second type can be omitted from the occupied slots. We assume a perfect sensing of the currently unused licensed spectrum and vacating the spectrum.

To achieve efficient spectrum utilisation, control messages are exchanged between the secondary base station and the secondary users through control channels. The control messages are associated with situations such as channel assignment, spectrum hand-off, etc. Similarly, control messages are also exchanged in similar situations between secondary users and the secondary base station through the control channels. If the control messages are not correctly received, the characteristics of some functions can be violated.

2.2   Cognitive Channel Model

In this paper, we consider a cognitive radio channel model in which a secondary transmitter attempts to send data to a secondary receiver with primary users present. The secondary users test channel activity. If the secondary transmitter selects its transmission when the channel is busy, the average power is $\overline{P}_1$ and the rate is r_1. When the channel is idle, the average power is $\overline{P}_2$ and the rate is r_2. We assume that $\overline{P}_1 = 0$ denotes the stoppage of the secondary transmission in the presence of an active primary user. Both transmission rates, r_1 and r_2, can be fixed or time-variant depending on whether the transmitter has channel side information or not. In general, we assume that $\overline{P}_1 < \overline{P}_2$. In the above model, the discrete-time channel input-output relation in the absence in the channel of the primary users is given by

$$y(i) = h(i)x(i) + n(i), \quad i = 1, 2, \ldots \tag{1}$$

where i is the symbol duration. If primary users are present in the channel, the discrete-time channel input-output relation is given by

$$y(i) = h(i)x(i) + s_p(i) + n(i), \quad i = 1, 2, \ldots \tag{2}$$

where $s_p(i)$ represents the sum of the active primary users' faded signals arriving at the secondary receiver $n(i)$ is the additive thermal noise at the receiver and is zero-mean, circularly symmetric, complex Gaussian random variable with variance $E\{|n(i)|^2\} = \sigma_n^2$ for all i.

We assume that the receiver knows the instantaneous lambda values $\{h(i)\}$, while the transmitter has no such knowledge. We construct a state-transition model for cognitive transmission by considering the cases in which the fixed transmission rates are lesser or greater than the instantaneous channel capacity values. In particular, the ON state is achieved if the fixed rate is smaller than the instantaneous channel capacity. Otherwise, the OFF state occurs.

Thus, we have the following four possible scenarios associated with the decision of channel sensing, namely [8]

1) channel is busy, detected as busy (correct detection),
2) channel is busy, detected as idle (miss-detection),
3) channel is idle, detected as busy (false alarm),
4) channel is idle, detected as idle (correct detection).

If the channel is detected as busy, the secondary transmitter sends with power $\overline{P}_1$. Otherwise, it transmits with a larger power, $\overline{P}_2$. In the above four scenarios, we have the instantaneous channel capacity, namely

$$C_1 = B \log_2(1 + SNR_1 \cdot z(i)) \quad \text{channel is busy, detected as busy} \tag{3}$$

$$C_2 = B \log_2(1 + SNR_2 \cdot z(i)) \quad \text{channel is busy, detected as idle} \tag{4}$$

$$C_3 = B \log_2(1 + SNR_3 \cdot z(i)) \quad \text{channal is idle, detected as busy} \tag{5}$$

$$C_4 = B \log_2(1 + SNR_4 \cdot z(i)) \quad \text{channel is idle, detected as idle} \tag{6}$$

where $z(i) = [h(i)]^2$, SNR_i for $i = 1, \ldots, 4$ denotes the average signal-to-noise ratio (SNR) values in each possible scenario.

The cognitive transmission is associated with the ON state in scenarios 1 and 3, when the fixed rates are below the instantaneous capacity values ($r_1 < C_1$ or $r_2 < C_2$). Otherwise, reliable communication is not obtained when the transmission is in the OFF state in scenarios 2 and 4. Thus, the fixed rates above are the instantaneous capacity values ($r_1 \geq C_1$ or $r_2 \geq C_2$). The above channel model has 8 states and is depicted in Fig. 2. In states 1, 3, 5 and 7, the transmission is in the ON state and is successfully realised. In the states 2, 4, 6 and 8 the transmission is in the OFF state and fails.

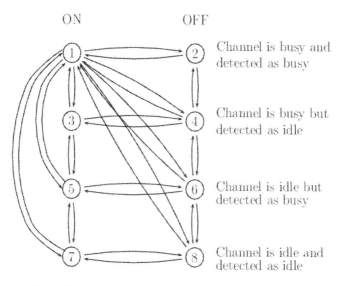

ON   OFF

Channel is busy and detected as busy

Channel is busy but detected as idle

Channel is idle but detected as busy

Channel is idle and detected as idle

Fig. 2. State transition model for the cognitive radio channel

2.3   Effective Capacity

The state-transition model above can be supported by service processes with the required QoS constraints identified through the notion of effective capacity. The effective capacity was defined by Wu and Negi [9] as the maximum constant arrival rate that a given time-varying rate in a given time-varying service process can support whilst meeting the QoS requirement specified by the QoS exponent theta. Hence, theta is defined as the decay rate of the tail distribution of the queue length Q, namely:

$$\lim_{q \to \infty} \frac{\log P(Q \geq q)}{q} = -q \tag{7}$$

where q is the queue length. For the large value of queue length, q_{max}, we have $P(W \geq q_{max}) \approx e^{-\theta q_{max}}$. Thus, the smaller θ corresponds to looser constraints and the larger θ indicates strict QoS constraints.

Thus, the smaller θ corresponds to looser constraints and the larger θ indicates strict QoS constraints.

3   Stochastic Game Formulation

There is a stochastic game among a number of players belonging to various classes. The stochastic game $\mathcal{G}$ is defined as a set of states denoted by $\mathcal{S}$, a set of actions described as $\mathcal{A}_1, \mathcal{A}_2, \ldots, \mathcal{A}_n$ - one for each player in the game. Each player selects a new state with a transmission probability determined by the current state and one action from each player, $\mathcal{T} : \mathcal{S} \times \mathcal{A}_1 \times \ldots \times \mathcal{A}_n \to PD(\mathcal{S})$. At each stage, each player attempts to maximise his expected sum of payoffs, namely: $E\{\sum_{j=0}^{\infty} \beta^j \eta_{i,t+j}\}$, where $\eta_{i,t+j}$ is the reward received j steps into the future by player i and β is the discount factor.

A secondary user can utilise unused spectrum bands belonging to L primary users. We assume that the bandwidth of licensed bands may be different, and each licensed band is partitioned into a set of adjacent channels with the same bandwidth. Thus, we can denote N_l channels in the primary user l's band. In our approach, when the primary user is active at time t in the l-th band, this is denoted by $P_l^t = 1$. Otherwise, the state is defined as $P_l^t = 0$.

Since the channel is modelled as a finite-state Markov chain (FSMC), the channel quality in terms of SNR of the lst band can be expressed by FSMC. Thus, the achievable gain of the licensed band depends on the primary usersŠ status ($P_l^t = 1$ when the primary user uses the lst band at any time t, otherwise $P_l^t = 0$). Thus, each state of the FSMC is jointly modelled by the pair (P^t, g_l^t), where g_l^t is the channel quality. The channel quality can take any value from a set of discrete values, i.e. $g_l^t \in \{SNR_1, \ldots, SNR_8\}$.

Consider the scenario with a two type of secondary users belonging to two classes. The actions of the secondary users from the first class of the secondary users can be defined as $\mathbf{a}^t = (a_{l,D_1}^t, a_{l,C_1}^t, a_{l,D_2}^t, a_{l,C_2}^t)$. The action a_{l,D_1}^t (or a_{l,C_1}^t) denotes that the secondary network will transmit data (control) messages at channels uniformly selected at time slot t. Next, the action a_{l,D_2}^t (or a_{l,C_2}^t) indicates that the secondary network will transmit data (control) messages in the a_{l,D_2}^t (or a_{l,C_2}^t) channel selected from the previously used channels without success.

Similarly, the action of the secondary users belonging to the second type of the secondary users is defined $\mathbf{a}_S^t = (a_{S,l,D_1}^t, a_{S,l,C_1}^t, a_{S,l,D_2}^t, a_{S,l,C_2}^t)$, where the action a_{S,l,D_1}^t (or a_{S,l,C_1}^t) denotes the secondary network will transmit data (control) messages at channels uniformly distributed at time slot t. Analogous, the action a_{S,l,S_2}^t (or a_{S,l,C_2}^t) denotes that the secondary users will transmit the data (control) messages in the a_{S,l,D_2}^t (or a_{S,l,C_2}^t) channel from previously used channel without success.

After defining the state at each stage, we may provide the state transition rule, namely assuming that secondary users should observe which channel has been occupied by secondary users. Based on these observations, the secondary users can define the pair $\{S_{l,D}^t, S_{l,C}^t\}$, where $S_{l,D}^t$ and $S_{l,C}^t$ denote data and control channel numbers being used by secondary users of the second class in the lst band observed at time slot t. We assume that the secondary users cannot be informed as to whether an idle channel is occupied or not by the secondary users from the second class. Thus, the number of idle channels that are not being engaged by the secondary users of the second class is not an observation by the secondary users from the first class.

Thus, at every time slot time t, the state of the stochastic game $\mathcal{G}$ is defined by $\mathbf{s}^t = \{s_1^t, s_2^t, \ldots, s_L^t\}$ where $a_l^t = (P_l^t, g_l^t, S_{l,D}^t, S_{l,C}^t)$ indicates the state associated with band $l (l \in \{1, \ldots, L\})$.

After defining the state at each stage, we may provide the state transition rule, namely

$$p(\mathbf{s}^t \mid \mathbf{s}^t, \mathbf{a}^t, \mathbf{a}_S^t) = \prod_{l=1}^{L} p(s_l^{t+1} \mid s_l^t, a_l^t, a_{l,S}^t) \tag{8}$$

Th transition probability $p(s_l^{t+1} \mid s_l^t, a_l^t, a_{l,S}^t)$ can be further expressed by

$$p(s_l^{t+1} \mid s_l^t, a_l^t, a_{l,S}^t) = p(S_{l,D}^{t+1}, S_{l,C}^{t+1} \mid S_{l,D}^t, S_{l,C}^t, a_l^t, a_{l,S}^t)$$
$$\times p(P_l^{t+1}, g_l^{t+1} \mid P_l^t, g_l^t) \qquad (9)$$

where the first term on the right side represents the transition probability of the number of secondary users of second type and data channels, and the second term denotes the transition of the primary user status and the channel conditions.

We will now consider the scenario with a two type of secondary users. After the all players choose their actions, the secondary users will transmit data and control messages in the selected channels and the secondary users belonging to the second type will intercept their channels. We assume that the same control messages are transmitted in all the control channels, and one correct copy of control information at time t is sufficient for coordinating the spectrum management in the next time slot.

We assume that the stage payoff of the secondary users maximizes the spectrum gain, namely

$$r(\mathbf{s}^t, \mathbf{a}^t, \mathbf{a}_S^t) = T(\mathbf{s}^t, \mathbf{a}^t, \mathbf{a}_S^t) \times (1 - p^{block}(\mathbf{s}^t, \mathbf{a}^t, \mathbf{a}_S^t)) \qquad (10)$$

where $T(\mathbf{s}^t, \mathbf{a}^t, \mathbf{a}_S^t)$ indicates the expected spectrum gain when not all control channels get intercept and $p^{block}(\mathbf{s}^t, \mathbf{a}^t, \mathbf{a}_S^t)$ denotes the probability that all control channels in all L bands are intercepted.

4   The Minimax-Q Learning to Obtain the Optimal Policy of the Stochastic Game

In this section, the minimax-Q learning for the secondary users to obtain the optimal policy of the stochastic game is presented.

In general, the secondary users treat the payoff in different stages differently. Then, the secondary users' objective is find an optimal policy that maximizes the expected sum of payoffs

$$\max E\{\sum_{t=0}^{\infty} \beta^t r(\mathbf{s}^t, \mathbf{a}^t, \mathbf{a}_S^t)\} \qquad (11)$$

for $\forall\, s \in \mathcal{S}$ **and** $\forall\, a \in \mathcal{A}$ **do**
  begin
    After receiving reward $r(\mathbf{s}^t, a^t, \mathbf{a}_S^t)$ for moving from $\mathbf{s}^t$ to $\mathbf{s}^{t+1}$ by taking action $\mathbf{a}^t$
    Let function $Q(\mathbf{s}^t, \mathbf{a}^t, \mathbf{a}_S^t)$ is given by Eq. (13)
    Update the optimal strategy $\pi^*(\mathbf{s}^t, \mathbf{a})$ by
      $\pi^*(\mathbf{s}^t) := \arg\max_{\pi(\mathbf{s}^t)} \min_{\pi(\mathbf{s}^t)} \sum_a \pi(\mathbf{s}^t, a^t) Q(\mathbf{s}^t, \mathbf{a}^t, \mathbf{a}_S^t)$
    Compute $V(\mathbf{s}^t) := \min_{\mathbf{s}_S(\mathbf{s}^t)} \sum_a \pi^*(\mathbf{s}^t, \mathbf{a}) Q(\mathbf{s}^t, \mathbf{a}, \mathbf{a}_S)$
      $\alpha^{t+1} := \alpha^t \cdot \mu;$
  end;

Fig. 3. The learning phase of the minimax-Q learning algorithm for two groups of secondary users

where β is the discount factor of the secondary user. In our approach, the policy of the secondary network is expressed by $\pi : \mathcal{S} \to \mathcal{PD}(\mathcal{A})$ and the policy of the secondary users of second type $\pi_S : \mathcal{S} \to \mathcal{PD}(\mathcal{A}_S)$, where $\mathbf{s}^t \in \mathcal{S}, \mathbf{a}^t \in \mathcal{A}, \mathbf{a}_S^t \in \mathcal{A}_S$. It is noticeable that the policy π^t at time t is independent of the states and actions in all previous states and actions. Then, the policy π is said to be Markov. If the policy is independent of time, the policy is said to be stationary.

In the stochastic game between the secondary users and the secondary users of second type is a zero-sum game, the equilibrium of each stage is the minimax equilibrium. To solve the game, we can use the minimax-Q learning method [10, 11]. The Q-function of stage t is defined as the expected discounted payoffs when the secondary users take action $\mathbf{a}^t$ and the secondary users of second type take the action $\mathbf{a}_S^t$. Then the Q-value in the minimax-Q learning of the game can be expressed as

$$Q(\mathbf{S}^t, \mathbf{a}^t, \mathbf{a}_S^t) = r(\mathbf{s}^t, \mathbf{a}^t, \mathbf{a}_S^t) + \beta \sum_{\mathbf{S}}^{t+1} p(\mathbf{s}^{t+1} \mid \mathbf{s}^t, \mathbf{a}^t, \mathbf{a}_S^t) V(\mathbf{s}^{t+1}) \qquad (12)$$

where $r(\mathbf{s}^t, \mathbf{a}^t, \mathbf{a}_S^t)$ is reward when making a transition from $\mathbf{s}^t$ to $\mathbf{s}^{t+1}$, $V(\mathbf{s}^{t+1})$ is the value of a state in the game of secondary users of second type. In order to obtain the state transition probability, we can modify the value of iteration and the Q-function according to [11], namely

$$Q(\mathbf{s}^t, \mathbf{a}^t, \mathbf{a}_S^t) = (1 - \alpha^t) Q(\mathbf{s}^t, \mathbf{a}^t, \mathbf{a}_S^t) + \alpha^t \{ r(\mathbf{s}^t, \mathbf{a}^t, \mathbf{a}_S^t) + \beta \cdot V(\mathbf{s}^{t+1}) \} \qquad (13)$$

where α^t denotes the learning rate decaying over time by $\alpha^{t+1} = \mu \cdot \alpha^t$ and $0 < \mu < 1$. Then, the learning phase of the used minimax-Q algorithm for the two groups of secondary users is given in Fig. 3.

5   Simulation Results

We conduct simulations to evaluate performance in QoS provisioning by a stochastic game. Firstly, we check the convergence of the minimax-Q learning algorithm and analyze the strategy of all secondary users for several stages.

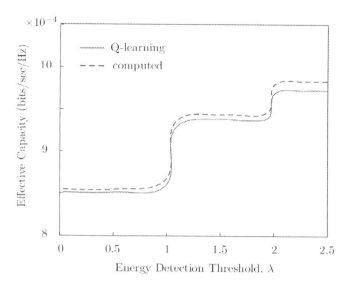

Fig. 4. Effective capacity as a function of the detection threshold value for the secondary users

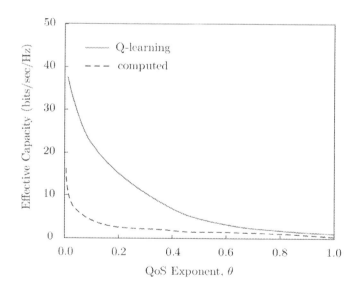

Fig. 5. Effective capacity versus QoS exponent θ for the secondary users

We assumed that the observation time is equal to 1 sec and the channel bandwidth is equal to 100 kHz. Moreover, we assumed that the QoS exponent is $\theta = 0.01$ and the average SNR values when the channel is detected correctly are $SNR_1 = 0$ dB and $SNR_4 = 10$ dB for busy and idle channels, respectively. In Fig. 4, we plot the effective capacity as a function of the detection threshold value λ. As we see in Fig. 4 the effective capacity is increasing with increasing the detection threshold value λ.

In Fig. 5, we plot the effective capacity as a function of the QoS exponent obtained for both classes of the secondary users under the assumption that the probability of false alarm is equal to 0 and the probability of detection is equal to 1 (perfect channel detection). We have seen that the effective capacity values are decreasing with increasing θ values.

6   Conclusion

In this paper, we studied the QoS provisioning transmission in cognitive radio networks. The impact upon the effective capacity of several system parameters, such as channel sensing duration, detection threshold, detection and false alarm probabilities, and the QoS parameters and transmission rates was investigated. Secondary users adapt their strategy on how to reserve and switch between control and data channels according to their observations about spectrum availability, channel quality, etc. By using the Q-learning algorithm in the stochastic game, secondary users learn the optimal policy that maximises the expected sum of discounted payoff sum, defined as the spectrum-efficient throughput.

References

[1]   Kang, X., Liang, Y.-C., Nallanathan, A.: Optimal Power Allocation for Fading Channels in Cognitive Radio Networks Under Transmit and Interference Power Constraints. In: IEEE Int. Conf. on Comm., pp. 3568–3572 (May 2008)

[2]   Attar, A., Nakhai, M.R., Aghvami, A.H.: Cognitive Radio Game: A Framework for Efficiency, Fairness and QoS Guarantee. In: IEEE Int. Conf. on Comm., pp. 4170–4174 (May 2008)

[3]   Newman, T.R., Barker, B.A., Wyglinski, A.M., Agah, A., Evans, J.B., Minden, G.J.: Cognitive Engine Implementation for Wireless Multicarrier Transceivers. Wiley Journal on Wireless Communications and Mobile Computing 7(9), 1129–1142 (2007)

[4]   Tsagkaris, K., Katidiots, A., Demestichas, P.: Neural Network-based Learning Schemes for Cognitive Radio Systems. Computer Communications 31, 3394–3404 (2008)

[5]   Kaniezhil, R., Kumar, C.D.N., Prakash, A.: Fuzzy Logic System for Opportunistic Spectrum Access using Cognitive Radio. UCSI International Journal of Computer Science 1(1), 703–709 (2013)

[6]   Tembine, H., Altman, E., El-Azouzi, R., Hayel, Y.: Evolutionary Games in Wireless Networks. IEEE Trans. on Systems, Man, and Cybernetics, Part B. Special Issues on Game Theory (2009)

[7]   Huang, J.-W., Krishnamurthy, V.: Dynamical Transmission Control. In: Zhang, Y., Guizani, M. (eds.) Game Theory for Wireless Communications and Networking. CRC Press, Boca Raton (2011)

[8]   Akin, S., Gursoy, M.C.: Effective Capacity of Cognitive Radio Channels for Quality of Service Provisioning. IEEE Trans. on Wireless Communications 9(11), 3354–3364 (2010)

[9]   Wu, D., Negi, R.: Effective Capacity: A Wireless A Wireless Link Model for Support of Quality of Service. IEEE Trans. on Wireless Communications 2(4), 630–643 (2003)

[10]   Watkins, C.J.C.H., Dayan, P.: Q-learning. Machine Learning 8, 279–292 (1992)

[11]   Littman, M.L.: Markov Games as a Framework for Multi-Agent Reinforcement Learning. In: Proc. 11th International Conference on Machine Learning, pp. 157–163 (1994)

An Adaptive Questionnaire for Automatic Identification of Learning Styles

Esperance Mwamikazi, Philippe Fournier-Viger, Chadia Moghrabi,
Adnen Barhoumi, and Robert Baudouin

Department of Computer Science, Université de Moncton, Canada
Department of Secondary Education and Human Resources, Université de Moncton, Canada
{eem7706,philippe.fournier-viger,chadia.moghrabi}@umoncton.ca,
bmhadnen@gmail.com, robert.baudouin@moncton.ca

Abstract. Learning styles refer to how a person acquires and processes information. Identifying the learning styles of students is important because it allows more personnalized teaching. The most popular method for learning style recognition is through the use of a questionnaire. Although such an approach can correctly identify the learning style of a student, it suffers from three important limitations: (1) filling a questionnaire is time-consuming since questionnaires usually contain numerous questions, (2) learners may lack time and motivation to fill long questionnaires and (3) a specialist needs to analyse the answers. In this paper, we address these limitations by presenting an adaptative electronic questionnaire that dynamically selects subsequent questions based on previous answers, thus reducing the number of questions. Experimental results with 1,931 questionnaires for the Myers Briggs Type Indicators show that our approach (Q-SELECT) considerably reduces the number of questions asked (by a median of 30 %) while predicting learning styles with a low error rate.

Keywords: adaptive questionnaire, association rules, neural networks, learning styles, Myers Briggs Type Indicator.

1 Introduction

Learning styles refer to how people acquire and process information [1, 2]. In education, knowing the learning styles of students is important because it allows to further personnalize interaction between teachers and students. It was shown in several studies that presenting information in a way that is adapted to the learning style of a learner facilitate learning [3].

Although several studies have been presented on how to perform learning style assessment, there are still several important limitations to current approaches. The first approach is used by e-learning systems that can provide adaptation according to the learning style of a learner. The approach consists of designing a software module that analyzes the learner's interactions with the system to detect the learning style [1, 4]. This approach has the benefit of being seamless for the learner. However, it suffers from a major drawback, namely that a random learning type is assigned initially to the

A. Moonis et al. (Eds.): IEA/AIE 2014, Part I, LNAI 8481, pp. 399–409, 2014.

learner and thus the system will initially guide and assist the learner according to that type. If the initial guess is incorrect, the system will thus interact with the learner according to the wrong learning style, which may have negative effect on learning. Furthermore, this interaction will continue until enough data is recorded to find the correct learning style [1].

The other main method for learning style assessment is to use a standardized questionnaire that a person has to fill out. A specialist then analyzes the answers to determine the correct learning style. The advantage of this approach is that the learning style of a person can be identified immediately. However, this approach suffers also from important limitations. First, questionnaires are usually very long. For example, the Myers-Briggs Type Indicator questionnaire discussed in this paper consists of more than 90 questions. This means that it is very time-consuming for a person to fill out the questionnaire. Second, long questionnaires have a negative effect on a person's motivation [4], which may lead to abandonning the test, skipping questions or answering falsely. This can furthermore provoke an incorrect learning style assessment, which may have undesirable consequences in future interactions [5]. For example, in the case of an e-learning system, if a learner does not answer the questionnaire correctly, the ensuing interactions with the system may be done according to a wrong learning style, which may have detrimental effect on learning. Third, using a questionnaire usually requires a specialist to analyze the learner answers and to determine the learning style.

In this paper, we address all the above limitations by presenting a novel learning style assessment approach, which takes the form of an adaptive electronic questionnaire. Our contributions are fourfold. First, the electronic questionnaire relies on an efficient algorithm PREDICT for predicting answers to upcoming questions based on associations between questions already answered and answers from previous users. Predicting answers allows skipping questions from the standardized questionnaire, thus reducing the number of questions to be answered by the learner. Second, the electronic questionnaire incorporates an efficient question selection algorithm Q-SELECT that analyzes associations between question answers to determine which questions should be asked first to minimize the number of questions asked when the aforementioned prediction algorithm is used. Third, once all questions have been answered or predicted, the electronic questionnaire uses a novel prediction algorithm to accurately predict a person learning style based on the answers and predicted answers. Fourth, we performed an extensive experimental study with 1,931 questionnaires for the assessment of the Myers-Briggs Type Indicator (MBTI). Results show that our approach reduces the number of questions presented to the user by a median of 30% while maintaining a low error rate in identifying the learning styles.

The rest of the paper is organized as follows. The Myers-Briggs Type Indicator model is presented in section 2. Section 3 discusses related work on adaptive questionnaires. In section 4 and 5, we respectively present the proposed electronic questionnaire and the experimental results. Finally, section 6 draws the final conclusions.

2   Myers-Briggs Type Indicator (MBTI)

A popular personality inventory that has been used for more than 30 years is the Myers-Briggs Type Indicator (MBTI). This is a self-report questionnaire that identifies personality types using Carl Jung's personality type theory. A four-letter code is used to describe each individual's personality. It uses choice items to classify individuals into dichotomous preferences. One can either be extraverted (E) or introverted (I); sensing (S) or intuitive (I); thinking (T) or feeling (F) and finally be either judging (J) or perceiving (P). Personality types are thus determined by the combination of these four dimensions. There are 16 four-letter codes that are possible (cf. Table 1). Descriptive outcomes of these codes or personality types help in one's classification [2, 6, 7].

Table 1. The sixteen Myers-Briggs Type Indicators

ISTJ	ISFJ	INFJ	INTJ
ISTP	ISFP	INFP	INTP
ESTP	ESFP	ENFP	ENTP
ESTJ	ESFJ	ENFJ	ENTJ

Each type describes tendencies and reflects variations in individual attitudes and styles of decision-making. The E-I dimension (extraverted-introverted) focuses on whether an individual's world attitude is outwardly-oriented to other objects and individuals, or it is internally-oriented. The S-N dimension (sensing-intuitive), on the other hand, describes the perceptual style of an individual. Sensing refers to attendance to sensory stimuli while intuition entails analyzing stimuli and events. Based on the T-F dimension, thinking encompasses logical reasoning together with decision processes. This dimension further shows that feeling has to do with a personal, subjective, and value oriented approach [6, 7, 8]. The J-P dimension involves either a judging attitude and quick decision-making or perception that demonstrates more patience and information gathering prior to decision-making.

Some of the preferences are dominant and others are auxiliary and can be influenced by other dimensions. For example, the J-P dimension influences the two function preferences: S or N versus T or F [6].

The MBTI has its limitations. Its theoretical and statistical imports are limited by the application of dichotomous choice items [8]. The large number of questions to be answered can discourage users and cause them to fill out the questionnaire without much attention. Reducing the number of questions in a questionnaire has been a way to increase its efficiency [4, 9].

3   Related Work

One major challenge in building a system that can adapt itself to a learner is giving it the capability of reducing the number of questions presented to the learner [4, 5, 11].

For instance, McSherry [9] reports that reducing the number of questions asked by an informal case-based reasoning system minimized frustration, made learning easier and increased efficiency.

Numerous researches on adaptive educational hypermedia systems have been conducted to minimize the number of questions asked to learners based on their capabilities and knowledge level [12 - 16] by using methods such as Item Response Theory [12, 14]. Questions are initially categorized by their difficulty. The score that a learner obtains for each completed section determines the difficulty level of the next questions that would be asked and whether some questions should be skipped. Nevertheless, few researches have attempted to measure the impact of reducing the number of questions on the correct identification of the learner profile. The AH questionnaire in [4] relies on decision trees to reduce the number of questions and to classify students according to the Felder-Silverman model of learning styles. Its experimental results with 330 students show that it effectively predicts the learning styles with high accuracy and limited number of questions.

Petri et al. [17] proposed EDUFORM, a software module for the adaptation and dynamic optimization of questionnaire propositions for profiling learners online. This tool, based on probabilistic Bayesian modeling and on Abductive reasoning, reduced the number of questionnaire propositions (items) by 30 to 50 percent, while maintaining an error rate between 10 to 15 percent. Experimental results have shown that a significant reduction in the numbers of proposition in the questionnaires was often accompanied by a correct classification of individuals.

Even though these studies have shown that it was possible to reduce the length of questionnaire using adaptive mechanisms, and in the case of [4] to apply it to learning styles, none of these studies have been done with the MBTI model of learning styles. Furthermore, our work differs from [4] in two important ways. First, our approach allows computing likely answers to unanswered questions and those are also taken into account to predict the learning style of a learner. Second, our proposal is based on the novel idea of exploiting associations between answers and questions to predict answers and skip questions (by mining association rules and using neural networks).

4   The Electronic Questionnaire

In this section, we present our proposed electronic questionnaire for the automatic assessment of learning style. It comprises three components: (1) an answer prediction algorithm, (2) a dynamic question selection algorithm and (3) an algorithm to accurately predict a person's learning style based on both user supplied and predicted answers.

4.1   The Answer Prediction Algorithm

Let there be a questionnaire such as that of the MBTI. Let $Q = \{q_1, q_2, \dots q_n\}$ be the set of multiple-choice questions from the *questionnaire*. Let $A(q_i) = \{a_{i,1} \dots a_{i,m}\}$ denote the finite set of *possible answers* to a given question q_i $(1 \leq i \leq n)$. Let $R = \bigcup_{i=1}^{n} A(q_i)$ be the set of all possible answers to all questions q_i ($1 \leq i \leq n$). A *set of answers* $U = \{u_1, u_2 \ldots u_k\}$ is a set $U \subseteq R$ where there does not exist integers a,b,x such that $u_a, u_b \in A(q_x)$ and such that a is different from b. A *completed set of answers* is a set of answers U such that $|U| = n$. A *partial set of answers* is a set of answers U such that $|U| < n$. An *empty set of answers* is a set of answers U such that $|U| = 0$. Given a set of answers U, a question q_x is an *unanswered question* if $A(q_x) \cap U = \emptyset$. Otherwise, q_x is an *answered question*. For a set of answers U and a set of questions Q, *Unanswered(U, Q)* denotes the *set of unanswered questions*, defined as $Unanswered(U, Q) = \{q_i \mid q_i \in Q \wedge A(q_i) \cap U = \emptyset\}$. Intuitively, a "set of answers" would be the filled-out answers in a questionnaire that might be supplied by a user. It could be completed, partial, or empty.

Problem of Answer Prediction. Let U be a partial set of answers. Let q_x be an unanswered question such that $A(q_x) \cap U = \emptyset$. The *problem of predicting the answer to q_x* is to determine the answer from $A(q_x)$ that the user would choose.

To address the above problem, we assume that we have a training set T of completed sets of answers. This set is used to build a prediction model that is then used to predict answers for any unanswered question. In a set of answers U, we use the term *predicted answer* to refer to an answer that was predicted by the prediction model.

Building the Prediction Model. To build the prediction model, we rely on association rule mining, an efficient and popular method to discover associations between items in sets of symbols, originally proposed for market basket analysis [18]. In our context, the problem of association rule mining can be defined as follows. Given the training set T, the *support* of a set of answers U is denoted as $sup(U)$ and defined as the number of completed sets of answers in T containing U, that is $sup(U) = |\{V \mid V \in T \wedge U \subseteq V\}|$. An association rule $X \rightarrow Y$ is a relationship between two sets of answers X, Y such that $X \cap Y = \emptyset$. The *support of a rule* $X \rightarrow Y$ is defined as $sup(X \rightarrow Y) = sup(X \cup Y) / |T|$. The *confidence of a rule* $X \rightarrow Y$ is defined as $conf(X \rightarrow Y) = sup(X \cup Y) / sup(X)$. The *lift of a rule* $X \rightarrow Y$ is defined as $lift(X \rightarrow Y) = sup(X \rightarrow Y) / (sup(X) \times sup(Y)/|T|^2)$. The *problem of mining association rules* is to find all association rules in T having a support no less than a user-defined threshold $0 \leq minsup \leq 1$ and a confidence no less than a user-defined threshold $0 \leq minconf \leq 1$ [18]. For instance, Figure 1 shows a set of completed answers T (left) and some association rules found in T for $minsup = 0.5$, $minconf = 0.5$ (right).

ID	sets of answers		ID	Rules	Support	Confidence
t_1	$\{a, b, c, e, f, g\}$		r_1	$\{a\} \rightarrow \{e, f\}$	0.75	1
t_2	$\{a, b, c, d, e, f\}$		r_2	$\{a\} \rightarrow \{c, e, f\}$	0.5	0.6
t_3	$\{a, b, e, f\}$	$\rightarrow$	r_3	$\{a, b\} \rightarrow \{e, f\}$	0.75	1
t_4	$\{b, f, g\}$		r_4	$\{a\} \rightarrow \{c, f\}$	0.5	0.6

Fig. 1. (a) A set of sets of answers and (b) some association rules found

To build the prediction model, in our experiments with the MBTI questionnaire, we used *minsup* = 0.15 and set *minconf* in the [0.75, 0.99] interval (the justification for these values are given in the experimental section). Choosing a high confidence threshold allows discovering only strong associations so that only those are used for prediction. Moreover, we also tuned the association rule mining algorithm to only discover rules of the form $X \rightarrow Y$ having a single item in the consequent, i.e. where $|Y|$ = 1. The reason behind such a choice is that we are only interested in predicting one answer at a time rather than multiple answers together.

Performing a Prediction. We now describe the algorithm for predicting the answer to an unanswered question q_z for a set of answers U, by using a set of association rules AR. Figure 2 shows the pseudocode of the prediction algorithm. It takes as input the question q_z, the current set of answers U and the set of association rules AR. The algorithm first initializes a variable named *prediction* that will hold the final prediction and a variable *highestMeasure* to zero (line 1 to 2). Then, the algorithm considers each association rule $X \rightarrow Y$ from AR such that the antecedent X appears in U and that the consequent Y contains an answer to q_z (line 3). For each rule, we calculate its usefulness for making a prediction, that we define as *measure* = $lift(X \rightarrow Y) * |X| - |Unanswered(U, Q)| / |X|$ (line 4). A larger value of this measure is considered better. In this measure, a lift higher than 1 means a positive correlation between X and Y, while a lift lower than 1 means a negative correlation. We multiply the lift by $|X|$ to give an advantage to rules matching with more answers from U over rules matching with fewer answers. The term $|Unanswered(U, Q)| / |X|$ is subtracted from the previous term so that previously predicted answers in X have a negative influence on the measure (to reduce the risk of accumulating error by performing a prediction based on a previous prediction). The algorithm then selects the answer with the highest measure (line 9) as the prediction and adds it to the set of answers U (line 9).

PREDICT(a question qz, a partial set of answers U, a set of association rules AR)

1.   *prediction* := *null*.
2.   *highestMeasure* := *0*.
3.   **FOR** each rule $X \rightarrow Y$ such that $X \rightarrow Y \in AR$, $Y \subseteq A(qz)$ and $X \subseteq U$
4.   *measure* = $lift(X \rightarrow Y) * |X| - |Unanswered(U, Q)| / |X|$.
5.   **IF** *measure* > *highestMeasure* **THEN**
6.   *highestMeasure* := *measure*.
7.   **END IF**
8.   **END FOR**
9.   **IF** *prediction* ≠ *null* **THEN** $U := U \cup \{prediction\}$.

Fig. 2. The answer prediction algorithm

4.2   The Question Selection Algorithm

We now describe Q-SELECT, the question selection algorithm of our electronic questionnaire that dynamically determines the order of questions. The pseudocode is given in Figure 3. The algorithm takes as input the set of association rules AR, previously extracted from the training set T. The algorithm first initializes the set of answers U for the current user to Ø. Then, the algorithm scans in one pass association rules AR to calculate dependencies of each question from Q. The *set of dependencies of a question* q_x is denoted as *dependencies*(q_x) and defined as the set of questions that can be used to predict an answer to q_x, i.e. *dependencies*$(q_x) = \{q_z \mid \exists X \to Y \in AR \land q_z \in X \land Y \subseteq A(q)\}$. A question q_x is said to be an *independent question* if no answer to that question can ever be predicted by the set of association rules AR, i.e. *dependencies*(q_x) $= \emptyset$. If there are independent questions, the algorithm starts by asking them to the user (line 3 to 4). The reason behind such a priority is that answers to independent questions cannot be predicted. But, their answers may be used to predict answers for other questions. Then, for each unanswered question q, the algorithm calls PREDICT (cf. Section 4.1) in an attempt to predict an answer for q (line 5).

After this loop, all independent questions and possible predictions have been exhausted. Next, the algorithm has to ask a question among the remaining unanswered questions. This is performed by a loop that continues until all questions have been answered (line 7). In this loop, the algorithm selects which question should be asked next. To make this choice, the algorithm estimates the number of questions that can be unlocked for each unanswered question if it was answered. The set of questions that a question q can unlock is denoted as *unlockable*(q) and defined as *unlockable*$(q) = \{q_z \mid \exists X \to Y \in AR \land \exists z \in A(q_z) \land z \in Y \land X \subseteq U \cup A(q) \land A(Q) \cap X \neq \emptyset\}$. The algorithm calculates this set for each unanswered question. This can be done by scanning the set of association rules once (line 8). Then the algorithm asks the question that can unlock the maximum number of questions according to the previous definition (line 9). Thereafter, for each unanswered question, the algorithm calls PREDICT to use the answer provided by the user to attempt to make a prediction (line 10). The WHILE loop then continue in the same way until no unanswered question remains. When the loop terminates, for each question q in Q, the set of answers U contains an answer from $A(q)$, which has either been answered by the user or predicted.

Q-SELECT(the set of questions Q from the questionnaire, association rules AR)
1. $U := \emptyset$
2. **SCAN** each association rules from AR to calculate dependencies for each question from Q.
3. **IF** there are independent questions **THEN**
4.   **ASK** all independent questions to the user. Add answers provided by the user to U.
5.   **FOR EACH** unanswered question q, **PREDICT**(q, U, AR).
6. **END IF**
7. **WHILE**$(|U| \neq |Q|)$
8.   **FOR EACH** unanswered question q, **CALCULATE** *unlockable*(q).
9.   **ASK** the question q such that $|unlockable(q)|$ is the largest among all unanswered questions.
10.   **FOR EACH** unanswered question q, **PREDICT**(q, U, AR).
11. **END WHILE**
12. **RETURN** U

Fig. 3. The question selection algorithm

4.3   The Learning Style Prediction Algorithm

We now describe how the electronic questionnaire automatically identifies the learning style of a user based on supplied and predicted set of answers.

The MBTI questionnaire evaluates each dimension (EI, JP, TF and SN) by a distinct subset of questions. Thus, we split the questionnaire into four sets of questions representing each dimension. A prediction algorithm is applied to each subset (dimension) to identify the individual's preference based on available answers. Finally, preferences in all four dimensions are combined to establish the learning style of the user.

Identifying the preference of a person in each dimension is essentially a classification problem. It is achieved, in our system, by a single layer feed-forward neural network, among the most common neural network architectures (it connects the input and output neurons directly, rather than connecting them through an intermediate layer). Neural networks are generally more accurate than other classifiers [9]. We trained a neural network for each dimension using 1,000 filled questionnaires (cf. experimental section). Thereafter, for each new user, the set of answers produced by the question selection algorithm (cf. Section 4.2) is used as input by the networks. The number of input neurons for each network is the number of questions for the corresponding dimension. The MBTI questionnaire uses 21 questions to assess the EI dimension, 23 questions for TF, 25 questions for SN, and 23 questions for JP. There is a single binary neuron as the output of each network because of the dichotomic nature of each dimension. Neural networks are built in MATLAB with the following parameters: activation function = TANSIG, performance function = MSE, number of iterations = 1000, the algorithm used for the training phase was TRAINLM with Goal = 0, Minimum gradient = $1e^{-10}$ and Max-fail = 6.

5   Experimental Results

A database of 1,931 MBTI completed questionnaires was provided by Prof. Robert Baudouin, an experienced specialist of the MBTI technique at the Université de Moncton. We used 1,000 samples for training and 931 for testing and evaluating the electronic questionnaire. The questionnaire is implemented using Java and MATLAB. The goal of the experiment was to measure how the elimination of questions influences the error rate. Since our questionnaire is dynamic and can eliminate a different number of questions for each questionnaire, the number of questions eliminated was measured using the median. Preliminary experimentation showed that *minsup* values lower than 0.15 did not increase accuracy. Thus, to vary the number questions eliminated (predicted), we instead varied the *minconf* threshold (in the [0.75, 0.99] interval). The error rate for a dimension is the number of questionnaire where the predicted preference is correct, divided by the total number of questionnaires in the test set.

Experimental results are shown in Table 2. The baseline error rates (when no questions are eliminated) are 3.7%, 4.9 %, 6% and 5% for the EI, SN, TF and JP dimensions, respectively. We limited our studies to error rates to no more than 12 %. The maximum number of questions that can be eliminated within this bound is shown in the last row of each column of Table 2. For EI, SN, TF and JP, the median number of questions eliminated is respectively 6, 9, 8, and 5, with an error rate of 9.9 %, 11.8 %, 12 % and 11.7 %. The combined median number of questions eliminated is 28, which represents 30.4 % of the MBTI questionnaire. It is important to note that the above numbers are medians. In many cases, individuals had more questions eliminated than the median. For example, Table 3 shows the distribution of questions eliminated for the TF dimension. Although the median is eight questions, nine questions were eliminated for 292 individuals, and less than eight questions were eliminated for only 231 individuals.

Given the error rates from Table 2, the probability of predicting incorrectly four preferences for a particular user is only 0.02%. The probability of predicting three erroneous preferences is 0.6 %. The probability of predicting two erroneous preferences is 6.66 %, while the probability of predicting one erroneous preference is 33 %, and the probability of a perfect prediction is 60 %. We note that the combined probability of having no errors or only one error is more than 92 %.

Table 2. Number of questions eliminated and the corresponding error rate

Median number of questions eliminated (predicted)	Error rate for EI	Error rate for SN	Error rate for TF	Error rate for JP
0	3.7 %	4.9 %	6 %	5 %
1	4.6 %	5,3 %	7,8 %	6 %
2	5.2 %	6,3 %	8,4 %	7.1 %
3	7.5 %	6,6 %	9,1 %	8.6 %
4	7.7 %	7,5 %	10 %	10 %
5	7.9 %	9 %	10,7 %	11.7 %
6	9.9 %	10,5 %	11,7 %	
7		11,1 %	11,7 %	
8		11,7 %	12 %	
9		11,8 %		

We compared the Q-SELECT algorithm results with those obtained by a C4.5 decision tree [19, 20]. Table 4 shows the comparable error rates of both methods for the highest number of questions eliminated by the Q-SELECT algorithm. It can be noticed that the error rates generated by the decision tree are two to four percent higher for each of the four preferences.

Table 3. Number of questions eliminated per questionnaire for the TF dimension

Number of questions eliminated (x)	Number of questionnaires (individuals)	Number of questionnaires in the [x, 10] interval
1	0	931 (100%)
2	11	931 (100%)
3	7	920 (99%)
4	3	913 (98%)
5	11	910 (98%)
6	14	899 (97%)
7	185	885 (95%)
8 (median)	348	700 (75%)
9	292,00	352 (38%)
10	60	60 (6 %)

Table 4. Comparative results for Q-SELECT and Decision Tree

	Error rate for EI (6 questions)	Error rate for SN (9 questions)	Error rate for TF (8 questions)	Error rate for JP (5 questions)
Q-SELECT	9.9 %	11.8 %	12 %	11.7 %
Decision Tree	12.35%	13.21%	16.54%	13.74%

6   Conclusion

Standardized questionnaires for learning style identification are long, time-consuming and require human intervention to determine an individual's learning style. To address this issue, we presented an adaptive electronic questionnaire. It incorporates an efficient answer prediction algorithm PREDICT for predicting answers to unanswered questions based on associations between answers. We also presented Q-SELECT, an algorithm that reorders questions and minimizes the number of those presented, based on associations between questions. Experimental results with 1,931 questionnaires filled for the Myers Briggs Type Indicators show that our approach considerably reduces the number of presented question.

The combined median number of questions eliminated is 28, which represents 30.4 % of the MBTI questionnaire. The combined probability of having no errors or only one error out of four preferences is more than 92 %. We also note that the Q-SELECT algorithm gave better results than the decision tree for all the preference types.

References

1. Graf, S.: Adaptivity in Learning Management Systems Focusing on Learning Styles. Ph.D. Thesis, Vienna University of Technology, Vienna (2007)
2. Felder, R.M.: Matters of Style. ASEE Prism 6(4), 18–23 (1996)

3. Felder, R.M., Brent, R.: Understanding student differences. J. of Engineering Education 94(1), 57–72 (2005)
4. Ortigosa, A., Paredes, P., Rodríguez, P.: AH-questionnaire: An adaptive hierarchical questionnaire for learning styles. Computers & Education 54(4), 999–1005 (2010)
5. García, P., Amandi, A., Schiaffino, S., Campo, M.: Evaluating Bayesian Networks' Precision for Detecting Students' Learning Styles. Computers & Edu. 49(3), 794–808 (2007)
6. El Bachari, E., Abelwahed, E.H., El Adnani, M.: Design of an Adaptive E-Learning Model Based on Learner's Personality. Ubiquitous Computing Comm. J. 5(3), 27–36 (2010)
7. Boyle, G.J.: Myers-Briggs Type Indicator (MBTI): Some psychometric limitations. Australian Psychologist 30(1), 71–74 (2009)
8. Francis, L.J., Jones, S.H.: The Relationship between Myers-Briggs Type Indicator and the Eysenck Personality Questionnaire among Adult Churchgoers. Pastoral Psychology 48(5), 377–386 (2008)
9. McSherry, D.: Increasing dialogue efficiency in case-based reasoning without loss of solution quality. In: Proc. 18th Intern. Joint. Conf. Artif. Intell., pp. 121–126 (2003)
10. Zhang, G.P.: Neural Networks for Classification: A Survey. IEEE Transactions on Systems, Man, and Cybernetics 30(4), 451–462 (2000)
11. Abernethy, J., Evgeniou, T., Vert, J.-P.: An optimization framework for adaptive questionnaire design. Technical Report, INSEAD, Fontainebleau, France (2004)
12. Baylari, A., Montazer, G.: Design a personalized e-learning system based on item response theory and artificial neural network approach. Expert Syst. Appl. 36(4), 8013–8021 (2009)
13. Papanikolaou, K.A., Grigoriadou, M., Magoulas, G.D., Kornilakis, H.: Towards New Forms of Knowledge Communication: the Adaptive Dimension of a Web-based Learning Environment. Computers & Education 39(4), 333–360 (2002)
14. Chen, C.M., Lee, H.M., Chen, Y.H.: Personalized e-learning system using item response theory. Computers & Education 44(3), 237–255 (2005)
15. Brusilovsky, P., Millán, E.: User models for adaptive hypermedia and adaptive educational systems. In: Brusilovsky, P., Kobsa, A., Nejdl, W. (eds.) Adaptive Web 2007. LNCS, vol. 4321, pp. 3–53. Springer, Heidelberg (2007)
16. Xanthou, M.: An Intelligent Personalized e-Assessment Tool Developed and Implemented for a Greek Lyric Poetry Undergraduate Course. Electron. Journal of e-Learning 11(2), 101–114 (2013)
17. Nokelainen, P., Niemivirta, M., Tirri, H., Miettinen, M., Kurhila, J., Silander, T.: Bayesian Modeling Approach to Implement an Adaptive Questionnaire. In: Proc. ED-MEDIA 2001, pp. 1412–1413. AACE, Chesapeake (2001)
18. Agrawal, R., Imieliński, T., Swami, A.: Mining association rules between sets of items in large databases. ACM SIGMOD Record 22(2), 207–216 (1993)
19. Quinlan, J.R.: C4.5: programs for machine learning, vol. 1. Morgan Kaufmann (1993)
20. Russell, S.J., Norvig, P.: Artificial Intelligence: A Modern Approach, 3rd edn. Pearson (2010)

Automatic Database Monitoring
for Process Control Systems

Hiromasa Kaneko and Kimito Funatsu

Department of Chemical System Engineering, Graduate School of Engineering,
The University of Tokyo, Hongo 7-3-1, Bunkyo-ku, Tokyo 113-8656, Japan
{hkaneko,funatsu}@chemsys.t.u-tokyo.ac.jp

Abstract. In industrial plants, soft sensors are widely used to predict difficult-to-measure process variables for process control. One of the problems of soft sensors is the degradation of the soft sensor models. The predictive accuracy decreases when the states in plants change. Many adaptive soft sensor models have been developed to reduce the degradation. Since the database management is required for those adaptive models, we previously proposed the database monitoring index (DMI) and the database managing method with the DMI. By judging whether new data should be stored in database or not compareing the DMI values, the amount of information can increase while controlling the number of data in the database. In this study, we proposed the automatic method determining the threshold of DMI with only training data. The model construction and data deletion are repeated while checking the DMI values. Through the analysis of simulation data and real industrial data, we confirmed that the proposed method can monitor the database appropriately and the predictive accuracy of the adaptive soft sensor models improve.

Keywords: Process control, Soft sensor, Database monitoring.

1 Introduction

In process control, soft sensors are widely used to predict process variables that are difficult to measure online [1]. An inferential model is constructed between the variables that are easy to measure online and those that are not, and an objective variable, $\mathbf{y}$, is then predicted using that model. Their use, however, is accompanied by some practical difficulties. One of these difficulties is the degradation of the soft sensor models. The predictive accuracy of soft sensors tends to decrease for several reasons, including changes in the state of the chemical plant, catalyzing performance loss, and sensor and process drift. This is called as the degradation of soft sensor models.

It is strongly desired to solve the degradation of a soft sensor model. To reduce the degradation, many adaptive soft sensor models have been developed [2]. These models can be categorized into three types, i.e. moving window (MW) models, just-in-time (JIT) models and time difference (TD) models [3]. MW models are constructed with data that are measured most recently; JIT models are constructed

A. Moonis et al. (Eds.): IEA/AIE 2014, Part I, LNAI 8481, pp. 410–419, 2014.
© Springer International Publishing Switzerland 2014

with data whose similarity with prediction data are higher than those of other data; and TD models are based on the time difference of **y** and that of explanatory variables, **X**.

However, there are no adaptive models having high predictive ability in all process states and the prediction accuracy of each adaptive model depends on a process state. Kaneko et al. categorized the degradation of a soft sensor model and discussed characteristics of adaptive models such as MW, JIT and TD models, based on the classification results, and confirmed the discussion results through the numerical simulation data and real industrial data analyses [3].

While the appropriate use of the MW and JIT models enables soft sensors to adapt to the changes of the relationship between **X** and **y**, there remain some problems for the introduction of soft sensors into practice. One of the problems is that reconstructed models have a high tendency to specialize in predictions over a narrow data range. Subsequently, when variations in the process variables occur, these models cannot predict the resulting variations in data with a high degree of accuracy. However if the model is not reconstructed frequently, the predictive ability of the model decreases due to the slow change of process states such as process and sensor drifts.

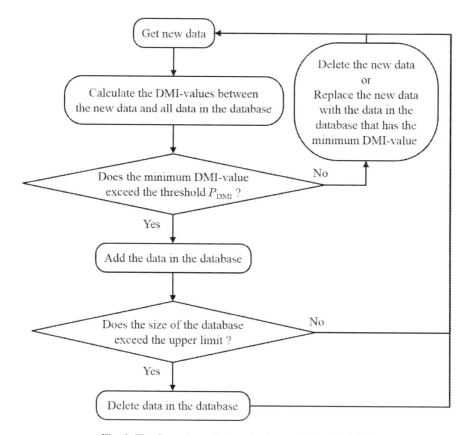

Fig. 1. The flow of monitoring database with the DMI [5]

Therefore, to construct adaptive models with high predictive accuracy for wide data range, we previously developed the database monitoring index (DMI) and the database managing method with the DMI [4,5]. The DMI is an index based on similarity between two data and is defined as the ratio of absolute difference of **y** devided by similarity of **X**. The more similar two data are, the smaller value the DMI has. The flow of monitoring database with the DMI is shown in Fig. 1. By storing only data having much information, the amount of information can increase while controlling the number of data in the database.

However, there is no method to determine the threshold of the DMI-values (P_{DMI}) above which the new data is stored in database. We therefore propose an automatic method determining the P_{DMI} with only training data. As long as the required prediction accuracy of the regression model is set, the P_{DMI} is automatically output by repeating the model construction and date deletion while checking the DMI-values.

To verify the effectiveness of the proposed method, we analyze simulation data where the relationship between **X** and **y** is nonlinear and data variation is small for a constant time. Then, the proposed method is applied to real industrial data of a distillation column. The management of databases using the DMI makes it possible for adaptive soft sensor models to adapt to rapid changes of process characteristics after long states of small variations in **X** and **y**. In this study, it is assumed that there are no abnormal data and no outliears in new measurement data. In practice, therefore, abnormal data and outliears must be detected first by using multivariate statistical process control methods [6].

2   Method

2.1   Database Management with DMI

The DMI for managing databases is defined between two data ($\mathbf{x}^{(i)}$, $y^{(i)}$) and ($\mathbf{x}^{(j)}$, $y^{(j)}$) as follows:

$$\text{DMI} = \frac{\left| y^{(i)} - y^{(j)} \right|^{a}}{\text{sim}\left(\mathbf{x}^{(i)}, \mathbf{x}^{(j)} \right)} \tag{1}$$

where sim($\mathbf{x}^{(i)}$, $\mathbf{x}^{(j)}$) is similarity between $\mathbf{x}^{(i)}$ and $\mathbf{x}^{(j)}$, and a is a constant. In this study, the similarity index is Gaussian kernel as follows:

$$\text{sim}\left(\mathbf{x}^{(i)}, \mathbf{x}^{(j)} \right) = \exp\left(-\gamma \left\| \mathbf{x}^{(i)} - \mathbf{x}^{(j)} \right\|^{2} \right) \tag{2}$$

where γ is a tuning parameter controlling the width of the kernel function. Therefore, the DMI used in this study is given as follows:

$$DMI = \frac{\left| y^{(i)} - y^{(j)} \right|^{a}}{\exp\left(-\gamma \left\| \mathbf{x}^{(i)} - \mathbf{x}^{(j)} \right\|^{2}\right)} \tag{3}$$

The DMI-values are large when two data of $\mathbf{X}$ and $\mathbf{y}$ are dissimilar, whereas the DMI-values are small when two data of $\mathbf{X}$ and $\mathbf{y}$ is similar. The weight of $\mathbf{y}$ for $\mathbf{X}$ can be changed by varying the a-value in eqs. (1) and (3). The DMI can quantify data similarity even for nonlinear and non-Gaussian processes.

The flow of the proposed database management is shown in Fig. 1. When new data is given, the DMI-values are calculated between the new data and all data included in the database. If the minimum value of the DMI-values exceeds the P_{DMI}, the new

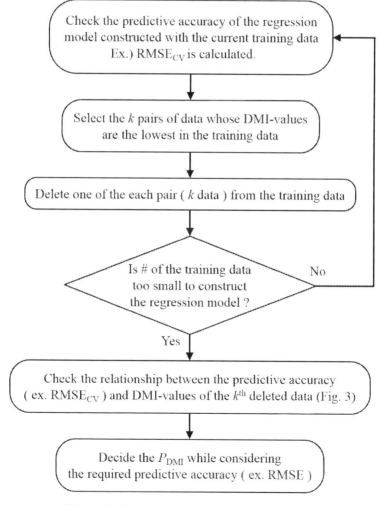

Fig. 2. The flow of automatic decision of the P_{DMI}

data is stored in the database. If not, the new data is not stored in the database or the new data replaces the data in the database that has the minimum DMI-value. Then, if the size of the database exceeds the upper limit, data is deleted from the database. If the current database has the necessary and sufficient data, it is reasonable that the oldest data is eliminated since the minimum DMI values for all the data in the database exceed the P_{DMI} at least and the information loss is not so different in the elimination of each data. In addition, the process state where the oldest data was measured would be different from the process state where the new data is measured.

2.2   Automatic Determination of the P_{DMI}

From Fig. 1, we need to decide the P_{DMI} first in order to manage database. The method determining the P_{DMI} with only training data and the required prediction accuracy is proposed in this study. The flow of the proposed method is shown in Fig. 2. The regression model is constructed with the current training data and the prediction accuracy of its model is checked. For example, an index of the prediction accuracy is the root-mean-square error of cross-validation and external validation ($\text{RMSE}_{\text{CVEX}}$) given as follows:

$$\text{RMSE}_{\text{CVEX}} = \sqrt{\frac{\sum_{i=1}^{n}\left(y^{(i)} - y_{\text{pred}}^{(i)}\right)^2}{n}} \qquad (4)$$

where n is the number of all training data, which means that the deleted training data are used as external data; $y^{(i)}$ is the y value of the ith data point; and $y_{\text{pred}}^{(i)}$ is the y value of the ith data point predicted with cross-validation or external validation. Then, the DMI values are calculated between all pairs of training data. The k pairs of data whose DMI values are the lowest sets are selected and one of the each pair is deleted from the training data, which means that k data are removed from the training data. Here, if plenty of training data remain, the check of prediction accuracy and the deletion of data are repeated while saving the DMI values of the removed data. If not, we check the relationship between the predictive accuracy of the model and the DMI values of the removed data for each loop. Fig. 3 shows the conceptual diagram for

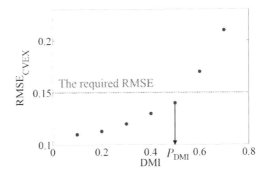

Fig. 3. The conceptual diagram for decision of the P_{DMI}

decision of the P_{DMI} when $RMSE_{CVEX}$ was used as an index of the prediction accuracy. The largest DMI value is set as the P_{DMI} while the $RMSE_{CVEX}$ does not exceed the requested RMSE. The database will be able to be managed not to exceed the required RMSE.

3   Results and Discussion

To verify the effectiveness of the proposed method, we analyzed simulation data and industrial distillation column data. The relationships between **X** and **y** have strong nonlinearity for the simulation data set.

3.1   Data in Which the Relationship between X and y is Nonlinear

The analysis using data where the relationship between **X** and **y** is nonlinear was performed to verify the performance of the proposed method. The relationship between **X** and **y** in the data was given as follows:

$$y = \sin(x_1)\cos(x_2) + 0.1x_1$$

(5)

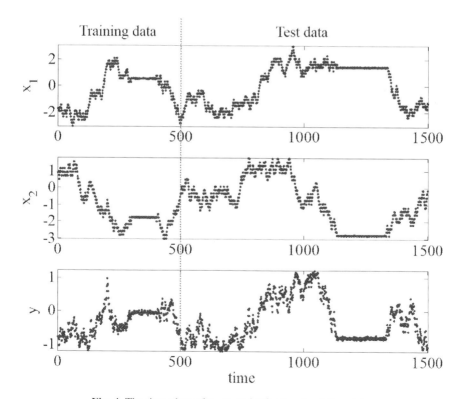

Fig. 4. The time plots of x_1, x_2 and **y** for the simulation data

Eq. (5) is described in the reference [7] as a test problem and $0.1x_1$ was added to the raw equation. The data of **X**-variables were generated to be randomly walked within ±3. Random numbers from the normal distribution with a mean of 0 and a standard deviation 0.01 were added to the **y**-variable. The time plots of x_1, x_2 and **y** for the simulation data are shown in Fig. 4. The data whose variation comes from only noise exist in the simulation data and this situation seems to be in Fig. 4. The first 500 data were used for training and the next 1,000 data were the test data.

The γ-value in eq. (3) was optimized with 5-fold cross-validation using support vector regression (SVR) [8] of Gaussian kernel. The DMI-values were calculated with the γ-value of 2^{-7} and eq. (3). Then, the k was set as 5; the required RMSE was set as 1.17×10^{-2} which is 1.05 times $RMSE_{CVEX}$ calculated with the first training data; and then, the P_{DMI} was determined through the flow of Fig. 2. Fig. 5 shows the plot of DMI and $RMSE_{CVEX}$. As the DMI values are high and the number of the deleted data is high, the $RMSE_{CVEX}$ are also high. 330 data were deleted from 500 training data and the P_{DMI} was 5.32×10^{-200}. The $RMSE_{CVEX}$ is 1.16×10^{-2}, which is lower than the required RMSE (1.17×10^{-2}).

Then we predicted **y** values with the online support vector regression (OSVR) model [9] which is one of the adaptive models. The details of OSVR are shown in the reference [10]. To consider measurement time of **y**, it was assumed that **y** values can be obtained after gaining time of 10, although **X**-values can be given in real time. When a **y** value is obtained, the DMI judges whether the new data should be stored in the database or not (Fig. 1).

Table 1 shows the prediction results of the OSVR models without database management and those with database management. The r_p^2 is the determination coefficient r^2 for test data and $RMSE_p$ is the RMSE for test data. The larger values of r_p^2 and the smaller values of $RMSE_p$ mean the more predictive accuracy of the model. From Table 1, the models with database management had more predictive ability than the model updating each time (without database management). The rate of update was 32.1% and, by using the P_{DMI} determined with the proposed method, data required to represent the nonlinear relationship between **X** and **y** could be properly selected and the OSVR model could predict **y** values with high prediction accuracy and low update frequency. We can say that the appropriate P_{DMI} for the database management can be decided using the proposed method.

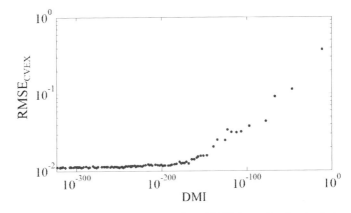

Fig. 5. The double logarithmic plot of DMI and $RMSE_{CVEX}$ for the simulation data

Table 1. The prediction results for the simulation data

	Without database management	With database manegement
r^2_p	0.986	0.996
$RMSE_p$	7.75×10^{-2}	4.36×10^{-2}
Rate of update	1.000	0.321

Fig. 6 shows the time plots of simulated and predicted **y** from 1300 to 1400. When the database was not managed and the model was updated each time, the prediction errors were large after time 1335 when the abrupt process change happened (Fig. 6a). This is because the model was constructed with only data including the small variation, and was specialized in that state. Thus, the model could not adapt to the next abrupt change. Meanwhile, when the database was managed with the flow of Fig. 1 and the model update with data in the small variation was avoided, the model could adapt to the abrupt variation after 1335 (Fig. 6b). We could confirmed that the appropriate selection of data required to the database with the DMI and P_{DMI} determined with the proposed method enables to predict wide range of **y** with high accuracy.

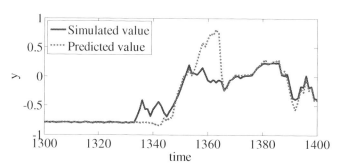

(a) Without database management

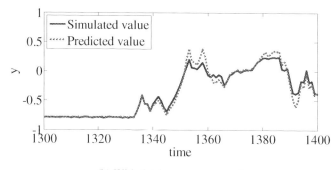

(b) With database management

Fig. 6. The time plots of simulated and predicted **y** from 1300 to 1400 for the simulation data

3.2   Distillation Column Data

We applied the proposed method to data obtained from the operation of a distillation column at the Mizushima Works, Mitsubishi Chemical Corporation. The **y**-variable is the concentration of the bottom product, and the **X**-variables are the 19 variables such as flow rate, temperature and pressure. The measurement interval of **y** was 30 min and **X**-variables are measured every minute. The details of the data set is given in the reference [5]. The OSVR method was used as the regression method as was the case in the simulation data analysis.

Fig. 7 shows the time plots of measured and predicted **y**. When the OSVR model was updated each time, the prediction errors were large in the relatively large variations after the small variations (see around time 1160 and 1300 in Fig. 7a). The model that specialized in the small variations could not adapt to the rapid time-varying variations occurring subsequently. Meanwhile, by managing the database with the P_{DMI} determined with the proposed method and the flow of Fig.1, the **y**-values were accurately predicted even in the rapid variations after the stable states (Fig. 7b). It was confirmed that the database can be appropriately managed with the P_{DMI} selected by using the proposed method and enables soft sensor models to predict **y** values for wide data range with high accuracy.

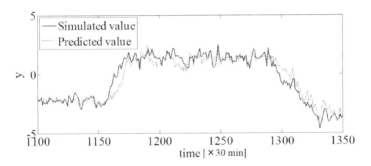

(a) Without database management

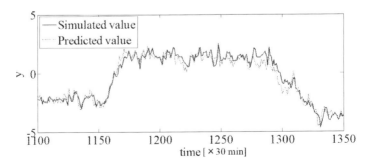

(b) With database management

Fig. 7. The time plots of measured and predicted **y** from 1100 to 1350 for the distillation columun data

4   Conclusion

To improve the prediction accuracy of the adaptive soft sensor models, the database is monitored with the DMI. In this paper, to determine the threshold of the DMI values (P_{DMI}) automatically, we proposed the method based on the model construction and the check of the predictive accuracy of the model while calculating the DMI values. As long as the required prediction accuracy is set, the proposed method can select the appropriate P_{DMI} for predicting **y**-values with high predictive accuracy. We analysed the simulation data set and the real industrial data set, and confirmed that the proposed method could decide the P_{DMI} with which adequate database monitoring could be performed and the prediction ability of the adaptive soft sensors could improve. The database managing method can be combined with not only the adaptive soft sensor models but also the process monitoring models that are updated or reconstructed with database including new measurement data. We believe that by managing the database and increasing the predictive accuracy of adaptive soft sensor models, chemical plants will be operated effectively and stably.

Acknowledgement. The authors acknowledge the support of Mizushima works, Mitsubishi Chemical Corporation, and the financial support of the Japan Society for the Promotion of Science (JSPS) through a Grant-in-Aid for Young Scientists (B) (No. 24760629) and Mizuho Foundation for the Promotion of Sciences.

References

1. Kadlec, P., Gabrys, B., Strandt, S.: Data-driven Soft Sensors in the Process Industry. Comput. Chem. Eng. 33, 795–814 (2009)
2. Kadlec, P., Grbic, R., Gabrys, B.: Review of Adaptation Mechanisms for Data-driven Soft Sensors. Comput. Chem. Eng. 35, 1–24 (2011)
3. Kaneko, H., Funatsu, K.: Classification of the Degradation of Soft Sensor Models and Discussion on Adaptive Models. AIChE J. 52, 1322–1334 (2013)
4. Kaneko, H., Funatsu, K.: Development of a New Index to Monitor Database for Soft Sensors. J. Comput.-Aided Chem. 14, 11–22 (2013) (in Japanese)
5. Kaneko, H., Funatsu, K.: Database Monitoring Index for Adaptive Soft Sensors and the Application to Industrial Process. AIChE J. (accepted)
6. Ge, Z., Song, Z., Gao, F.: Review of Recent Research on Data–based Process Monitoring. Ind. Eng. Chem. Res. 52, 3543–3562 (2013)
7. Li, G., Aute, V., Azarm, S.: An Accumulative Error Based Adaptive Design of Experiments for Offline Metamodeling. Struct. Multidiscip. O. 40, 137–155 (2010)
8. Bishop, C.M.: Pattern Recognition and Machine Learning. Springer, New York (2006)
9. Ma, J., Theliler, J., Perkins, S.: Accurate On-line Support Vector Regression. Neural Comput. 15, 2683–2703 (2003)
10. Kaneko, H., Funatsu, K.: Adaptive Soft Sensor Model Using Online Support Vector Regression with the Time Variable and Discussion on Appropriate Hyperparameters and Window Size. Comput. Chem. Eng. 58, 288–297 (2013)

Fuzzy Logic Guidance Control Systems for Autonomous Navigation Based on Omnidirectional Sensing

Danilo Cáceres Hernández, Van-Dung Hoang,
Alexander Filonenko, and Kang-Hyun Jo

Intelligent Systems Laboratory, Graduate School of Electrical Engineering,
University of Ulsan, Ulsan 680-749, South Korea
{danilo,hvzung,alexander}@islab.ulsan.ac.kr, acejo@ulsan.ac.kr
islab.ulsan.ac.kr

Abstract. Sensing in an unknown environment is one of a few challenges faced by fully autonomous navigation. Getting a head estimation with respect to the world coordinate system can definitely be a difficult task. In this paper, the authors propose a real time fuzzy logic application for autonomous navigation based on omnidirectional sensing. First, it was proposed to extract the longest segments of lines from the edge frame. Second, RANSAC curve fitting method was implemented for detecting the best curve fitting given the data set of points for each line segment. Third, the set of intersection points for each pair of curves were extracted. Fourth, the DBSCAN method was used in estimating the VP. Finally, to control the mobile robot in an unknown environment, a fuzzy logic controller by using the vanishing point (VP) was implemented. Preliminary results were gathered and tested on a group of consecutive frames. These specific methods of measurement were chosen to prove their effectiveness.

Keywords: Fuzzy logic, Omnidirectional sensing, Vanishing Point, RANSAC curve fitting, DBSCAN.

1 Introduction

Autonomous ground navigation is still facing important challenges in the field of robotics and automation due to the uncertain nature of the environments, moving obstacles, and sensor fusion accuracy. Therefore, for the purpose of ensuring autonomous navigation and positioning along the environments aforementioned, a visual based navigation process is implemented by using an omnidirectional camera and fuzzy logic control. First, in the sensing task, based on the perspective drawing theory, the VP was estimated. In the case of omnidirectional scenes, parallel lines are projected as curves that converge and disappear into the horizon. Therefore, in order to address the challenges of VP detection using omnidirectional scenes, the authors decided to present an iterative VP detection

A. Moonis et al. (Eds.): IEA/AIE 2014, Part I, LNAI 8481, pp. 420–429, 2014.
© Springer International Publishing Switzerland 2014

based on curve fitting and clustering approaches. In fact, there are several approaches for estimating VP in the field of Intelligent Systems. However, those approaches are based on the constraint that the estimate vanishing point is localized in the front of the designed system. In order to decrease these limitations, omnidirectional camera systems were used. The authors are proposing to extract the cluster which contains the large amount of points located in either the front most or rear most part of the mobile robot systems. Second, based on the cognitive process of human decision a fuzzy logic controller (FLC) was implemented. To this end, the main contributions of the presented method are:

– Implementation of real time RANSAC curve fitting algorithm for VP detection.
– Implementation of DBSCAN, due that the number of cluster in the image are not specified, for VP estimation.
– Implementation of fuzzy logic controller using omnidirectional sensing.

The rest of this paper is structured as follows: (2) Related Works (3) Proposed Method, (4) Experimental Result, (5) Conclusions and Future Works.

2   Related Works

In the field of autonomous navigation systems, efficient navigation, guidance and control design are critical in averting current challenges. When referring to the need for estimation of rotation angle for automatic control, one VP plays an important role. By detecting VP in the 2D image, autonomous unmanned systems are able to navigate towards the detected VP. Several approaches for VP detection can be used for detection, for example, Hough transform (HT) [1], RANSAC algorithm [2], dominant orientation [3], the equivalent sphere projection [4], and lastly polynomial system solver [5] were all during experimentation. On the other hand, autonomous guidance and control algorithms based on fuzzy control approaches has been introduced by several researches since Fuzzy Set theory was introduced in [6] by L. A. Zadeh. In the case of autonomous navigation fuzzy logic sensor-based control have been widely implemented using reflective sensors such as laser range finder and ultrasonic devices. In [7,8] authors propose to use ultrasonic sensors for path planning task as well autonomous mobile navigation. An infra-red solutions have been proposed by the authors in [9]. Finally, authors in [10] uses a Laser Range Finder (LRF) for the task of autonomous mobile navigation. However, in the case of visual sensing there are a few jobs throughout the years. For example, authors in [11] uses six CMOS NTSC video cameras (representing a 180 deg. field of view) for a Real-Time Autonomous Rover Navigation System. The processing time was less than 2 seconds CPU time. In [12], authors presented a system based on a BW camera and a conical mirror for a vision-based localization with a rate of 12 cycles per second.

3   Proposed Method

Essentially, the proposed method consists mainly of extracting the information surrounding the ground plane. Frames were extracted in short time intervals that started just before the earliest detection from the video-capture sequence. In this section, the proposed algorithm has two steps: (1) omnidirectional sensing, (2) controller design.

3.1   Omnidirectional Sensing

Extracting Line Segments. Road scenes can be described as structures that contain lane marking, soft shoulders, gutters, or a barrier curb. Therefore, in 2D images these features are represented as a set of connected points. The main idea of this section consisted of extracting the longest line segments around the ground by applying the canny edge detector from a group of consecutive frames captured. After the edge detection step, the subsequent task was to remove the smallest line segments by extracting the longest line segments after applying canny edge detection. This was achieved by using a 3-chain rule for each possible curve candidate. From the tracking algorithm, the authors were able to extract the basic information such as: length, number of point per line segments, as well as the pixel position location of the points of each line segments into the image plane. As a result, extraction from an image sequence the set of longest line segments was completed.

Curve Fitting. At this point the set of line segments in the image plane are known. Hence, the new task consists of defining the function for each line segment. The process starts at every iteration by selecting a set of three random points for each line segment. This will help to describe the function that determines the curvature of the lines. As far as we know, in numerical analysis there are various approaches for solving a polynomial system. For example, in [4] authors use polygonal approximation to extract that lines segments. On the other hand, in [5] authors use an Auzinger and Stetter method. In that sense, the main contribution of the authors work is to estimate VP in real time processing, required by autonomous navigation. To this end, a circle geometry was implemented after a comparison with Lagrange interpolating polynomials has shown a better performance. The idea relies on the concept of parallel lines projected into the image as a circle following the idea of sphere projection model. In order to define the curve model by the three given points, researchers relied mainly on circle geometry. The basic algorithm computed was the perpendicular bisector of: $(P_1(x_1, y_1), P_2(x_2, y_2))$ and $(P_1(x_1, y_1), P_3(x_3, y_3))$, the center of the circle is determined by solving the perpendicular bisectors equation. The radius is the distance from the computed center of a candidate circle to $P_1(x_1, y_1)$. Once the polynomials have been disclosed, the next step consists of selecting the appropriate curve fit model by finding the function that contains the largest number of inliers. In other words, by using RANSAC the curve model that has the largest number of points based on the given data set becomes the points that are extracted. As result, the set of curves are clearly defined. The next step entails the extraction of the data set of intersection point by finding the same for each pair of defined curves at the point where they intersect, see Fig.1.

Cluster Extraction by DBSCAN. Given the data set of intersection point from the previous step, the algorithm should be able to extract clusters. From figure 1, it is clear that the projected data into the image plane give us vital information about the data set points, as follows:

- The data does not depict a well-defined shape.
- Presence of noise in the data due to the lack of a pre-processing model for road analysis. To compensate, all line segments are considered as a part of the road.
- The number of cluster could not be described in advance.

Considering the above mentioned occurrences, among the various clustering algorithms it was proposed to use the DBSCAN algorithm. Shown in [13]; an unsupervised method was used, due to the algorithm having achieved a good performance with respect to some other algorithms. In essence, the main idea of DBSCAN is that for each point of a cluster, the neighborhood of a given radius has to contain a minimum number of points. The DBSCAN algorithm depends mostly on two parameters:

- Eps: number of points within a specified radius.
- MinPts: minimum number of points belonging to the same cluster.

After the candidate clusters are formed, the idea is to define the cluster that contains the largest amount of points. Then, the centroids of the cluster are calculated. It is important to remark that one of the main advantages of using omnidirectional cameras lies in the fact that the possibility to extract clusters which are located either in the front most or rear most part of the mobile robot systems, thereby helping the estimation of the VP. Thus, it is of the utmost importance in determining the VP in frames which are affected by an illumination distribution, such as strong sunlight or dark shadows. On the other hand, once the VP has been extracted, the angle between (θ) the line formed by the VP and the center point of the omnidirectional camera, in respect to the x-axis of the coordinate system (which passes through the center point of the omnidirectional camera) is now able to be accurately computed, see Fig.2. Finally, Fig.3 shows the Frame-rate image sequences at a resolution of 160x146 pixels for the omnidirectional sensing step.

3.2 Controller Design

Based on the cognitive process of human decision, the main idea of this step consist in track a reference trajectory given by the VP by using FLC. First

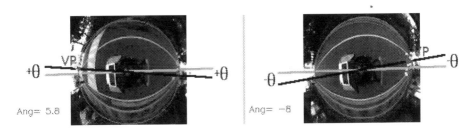

Fig. 1. Coordinate system. The green line shows to the x-axis of the coordinate system. The line in black is the line formed by the VP and the center point of the omnidirectional camera. Theta (θ) is the angle between the line formed by the VP and the center point of the omnidirectional camera, in respect to the x-axis. The point in blue represent the vanishing point. Points in other colors show the detected set of clusters in the current frame.

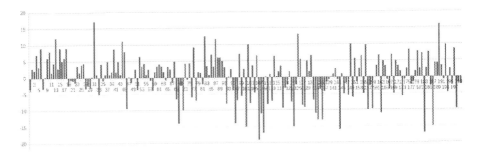

Fig. 2. Angle frame rate

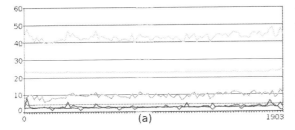

Process (msec.)	UoU
Capture	2.62
Preprocessing	4.56
Line extracting	23.04
RANSAC	9.50
DBSCAN	3.27
Total	42.99

(a)   (b)

Fig. 3. Frame-rate image sequences at a resolution of 160x146.(a) Data in blue sky represent the total running time. Data in yellow represent the line extraction. Data in green represent the curve fitting by using RANSAC. Data in brown represent the vanishing point estimation by using DBSCAN. Data in orange represent the preprocessing time. Data in blue represent the image capture process. (b) Average time for main phase of processing.

at all, the mobile robot used in this work has four independently driven DC motors, controlled by its own driver. Control signals are generated by the 16 MHz micro controller. The robots itself is a heavy machine with located high center of gravity and it can overturn when it tries to turn while moving with high speed. There is a need to set limits on the speed of rotation depending on the linear speed of the robot, see Fig. 4. To comply these conditions and make the robot fully autonomous a decision system which controls all the movements must be implemented. Using the fuzzy controller as a human-like decision system can be one of the most natural and understandable ways of doing that.

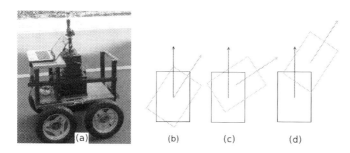

Fig. 4. Linear speed of the robot. (a)Mobile robot. (b) When the robot turns without moving forward or backward, it is possible to maximize the speed of rotation. (c)In slow linear movement the rotation speed should be decreased by stopping wheels on one side. (d) When the robot is moving fast, it is dangerous to stop any wheels. Motors on one side should rotate slower than on another instead

Control System Scheme. The robot can be controlled both by a computer and by a mobile device as shown at Fig. 4. The control signal from the mobile device is delivered via Bluetooth. Tests have shown the effective control range is 100 meters. The graphical interface allows intercepting control from the computer to prevent accidents, see Fig. 5. The signal from the computer is sent by a ZigBee wireless network. The computer itself is mounted on a robot but it can be installed separately since this type of communication allows controlling the robot in a long range [14].

Fuzzy Decision System. The whole robot control system can be considered as a closed-loop one but in a Fuzzy control part (mamdani type) which is operated by the microcontroller it is the open-loop system. All the calculations of the new input parameters are done in the omnidirectional visual sensing step. The data protocol is following: *Angle,Speed#. Where ANGLE is the value from 0 to 90. Values from 1 to 45 are considered as a turning right angle. Values from 46 to 90 are considered as a turning left angle and calculated: a = Angle 45, if Angle > 45, where a is the left rotation angle and ANGLE is the value received by a ZigBee wireless module.The SPEED value is sent by a characters and deffuzified to integer values. There are two output signals each of them control torque in

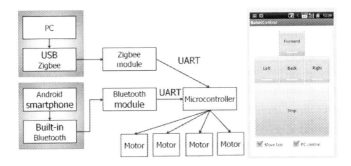

Fig. 5. Control system design and the graphical user interface

percent for left and right motors. In the real program we simplify our calculations by using the central values for the output and using pre-calculated table for the membership functions. Otherwise we cannot guarantee micro-controller works fast enough. If the torque less than 50 percent, the micro-controller sends a signal to a low speed pin. Otherwise, it sends a signal to the high speed pin of the motor driver. The control surface is pre-calculated too to decrease the processing time which is crucial for the micro-controller which operates with a fuzzy control, communication, and other systems in real time. NL, NS, ZE, PS, PL are negative large, negative small, zero error, positive small, positive large values respectively. Zero error value is kept is narrow boundaries to prevent oscillating of the robot due to small error. As a SPEED value the robot receives discrete values MBF, MBS, S, MFS, MFF (move backward fast, move backward slow, stop, move forward slow, move forward fast respectively). Output membership functions for left and right motors are the same. Figure 6 show the input/output membership functions.The fuzzy controller rule-base consists of 25 rules as shown in table 1. Due to discrete SPEED input the control surfaces are not smooth. Figure 7 shows the step results for angles from -180 to 180. Current fuzzy decision system allows preventing overturn of the robot. This simplified control scheme does not guarantee the smoothness of movement. To address this problem, motors drivers should be upgraded to control motor speed smoothly.

Table 1. FUZZY CONTROLLER RULE-BASE

Right/Left	NL	NS	Z	PS	PL
MBF	BF/BS	BF/BS	BF/BF	BS/BF	BS/BF
MBS	BF/S	BS/S	BS/BS	S/BS	S/BF
S	BF/FF	BS/FS	S/S	FS/BS	FF/BF
MFS	S/FF	S/FS	FS/FS	FS/S	FF/S
MFF	FS/FF	FS/FF	FF/FF	FF/FS	FF/FS

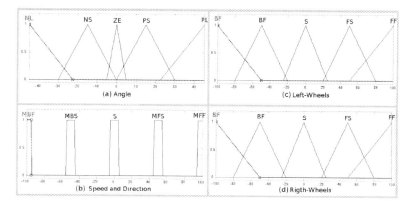

Fig. 6. Membership functions. (a) Angle input. (b) Speed input. (c) Left motors output. (d) Right motors output.

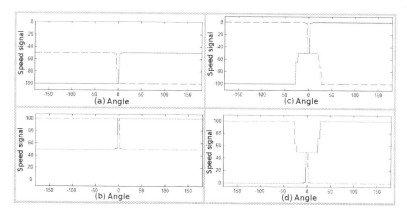

Fig. 7. Simulation results for angles from -180 to 180. (a) Fast backward output. (b) Fast forward output. (c) Slow backward output. (c) Slow forward output.

4   Experimental Result

In this section, the ending results of the experiment will be introduced. All of the experiments for the omnidirectional sensing were done on Pentium Intel Core 2 Duo Processor E4600, 2.40 GHz, 2 GB RAM. Implementation was done in C++ under Ubuntu 12.04. The algorithm used a group of 1,933 frames taken at the University of Ulsan (UOU), South Korea with a frame resize to 160x146 pixels. In the case of Fuzzy Logic control the implementation was done on ATmega 128A microcontroller connected to four independent driven DC motors. On the other hand, in order to prevent accidents, during the test a mobile device with built-in Bluetooth module and Android OS 2.2+ was used as an emergency collision avoidance control. The depicted results are to the implementation of the RANSAC curve fitting, DBSCAN and Fuzzy logic.

5 Conclusions and Future Works

The preliminary results of the proposed method presented relevant information for finding the VP estimation in road scenes. Specified imaging results show the performance of our proposed work over a set of possible scenarios. The proposed algorithm works well despite the problems given by the illumination distribution, strong sunlight, and dark shadows. On the other hand, the vanishing point has an error in places where single or double continuous white or yellow lines along the center of the road lines does not appear at all due to the following: weather conditions, wildlife, pavement deterioration or where turn lanes are not marked in places like junctions, intersection, etc. Part of this discrepancy is due to the fact that picture looses a lot of detail during resizing. The continuous white lines along the road segments were not extracted from the binary image obtained after applying the Canny edge detector with a length $l \geq 25$, where l is the Euclidean distance between the endpoints of the line segments. As a result, the VP had to be extracted using the spatial density information from the side strip areas (bicycle and/or planter) or sidewalks. Table 2 shows the time measuring performance of our proposed algorithm. The FLC is designed to control a robot in existing hardware restrictions. It can address the overturn problem which may happen if diverse turning speeds are not applied. The small gap at the zero angle point prevents the robot from oscillating due to small signal errors got in an image processing part. Introducing the smooth changing of motor torque will give more precise speed control with minor changes membership functions. Finally,the fuzzy logic guidance control algorithm show a satisfactory performance, however in order to improve the robustness, there will be improvements to the performance and the experiment by using road model features, surface as well path planning approaches. Regarding autonomous navigation, guidance and control design this result reinforces the point of usage of a set of sensor (GPS, IMU, LRF, and online interactive maps) for dealing with this problem in real time.

Results show satisfactory performance of the proposed control systems and indicate that the modified fuzzy controller manages integrated vehicle control systems more effectively compared to standard fuzzy logic controller.

Acknowledgments. This work was supported by the National Research Foundation of Korea(NRF) Grant funded by the Korean Government(MOE) (2013R1A1A2009984)

References

1. Ebrahimpour, R., Rassolinezhad, R., Hajiabolhasani, Z., Ebrahimi, M.: Vanishing point detection in corridors: using Hough transform and K-means clustering. IET Computer Vision J. 6, 40–51 (2012)

2. López, A., Cañero, C., Serra, J., Saludes, J., Lumbreras, F., Graf, T.: Detection of Lane Markings based on Ridgeness and RANSAC. In: Proceeding of the 8th International IEEE Conference on Intelligent Transportation Systems (ITSC 2005), Vienna, Austria, pp. 733–738 (2005)
3. Miksik, O., Petyovsky, P., Zalud, L., Jura, P.: Detection of Shady and Highlighted Roads for Monocular Camera Based Navigation of UGV. In: IEEE International Conference of Robotics and Automation (ICRA 2011), Shanghai, China, pp. 4844–4849 (2011)
4. Bazin, J.-C., Pollefeys, M.: 3-line RANSAC for Orthogonal Vanishing Point Detection. In: in IEEE/RSJ International Conference on Intelligent Robots and Systems (IROS 2012), Algarve, Portugal, pp. 4282–4287 (2012)
5. Mirzaei, F.M., Roumeliotis, S.I.: Optimal Estimation of Vanishing Points in a Manhattan World. In: IEEE International Conference on Computer Vision (ICCV 2011), Barcelona, Spain, pp. 2454–2461 (2011)
6. Zadeh, L.A.: Fuzzy Sets*. DBLP Information and Control 8, 338–353 (1965)
7. AbuBaker, A.: A Novel Mobile Robot Navigation System Using Neuro-Fuzzy Rule-Based Optimization Technique. Research Journal of Applied Sciences, Engineering and Technology 4(15), 2577–2583 (2012)
8. Al-Din, M.S.N.: Decomposed Fuzzy Controller for Reactive Mobile Robot Navigation. International Journal of Soft Computing and Engineering 2(4), 140–149 (2012)
9. Muthu, T., Thierry Gloude, R., Swaminathan, S., Satish Kumar, L.: Fuzzy Logic Controller for Autonomous Navigation. In: Proceeding of 12th International Conference on Ceramic Processing Science (ICCPS 2012), Oregon, USA, pp. 81–92 (2012)
10. Khatoon, S., Ibraheem: Autonomous Mobile Robot Navigation by Combining Local and Global Techniques. International Journal of Computer Applications 37(3), 1–10 (2012)
11. Howard, A., Seraji, H.: A Real-Time Autonomous Rover Navigation System. In: Proceeding of 3rd International Symposium on Intelligent Automation and Control, World Automation Congress (WAC 2000), Hawaii, USA (2000)
12. Zhang, J., Knowll, A., Schwert, V.: Situated Neuro-Fuzzy Control for Vision-Based Robot Localisation. Journal of Robotics and Autonomous Systems 28(1), 71–82 (1999)
13. Ester, M., Kriegel, H.-P., Sander, J., Xu, X.: A Density-Based Algorithm for Discovering Clusters in Large Spatial Databases with Noise. In: Proceeding of 2nd International Conference of Knowledge Discovery and Data Mining (KDD 1996), Oregon, USA, pp. 226–231 (1996)
14. Filonenko, A., Fei, Y., Vavilin, A., Jo, K.-H.: Self-configuration for surveillance sensor network. In: Proceeding of 9th International Conference on Ubiquitous Robots and Ambient Intelligence (URAI 2012), Daejeon, Korea, pp. 425–428 (2012)

Production Planning and Inventory Control
in Automotive Supply Chain Networks

Ashkan Memari[1], Abdul Rahman Bin Abdul Rahim[2], and Robiah Binti Ahmad[2,*]

[1] Faculty of Mechanical Engineering, Universiti Teknologi Malaysia
81310 Skudai, Johor Bahru, Malaysia
ashkan.eng@gmail.com
[2] UTM Razak School of Engineering and Advanced Technology
UTM Kuala Lumpur
54100 Jalan Semarak, Kuala Lumpur, Malaysia
rahmanar@utm.my
robiah@ic.utm.my

Abstract. This paper addresses a non-linear optimization model by integrating production planning and inventory control in the automotive industry at the strategic and operational level. In order to provide an effective modeling, we developed a framework to integrate manufacturing system and suppliers within an automotive supply chain network. The numerical experiments demonstrate the efficiency of the proposed model on minimization of total delivery cost and due date delivery.

Keywords: Optimization Modeling, Production Planning, Inventory Control, Supply Chain.

1 Introduction

In today's global market, supply chain management (SCM) has provided several advantages for companies' strategy in order to improve their competitiveness. Study on supply chain (SC) behavior is valuable for understanding causal effects and vari-ous or even extreme scenarios. Inventory control is one of the vital aspects in compa-nies supply chain network (SCN) in order to promote their efficiency. Since the flow of inventories is at the heart of each company, the role of inventory controlling tools is unavoidable in every SCN. Just-in-Time (JIT), Material Requirements Planning (MRP) and its modified version Manufacturing Resource Planning (MRPII) are well-known and the most powerful controlling tools that have a substantial effect upon the failure or success of an entire manufacturing system [2]. In this regard, the important characteristics of material flow are directed toward planning and controlling by MRP and JIT.

However, a pure MRP or JIT production system are rarely exists in practice. Even in pioneers of JIT systems like Toyota Japan, production smoothing is planned by the master production schedule (MPS), and the MRP is followed based on the MPS by

---

* Corresponding author.

A. Moonis et al. (Eds.): IEA/AIE 2014, Part I, LNAI 8481, pp. 430–439, 2014.
© Springer International Publishing Switzerland 2014

applying Bill of Material (BOM). This type of hybrid MRP/JIT planning has been especially adopted by automobile manufacturers [1]. In practice, both distribution and manufacturing systems enclose features of pull or push to varying degrees regardless of the identity of system [2].

Previous studies have been demonstrated the solely implementation of JIT or MRPII cannot reduce deliveries costs [1]. Specially, when suppliers are located almost faraway from manufacturers plant, raw materials and parts should be delivered on time to manufacturer. To do so, in our real world automotive manufacturer case study, we propose a non-linear mathematical model in a hybrid MRPII/JIT manufacturing system to overcome this issue. In the first level of the proposed model, Economic Orders Quantity (EOQ) for raw materials and parts are calculated based on JIT philosophy with respect to numbers of pre-determined suppliers. The objective of the second level is to reduce the delivery tardiness by considering the optimized delivery time, which determined by customers.

2 Literature Review

In the most basic definition, perhaps the central notion of mutual exclusivity of MRP and JIT systems is "push" and "pull" terminology, which used to interpret the control strategy of information and material flow [3]. MRP is a manufacturing plan-ning and control strategy that determines a schedule for the deterministic demand items from the stochastic demand within the BOM explosion process, netting and offsetting [4]. Whilst, JIT production is relatively younger in terms of use and simply means producing with the right amount of material at the right place and at the right time. However, in JIT implementation, there is a lack of the support of a standardized software package due to its initial detachment from information technologies [3].

In the majority of studies in the area of MRP and JIT systems, the assessment is not just comparison among pull and push control principles, however, the hybrid control principles versus pure pull or pure push strategies are compared in order to search the advantages of the optimized hybrid system. Analytical hybrid MRP/JIT studies discuss in smaller groups and they originate with the premise that integrated control is a better approach inherently in comparison with simpler control techniques. In these types of studies, changes in hybrid systems and in the manufacturing parameters tested by focusing on how hybrid system is fitted to the manufacturing environment exclusively rather than the advantages would be provided by this integration.

Performance comparison of different control techniques is provided by complex mathematical analysis and simulation tools in manufacturing environments. This means are able to compare manufacturing system performance ranging from simpler systems with few processes to more complex manufacturing environments such as multistage production environments with multiple lines as well as parallel configurations. Optimization of these manufacturing environments out-performs simpler control principles in the various dimensions of inventory performance and service level.

Hirakawa et al. [5] and Takashi et al. [6] applied mathematical model for multistage production-inventory system with different inventory level and production quantities at each stage as the performance criteria. As a result, hybrid systems were reported have smaller variability in both studies. In other words, changes in demand fluctuation do not change this fact; however, they affect the best configuration of the hybrid system and the place of push-pull boundary. Moreover, combination between pull in downstream and push at upstream make greater results compared to the reverse case.

An optimization model of a horizontally integrated push–pull hybrid production system in a serial production line constructed by Cochran [7]. In this study, pull-push control to production systems were found more efficient compare with all-push and all-pull systems. Furthermore, Salum [8] study the application of push–pull control to production systems by comparing the dual-resource constrained/push–pull controlled system with the dual-resource constrained/Kanban system. Hodgson [9] applied the Markov Decision Process (MDP) as the alternative for production system in a push–pull multistage production/inventory system in his study. Literatures in push-pull control underlie various case studies such as telecommunication industry [10] and semiconductor industry [11].

One of the most important issues, which faces by many manufacturers, is the customers' order scheduling. Furthermore, delivery of material and parts should be started as long as their processing is done to escape tardiness which may lead to customer dissatisfaction. As a result, between the two types of costs such as delivery cost and the costs caused by tardiness is often a trade off [12].

The problem material delivery tardiness scheduling without engaging suppliers into consideration in hybrid MRP/JIT has been studied by many studies [4-11]. Delivery tardiness is a significant issue since it may lead to impose penalties for manufacturers. In this study, a mathematical model based on tardiness control at supplier is developed for minimizing delivery tardiness and delivery costs as well. This model can be solved by a commercial solver to find local optimums. We use Matlab 2011 to run the proposed model.

3   Modeling

The objectives of MRP and JIT integration in the supply chain were explained in Section 2. In this study, our objective is to find the minimum values for the delivery due date of raw materials and semi-finish parts. Also, this model focuses on minimizing late delivery cost, effective JIT implementation, Kanban and planning on the MRPII information feedback. Moreover, it may have an impact on the bullwhip effects and relevant indirect delivery costs. The proposed model includes three nonlinear pro-gramming models that are applied in two stages of the whole supply chain (supplier and manufacturer) based on two types of piston in a spare parts automotive manufacturer. Figure 1 shows the BOM for these two types of products. We assumed the internal logistics system is in its optimum state. Furthermore, it assumed order batches are fixed for each order period and the procurement process includes ordering, quality control, loading, transportation and set up. To develop the optimization model, we apply non-linear programming based on dynamic time series with stochastic variables. The producer of the modeling approach in our study is based on the developed framework in figure 2.

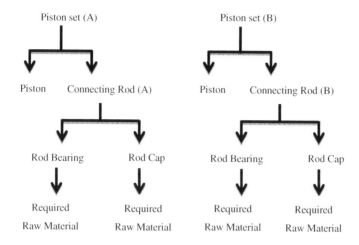

Fig. 1. BOM For Selected Case Study

3.1   Model Description and Formulation

C_i	Capacity of supplier i
q	Quantity of parts and raw materials in BOM
P	Planning horizon/day
d	Demand growth rate
$\overline{Q}_p$	Average amount of required material for procurement part p
$\overline{t}_p$	Average time between consecutive deliveries
DT(m)	Delivery time
O	Amount of operations in ordering process
$\overline{S}_q$	Average procurement time by supplier
$\overline{U}_{qo}$	Average time per unit for preparation of part p
$E(G_{pm}(Q^*))$	Final expectation
TIC	Total inventory cost
DC	Delay cost
m	Number of suppliers that supply part p
Bi	Order batches quantity in Kanban system
Z	Order center
TS_q	Total satisfied orders
DH	Demand in horizon plan
σ_q	Fixed order price
I_q	Inventory holding cost
φ_q	Fixed purchase cost
θ_q	Inventory shortage cost (late delivery)
τ_q	Deviation cost of the delivery cost, the optimum value is 1.
w	$w = \sigma_q + \varphi_q$
w_i	Q* division rate between two suppliers

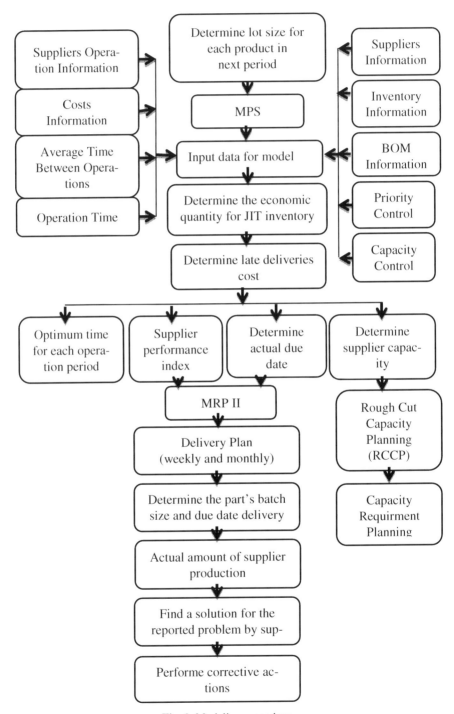

Fig. 2. Modeling procedure

Based on the developed framework, equation 1 calculates the actual parts delivery time. $E\ (G_{pm}\ (Q^*)) = S\%\ (G_{pm}\ (Q^*))$ calculates expected arrival time for the required parts in final product assembly in production line while considering delay time for the other parts [13]. To obtain production capacity we apply equation (2) [14]. In this problem material capacities, parts orders and amount of inventory that should be available in the specific planning horizon are determined.

$$DT(m) = \frac{\text{supplier performance for each operation}}{\text{supplier performance for total operations}} \times \text{supply average time} \tag{1}$$

$$DT(m) = \sum_{q=1}^{q} \sum_{o=1}^{oq} \varphi_{qom} \times \left[\frac{\frac{P}{\bar{t}Q_q^*}}{\sum_{q=1}^{q} \sum_{o=1}^{oq} \varphi_{qom} \frac{P}{\bar{t}Q_q^*}} \right] \times \left[S\%\ (Gpm\ (Q\ *)) + Q_q^* \overline{U}_{qo} \right] \tag{2}$$

$\begin{cases} \varphi_{qoz} = 1, & \textit{If operation o is done to fulfill the order of part q in center z} \\ \varphi_{qoz} = 0, & \textit{otherwise} \end{cases}$

$\begin{cases} \varphi_{qom} = 1, & \textit{If operation o is done to fulfill the order of part q by} \\ & \qquad\qquad \textit{supplier m} \\ \varphi_{qom} = 0, & \textit{otherwise} \end{cases}$

$\begin{cases} \varphi_{qoz \rightarrow m} = 1, & \textit{If operation o is done to fulfill the order of part q at} \\ & \textit{cnter z and next operation will be done by supplier m} \\ \varphi_{qoz \rightarrow m} = 0, & \textit{otherwise} \end{cases}$

$$\tag{3}$$

The amount of satisfied demand in specific planning horizon, are obtained by equation (3) [13]. Furthermore, the order placing period is calculated by equation (4)[13].

$$S = [DT(m)] \times \frac{DH}{P} \tag{4}$$

$$TS_q = \left[\sum_{q=1}^{q} \sum_{o=1}^{oq} \varphi_{qoz} \leftrightarrow m \frac{P}{\bar{t}_p Q^*} (\overline{S}_{qo} + Q_q^* \overline{U}_{qo}) \right] \Big/ \left[\sum_{q=1}^{q} \sum_{o=1}^{oq} \varphi_{qoz} \rightarrow m \frac{P}{\bar{t}_q Q_q^*} \right] \tag{5}$$

Ordering information in the current system is acquired with respect to equation 3 and 4. In this step of determining the optimum value of the parts to be received, firstly, we calculate economic order quantity of a supplier with respect to on time delivery approach (equation (5)) [14,15] and second this relation is measured for two suppliers in equation (6) [16,17] by considering stochastic conditions. The obtained output from equation 6 is applied as an input for MRP, delivery time determination model and operation length.

$$E(I_q) = I \int_0^\infty \int_0^\infty \left(\frac{Q^2}{2} + rQ - JQ - lQ + K_{1j}Q\right) Q_T(j)Q_L(l)\,d.jdl$$
$$+ (I + \theta) \int_0^\infty \int_0^\infty \frac{1}{2}(l - r)^2 f_j(j) f_L(l)d.jdl$$
$$+ (I + \theta) \int_0^{r+w_1Q} \int_{r+w_1Q-1}^\infty \frac{1}{2}(l + j - r - w_1Q)^2 q_j(j)q_L(l)d.jdl$$
$$+ (I + \theta) \int_{r+w_1Q}^\infty \int_0^\infty \left(l_j + \frac{J^2}{2} - w_{1j}Q - tj\right) w_j(j)w_L(l)d.jdl$$

(6)

$$TIC = \frac{B}{Q}(1 + d) + E(I) + \tau_1 + \tau_2$$

(7)

$$E(I) = (I + \theta) \int_r^\infty \frac{(r-l)^2}{2Q} f(l/\tau)dl + I \int_0^\infty r - l + \frac{Q}{2}) \frac{f(l/\tau)}{\tau} + \tau_i$$

(8)

$$TIC = \frac{B}{Q}(1 + d) + E(I) + c\left(\frac{1}{\tau} - 1\right)$$

(9)

$$s.t \, Q \le CA_m$$

(10)

The objective in the second level of modeling is to reduce cost and obtained time. To do so, mean and standard deviation of delay time is needed. These considerations are provided by formula (7) [18] and (8) [19] for every product. In addition, the achieved results for Q* from this function will be applied in the final delivery time model (equation 9) [20].

$$E(G_p) = \sum_{p=1}^p E\left(G_{pm}(Q^*)\right) \varphi_{pom} + Q^*\overline{U}_{qo}$$

(11)

$$V(W_f) = \sum_{o=1}^{op} V\left(W_{qm}(Q^*)\right) \varphi_{qom} + \sum_{o=1}^{op} Q_p^* S_{Iqo}^2$$

(12)

$$Clostsale = \sum_{i=1}^{\acute{p}} lostsale_p$$

(13)

$$Choldingcost_{dp} = \sum_{i=1}^p \sum_{i=1}^{op} Q_{pom \to z}^* \times Holding \, Cost$$

(14)

$$MinDC = \sum_{p=1}^{p} (V_p(G_p)/E_p(P_q(m))) \times Choldingcost_{dp} \times Clostsale$$
$$+ (E_p(G_p)/E_p(P_z(m))) \times Choldingcost_{dp}$$

(15)

$$MinDC = \sum_{p=1}^{p} \frac{b_p}{x_p} \times \frac{c_p}{x_p}$$

$$x_p \le \max(x_2, x_3)$$

$$\max(x_1, \dots, x_n, x_{n-1}, x_p) \le x_1$$
$$(x_1, \dots, x_n, x_{n-1}, x_p) > 0$$

(16)

4   Results and Discussion

To solve this problem, equations 1-4 calculate the value of current system parameters. Then, the optimum values of these parameters are obtained from equations 5-9. In the case study under consideration the amount of annual production for piston set (A) and piston set (B) are 2000 and 3000 respectively. Annual Re-Order Point (ROP) is announced 3-4 months before planning horizon. Table 1 shows the collected data for the corresponding case study. Also, the developed framework in figure 2 summarizes the require steps for solving this problem.

Table 1. Collected Data

	Basic Parts Data			Inventory Data		
Parts Name	Required Amount	Produced/ Purchased	Current Inventory	Allocated Inventory	Shipped orders	Delivery Time/day
Piston Set (A)	2000	Produced	100	0	0	0
Piston Set (B)	3000	Produced	900	0	0	0
Piston	17000	Purchased	150	0	200	200
Connecting Rod	5000	Purchased	300	0	1200	700
Rod Bearing	5000	Purchased	1200	500	0	0
Rod Cap	5000	Purchased	700	100	0	0
RB raw material	5000	Purchased	0	0	100	500
RC raw material	5000	Purchased	0	0	0	0

Table 2. Model Outputs (Percentage of Customers response)

Items / Parts	minimum due date delivery/per unit per day	Actual due date delivery/per unit per day	Percentage of improvement in due date delivery	Performance index for each supplier
Piston set (A)	2	36	85%	0.18849
Piston set (B)	3	5	40%	0.03278
Piston	1	0.84	0	0.41146
Connecting Rod	1	0.73	0	0.09778
Rod Bearing	2	3.38	41% ،	0.45612
Rod Cap	1	9.9	90%	0.065921
Rod Bearing raw material	2	9.5	60%	0.063511
Rod Cap raw material	2	0.16	0	0.05424

Table 3. Model Outputs (Economical Aspects)

Costs / Parts	Piston set (A)	Piston set (B)
Minimum total delay cost	43410	64387
Total delay cost	267049	276472
Improvement percentage	83.74%	76.71%

The numerical experiments demonstrate the benefit of the developed framework.

5 Conclusion

The proposed model optimizes due date delivery with respect to suppliers' performance index which can aid suppliers to better adjust production planning and ordering. Furthermore, the model enables manufacturers to respond quickly to customer requirements, enhance productivity, reduce material levels, reduce tardiness, holding cost reductions, leading to the potential benefit of long term contracts with suppliers.

Acknowledgment. The authors would like to thank Ministry of Higher Education (MOHE) Malaysia and Universiti Teknologi Malaysia (UTM) under Grant University Project Scheme Vot 08J28 for the financial support provided throughout the course of this research.

References

1. Monden, Y.: Toyota production system: an integrated approach to just-in-time, 4th edn. CRC Press (1998)
2. Pyke, D.F., Cohen, M.A.: Push and pull in manufacturing and distribution systems. Journal of Operations Management 9(1), 24–43 (1990)

3. Hopp, W.J., Spearman, M.L.: To pull or not to pull: what is the question? Manufacturing & Service Operations Management 6(2), 133–148 (2004)
4. Orlicky, J.A.: Material requirements planning: the new way of life in production and inventory management. McGraw-Hill, Inc. (1974)
5. Hirakawa, Y., Hoshino, K., Katayama, H.: A hybrid push/pull production control system for multi-stage manufacturing processes. In: Achieving Competitive Edge Getting Ahead Through Technology and People, pp. 341–346. Springer London (1991)
6. Takahashi, K., Soshiroda, M.: Comparing integration strategies in production ordering systems. International Journal of Production Economics 44(1), 83–89 (1996)
7. Cochran, J.K., Kaylani, H.A.: Optimal design of a hybrid push/pull serial manufacturing system with multiple part types. International Journal of Production Research 46(4), 949–965 (2008)
8. Salum, L., Araz, Ö.U.: Using the when/where rules in dual resource constrained systems for a hybrid push-pull control. International Journal of Production Research 47(6), 1661–1677 (2009)
9. Hodgson, T.J., Wang, D.: Optimal hybrid push/pull control strategies for a parallel multistage system: Part I. The International Journal of Production Research 29(6), 1279–1287 (1991)
10. Perdaen, D., Armbruster, D., Kempf, K.G., Lefeber, E.: Controlling a Re-entrant Manufacturing Line via the Push–Pull Point. In: Decision Policies for Production Networks, pp. 103–117. Springer London (2012)
11. Boukerche, A., Dash, T., Pinotti, C.M.: Performance analysis of a novel hybrid push–pull algorithm with QoS adaptations in wireless networks. Performance Evaluation 60(1), 201–221 (2005)
12. Mahdavi Mazdeh, M., Rostami, M., Namaki, M.H.: Minimizing maximum tardiness and delivery costs in a batched delivery system. Computers & Industrial Engineering 66(4), 675–682 (2013)
13. Vandaele, N.J., Lambrecht, M.R.: Reflections on the use of stochastic manufacturing models for planning decisions. In: Stochastic Modeling and Optimization of Manufacturing Systems and Supply Chains, pp. 53–85. Springer US (2003)
14. Lambrecht, M.R., Ivens, P.L., Vandaele, N.J.: ACLIPS: A capacity and lead time integrated procedure for scheduling. Management Science 44(11-Part-1), 1548–1561 (1998)
15. Choi, J.W.: Investment in the reduction of uncertainties in just-in-time purchasing systems. Naval Research Logistics (NRL) 41(2), 257–272 (1994)
16. Barnes-Schuster, D., Bassok, Y., Anupindi, R.: Optimizing delivery lead time/inventory placement in a two-stage production/distribution system. European Journal of Operational Research 174(3), 1664–1684 (2006)
17. Mohebbi, E., Hao, D.: When supplier's availability affects the replenishment lead time— An extension of the supply-interruption problem. European Journal of Operational Research 175(2), 992–1008 (2006)
18. Mohebbi, E.: Supply interruptions in a lost-sales inventory system with random lead time. Computers & Operations Research 30(3), 411–426 (2003)
19. Van Nieuwenhuyse, I., Vandaele, N.: The impact of delivery lot splitting on delivery reliability in a two-stage supply chain. International Journal of Production Economics 104(2), 694–708 (2006)
20. Ayers, J.B. (ed.): Handbook of supply chain management. CRC Press (2002)

Dynamic Scripting with Team Coordination in Air Combat Simulation

Armon Toubman[1,2], Jan Joris Roessingh[1], Pieter Spronck[2],
Aske Plaat[2], and Jaap van den Herik[2]

[1] National Aerospace Laboratory,
Department of Training, Simulation, and Operator Performance
Anthony Fokkerweg 2, 1059 CM Amsterdam, The Netherlands
{Armon.Toubman,Jan.Joris.Roessingh}@nlr.nl
[2] Tilburg center for Cognition and Communication (TiCC), Tilburg University,
P.O. Box 90153, 5000 LE Tilburg, The Netherlands
{p.spronck,aske.plaat,jaapvandenherik}@gmail.com

Abstract. Traditionally, behavior of Computer Generated Forces (CGFs) is controlled through scripts. Building such scripts requires time and expertise, and becomes harder as the domain becomes richer and more life-like. These downsides can be reduced by automatically generating behavior for CGFs using machine learning techniques. This paper focuses on Dynamic Scripting (DS), a technique tailored to generating agent behavior. DS searches for an optimal combination of rules from a rule base. Under the assumption that intra-team coordination leads to more effective learning, we propose an extension of DS, called DS+C, with explicit coordination. In a comparison with regular DS we find that the addition of team coordination results in earlier convergence to optimal behavior. In addition, we achieved a performance increase of 20% against an unpredictable opponent. With DS+C, behavior for CGFs can be generated that is more effective since the CGFs act on knowledge achieved by coordination and the behavior converges more efficiently than with regular DS.

Keywords: computer generated forces, machine learning, air combat.

1 Introduction

Military organizations are increasingly using simulations for training purposes. Simulations are cheaper, safer, and more flexible than training with real equipment in real-life situations [1, 2]. In military simulations, the roles of allies and adversaries are performed by computer generated forces (CGFs).

Traditionally, the behavior of CGFs is scripted [3]. Production rules—rules that map conditions to actions—are manually crafted to suit specific (types of) CGFs. In complex domains, such as that of air combat, this leads to complex scripts and requires availability of domain expertise. These scripts then produce rigid behavior, because it is impossible to account for all situations that CGFs might encounter during simulations.

A. Moonis et al. (Eds.): IEA/AIE 2014, Part I, LNAI 8481, pp. 440–449, 2014.
© Springer International Publishing Switzerland 2014

Artificial Intelligence techniques may provide a solution by automating the process of generating CGF behavior. Automation bypasses the requirement of expertise availability and shortens the time needed to generate the behavior. Various efforts have been made at realizing automatic generation of CGF behavior [4, 5].

At the National Aerospace Laboratory (NLR) in the Netherlands, CGF research aims to generate behavior for air combat training simulations. The focus of recent work has been to generate behavior through the use of cognitive models, and optimizing these models with machine learning (ML) techniques such as neural networks and evolutionary learning [3], [6]. In this paper, we diverge from the earlier approach of using cognitive models by applying ML directly to the generation of behavior.

The envisaged new ML technique should satisfy at least four conditions to be suitable for our domain. First, the technique should provide *transparent behavior models* as a result. Techniques such as neural networks are opaque in the sense that the resulting models are hard to relate to the behavior they produce. The new technique should produce understandable models that are manually editable and reusable by training instructors. Second, the new technique should be *scalable* to the domain of air combat with team missions. The scope of the mentioned research with cognitive models [3], [6] was not scalable, since it was limited to a single learning agent. Third, the chosen machine learning technique should *converge to practically usable behavior* in a timely fashion, to allow rapid development of new training scenarios. Fourth, the ML technique must be able *to learn robust behavior*. Since the CGFs will be used for training humans, the CGFs should have good performance against a variety of tactics.

Dynamic Scripting (DS) is a reinforcement learning technique specifically designed to satisfy requirements similar as the ones stated above [7]. While DS has been used with teams, no attention has been given to the explicit coordination of teams using DS. In this paper, we present a technique based on DS called DS+C, which enables team coordination using DS through direct communication between agents. We compare the performance of a team using DS with and without coordination. The main contributions of this paper are that we (1) present explicit coordination in DS and (2) show, using an existing combat simulator, experimental evidence that coordination leads to faster convergence to optimal behavior.

The course of the paper is as follows. Section 2 describes the Dynamic Scripting method. Section 3 describes our method of team coordination. Section 4 describes a case study. Section 5 shows the results. Finally, the paper is concluded by a discussion in Section 6 and a conclusion in Section 7.

2   Dynamic Scripting Method and Related Work

Dynamic Scripting is an online learning technique based on reinforcement learning. It was introduced by Spronck et al. [7] to address certain requirements for adaptive game AI in commercial video games, such as "easily interpretable results" and "reasonably successful behavior at all times." These requirements are also applicable in the domain of military training, where quality controls such as transparent results and robust behavior are important.

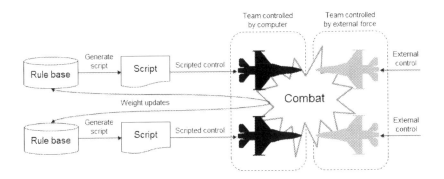

Fig. 1. Dynamic Scripting in the context of two learning agents in the air combat domain

In DS, the learning process works as follows. The learning agent has a rule base with behavior rules. The DS algorithm selects a set of rules through weighted random selection. The selected rules together form a script that governs the behavior of the agent during a trial with one or more other agents. After each trial, the weights of the rules that were activated in the encounter are updated. The learning process is illustrated in Fig. 1.

In the original DS experiments [7] team behavior was a result of emergence, guided by a fitness function which rewarded team victories as well as individual success. However, to make sure that CGFs act conforming to the training goals of a particular air combat training simulation, more control over the team members' actions is required. Such intensive control can be formalized by coordination rules.

There are two general methods of team coordination: centralized and decentralized coordination [8]. With centralized coordination, one agent may direct the actions of a team. With decentralized coordination, all agents in a team may influence each other's actions by sharing information through some form of communication.

In this paper, we have chosen to implement decentralized control, because of its straightforward implementation. In terms of DS, decentralized control translates to each agent having their own rule base with its own weights and generated scripts. Coordination is achieved through communication. However, adding communication to multi-agent systems in general is not trivial [9]. For this reason, we attempted to fit the communication (and therefore also the coordination) explicitly into the DS mechanism.

3   Dynamic Scripting with Team Coordination

We implemented team coordination in DS through communication between agents, resulting in a technique which we called DS+C. In brief, the technique works as follows. Each agent starts with certain rules in their rule bases that are activated when particular messages are received. Whenever an agent activates a rule, it sends a message to its teammates describing its actions. The description of actions should not be too narrow; otherwise no match will occur during trials. The DS+C algorithm decides which actions in response to the messages are valuable.

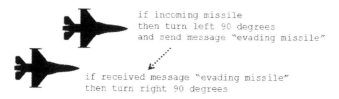

```
        if incoming missile
        then turn left 90 degrees
        and send message "evading missile"


        if received message "evading missile"
        then turn right 90 degrees
```

Fig. 2. Illustration of the communication applied in DS+C. Messages sent by one agent trigger rules in another agent

In more detail, the communication scheme consists of three parts. The first part is an addition to existing behavior rules: each rule, when activated by an agent a, now also sends a message from agent a to every agent b in the same team. This message contains the nature of the actions described by the rule. The second part is a new component for the agents. Agent b stores the messages received during the activation of rules by its teammates until b has processed its own rules. The third part handles the processing of the received messages. For each agent, rules (i.e., the 'coordination rules') are added to its rule base that will lead to new behavior after aforementioned messages have been received. Together, these parts form a robust communication system that will remain functioning even when the recombination of behavior rules detects conflicting messages. The communication principle is shown in Fig. 2.

It must be emphasized that the form of coordination as described above is completely rule-based. The coordination rules undergo the same selection process as all other behavior rules. In other words, by expressing the coordination as rules, DS+C will learn which messages are relevant and how they should be acted upon. The rule selection part of the DS+C algorithm will include or exclude a subset of these coordination rules in scripts based on their added value.

4   Case Study and Experimental Setup

In order to test the suitability of the approach, it has been applied in the domain of air combat. In this domain, agents must exhibit realistic tactical behavior in order to increase the value of simulation training for fighter pilots.

We have taken a 'two versus one' combat engagement scenario as our testing ground. The scenario is illustrated in Fig. 3. Two 'blue' fighters (virtual pilots controlling fighter planes), i.e., a 'lead' together with its 'wingman' attempt to penetrate the enemy airspace. In more detail, the 'blue' formation seeks an engagement with a 'red' fighter that defends a volume of airspace, by flying a so-called Combat Air Patrol (CAP) pattern. The

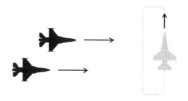

Fig. 3. Diagram of the scenario used in the case study. The 'blues' (left) fly towards the 'red' (right). Red is flying a CAP.

'blue' mission is considered successful (a win) if 'red' is eliminated, and is considered unsuccessful (a loss) if one or both of the 'blue' aircraft are eliminated, after which the 'blue' mission will be aborted. 'Rules of Engagement' for the 'red' fighter dictate that it will intercept fighter aircraft that fly in its direction.

The behavior of the 'blue' agents is governed by scripts generated by the newly implemented DS+C. The rule bases of the 'blue' agents contain three sets of rules. The first set consists of default rules. The default rules define basic behavior, on which the agents can fall back if no other rules apply. The rules are included in every script, and their weights will not be changed by the DS+C process. The rules also define the 'missions' of the agents; for instance, the 'blues' have default rules that let them fly to 'red' in formation, while 'red' has default rules that let it fly its CAP. The second set consists of general rules for air combat. These rules are based on domain knowledge, although highly simplified to illustrate the principles. Two instances are 'if I see an enemy on my radar, I lock this enemy with my radar' and 'if the enemy is locked by my radar, I fire a missile'. The third set consists of coordination rules. In the case of DS+C, these are the rules that produce behavior in response to the reception of certain messages. However, in the case of regular DS, these rules are 'filler' rules; rules that cannot be activated and therefore produce no behavior. These 'filler' rules were added to keep the sizes of the rule bases constant between the DS and DS+C, thus providing a fair comparison. The scripts generated by DS+C consist of 6 rules, to which the default rules were added. All rules started with a weight of 50. In total, the rule bases had 31 rules each.[1]

The 'red' agent used three basic tactics, implemented as three static scripts. The three tactics are called *Default*, a basic CAP where 'red' fires on enemies it detects; *Evading*, like *Default* but with evasive maneuvers; and *Close Range*, like *Default* but only firing from close range. These three tactics each had alternative versions in which 'red' would start the engagement from flying the CAP in the clockwise direction, rather than the counter-clockwise direction. To test whether the 'blues' would be able to learn generalized behavior, 'red' was given a composite tactic that consists of the three basic tactics plus their alternative versions. With this seventh tactic (henceforth called mixed tactics) 'red' randomly selects one of the six basic tactics and uses that tactic until it loses, at which point it would select a new tactic at random.

The performance of the 'blues' in a trial is measured using the following fitness function:

$$fitness = (0.25 + (0.5 * winner)) + \quad 0.125 * speed + 0.125 * resources \quad (1)$$

In Eq. 1, *winner* is 1 if the 'blues' won, while it is 0 if they lost; *speed* is 1 minus the ratio of the maximum duration of a trial and the actual duration to complete the trial; and *resources* is a value between 0 and 1 based on the number of missiles spent in the trial (the idea is to learn to defeat the opponent using the least number of

---

[1] Descriptions of the rules are omitted for brevity, but will be published in a technical report.

missiles). The fitness function is used to calculate the adjustments to the weights as follows:

$$adjustment = \max(50 * ((fitness * 2.0) - 1.0), -25) \qquad (2)$$

The constants in these equations represent the balance between reward and punishment; for example, the constant -25 in equation (2) is the maximum negative adjustment after a loss, such that the associated rules with an initial weight of 50 still have some selection probability in a subsequent trial.

With DS+C, agents have additional rules in their rule bases, which lead to a larger number of possible scripts. In practice, this would lead to more trials needed to converge to successful behavior. However, since additional rules provide more options to the agents, there are also more possibilities to find optimal behavior, even in a rapid way. Below, we compare the performance of DS+C to that of regular DS. To do so, we first define performance in terms of efficiency (learning speed) and effectiveness (combat results). We define effectiveness as the mean win/loss ratio during a learning episode. It is difficult to define the efficiency of the DS algorithm, because it is hard to establish precisely when stationary performance, i.e., no further improvement takes place, is reached during learning. Both the DS algorithm and the agent environment are stochastic by nature. Therefore it is unlikely that DS converges to a single winning script. It is more likely that there is a set of sufficiently successful scripts available for a variety of situations.

To cope with the inherent variations in the learning process, we define the Turning Point (TP) measure TP(X) (based on Spronck's TP measure [7]) as the trial after which the 'blues' have won X percent of the last 20 trials. The window size (20 trials) was chosen to allow for a sufficient number of evaluation points during a learning episode (in this case 250 trials). X thus represents the chance that a winning script will be selected at that point. An early TP now represents a more efficient learning process, while a late TP represents a less efficient learning process.

Two series of experiments were run. In the first series, the 'blues' used the regular DS algorithm. In the second series, the 'blues' used DS+C. 'Red' used one of the seven tactics described in this section. The results of the experiments are described in the next section.

5   Results

For each basic tactic used by 'red', results were averaged over ten learning episodes, a learning episode representing the learning process of the 'blue' agents from zero to 250 trials (encounters). In the case of the mixed tactics, results were averaged over one hundred learning episodes to reduce noise (and thus improving the chances to observe a difference between DS and DS+C agents).

The average TPs at different percentages (50%, 60%, 70% and 80%) were calculated against each of the tactics of 'red'. For the mixed tactics, the TPs were compared using independent two-sample t-tests. Learning curves (Fig. 4) were created using a rolling average (with a window size of 20 trials) of the win/loss ratio. Additionally, the weights of all rules were recorded to check to what extent coordination rules were selected by the DS+C agents.

Table 1. TPs of DS and DS+C against mixed tactics (averaged over one hundred episodes) and the basic tactics (aggregated results, ten episodes per tactic)

		TP(50%)		TP(60%)		TP(70%)		TP(80%)	
Tactics of 'red'	DS	μ	σ	μ	σ	μ	σ	μ	σ
Mixed	DS	83.8	78.1	94.5	78.9	110.5	78.4	129.9	79.1
Mixed	DS+C	48.4	48.4	60.9	49.6	75.8	55.5	103.9	69.7
Basic (aggregated)	DS	55.8	56.7	66.1	57.8	87.3	66.5	122.4	82
Basic (aggregated)	DS+C	34.8	31.3	48	42.7	65	61.3	90.7	80.6

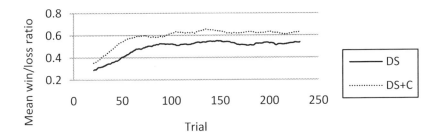

Fig. 4. Rolling mean (window size twenty) of win/loss ratio of the 'blues' against mixed tactics, with DS and DS+C. Ratios are averaged over one hundred learning episodes.

Table 1 shows the TPs of DS and DS+C against the mixed tactics. DS+C agents generally reached all TPs (50%, 60%, 70%, 80% wins) earlier than DS agents did. Note that the standard deviation in TP generally has the same order of magnitude as its mean. Independent two-sample two-tailed t-tests show that against the mixed tactics at TP(50%), the mean TPs are achieved significantly earlier using DS+C ($t = 3.85$, $p = 0.00016$) at the $a = 0.05$ significance level. The same holds for TP(60%) ($t = 3.60$, $p = 0.00039$), TP(70%) ($t = 3.60$, $p = 0.00039$), and TP(80%) ($t = 2.46$, $p = 0.015$).

In contrast with the performance against opponents that employed mixed tactics, TPs for DS+C agents were generally achieved later against the basic tactics. The learning curve of DS and DS+C against the mixed tactics is shown in Fig. 4. Both DS and DS+C agents seem to have passed a point of inflection after around 100 trials. After the first 100 trials, DS and DS+C maintain a mean win/loss ratio of 0.53 and 0.63, respectively. The mean percentage difference between the learning curves is 20.3%, with DS+C agents clearly outperforming DS agents during the entire learning process.

Log traces show that the coordination rules were selected and activated multiple times. This means that according to the DS algorithm, the coordination rules had added value. Considering the final weights of the rules, it can be observed that some of the coordination rules received high weights. The 'blue lead' favored one rule in particular, with a mean final weight of 178.6. This rule stated 'if I receive a message that my wingman is evading an enemy, turn approximately towards the enemy'. Interestingly, the 'blue wingman' mainly favored two rules, with mean final weights of

103.8 and 106.6. Both rules made the 'wingman' perform an evasive action when it received a message that the 'lead' was trying to avoid being detected by 'red'.

In the case without coordination, the rule that stood out most was the so-called 'beam' maneuver (flying perpendicular to an enemy's radar to avoid detection). This rule received high weights from the 'blue lead' (386.7) and the 'blue wingman' (323.5). A final interesting observation is that in all cases the agents preferred firing from a greater distance. This had the obvious advantage that it would be hard for 'red' to hit a 'blue', but would also diminish the chances of the 'blues' to make a hit on red.

6   Discussion

In this paper, we have presented a method for team coordination through communication using DS, called DS+C. The method was tested in a (simulated) air combat environment, in which a team of two learning agents had to learn how to defeat an opponent. Over a large set of experiments, DS+C showed clear advantages over traditional DS for multi-agent reinforcement learning. Throughout the learning process of 250 trials (an episode) DS+C agents won more often than DS agents from opponents that frequently change their tactics. On the basis of a decentralized coordination scheme, DS+C agents are able to develop more successful and more robust tactics against a less predictable opponent. Coordination in multi-agent systems is an extensively researched topic, with many issues and learning opportunities [9]. From the literature we know that other authors have found that the addition of coordination to a multi-agent system does not automatically lead to increased performance [10].

To judge the relative efficiency of DS and DS+C, we defined the TP(X) measure, based on the TP measure from [7]. DS+C agents reached the TP(X) at 50%, 60%, 70% and 80% significantly earlier than regular DS did, against an opponent with *mixed* tactics. From the fact that DS+C reached these 'milestones in learning' earlier than DS did, we may conclude that DS+C agents learn more efficiently than DS agents.

Looking at the learning curves shown in Fig. 4, it can be observed that DS+C agents generally maintain a higher win/loss ratio than DS agents throughout the learning process, against opponents that employed mixed tactics. Therefore, we may provisionally conclude that in this case, DS+C agents are not only more efficient in their learning process, but also more effective than DS agents, after training.

The higher performance of DS+C should probably be attributed to the addition of more evasive rules to the rule bases. Since the 'blues' would lose if only a single 'blue' was hit, cautious behavior was rewarded. This can be seen in the high weights that several evasive rules received. Also, because the coordination rules were proven to be valuable, it is also easy to explain the faster convergence on optimal scripts, since the DS+C agents simply had more good options available. However, the coordination rules were not intentionally biased towards evasion, and it remains possible that more aggressive rules would have a similar effect.

Against the opponent with a basic tactic, the picture is slightly different. As can be seen in Table 1, DS+C agents also reached the TPs earlier against opponents with a basic tactic. However, the TPs against the basic tactics were achieved relatively early for both DS and DS+C. Surprisingly, against the *Close Range* tactic, DS achieved earlier TPs than DS+C did. We hypothesize that if DS was able to rapidly find optimal behavior against this tactic of 'red', then the additionally included coordination rules for DS+C only hindered the convergence to successful rules, resulting in later TPs. Additionally, there seemed to be a trend of both DS and DS+C having later TPs against the alternative (reverse direction) versions of tactics (see Section 4). Additional experiments, in which the formation of the 'blues' was mirrored, also led to later TPs, when the opponent employed a non-reversed tactic. While this can be considered an artefact, it is also an indication that the spatial configuration of a formation of cooperating aircraft is a relevant factor in air combat.

Table 1 shows that the means and standard deviations of the TPs generally had the same order of magnitude. Each learning episode starts with a rule base in which each rule has an equal weight. There are nevertheless two sources of variance when averaging TPs over episodes. The first source is the stochastic sampling of the rule base by the DS algorithm. The second source is stochastic variation in the simulation environment (e.g., radar detection probability and missile kill probability). These sources cause stochastic variations in win/loss ratio and hence stochastic differences between episodes. Note that the first source of variance is non-stationary, in the sense that the distribution of weights in the rule base continuously changes during an individual episode, and eventually diminishes after a subset of relatively successful rules are identified by the algorithm.

The high weights of both general and coordination rules promoting 'evasion' were likely caused by the fact that 'blue' would lose the trial if only one of the two 'blues' was hit. Thus, 'blue' was relatively vulnerable. At the same time, the two 'blues' together had more missiles at their disposal than red, thereby promoting the 'distant firing' rules as well, overall resulting in a low risk strategy.

7   Conclusions and Future Work

From the experimental results given above we may conclude that the difference in performance against mixed tactics is the most interesting outcome: it shows that DS+C agents are better able to generalize their behavior against unpredictable enemies than DS agents.

The next step is to expand the scenario and investigate the use of DS+C with more agents, both friendly and enemy. Further work could investigate how existing extensions to DS, such as performance enhancements [7] and extensions leading to variety in the learned behavior [11] would interact with DS+C. In the future it could also be investigated which communication is most effective against specific enemy tactics, or if a centralized coordination method would offer any benefits over the currently used decentralized method. Research in these directions will further improve CGF behavior and the effectiveness of training simulations.

Acknowledgments. LtCol Roel Rijken (Royal Netherlands Air Force) provided the first version of the simulation environment used in this work. The authors also thank Pieter Huibers and Xander Wilcke for their assistance with the simulation environment.

References

1. Laird, J.E.: An exploration into computer games and computer generated forces. In: Eighth Conference on Computer Generated Forces and Behavior Representation (2000)
2. Fletcher, J.D.: Education and training technology in the military. Science 323, 72–75 (2009)
3. Roessingh, J.J., Merk, R.-J., Huibers, P., Meiland, R., Rijken, R.: Smart Bandits in air-to-air combat training: Combining different behavioural models in a common architecture. In: 21st Annual Conference on Behavior Representation in Modeling and Simulation, Amelia Island, Florida, USA (2012)
4. Benjamin, P., Graul, M., Akella, K.: Towards Adaptive Scenario Management (ASM). In: The Interservice/Industry Training, Simulation & Education Conference (I/ITSEC), pp. 1478–1487. National Training Systems Association (2012)
5. De Kraker, K.J., Kerbusch, P., Borgers, E.: Re-usable behavior specifications for tactical doctrine. In: Proceedings of the 18th Conference on Behavior Representation in Modeling and Simulation (BRIMS 2009), Sundance, Utah, USA, pp. 15–22 (2009)
6. Koopmanschap, R., Hoogendoorn, M., Roessingh, J.J.: Learning Parameters for a Cognitive Model on Situation Awareness. In: The 26th International Conference on Industrial, Engineering & Other Applications of Applied Intelligent Systems, Amsterdam, the Netherlands, pp. 22–32 (2013)
7. Spronck, P., Ponsen, M., Sprinkhuizen-Kuyper, I., Postma, E.: Adaptive game AI with dynamic scripting. Mach. Learn. 63, 217–248 (2006)
8. Van der Sterren, W.: Squad Tactics: Team AI and Emergent Maneuvers. In: Rabin, S. (ed.) AI Game Programming Wisdom, pp. 233–246. Charles River Media, Inc. (2002)
9. Stone, P., Veloso, M.: Multiagent systems: A survey from a machine learning perspective. Auton. Robots. 8, 345–383 (2000)
10. Balch, T., Arkin, R.C.: Communication in reactive multiagent robotic systems. Auton. Robots 1, 27–52 (1994)
11. Szita, I., Ponsen, M., Spronck, P.: Effective and Diverse Adaptive Game AI. IEEE Trans. Comput. Intell. AI Games 1, 16–27 (2009)

Proposal of Design Process of Customizable Virtual Working Spaces

Darío Rodríguez and Ramón García-Martínez

Information Systems Research Group, National University of Lanus, Argentina
PhD Program on Computer Science, National University of La Plata, Argentina
rgarcia@unla.edu.ar

Abstract. The evolution of communications based on Internet technology allows considering the development of Virtual Workspaces. Recently, modeling formalisms have been proposed to specify the interactions among the various members of a workgroup interacting through a virtual space. Moreover, depending of the nature of the tasks developed by the workgroup, not all the communication resources based on internet are necessary. In this context, this paper introduces a design process of customizable virtual work spaces. The proposed process specifies the components of the virtual workspace architecture necessary to support the workgroup task. The process leads the specification based on the modeled interactions among members.

Keywords: virtual workspace, design process, modeling human interactions.

1 Introduction

Collaborative work is based on communication and exchange of information among individuals in order to develop a physical or conceptual object [1]. Systems within the paradigm Computer Supported Cooperative Work (CSCW) constitute an approach to facilitate group work processes oriented to developing conceptual objects. The interaction activities among group members related to development of object are mediated by communication resources based on Internet technology.

It has been proposed [2] that there are three types of conceptual frames for developing CSCW systems:

[a] Development ad-hoc, in which systems are built in a completely adapted way to the specific problem to which it is intended to support, this has been, until now, the usual trend in creating groupware systems.

[b] The use of programming toolkits, which provide a higher level of programming abstraction through functions and APIs (Application Programmer Interface).

[c] The development of CSCW systems based on components that allows the construction of CSCW systems using predefined building blocks that can be reused and combined in different ways.

This paper is related to first type of conceptual frames for developing CSCW systems, to which we call Customizable Virtual Working Spaces (CVWS).

A. Moonis et al. (Eds.): IEA/AIE 2014, Part I, LNAI 8481, pp. 450–459, 2014.
© Springer International Publishing Switzerland 2014

In [3] is introduced a set of formalisms to deal with the modeling of aspects of group dynamics such as interactions among group members, and agreements over responsibilities of each member related to development of certain conceptual objects. The set of interaction modeling formalisms among group members within a virtual collaborative work space is briefly describe as follows:

- *Table Concept-Category-Definition*: Its function is to represent the factual knowledge of the conceptual model of group dynamics. This table introduces, in lexicographic order, the concepts that are going to be used in other formalisms specifying the category and giving the concept definition. There are three categories: actor (person), interaction and object.

- *Interaction Cases and Interaction Group Diagrams*: The modeling of the interactions among actors is made using two formalisms: [a] Interaction Cases and [b] Interaction Group Diagrams. An Interaction Case captures interactions between two actors. In particular, the reflection is a case of interaction of an actor with himself. An Interaction Group Diagram provides, in an integrated way, interactions among all actors considered in the modeling process.

- *Interaction Procedures*: The procedures describe the composition of interactions among the actors made for the development of an object. To express the procedures that actors can perform on the objects, is proposed to use predicates of order N.

- *Sequence Diagram of Group Dynamics*: It is used to express the group dynamics among the actors in the timeline imposed by the procedures of interaction. The formalism is called the Sequence Diagram of Group Dynamics.

- *Diagram of Conceptual Object Development*: Virtual spaces dedicated to collaborative work are intended to facilitate mediation inside teams whose members are not physically contiguous, and have to develop a conceptual object (for example: research, project development, software, thesis plan, technical articles, reports, among others). The modeling of interactions in virtual spaces dedicated to collaborative work should help to specify the interactions among group members, and the developing work stages of the conceptual object that the collaborative working team is carrying on. The virtual space for collaborative work must satisfy the requirement of keeping and documenting the different versions of the conceptual object that is being developed by the collaborative working team; leaving a record of the evolution from the agreement among the members of the working group since initial specifications of the conceptual object until its final stage development. These diagrams are digraphs with two types of nodes: the "conceptual objects" which will be denoted with circles and the "transformations" that will be denoted by rectangles. The "transformation" represents the action that must to be performed to make evolve the "conceptual object" from a level of development into another.

This paper is structured as follows: the definition of the problem of customizable virtual working spaces is presented (section 2), a design process for this type of spaces is proposed (section 3), a concept proof to illustrate the application of the design process is given (section 4), and preliminary conclusions and future research work are presented (section 5).

2   Definition of the Problem

Several authors [4-8] have pointed out that the current state of conceptual modeling of work group is characterized by the following limitations:

- Lack of conceptual models that adequately specify the interactions related to the development of group activities supported by virtual workspaces.

- Lack of processes that allow deriving the architecture of the virtual space designed for the particular needs of a workgroup, from conceptual models which specify the interactions among its members.

With regard to the first limitation, since 2009, authors have been working on tools for interaction modeling among persons and analysis and design of virtual working spaces [3,9]. Regarding the second limitation, in this paper is proposed a preliminary solution to the problem of defining a design process for customizable virtual workspace, with emphasis on identifying the components of its architecture.

3   Proposal of Design Process

We propose a Design Process of Customizable Virtual Working Spaces (CVWS) defined by two phases: Conceptualization Phase of CVWS and Modelling Phase of CVWS.

The Conceptualization Phase of CVWS has the goal of transforming a description of the activities within workspace (emphasizing necessary interactions among the members to deal with the tasks), into the interaction modelling formalisms previously indicated (see section 1). Two activities are performed: "Conceptualization of interactions" and "Specification of CVWS functionalities". The activity of "Conceptualization of Interactions" has as input the description of the workspace and generates as output a conceptual description formalized through artefacts: Table Category-Concept-Definition, Interaction Cases and Interaction Group Diagrams, Interaction Procedures, Sequence Diagram of Group Dynamics, and Diagram of Development of Conceptual Object. The activity of "Specification of CVWS Functionalities" has as input the conceptual artefacts that give a formalized description of the workspace, and generates as output the list of functionalities that CVWS has to support.

The Modelling Phase of CVWS has the goal of deriving the architecture of CVWS from the modelling formalisms obtained in the first phase. Two activities are performed: "Component Selection of CVWS" and "Modelling the architecture of CVWS". The activity of "Component Selection of CVWS" takes as input the list of functionalities identified in the previous activity and generates as output a list of components of CVWS. The activity of "Modelling the Architecture of CVWS" takes as input the list of components of the artefacts of interaction modelling, and generates as output architectural model of CVWS.

The design process [10] of customizable virtual working spaces (CVWS) is summarized in Figure 1.

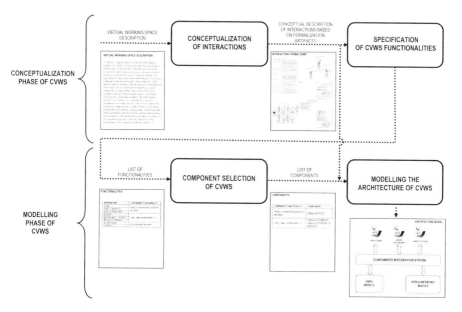

Fig. 1. Design Process of Customizable Virtual Working Spaces

4   Concept Proof

To illustrate the proposed Design Process is provided a proof of concept based on a case brought in [7]. The situation described in the case is based on developed interactions within a virtual space during the thesis plan review of a master´s degree student made by a PhD degree student (co-director of the master's thesis) under supervision of a senior researcher (director of the master's thesis and doctoral´s thesis). The case "Review of Master's Thesis Plan" is described in the following bit of text:

"...Master's degree student sends the PhD degree student, his master's thesis plan developed previously. PhD degree student reviews the plan and made the corrections and comments that he considers relevant and then send them to master´s degree student. He appropriates the corrections and comments to continue working on his master's thesis plan. Once the PhD degree student believes that the version of the master's thesis plan has not problems, forward it to senior researcher asking for his overseeing of the final version of master´s thesis plan. Senior researcher oversees the corrections made by the PhD degree student. As a result of overseeing, he can send comments which may include observations about the

correction made and/or to make further corrections to be introduced in master's thesis plan. Upon receiving these comments, the PhD degree student appropriates these and forwards them to master's degree student for his appropriating also, allowing in this way the generation of new versions of the document ..."

4.1   Conceptualization Phase of CVWS

In this section is presented the results of activity "Conceptualization of interactions" (section 5.1.1) and activity "Specification of CVWS functionalities" (section 5.1.2) for case "Review of Master's Thesis Plan".

4.1.1   Activity: Conceptualization of Interactions

In the proposed case are identified: three actors, an object, eight interactions. These are shown in Table Category-Concept-Definition illustrated in Table 1.

Table 1. Table Concept-Category-Definition of case "Review of Master's Thesis Plan"

Concept	Category	Definition
INCORPORATE	INTERACTION	Actor "A" incorporates the received information in the document and / or comments in it.
PhD STUDENT	ACTOR	Professional who has a master degree or academic equivalent and is making a career of doctoral degrees, scientific production of national importance, with a history of co-management of R&D, with expertise in co-management of in human resources training at level of master degree, specialization degree, and accreditation of being investigator category III or IV of the Argentine Ministry of Education.
SEND	INTERACTION	Actor "A" sends to actor "B" a document or information.
SEND COMMENTS	INTERACTION	Actor "A" sends Actor "B" the comments on the results of overseeing carried out, this may include observations about the correction made and/or further corrections to make.
SEND CORRECTION	INTERACTION	Actor" A" sends to actor "B" the result of the review and correction of the document with its observations.
SENIOR RESEARCHER	ACTOR	Professional with a PhD degree or academic equivalent, with scientific production of international importance, with background in project management of R & D, with background in human resources training at the doctoral level, master degree, and grade, and accreditation of being investigator category I or II of the Argentine Ministry of Education.
MASTER STUDENT	ACTOR	Professional with grade title and who is making a master degree, with national scientific production, with a history of collaboration in the development of human resources at grade level, and accreditation of being investigator category IV or V of the Argentine Ministry of Education.
THESIS PLAN	OBJECT	Document referred to student´s research project who is carrying out to earn a PhD, master's, specialty or grade degree.
REVIEW	INTERACTION	The actor reviews the document and states his comments (in case needed) but without doing any correction.
REVIEW AND CORRECT	INTERACTION	The actor revises and corrects the document with indication of his comments and corrections (if it was necessary).
REQUEST OVERSEE	INTERACTION	Actor "A" asks oversee of review / corrections on a document generated by a third actor. Overseeing will be made by actor "B".
OVERSEE	INTERACTION	Actor "A" oversees the reviews or corrections made by an actor "B" on a document that has been sent previously to him by a third actor.

From persons and interactions introduced in Table Category-Concept-Definition, interaction cases that are part of the group interaction diagram which is shown in Figure 2 are identified.

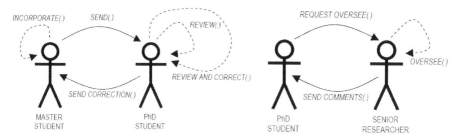

Fig. 2a. Interaction case between Master Student and PhD Student

Fig. 2b. Interaction case between PhD Student and Senior Researcher

From persons and interactions introduced in Table Category-Concept-Definition, interaction cases that are part of the group interaction diagram which is shown in Figure 3 are identified.

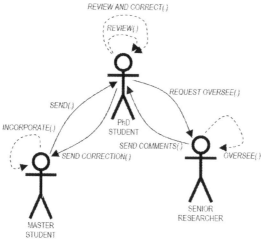

Fig. 3. Group interaction diagram among Master Student, PhD Student and Senior Researcher

The group dynamics developed among actors through the timeline, expressed through the interactions identified in the case of concept proof, is shown in the Sequence Diagram of Group Dynamics in Figure 4. The conceptual object identified is "Master Thesis Plan" and the Diagram of Conceptual Object Development is shown in Figure 5.

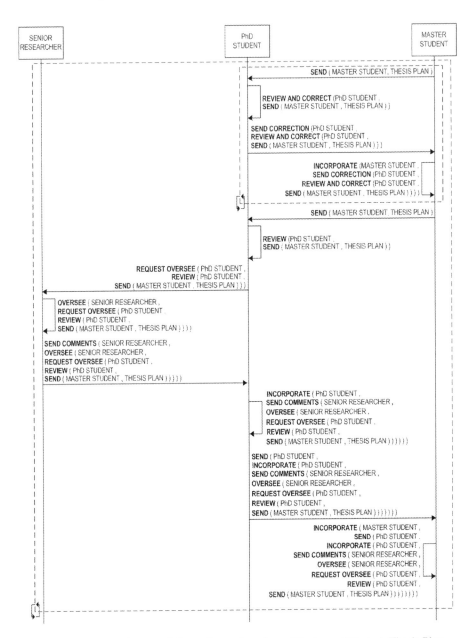

Fig. 4. Sequence Diagram of Group Dynamics of case "Review of Master's Thesis Plan

4.1.2 Activity: Specification of CVWS Functionalities

Based on the information contained in the Table Concept-Category-Definition the subset of interactions may be built and the functionalities that serve to each interaction are identified. It may happen that several interactions may be satisfied by the same functionality. For the case study, the relation interaction-functionality is presented in Table 2.

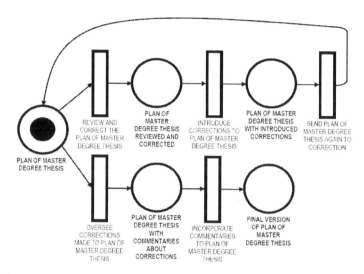

Fig. 5. Diagram of Conceptual Object Development for case "Review of Master's Thesis Plan

4.2   Modelling Phase of CVWS

In this section is presented the results of activity "Component Selection of CVWS" (section 5.2.1) and activity "Modelling the architecture of CVWS" (section 5.2.2) for case "Review of Master's Thesis Plan".

4.2.1   Activity: Component Selection of CVWS

Based on the results in Table 2, components that give satisfaction to each functionality, are identified. For the concept proof, the relation Functionality-Component is presented in Table 3.

Table 2. Relation Interaction-Functionality

INTERACTION	FUNCTIONALITY
INCORPORATE	No component required
REVIEW AND CORRECT	
REVIEW	
SEND	Ability to transmit documents in real time
SEND COMMENTS	
SEND CORRECTION	
REQUEST OVERSEE	Carry video conferences 1-1
OVERSEE	

Table 3. Relation Functionality-Component

COMPONENT FUNCTIONALITY	COMPONENT
Ability to transmit documents in real time	EMAIL MODULE
Carry video conferences 1-1	WEB-CONFERENCE MODULE PERSON TO PERSON

4.2.2   Activity: Modelling the Architecture of CVWS

The Architecture of Virtual Working Space is modelled based on the results in Table Relation Component-Functionality, Interaction Group Diagrams, Sequence Diagram of Group Dynamics, and Diagram of Development of Conceptual Objects.

The description of the case shows that it is not necessary that the "Master Thesis Plan" object needs to be in the virtual working space. This object may be shared through e-mail module and web-conference module. The selected modules are integrated through the "components integration system". For the concept proof, the model architecture of customized virtual working space is presented in Figure 6.

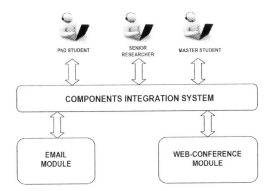

Fig. 6. Model architecture of customized virtual working space of the concept proof

5   Conclusions

Work in groups is one of the usual labour strategies that may be mediated by Internet technology. Virtual workspaces arise as a possibility to establish working groups in which persons are not physically contiguous or have difficulty to share the same real space.

In this context, this paper presents a design process for customizable virtual working spaces that require to be strictly adjusted to the needs defined by the nature of task developed by the work group.

The proposed design process, which falls within the type of production processes by project, allows the design of the virtual space architecture in which the virtual work will take place. This design is based on the formalization of the interactions among the members of the working group.

To consolidate the results presented in this paper, the following research works have been started up [11]:

[a] The refinement of the specification of the procedure steps of derivation to obtain the model architecture of CVWS from the modelling formalisms of group members interactions and tasks.

[b] The development of a prototype configuration of CVWS component-based and a prototype tool to support the process of formalizing interactions.

[c] The development a working environment that integrates the developed prototypes.

[d] Explore the validity of the Design Process of Customizable Virtual Working Spaces proposed in this paper in the following cases: (i) CVWS for Arquitects team working in building design, and (ii) CVWS for Software Engineers team working in software development.

Acknowledgments. The research reported in this paper was partially funded by Project UNLa-33A166 of the Secretary of Science and Technology of National University of Lanus (Argentina), and sponsored by the Department of Research and Development of Staffing IT Software & Services.

References

1. Grudin, J.: Computer-Supported Cooperative Work: History and Focus. IEEE Computer 27(5), 19–26 (1994)
2. Molina, A., Redondo, M., Ortega, M.: A Review of Notations for Conceptual Modeling of Groupware Systems. In: Macías, J., Granollers, A., Latorre, P. (eds.) New Trends on Human-Computer Interaction, pp. 1–12 (2009) ISBN 978-1-84882-351-8
3. Rodriguez, D., Ramon Garcia-Martinez, R.: A Proposal of Interaction Modelling Formalisms in Virtual Collaborative Work Spaces. Lecture Notes on Software Engineering 2(1), 76–80 (2014) ISSN 2301-3559
4. Giraldo, W., Molina, A., Collazos, C., Ortega, M., Redondo, M.: Taxonomy for Integrating Models in the Development of Interactive Groupware Systems. Journal of Universal Computer Science 14(19), 3142–3159 (2008) ISSN 0948-695X
5. Molina, A., Redondo, M., Ortega, M.: Evolution of an E-Learning Environment Based on Desktop Computer to Ubiquitous Computing. In: Proceedings of 34th ASEE/IEEE Frontiers in Education Conference (2004)
6. Molina, A., Redondo, M., Ortega, M.: A System to Support Asynchronous Collaborative Learning Tasks Using PDAs. Journal of Universal Computer Science 11(9), 1543–1554 (2005) ISSN 0948-695X
7. Molina, A., Redondo, M., Ortega, M., Hoppe, U.: CIAM: A Methodology for the Development of Groupware User Interfaces. Journal of Universal Computer Science 14(9), 1435–1446 (2008) ISSN 0948-695X
8. Molina, A., Redondo, M., Ortega, M.: A Review of Notations for Conceptual Modeling of Groupware Systems. In: Macías, J., Granollers, A., Latorre, P. (eds.) New Trends on Human-Computer Interaction, pp. 1–12 (2009) ISBN 978-1-84882-351-8
9. Rodríguez, D., Bertone, R., García-Martínez, R.: Collaborative Research Training Based on Virtual Spaces. In: Reynolds, N., Turcsányi-Szabó, M. (eds.) KCKS 2010. IFIP AICT, vol. 324, pp. 344–353. Springer, Heidelberg (2010)
10. Curtis, B., Kellner, M., Over, J.: Process Modelling. Communications of the ACM 35(9), 75–90 (1992)
11. Fields, B., Merrian, N., Dearden, A.: DMVIS: Design, Modelling and Validation of Interactive Systems. In: Design, Specification and Verification of Interactive Systems. Springer (1997)

Develop and Evaluate the Mobile-Based Self-Management Application for Tele-Healthcare Service

Hao-Yun Kao[1], Yi-Ting Cheng[1], and Yi-Kuang Chien[2,*]

[1] Department of Healthcare Administration and Medical Informatics,
Kaohsiung Medical Universi-ty, 100, Shih-Chuan 1st Road, Kaohsiung, 80708, Taiwan, ROC
[2] Tele-healthcare Center, Kaohsiung Medical University hospital, 100, Shih-Chuan 1st Road, Kaohsiung, 80708, Taiwan, ROC

Abstract. The problem of the aging population and its likely impact on the provision of healthcare systems has been increasing. This study presents a conceptual model to explore and investigate the Mobile-based technology and the business model in enhancing chronic disease care performance. Some scales to measure Mobile-based technology and the business model were developed and validated; then they were used to collect survey data from respondents who had experience. In addition, this study has verified the entrepreneurial alertness as the moderator and influenced the relationship between Mobile-based technology and the business model. The empirical results support the proposed model and can potentially be used in advance of Mobile-based technology adoption for contemporary hospitals using the proper business model, and sensing the change for chronic diseases care. Among key managerial implications, hospital administrators must focus on alerting environmental change then creating reorganized capacities for Mobile-based technology implementations to ensure better care performance for chronic diseases.

Keywords: Mobile-based Technology, Business Model, Performance, Healthcare.

1   Introduction

An aging population structure leads to a considerable demand for healthcare [1], in turn, continued advances in science and technology and general improvements in environmental and social conditions have increased life expectancy around the world. As a result, the world's population is aging. The problem of the aging population and its likely impact on the provision of health care systems has been increasing. With an aging population, the proportion of patients needing complex care for one or more chronic illnesses has increased [2]. Thus, it is important to integrate medical care resources to achieve resource sharing. Hospitals aren't only trying to prevent diseases reaching critical conditions rapidly, but also need to manage efficiently and effectively.

The application of real-time Tele-healthcare has been focused on home-based health monitoring, which is an extension of in-hospital services via information

---

* Corresponding author.

A. Moonis et al. (Eds.): IEA/AIE 2014, Part I, LNAI 8481, pp. 460–469, 2014.

communication technology (ICT) [3]. Care devices based on ICT include real-time visual and auditory contacts and relationships between caregivers and users [4]. Dramatic increases in the numbers of chronically ill patients in the face of shrinking provider numbers and significant cost pressures mean that a fundamental change is required in the process of care. Organizational and societal changes, such as cost reduction policies and an aging population, are the main driving forces for the development of Tele-healthcare [5]. We need to identify patient management approaches that would ensure appropriate home-based health monitoring and treatment of patients while improving the care performance involved in the process.

Novel methods need to be developed if health care systems are to manage the increasing level and complexity of diseases. The development of a real-time monitoring health care service model with the intervention of ICT has become a research priority [6]. Mobile-based technology can now overcome these challenges. Users in rural areas are increasing rapidly because of the availability and affordability of the technology. Some of these models establish conditions that foster the adoption of enabling technologies, including remote patient management technology (RPMT), and adoption rates are now beginning to accelerate [7,8].

Meanwhile, a body of research has been under way to develop better chronic illness care through the service model. The service model had been developed from academic medical center in Taiwan on how to develop and evolve chronic disease management, which has been designed to improve outcomes and reduce spending in long-term care where there are still limited understandings [5]. As an affordable and accessible means of communication, rural communities are realizing the potential of mobile technology to create economic opportunities and strengthen social networks. Mobile technology effectively reduces the "distance" between individuals and expert caregivers, making the sharing of information and knowledge easier and more effective. It is hoped that the development of this Android-based application can be widely implemented in the near future.

This study describes the development of an android-based self-management application (I-health App) based on a novel developmental methodology, namely the design science research methodology (DSRM). This method consists of six major processes: identify problem and motivation, define solution objectives, design and development, demonstration, evaluation, and communication [9-11]. Patients are generally satisfied with Tele-healthcare, but they prefer a combination of Tele-healthcare with conventional health care service [5]. Consequently, this research extends understanding through increasingly important issue which applied mobile technology to enhance the benefits in a service model [12]. The study integrates the various strands of research and provides a common ground from which further work can proceed. We developed and tested the systems to answer the above question that based on design science methodology from an Academic Medical Centre and its affiliated facilities in Taiwan.

The next section describes the theoretical foundations and the re-search hypotheses in this research. Then, we describe our research method and the process through which we gathered data. Finally, we present the analyses and results of our research and discuss their implications for future research and practice.

2   Methodology

2.1   Design Methodology

In recent years, several researchers succeeded in bringing design science (DS) into the IS research community, successfully making the case for the validity and value of design science as an IS research paradigm and actually integrating design as a major component of research [13]. The design science research methodology is rooted from engineering and the sciences of the artificial [14]. It is a problem-solving guideline that aims at creating innovations with effective and efficient analysis, design, implementation, management, and use of information systems [15]. Hevner et al. (2004) argues that any de-sign artifact should rely upon the application of rigorous methods in its construction. IS researchers have not focused significantly on the development of a consensus process and mental model for DS research [10].

Design science research methodology (DSRM) was developed in engineering [16,17], with Eekels and Roozenburg (1991) raising the need for a common DSRM [18]. Archer's methodology focuses on one kind of DS research, with building system instantiations as the research outcome [19]. Archer believed that design could be codified, even its creative aspects, and his industrial engineering research outcomes reflect his views on re-search methodology. His work included a purpose-oriented designs for hospital beds and for the mechanisms that prevented fire doors from being propped open.

Following this direction, several researchers have succeeded in bringing design research in the information systems (IS) research community, successfully making the case for the validity and value of design science (DS) as an IS research paradigm [10,20,17] and actually integrating design as a major component of research [21]. The DSRM presented here incorporates principles, practices, and procedures required to carry out such research and meets three objectives: it is consistent with the prior literature, it provides a nominal process model for doing DS research, and it pro-vides a mental model for presenting and evaluating DS research in IS [13]. The DS process includes six activities: problem identification and motivation, definition of the objectives of a solution, design and development, demonstration, evaluation, and communication.

2.2   The Service Model

This research project was carried out from January, 2012 to August, 2013. It is methodologically based on the project development life cycle, and an adoption of the waterfall model. Regarding the healthcare framework, we use home self-care as the basic unit of tele-healthcare, and the healthcare professionals would bridge the six kinds of services in this model, Including (1) 24-hour real-time health status tracing and monitoring (2) Emergency care referral and health consultation (3) Home visits (4) Return visit scheduling (5) Prescriptions delivery (6) Social welfare services application. Once the services model is mature, the expansion of the equipment measuring physiological information will be considered. The recorded physiological

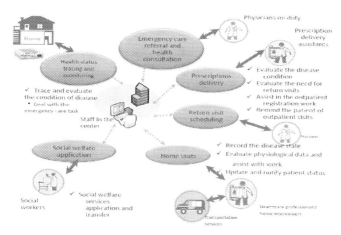

Fig. 1. Service model in Academic Medical Center

information of each patient will be turned into a continuous report and be sent via the information sys-tem (IS) to healthcare professionals for monitoring purposes. A simplified service model is displayed in Figure 1.

2.3   System Architecture and Software Development

The implemented mobile Tele-healthcare architecture includes: a home box connected to BP device, an Android OS Smartphone associated with pervasive computing, human computing interfacing in Smartphone and a cloud-based Tele-healthcare information system.

The system is able to perform data monitoring and management of patient vital signs and daily activity, materializing an effective inter-face between clinical staff and remotely assisted patients. The implemented system is presented in Figure 3 including Bluetooth and wireless internet protocols associated with data communication. The Smartphone materializes the main computing platform and data storage associated with the assessed patient. A distributed data processing is implemented in the system. The primary processing, including the physiological parameters calculation, is performed on the connected BP device on home box. The intermediate processing and data representation are performed at the Smartphone level, while the advanced data processing and database management is per-formed by the cloud-based Tele-healthcare information system.

This application is designed with the Android software stack produced by Google. Android is an open source framework designed for mobile devices. It packages an operating system, middleware, and key programs [22]. The Android SDK provides libraries needed to interface with the hardware at a higher level and make/deploy An-droid applications [23]. The application is written in Java and use SQL databases to store persistent data. We choose this platform as opposed to others because of the ability to easily thread background running processes and compatibility with other Android devices.

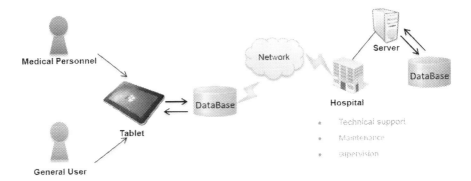

Fig. 2. System Architecture of mobile-based self-management application

2.4   Requirement Modelling and Implements

The objective of this stage is to acquire and model the Tele-healthcare activities (i.e., the work flow) and its associated knowledge. The steps included acquisition and classification. There-fore, we utilized the unified model language (UML) approach to capture and organize requirement via interview with providers. Finally, the knowledge of Tele-healthcare was generalized from two approaches: use case and activity diagram in modeling as shown in figure 4. The use case is valuable in goal modeling. A goal-based use-case approach was proposed as a way to extend the use-case approach [24]. The activity diagram is used to model the activity flow controlled by conditions and connections. A drawing represents the input and output visual layout of an activity. The data glossary is used to describe the activity data information [25].

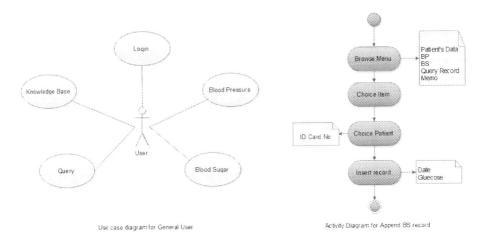

Fig. 3. Use case and Activity Diagram

An activity diagram represented using UML is commonly used today in modeling business process flow [26]. It includes several elements: Activity, Start Activity, End Activity, Transition, Fork, and Branch, Merge, and Join. The first three elements encompass the activity. The fourth element represents the connection. Merge and Join rep-resent the precondition. Fork and Branch represent the post condition. Although the activity diagram can easily represent a scenario, it cannot effectively represent detailed input and output information for each activity. Therefore, to support this, other tools must be added. A drawing is a way of effectively expressing input and output information, such as the title, presentation position, lines, figures, and tables; these are widely used in systems analysis and design. One data glossary record format might contain the Data_type, Origin_type, Source_field, and Computing_rules that describe the data fields that make up an activity.

3   Results

This study follows the above DSRM activities to develop and implement an application for self-management of a tele-healthcare service model in southern Taiwan.

3.1   Problem Identification and Motivation

Motivated to secure the advantages of self-management application, Academic medical center sought to implement a database to collect and support the development of tele-healthcare service model. The tele-healthcare center can use self-management information systems to improve quality of care, user satisfaction, and operational efficiencies. The application allows users to collect and represent data from devices for self-management. Therefore, it was designed to collect information regarding the progress of health management processes, providing significant benefits for users and caregivers.

3.2   Define the Solution Objectives

The objective of the project was to develop an artifact referred to as the android-based self-management application based on a novel developmental methodology, called the design science research methodology (DSRM). The major challenges to implementation included the collect the vital sign from devices through blue tooth, the diversity of objectives for which reports were generated, and the need to conform to the requirement of users and call center. The application provides a rich environment to promote the improvement of self-management capabilities, with the long-term goal of monitoring and improving the effectiveness of self-health management processes.

3.3   Design and Development

Figure 3 illustrates the system architecture, which consists mainly of data sources, devices and tools. The system was developed using Android SDK, and connected to devices via BT or wireless technology. Given these data sources, the application had developed which based on UML system analysis methodology and examples as shown in Fig. 4-7. These included the use case diagrams and activity diagrams for developing the user interface.

Finally, the application had developed with maximizing usability. The user interface to the users and caregivers was designed to present graphs and tables. This interface was chosen to assist users with presenting and reporting the results of queries in a simplified viewing format as shown in figure 5. In addition, discrepancies in data combinations are highlighted to instantly alert users and caregivers of abnormal indicators and needed to notice in self-management.

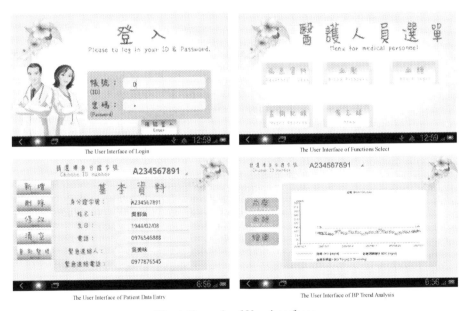

Fig. 4. Example of User interfaces

3.4   Usability Testing

We conducted the usability testing to collect the feedback from participating users. According to international standards, usability refers to effectiveness and efficiency to achieve specified goals and user satisfaction. Usability testing is typically conducted as a standard method for usefulness. Nielsen proposes five attributes of usability: easy to learn, efficient to use, easy to remember, minimal errors, and pleasant to use. In this study, we use completeness, correctness (i.e., Effectiveness), and usefulness as the testing criteria. The objective of usability testing is to explore the strengths and weaknesses of the application, and to improve usability. This study adopted the

Post-Study System Usability Questionnaire (PSSUQ) proposed by Lewis (1995) to assess the usability and satisfaction for information system use [27]. The contents of PSSUQ were constructed to measure to usefulness, ease of learning, information quality, and interface quality, with the evaluation constructs listed in Table 2. The PSSUQ survey was used to obtain feedback from users who had completed the specific tasks. The questions were answered using a scale from 1 to 7 (1 – Strongly Dis-agree to 7 – Strongly Agree).

Usability testing reflects the five constructs of system usefulness, ease of learning, information quality, interface quality, and overall satisfaction. I-health App was found to assist users in collecting and monitoring indicators, thus increasing work efficiency. The results show a clear need to collect and integrate data across the different departments. For example, using report patterns applied with the advantages of I-health APP. The results also indicate that ease of use and interface usability are critical factors in improving work efficiency and streamlining the debugging process.

Table 1. PSSUQ measurement items and results

Measurement	Results (Mean ± SD)
System usefulness	5.13±1.58
Ease of learning	5.45±1.41
Information quality	5.11±1.45
Interface quality	5.35±1.24
Overall satisfaction	5.48±1.52

4   Discussion and Conclusions

This study proposed an information system which based on the de-sign science perspective through a rigorous system analysis and related methodology. Our study results not only support the previous findings on the effectiveness of mobile-based services but develop an application to enhance the results. The study possesses several notable strengths as follows:

First, mobile-based technologies offers a means of making care more affordable, which has been shown to support patient self-management, co-operate with caregivers, shift responsibilities for nonclinical providers, and reduce the use of emergency services for hospitals. Because transformative technologies (e.g. mobile-based) offers major opportunities to advance national goals of improved quality and efficiency in health care, it is important to understand their evolution, the experiences of early adopters, and the proper business models that may support their deployment.

Second, the service model could effectively monitor the risk of dis-ease occurrence when patients were at home and provide an emergency transfer instantly. Moreover, compared with traditional care service models where only passive information was provided, we provided new services and its application in which healthcare professionals could intervene by mobile-based technology.

In this study, we provided an application they base on IT-intervened service model, which could extensively improve the efficiency and effectiveness of patient management in the future. The resulting mobile-based technology and evolving new development tools (e.g. Android-based Tablet and Smartphone) can then guide the relevant activity in relation to develop information systems. The healthcare professional can also take advantage of such a profile to assist in clinical decisions by evaluating the levels and development needs.

From the viewpoint of managerial implications, our findings have important implications for users involved in efforts to introduce mobile-based application into their life. The development and setup of the mobile-based platform are the core of this service model. In the future, with the application in the proper service model, the integration of the available healthcare services will become easier, and the elasticity and efficiency of innovative services will increase. While numerous advocates have pre-scribed such a collective responsibility as a normative guideline, our research provides empirical support for this prescription.

The findings are consistent with a previous study which proposed "Chronic Care Model". Our work provides judicious knowledge to researchers and practitioners interested in learning how hospitals facilitate more effective mobile-based technology adoption to today's e-health environments. We hope that our design science perspective and findings will stimulate and encourage more research into this field.

Acknowledgement. This research is supported by Grant from National Science Council (NSC102-2815-C-037-008-E) and Kaohsiung Medical University affiliated with National Sun Yat-sen University (NSYSUKMU 103-P030)

References

1. Reinhardt, U.E.: Does the aging of the population really drive the demand for health care? Health Aff (Millwood) 22(6), 27–39 (2003)
2. Larson, E.B., Reid, R.: The patient-centered medical home movement. JAMA: the Journal of the American Medical Association 303(16), 1644–1645 (2010)
3. Yousef, J., Lars, A.: Validation of a real-time wireless telemedicine system, using bluetooth protocol and a mobile phone, for remote monitoring patient in medical practice. Eur. J. Med. Res. 10(6), 254–262 (2005)
4. Huang, J.-C.: Innovative health care delivery system—A questionnaire survey to evaluate the influence of behavioral factors on individuals' acceptance of telecare. Comput. Biol. Med. (2013)
5. Botsis, T., Demiris, G., Pedersen, S., Hartvigsen, G.: Home telecare technologies for the elderly. J. Telemed. Telecare 14(7), 333–337 (2008)
6. Clarke, M., Thiyagarajan, C.A.: A systematic review of technical evaluation in telemedicine systems. Telemed e-Health 14(2), 170–183 (2008)
7. Coye, M.J., Haselkorn, A., DeMello, S.: Remote patient management: technology-enabled innovation and evolving business models for chronic disease care. Health Aff (Millwood) 28(1), 126–135 (2009)
8. Singh, R., Mathiassen, L., Stachura, M.E., Astapova, E.V.: Dynamic capabilities in home health: IT-enabled transformation of post-acute care. Journal of the Association for Information Systems 12(2), 2 (2011)

9. Hevner, A.R.: The three cycle view of design science research. Scandinavian Journal of Information Systems 19(2), 87 (2007)
10. Hevner, A.R., March, S.T., Park, J., Ram, S.: Design science in Information Systems research. Mis. Quart. 28(1), 75–105 (2004)
11. March, S.T., Storey, V.C.: Design Science in the Information Systems Discipline: An Introduction to the Special Issue on Design Science Research. Mis. Quart. 32(4), 725–730 (2008)
12. Paré, G., Jaana, M., Sicotte, C.: Systematic review of home telemonitoring for chronic diseases: the evidence base. Journal of the American Medical Informatics Association 14(3), 269–277 (2007)
13. Peffers, K., Tuunanen, T., Rothenberger, M.A., Chatterjee, S.: A design science research methodology for Information Systems Research. J. Manage Inform. Syst. 24(3), 45–77 (2007), doi:10.2753/Mis0742-1222240302
14. Simon, H.A.: The sciences of the artificial. MIT Press (1996)
15. Deming, D.: Design, science and naturalism. Earth-Sci. Rev. 90(1-2), 49–70 (2008), doi:10.1016/j.earscirev.2008.07.001
16. Hoffman, R.R., Roesler, A., Moon, B.M.: What is design in the context of human-centered computing? IEEE Intelligent Systems 19(4), 89–95 (2004)
17. Walls, J.G., Widmeyer, G.R., El Sawy, O.A.: Building an information system design theory for vigilant EIS. Information Systems Research 3(1), 36–59 (1992)
18. Eekels, J., Roozenburg, N.F.M.: A methodological comparison of the structures of scientific research and engineering design: their similarities and differences. Design Studies 12(4), 197–203 (1991), doi:10.1016/0142-694x(91)90031-q
19. Archer, L.: Systematic method for designers. In: Cross, N. (ed.) Developments in Design Methodology, pp. 57–82. Wiley & Sons, Chichester (1984)
20. March, S.T., Smith, G.F.: Design and natural science research on information technology. Decision Support Systems 15(4), 251–266 (1995)
21. Nunamaker, J.F., Chen, M., Purdin, T.D.M.: Systems Development in Information Systems Research. J. Manage. Inform. Syst. 7(3), 89–106 (1990)
22. Developers, A: What is android? (2011), http://developerandroid.com/guide/basics/what-is-android.html2
23. Rogers, R., Lombardo, J., Mednieks, Z., Meike, B.: Android application development: Programming with the Google SDK. O'Reilly Media, Inc. (2009)
24. Lee, J., Xue, N.-L.: Analyzing user requirements by use cases: A goal-driven approach. IEEE Software 16(4), 92–101 (1999)
25. Wu, J.-H., Shin, S.-S., Heng, M.S.: A methodology for ERP misfit analysis. Inform. Manage-Amster. 44(8), 666–680 (2007)
26. Schambach, T.P., Walstrom, K.A.: Systems development practices: Circa 2001. Journal of Computer Information Systems 43(2), 87–92 (2003)
27. Lewis, J.R.: IBM computer usability satisfaction questionnaires: Psychometric evaluation and instructions for use. International Journal of Human-Computer Interaction 7(1), 57–78 (1995), doi:10.1080/10447319509526110

Multiobjective Optimization of Microchannels with Experimental Convective Heat Transfer Coefficient of Liquid Ammonia

Normah Mohd-Ghazali[1,*], Oh Jong-Taek[2], Nuyen Ba Chien[2], Kwang-Il Chi[2], Nor Atiqah Zolpakar[1], and Robiah Ahmad[3]

[1] Faculty of Mechanical Engineering, UniversitiTeknologi Malaysia
81310 Skudai, Johor Bahru, Malaysia
[2] Dept. Refrigeration & Air Conditioning Engineering, Chonnam National Univ., San 96-1, Dunduk-Dong, Yeosu, Chonnam 550-749, Republic of Korea
[3] UTM Razak School of Engineering and Advanced Technology, UTM KualaLumpur
54100 JalanSemarak, Kuala Lumpur, Malaysia
normah@fkm.utm.my

Abstract. A multi-objective optimization on a system of microchannels with environmentally friendly liquid ammonia is presented. Further, comparative studies were done on two approaches of obtaining the convective heat transfer coefficient necessary for the procedure; from the conventional Nusselt number correlation and that from experimentally measured data. The thermohydrodynamic performance of the coolant agrees well with theory with a higher resistance associated with the experimentally obtained data due to overall contributions from experimental apparatus generally not considered in mathematical representations of actual processes. The study shows that the pairing of a fast and simple evolutionary algorithm method as MOGA with experimental data is a powerful combination when new coolants are being explored for replacements in current systems. The results would be useful in providing the trends and patterns needed to evaluate the potentials of new coolants in microchannels.

Keywords: Multi-objectiveOptimization, Microchannels, thermohydrodynamic performance, new coolants.

1 Introduction

A microchannel heat sink (MCHS) has proven its application in the micro-electro-mechanical systems (MEMS) of very-large-integrated-system (VLSI) of electronic packages [1]. The series of parallel microchannels machined above or below the substrate of electronic chips have demonstrated to be effective in removing the high heat flux generated. Increasing demand for efficient miniaturized heat sinkshas focused research into optimization in the design of the microchannelswhich can improve the performance of the system, comprehensive, as well as flexible within the constraints

---

* Corresponding author.

A. Moonis et al. (Eds.): IEA/AIE 2014, Part I, LNAI 8481, pp. 470–478, 2014.
© Springer International Publishing Switzerland 2014

given. The thermal resistance and pressure drop model has continuously been used as the performance criteria for a MCHS,both of which are conflicting to each other [2-7]. Studies to improve the performancehave included variations in the microchannel geometry, materials, coolant flow conditions and coolant types [8]. Concerns over the environmental impact of hazardous refrigerants encourage research into more environmentally friendly coolants that can perform as good as the current coolants. Correlations for the Nusselt number which subsequently gives the convective heat transfer coefficient (h) necessary to determine the convective thermal resistance have been developed for easily available coolants in a MCHS or for macro channels. Depending on the flow regime, the correlations were developed under experimentally controlled and specific conditions with a particular coolant. Many of the optimization schemes based on these correlations were then completed experimentally and/or numerically with the outcome of a local minimum/maximum of the objectives depending on the discrete values of the parameters that have been varied.The exercise is time-consuming and limited [9-12]. Khan et al. [13] overcame the complexity and time consuming issue of previous optimization techniques with genetic algorithm (GA) but their system was a water-cooled MCHS and comparisons were completed with experimental data of Tuckerman and Pease of 1981[14].Although the use of theoretical values in the optimization schemes have shown good agreement with the results using measured parameters, it is desirable to have available experimental values to represent the actual applications.

This study investigates the application of the simple multi-objective optimization algorithm (MOGA)in the optimization of the microchannel aspect ratio and the mircochannel wall to channel width ratio of rectangular microchannels with ammonia liquid as coolant. For this investigation of the comparison between the optimized output using theoretical and experimental data, the simple MOGA is believed to be adequate to showcase the importance of the usage of the latter in exploring the potential performance of new coolants being considered for a real system such as the MCHS [15]. Ammonia has zero global warming potential (GWP) and zero ozone depletion potential (ODP) which means it is environmentally friendly [16].Experimental data on liquid ammonia is scarce particularly those completed on microchannels. In the present study, the convective heat transfer coefficient is obtained through a Nusselt number correlation, developed for water with the porous medium model, and through experimental data for liquid ammonia. No known correlation for microchannels developed for ammonia is available presently. Although this optimization is performed on a MCHS specific for MEMS generally found in VLSI applications, the algorithm andresults could be used in promoting ammonia as an alternative coolant in other microchannelcooling.

2   Mathematical Model Formulations

The system of microchannels considered is schematically represented in Fig. 1. Heat flux is assumed to be generated from the bottom of the heat sink system with the top being covered by an adiabatic plate.

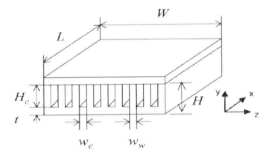

Fig. 1. Schematic of microchannels system

The total thermal resistance representing the performance of the MCHS is given by:

$$R_{tot} = R_{cond} + R_{capa} + R_{conv} \tag{1}$$

where R_{cond}, R_{capa}, and R_{conv} are the conductive, capacitive and convective thermal resistance, respectively, assuming that the resistances are in series. The heat flux is first resisted by the substrate thickness, R_{cond}, then by the coolant capacitance, R_{capa}, and finally by convective heat transfer, R_{conv}, as the coolant flows out of the MCHS.Many thermal resistance calculations are available in the literature but themethodology offered by Wen and Choo [17] is chosen as the most suitable for the system in the current study. The thermal resistance has been reformulated to result in only two independent variables and the constrictive thermal resistance omitted due to its insignificant contribution to the total resistance. The convective thermal resistance from the experimental data is given by

$$R_{conv} = \frac{1}{h_{av}} \times \frac{1+\beta}{1+2\alpha \cdot \eta} \times \frac{1}{LW} \tag{2}$$

where h_{av} is calculated from

$$h_{av} = \frac{Nu \cdot k_f}{D_h} \tag{3}$$

The values for k_f and D_h have been taken from experimental measurements and Nu is 4.36 for laminar flow in a circular microchannel.The optimization based on experimental data has the h_{av} determined directly from Eqn. (3) with k obtained from REFROP 8 using the measured temperatures of ammonia liquid along the 2-meter 3 mm internal diameter microchannel at the saturation temperature of $T_{sat} = 10°C$.K-type thermocouples were positioned at 19 locations along the test section, each location 100 cm apart. At each location, three temperature measurements were taken inside the microchannel; at the top, middle, and bottom wall. Since the test system was connected to a data acquisition system, the thermo physical properties were determined using REFROP 8 based on the average temperature at each temperature-measuring station for laminar liquid flow portion. Downstream of the long micro-channel, two-phase flow occurs due to the heat flux received by the coolant. The optimization scheme was performed for four cases for comparison; for theoretical properties with h_{av} using Eqn. (5), and values for properties from the experiments completed at three different heat fluxes; 15, 30, 60 kW/m² , denoted by experiment C, B, and A respectively. For this investigation, the $L \times W$ for the MCHS is 1×1 cm with a thickness of 0.213 mm and height of each channel fixed at 0.32 mm. A schematic of the details of the experimental rig with the circular test section is shown in Fig. 2.

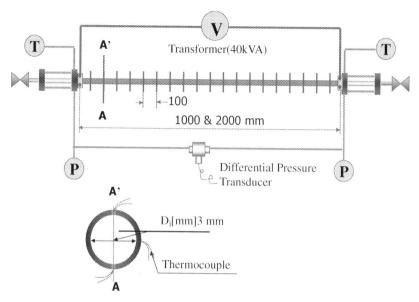

Fig. 2. Circular test section schematics

The thermal resistance due to convection from the theoretical calculations is given by Kim and Kim [18]as:

$$R_{conv} = \frac{1}{Nu \cdot k_f} \times \frac{1+\beta}{1+2\alpha \cdot \eta} \times \frac{2\alpha w_c}{1+\alpha} \tag{4}$$

where

$$Nu = 2.253 + 8.164 \left(\frac{\alpha}{\alpha+1}\right)^{1.5} \tag{5}$$

The final total thermal resistance is expressed in terms of the channel aspectratio (α) and the wall width to channel width ratio (β),

$$\alpha = \frac{H_c}{w_c} \tag{6}$$

$$\beta = \frac{w_w}{w_c} \tag{7}$$

The second objective in this optimization scheme is the minimization of the pressure drop where the heat sink pressure drop offered by Kleiner et al. [6] is modified in terms of α and β,

$$\Delta p_{tot} = f_{hs} \frac{L}{D_h} \rho_f \frac{v_{mf}^2}{2} + \left(1.79 - 2.32 \left[\frac{1}{1+\beta}\right] + 0.53 \left(\frac{1}{1+\beta}\right)^2\right) \rho_f \frac{v_{mf}^2}{2} \tag{8}$$

Generally, the pumping power is of concern, defined as,

$$Pp = \Delta p_{tot} \times G \tag{9}$$

where G is the volumetric flow rate of the coolant. Thus, Eqn. (9) is used as the second objective function instead of Eqn. (8).

3   Multi-objective Optimization of a MCHS

Pareto optimization is a multi-criteria optimization that allows for the optimization of a vector of multiple criteria. The shape of the Pareto surface indicates the nature of the trade-off between the different objective functions. These points are called non-inferior or non-dominatedpoints. The optimization of a multi-objective function can be mathematically defined as

$$\min \bar{f}(\bar{x}) = \left[f_{1(\bar{x})}, f_{2(\bar{x})}, \dots, f_{n(\bar{x})}, \right]$$

subject to

n:number of the objective functions.

$\bar{x}$: vectors contain design variables.

The multi-objective optimization of MCHS using GA is summarised in Fig. 3.

In the current study, the number of objective functions, n, is two and the design variables are also two as described by Eqn. (6) and Eqn. (7). The first and second objective functions are re-written as follows

$$f1 = R_{tot} = \left(\frac{t}{k_w} + \frac{t_{cov}}{k_{cov}}\right) + \frac{1+\beta}{\pi k_w} ln\left(\frac{1}{\sin\frac{\pi\beta}{2(1+\beta)}}\right)\alpha H_c + \frac{1}{h_{av}}\frac{1+\beta}{1+2\,\alpha\,\eta} + \frac{L}{c_{pf}\,\mu_f}\frac{2}{Re}\frac{1+\beta}{1+\alpha} \quad (10)$$

$$f2 = P_p = \Delta p_{tot} \times G \quad (11)$$

where

$$\Delta p_{tot} = f_{hs}\frac{(1+\alpha)L}{2H_c}\rho_f\frac{v_{hs}^2}{2} + \left(1.79 - 2.32\left(\frac{1}{1+\beta}\right) + 0.53\left(\frac{1}{1+\beta}\right)^2\right)\rho_f\frac{v_{hs}^2}{2} \quad (12)$$

Table 1 lists the parameters and their values used in this MOGA scheme

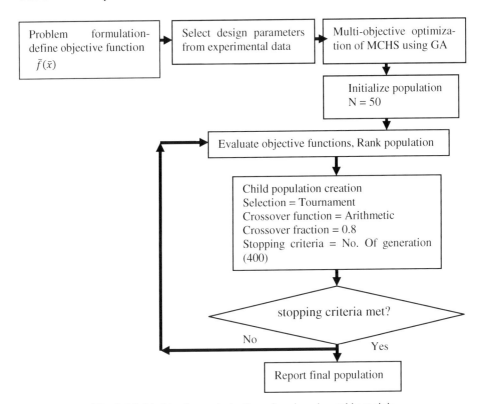

Fig. 3. Multi-objective optimization of a microchannel heat sink

Table 1. Values of parameters used in the Pareto optimization

Parameters	Theoretical	Expt. A	Expt. B	Expt. C
k_f (W/m.K)	0.5305	0.5066	0.5173	0.519
C_{pf} (J/kg.K)	4675	4727.7	4701.7	4698
μ_f (kg/m.s)	153.5×10^{-6}	141.6×10^{-6}	146.9×10^{-6}	147.8×10^{-6}
H_c (mm)	320	320	320	320
h_{av} (W/m^{2}K)	f(Nusselt)	754.32	751.882	736.24

The design limits set for the parameters to be optimized are $1<\alpha<5$ and $0.01<\beta<0.1$ with the height of each channel fixed at 320 μm. These constraints have been based on the water-cooled MCHS tested by Tuckerman and Pease [14] due to the absence of similar studies on liquid ammonia. The population size is set at 100 with selection function used is tournament. The freedom of placing arbitrary constraints with multiple objectives that are conflicting with each other using the simple evolutionary algorithm of MOGA is attractive in this investigation of the thermal performance of microchannels with the potential coolant liquid ammonia.

4   Results and Discussion

Results of the Pareto optimization is shown in Fig. 4 for both approaches of obtaining the convective heat transfer coefficients. A total of 35 optimal solutions is produced for each optimization. As expected, increasing objective1 decreases objective 2 due to their conflicting effects. The optimized conditions for h_{expC} occur over a wider range compared to that for h_{expA}. Although minimal thermal resistance is often the desired outcome, the available solutions provide flexibility for the application of a MCHS depending on which objective is of outmost priority.

Fig. 5 shows the optimized microchannel aspect ratio (α) against the thermal resistance. The patterns and trends of decreasing thermal resistance (R_{th}) with increasing α, for both approaches of obtaining the convective heat transfer coefficient are the same with the R_{th} for the experimental h having much higher values at the same optimized α. Wider channels increases the thermal resistance as the capacitance as well as the convective resistances increases. Increased channel width occurs when α increases at a fixed channel height. The optimization results agree well with theoretical expectation.

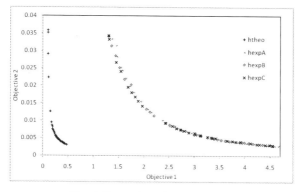

Fig. 4. Pareto front for all cases

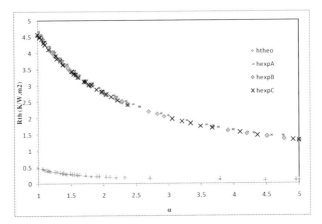

Fig. 5. Thermal resistance, R_{th}, against microchannel aspect ratio, α

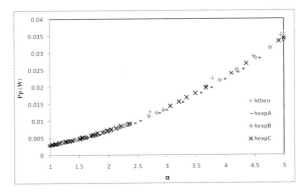

Fig. 6. Pumping power, Pp, against microchannel aspect ratio, α

The higher thermal resistance associated with the experimental data is expected since resistances during experiments come from all apparatus and the environment which have not and cannot always be represented.

Furthermore, the data for liquid ammonia, for h_{expA}, h_{expB}, and h_{expC}, were determined from experimentally measured temperatures for circular microchannelswhilst the theoretical h_{th} utilized the correlation formulated from the porous medium model with liquid water.

The effect of increasing the aspect ratio against the pumping power is shown in Fig. 6. Interestingly, at each pumping power, the optimized aspect ratio, α, is almost the same for all methods. Thus, under optimized conditions for the same pumping power, the actual minimum thermal resistance could be higher by as much as a factor of 8 over the theoretical minimum thermal resistance. The results of this study has indicated that the pairing of experimental data and the MOGA optimization, can be expected to be a powerful and reliable tool in predicting the expectation in the thermal and hydrodynamic performance of a MCHS with new coolants such as shown here with liquid ammonia. This is of course only the outcome from using just one algorithm, the simple MOGA. There could be a better algorithm yet to be investigated;

NSGA-II, GAM, and SPEA2 although MOGA should be adequate for a preliminary investigation such as the present one. A comprehensive optimization could be done with evolutionary algorithm unlike the limited optimization procedure associated with numerical simulation and repeated series of experiments. This is especially useful when new coolantsare being explored and assessed as potential replacements in the current systems.

5   Conclusion

The application of the multi-objective genetic algorithm (MOGA) on the optimization of a microchannel heat sink used in MEMS generally found in very-large-system integration (VLSI) applications has been completed. Theoretical correlation and actual experimental data have been used to look at the thermo hydrodynamic performance of environmentally friendly liquid ammonia, a coolant being considered as a replacement in microchannels as well heat exchangers of the refrigeration and air-conditioning industry. The results show the potential of the powerful and possibly necessary combination of experimental data and analytical optimization scheme such as MOGA to identify optimized parameters when there are conflicting multi-objectives such as the thermal resistance and pumping power in the study of new environmentally friendly refrigerants. Knowing what is possible under optimized conditions provides designers and manufacturers options to produce the most efficient systems.

Acknowledgments. The authors would like to thank the ASEAN University Network (AUN) and Republic of Korea (ROK) under the Exchange Fellowship Programme, the Department of Refrigeration and Air-Conditioning of the Chonnam National University of Korea and the UniversitiTeknologi Malaysia (UTM) for the grant and assistance throughout the course of this research.

References

1. Kandlikar, S.G., Grande, W.J.: Evolution of microchannel flow passages-thermohydraulic performance and fabricationtechnology. Heat Transfer Eng. (24), 3–17 (2003)
2. Kleiner, M.B., Kuhn, S.A., Haberger, K.: High performance forced air cooling scheme employing microchannel heat exchangers. IEEE Transactions on Components, Packaging, and Manufacturing Technology, Part A 20(4), 795–804 (1995)
3. Zhimin, W., Fah, C.K.: Optimum thermal design of microchannel heat sinks. In: Proceeding of the 1997 1st Electronic Packaging Technology Conference, EPTC, Singapore, pp. 123–129 (1997)
4. Iyengar, M., Garimella, S.: Design and optimization of microchannel cooling systems. IEEE Transactions, 54–62 (2006)
5. Hu, G., Xu, S.: Optimization design of microchannel heat sink based on SQP Method and numerical simulation. In: Proceeding of 2009 IEEE International Conference on Applied Superconductivity and Electromagnetic Devices, China, pp. 89–92 (2009)

6. Escher, W., Brunschwiler, T., Shalkevich, N., Shalkevich, A., Burgi, T., Michel, B., Poulikakos, D.: On the cooling of electronics with nanofluids. Journal of Heat Transfer (133), 1–11 (2011)
7. I.,S.R.: Nanofluid as a coolant for electronic devices (cooling of electronic devices). International Journal of Thermal Sciences (32), 76–82 (2012)
8. Ahmed, M.A., Normah, M.G., Robiah, A.: Thermal and Hydrodynamic Analysis of Microchannel Heat Sinks: A Review. Renewable and Sustainable Energy Reviews (21), 614–622 (2013), doi:10.1016/j.rser.2013.01
9. Philips, R.J.: Microchannel heat sinks. In: Bar-Cohen, A., Kraus, A.D. (eds.) Advances in Thermal Modeling of Electronic Components and Systems, vol. 2, pp. 109–184. ASME Press, New York (1990)
10. Knight, R.W., Goodling, J.S., Gross, B.E.: Optimal thermal design of air cooled forced convection finned heat sinks – experimental verification. IEEE Transaction on Components, Hybrids and Manufacturing Technology, 602–212 (1992)
11. Kleiner, M.B., Kuhn, S.A., Haberger, K.: High performance forced air cooling scheme employing microchannel heat exchangers. IEEE Transactions on Components, Packaging, and Manufacturing Technology, Part A 20(4), 795–804 (1995)
12. Fedorov, A.G., Viskanta, R.: Three-dimensional conjugate heat transfer in the microchannel heat sink for electronic packaging. Int. J. Heat Mass Transfer 43, 399–415 (2000)
13. Khan, W.A., Kadri, M.B., Ali, Q.: Optimization of microchannel heat sinks using genetic algorithm. Heat Transfer Engineering 34(4), 279–287 (2013)
14. Tuckerman, D.B., Pease, R.F.: High performance heat sinking for VLSI. IEEE Electron Devices Lett. EDL-2, 126–129 (1981)
15. Adham, A.M., Mohd-Ghazali, N., Ahmad, R.: Multi-objective Optimization Algorithms for Microchannel Heat Sink Design. In: Ali, M., Bosse, T., Hindriks, K.V., Hoogendoorn, M., Jonker, C.M., Treur, J. (eds.) Contemporary Challenges & Solutions in Applied AI. SCI, vol. 489, pp. 169–174. Springer, Heidelberg (2013)
16. http://www.iiar.org (visited September 19, 2013)
17. Wen, Z., Choo, F.K.: The optimum thermal design of microchannel heat sinks. In: Proceedings of the 11th IEEE Electronic Packaging Technology Conference, Singapore, pp. 123–129 (1997)
18. Kim, S.J., Kim, D.: Forced convection in microstructures for electronic equipment cooling. J. Heat Transfer (121), 639–645 (1999)

Multi-objective Dual-Sale Channel Supply Chain Network Design Based on NSGA-II

Chia-Lin Hsieh[1,*], Shu-Hsien Liao[2], and Wei-Chung Ho[2]

[1] Department of Statistics and Actuarial Science, Aletheia University, No. 26, Chenli Street, Danshuei District, Taipei 251, Taiwan, ROC
[2] Department of Management Sciences and Decision Making, Tamkang University, No. 151, Yingjuan Road, Danshuei District, Taipei 251, Taiwan, ROC

Abstract. In this study, we propose a two-echelon multi-objective dual-sale channel supply chain network (DCSCN) model. The goal is to determine (i) the set of installed DCs, (ii) the set of customers the DC should work with, how much inventory each DC should order and (iv) the distribution routes for physical retailers or online e-tailers (all starting and ending at the same DC). Our model overcomes the drawback by simultaneously tackling location and routing decisions. In addition to the typical costs associated with facility location and the inventory-related costs, we explicitly consider the pivotal routing costs between the DCs and their assigned customers. Therefore, a multiple objectives location-routing model involves two conflicting objectives is initially proposed so as to permit a comprehensive trade-off evaluation. To solve this multiple objectives programming problem, this study integrates genetic algorithms, clustering analysis, Non-dominated Sorting Genetic Algorithm II (NSGA-II). NSGA-II searches for the Pareto set. Several experiments are simulated to demonstrate the possibility and efficacy of the proposed approach.

Keywords: Supply chain management, Integrated supply chain design, Dual sale channel, Multiple objective evolutionary algorithm, NSGA-II.

1 Introduction

In general prospective, there are two streams of research solving the integrated supply chain network (SCN) problem, one stream of study is based on the concept of the Location-Allocation Problem (LAP), and the other stream is based on the Location-Routing Problem (LRP). The LRP is defined to solve a facility location problem, but in order to achieve this we simultaneously need to solve a vehicle routing problem. The main difference of the LRP from the LAP is that, once the facilities have been placed, the LRP requires the visitation of demands nodes through tours, where the latter assumes straight-line or radial trips between the facilities and respective customers. The LRP considers three main decisions of difference levels simultaneously: location of depots - strategic level; allocation of customers to depots - tactical level and the routes to visit these customers - operational level. The interdependence between these decisions has

---

* Corresponding author.

A. Moonis et al. (Eds.): IEA/AIE 2014, Part I, LNAI 8481, pp. 479–489, 2014.
© Springer International Publishing Switzerland 2014

been noticed by researchers long ago. Due to the complexity of both location and routing problems, they have been traditionally solved separately [1] and have made the proposed models too simple and led to sub-optimality.

In last few years, the advent of e-commerce (EC) has made retailing more complicated and more competitive. New channels for supply chains have attracted much interest. Since the internet made on-line shopping easy, it has become an important *internet-enabled* channel as well. Dual-channel supply chain design (DCSCN) is becoming more common. In DCSCN, customers select the channel through which to buy goods, so dual channels mean more shopping choices and potential cost savings to customers. Therefore, several models addressing these issues are developed. Especially, on-time delivery relies heavily on effective vehicle routing once the merchandise is out the supplier's door and on its way to the customer. The LRP has become more complicated in a B2C environment in dual-channel supply chains.

2   Literatures Reviews

In the last two decades, many LRP models have been proposed in the literature. Most of them are related to a simple distribution network with two layers (depots and customers) and are solved by either exact or heuristic solution methods. Only few exceptional studies addressed more complex distribution network design problems. [2] developed a four-tier integrated LRP made up of four layers (plants, central depots, regional depots and customers), with the aim of defining the number and the location of the different types of facilities. [3] proposed a three-layer distribution logistics model for the conversion from brick-and-mortar to click-and-mortar retailing by a static one-period optimization model. [4] considered four layer supply chains similar to [2]. A heuristic algorithm based on LP-relaxation was proposed. Research on dual channel environment problem is relatively rare. [5] considered ordering and allocation policies for multi-echelon systems with two sales channels. [6] reviewed an inventory equilibrium performance or the inventory control policy within dual sale channel. There are also very limited researches addressing the retail/e-tail routing operations in dual sale channel. [7, 8] both agreed that a quick-response vehicle dispatching system is more necessary in the B2C environment than in B2B. [3] solved a three-tier static location-routing-based problem that embraces the clicks-and-bricks strategy in their retail operations. Multi-objective optimization problems in SCN have been considered by different researchers in literature [9,10]. The evolutionary algorithms have been validated to have better computational efficiency on solving the optimization problems for SCN. In the last decade, there has been a growing interest to adopt Multi-objective evolutionary algorithms (MOEAs), such as Non-dominated Sorting Genetic Algorithm II (NSGA-II), to solve a variety of multi-objective SCN problems [11]. Through MOEAs, decision-makers can obtain Pareto optimal solutions.

From the survey, some innovative research aspects that are noteworthy have been incorporated in our research work. This study incorporates two streams of SCN research, LAP and LRP, to solve an integrated DCSCN problem. We propose a nonlinear mix-integer Multi-Objective Location-Routing model with multiple objectives so as to minimize the total location cost and the routing cost simultaneously. We also provide a decision making approach via NSGA-II which is employed as a "filter" to approximate a set of Pareto-optimal solutions. Up to now, very few studies have applied similar problem-solving approaches in the same research context.

3   Problem Statement and Formulation

3.1   Problem Description and Assumptions

In this paper, we consider a multi-objective two-echelon DCSCN problem (see Fig 1.) that consists of a vendor with a warehouse at the top echelon, multi-distribution centers (DCs) in the middle echelon and the retailers from dual sale channels (either traditional or internet-enabled channels) at the bottom echelon. Our problem incorporates the Vehicle Routing Problem with Time Windows (VRPTW) which is the problem of designing least cost routes from one depot (say DC) to a set of geographically scattered points of e-retailers. The routes is designed in such a way that each point is visited only once by exactly one vehicle within a given time interval; all routes start and end at the same DC, and the total demands of all e-retailers on one particular route must not exceed the capacity of the vehicle. In addition, for each DC, there are two different delivery policies. A point-to-point policy is adopted for the shipment between DCs and retailers for the traditional channel. However, DCs have to quickly response to the online customer's requirements through the internet. A home delivery services guarantees that shipment should arrive during designated time window. In addition to the typical costs associated with LAP, we explicitly consider the pivotal routing costs between the DCs and their assigned customers incurred from VRP. Two objectives are provided to minimize the total facility location and the inventory-related costs in LAP as well as to minimize the total routing costs in VRP. Our problem then is modeled as a multi-objective nonlinear integer program.

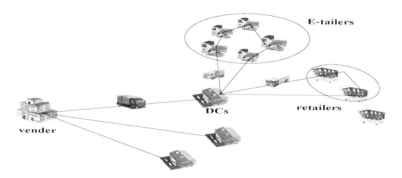

Fig. 1. Graphic representation of DCSCN model

The following assumptions are used throughout the whole paper. The product is always available to customers throughout both channels. The product price is identical for both channels. The system receives orders from both channels according to customers' preferences. The demands from both channels at each DC occurred randomly and are identically independent and normally distributed. The centralized inventory policy under the vendor managed inventory (VMI) mode is considered where the vendor is responsible for the safety stock pooled at DCs. At any DC j, we assume a continuous inventory revision, and a (Q_j, r_j) policy to meet a stochastic demand pattern. That is, when the inventory level at DC j falls to or below a reorder point r_j, a fixed quantity Q_j is ordered to the vendor. Each order is fulfilled and delivered by only a specific DC but the assignment of e-tailer/retailers to a DC is known a priori. For e-tailers, the last-mile home delivery within time windows is adapted to fulfill the quick response requirement. The vendor storage capacity is unlimited but each DC has capacity restriction for retailers but not for e-tailers due to the fact that the requirement of e-tailers is relatively small as compared to retailers. For retailers, therefore, the assignment rule is based on the DC's capability and distance coverage; for e-tailers, the routing distance is the only concern. Each DC possesses two types of vehicles' capacities for dual sale channels. Vehicles' capacities in the same channel are the same, and fleet type is homogeneous but the inter-dispatch shipping is prohibited. In addition, we integrate three decisions in a mathematical model under the aforementioned assumptions.

— *Location and allocation decisions*: how many DCs to locate, where to locate the opened DCs, and how to allocate the e-tailer /retailers to them.
— *Routing decisions*: how to build the vehicles' routes starting from an opened DC to serve its customers.
— *Inventory decisions*: how often to reorder, what quantity to replenish for each order at a DC from each retailer.

3.2   Mathematical Model

Before presenting the model, we depict the notation used throughout the paper.

Indices. j is an index set of potential DCs ($j \in$ J). i is an index set for retailers ($i \in$ I). n is an index set for e-tailers ($n \in$ N). r is an index set of all routes (vehicles) ; $\forall r \in$ R. v is an index set of vehicles ($v \in$ V). M is a merged set of e-tailers and potential DCs (N∪J). P is a merged set of e-tailers and potential DCs.

Decision Variables. Q_j is the order quantity at DC j. Y_j is a binary variable to decide if DC j is opened. X_{ji} is a binary variable to decide if retailer i is assigned to DC j. W_{jn} is a binary variable to decide if e-tailer n is assigned to DC j. R'_{nh} is a binary variable to decide if node n precedes node h in the route r. F^v_{st} is a binary variable to decide if node s precedes node t in the route v. M_r is an auxiliary variable for sub-tour elimination constraints in route r.   if DC j is opened; 0 otherwise see if RBC i is chosen or not. v_j is a binary variable if CBC j is opened or not.

Model Parameters. B is the number of e-tailers contained in set N, i.e. B = |N|. d_i is the mean of annual demand at retailer i. u_n is the mean of annual demand at e-tailer n . δ_i is the standard deviation of annual demand at retailer i. δ_n is the standard deviation of annual demand at e-tailer n. f_j is the annual fixed cost for opening and operating DC j. rc_j is the unit transportation cost between the vendor and DC j. tc_{st} is the unit routing cost between node s and node t; $\forall$ s, t $\in$ I∪J. vc_{nh} is the unit routing cost between node n and node h; $\forall$ n , h $\in$ N∪J. a^r_n is the earliest time of route r to serve e-tailer n. b_n^r is the latest time of route r to serve e-tailer n. t_n^r is the specified arrival time of route r for e-tailer n. s_j is the inventory holding cost per unit time (annually) at DC j. o_j is the inventory ordering cost per order to the vendor from DC j. β is the weight factor associated with routing cost. Θ is the weight factor associated with inventory cost. ζ_j is the average lead time in days to be shipped to DC j from the vendor. z_α is the left α-percentile of standard normal random variable Z.

According to the mentioned notations and assumptions, we formulate a multi-objective mixed-integer programing model as follows.

$$\min \quad \sum_{j \in J} f_j \times Y_j + \sum_{j \in J} (o_j \times \frac{\sum_{i \in I} \sum_{n \in N} (d_i \times X_{ji} + u_n \times W_{jn})}{Q_j})$$
$$+ \theta \times \{ \sum_{j \in J} s_j [{Q_j}/{2} \times Y_j + z_{1-\alpha} (\sum_{i \in I} \delta_i \sqrt{\zeta_j \times X_{ji}} + \sum_{n \in N} \delta_n \sqrt{\zeta_j \times W_{jn}})] \} \tag{1}$$

$$\min \quad \beta \times [\sum_{j \in J} \sum_{i \in I} \sum_{n \in N} rc_j \times (d_i \times X_{ji} + u_n \times W_{jn})$$
$$+ \sum_{n \in M} \sum_{h \in M} \sum_{v \in V} vc_{nh} \times R_{nh}^r + \sum_{s \in P} \sum_{t \in P} tc_{st} \times d_i \times F_{st}^v] \tag{2}$$

subject to :

$$\sum_j X_{ji} = 1 \tag{3}$$

$$X_{ji} \leq Y_j \quad , \tag{4}$$

$$\sum_{h \in M} R_{nh}^r = 1 \tag{5}$$

$$\sum_{t \in P} F_{st}^v = 1 \tag{6}$$

$$M_n - M_h + (B \times R_{nh}^r) \leq B - 1 \tag{7}$$

$$\sum_{h \in M} R_{nh}^r - \sum_{h \in M} R_{hn}^r = 0 \tag{8}$$

$$\sum_{j \in J} \sum_{n \in N} R_{jn}^r \leq 1 \tag{9}$$

$$-W_{jn} + \sum_{h \in M} (R_{nh}^r - R_{jh}^r) \leq 1 \tag{10}$$

$$a_n^r \times W_{jn} \leq t_n^r \leq b_n^r \times W_{jn} \tag{11}$$

$$X_{ji} \in \{0,1\} \quad Y_j \in \{0,1\} \quad W_{jn} \in \{0,1\} \quad R_{nh}^r \in \{0,1\} \tag{12}$$

The objective function Eq. (1) minimizes the facility location and inventory-related costs in LAP. The first term indicates the *facility operating* cost of DCs; the second term considers the *dual channel ordering* cost and the last one is the *holding cost at* DCs, including working inventory cost and safety stock cost. The objective function

Eq. (2) minimizes the transportation cost in VRP. The first term indicates the *inbound transportation cost* from the vender to DCs; the second and the third terms refer to the *outbound routing costs* incurred by the orders of *retailers* and *e-tailers* respectively. We split the total cost into these objectives for the sake of concerning the association of costs incurred between LAP and VRP. In Eq. (1) and Eq. (2), β and θ denote the weights of different scenarios corresponding to their impacts on inventory and routing factors, respectively. Eq. (3) restricts a retailer to be serviced by a single DC. Eq. (4) states that retailers can only be assigned to open DCs. Eq. (5) ensures that each e-tailer is assigned on exactly one vehicle route at a time. Eq. (6) ensures each retailer is placed on only one vehicle at a time. Eq. (7) is the sub-tour elimination constraint which guarantees each tour must contain a DC from which it originates, i.e. each tour must consist of a DC and some e-tailers. Eq. (8) carries out the flow conservation saying that whenever a vehicle enters an e-tailer or DC node, it must leave again and ensuring that the routes remain circular. Eq. (9) implies that only one DC is included in each route. In Eq. (10), the e-tailer is assigned to the DC only if a specific route starts its trip from the DC. Eq. (11) ensures the DC delivery service meets the e-tailer's time requirement. Eq. (12) enforces the integrality restrictions on the binary variables. Since the order quantity Q_j in Eq(1) is convex in $Q_j > 0$, the optimal order quantity Q_j^* is obtained by differentiating Eq. (1) with respect to Q_j.

$$Q_j^* = \sqrt{\frac{2 \times o_j \times (\sum_{i \in I} d_i \times X_{ji} + \sum_{n \in N} u_n \times W_{jn})}{s_j}} \tag{13}$$

4   Solution Methodologies

Our proposed model combines the location-allocation problem (LAP) and the multi-depot vehicle routing problem (VRP) in dual sales channel environments that results in NP-hard. Due to the complexity of the problems that exact methods can only tackle relatively small instances, as an alternative, a heuristic procedure is applied. Fig. 2 depicts the solution scheme of our heuristic procedure.

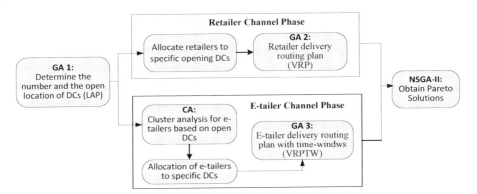

Fig. 2. The solution scheme of proposed heuristic procedure for DCSCN

As we can see in the heuristic procedure in Fig. 2, genetic algorithms (GA) and cluster analysis (CA) are integrated to solve our DCSCN model. The heuristic procedure is decomposed into LAP and VRP stages. In the LAP stage, a genetic-based heuristic procedure (GA1) is first applied to determine the number, location of DCs, assignment of specific retailers to each of DC. In the VRP stage, the procedure is then decomposed into two phases: retail channel phase and e-tail channel phase. The former mainly arranges retailer' delivery routing plan between each opened DC to their allocated retailers by a second genetic-based heuristic (GA2), the latter clusters the e-tailers based on open DCs and also makes delivery routing by time-windows by a hybrid heuristic via a K-means cluster analysis (CA) and a genetic algorithm (GA3). Vis those heuristics, we obtain all costs incurred in the proposed DCSCN model problem. Subsequently, NSGAII is adopted to search for the Pareto solutions.

— GA1 for LAP: the major task of this procedure (GA1) is to determine the number of potential DCs will be opened and the allocation of downstream retailers to specific opening DCs, the solution is encoded in a binary string of length |J| (the number of DC and $\forall j \in J$).
— GA2 for VRP: this procedure is to decide the retailer's delivery routing plan for DCs.
— CA for e-tailers: this procedure is to classify e-tailers into k groups according to the number of open DCs given priori by k-means. After clustering, the DC-Group *allocation* procedure is performed to allocate each opening DC to one of the groups based on the shortest distance between DCs and the group centroids. Due to DC capacity restrictions, it is allowed for a specific group to select the secondary closest DC, if its closest DC cannot afford sufficient capacity for all e-tailers in the same group, until every group has been assigned.
— GA3 for e-tailer's VRPTW: this procedure is to determine the e-tailer's delivery routing plan within time windows for DCs. The process of GA3 is quite similar to GA2 except for delivery time requirements. In practice, on-line delivery service allows its customers to choices favorite time periods to receive orders instead of exact arriving time. For this reason, we randomly divide each group of e-tailers into three sub-groups with respect to different time requirements of delivery service.

4.1   NSGAII for Pareto Solutions

NSGA-II [12] is one of the best techniques for generating "good" solutions in MOEAs in which two primary goals should be achieved: (i) convergence to a Pareto-optimal set, and (ii) maintenance of population diversity in a Pareto-optimal set. First of all, for each solution in the population, one has to determine how many solutions dominate it and the set of solutions to which it dominates. Then, it ranks all solutions to form non-dominated fronts according to a *non-dominated sorting* process, hence, classifying the chromosomes into several fronts of non-dominated solutions. To allow for diversification, NSGA-II also estimates the solution density surrounding a particular solution in the population by computing a *crowding distance* operator. During selection, a crowded-comparison operator considering both the non-domination rank of an individual and its crowding distance is used to select the offspring, without losing good solutions (*elitism* strategy). However, the crossover and mutation operators remain the same as usual.

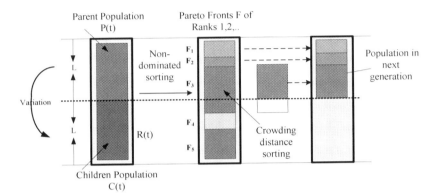

Fig. 3. NSGA-II solution scheme

Based on NSGA-II, a hybrid evolutionary algorithm is proposed for our model. The solution scheme is graphically represented as in Fig. 3. This algorithm starts by generating a random population P(1) of size L. For each chromosome in P(1), the algorithm evaluates its costs using the encoded solution expressions. Then, it applies *non-dominated sorting* on P(1) and assigns to each chromosome a front to which it belongs. Next, the algorithm applies binary tournament selection (to form the crossover pool), crossover, and mutation operators to generate the children population C(1) of size L. After that, a combined population R(1)=P(1)∪C(1) of size $2L$ is sorted according to the *elitism strategy* aforementioned . Therefore, a new parent population P(2) is formed by adding solutions from the first front till the size exceeds L. Once initialized, the algorithm repeats for T generations.

5   Numerical Experience

To evaluate the performance of the DCSCN consisting of LAP and VRP issues, we provide some computational experiments. For the best of our knowledge, there are no similar instances in the public domain, nor have any benchmarking available in previous studies. To explore DCSCN, we developed a test problem by generating problem instances with 25 potential DCs, 100 retailers and 500 e-tailers in a square of 50 distance units of width. For simplicity, Euclidean distance is used for measuring distribution distances. For the hybrid GA implementation, we used the following input parameters: population size = 100; maximum number of generations = 200; cloning = 20%; crossover rate = 80%; mutation rate varies from 5% to 10% as the number of generations increases. The approach program was coded in MATLAB. In Fig. 4, we represent the solution evolutionary process of our optimization scheme visually from a variety of feasible solutions to a non-dominated solution set through NSGAII. The non-dominated solution set of DCSCN is obtained by applying NSGAII is illustrated in Table 2, where 30 alternatives of non-dominated solutions are listed. Each alternative contains the number of opening DCs, the operation cost in LAP (Z_1) as well as the transportation cost in VRP (Z_2).

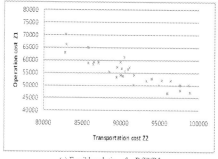
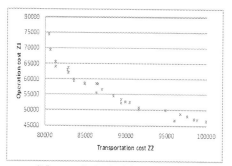

(a) Feasible solutions for DCSCN   (b) Pareto non-dominated solutions through NSGA-II

Fig. 4. The solution evolutionary process of optimization scheme

Table 1. Non-dominated solution set from NSGAII

Alternative	# of Open DCs	Operation cost in LAP (Z1)	Transportation cost in VRP (Z2)	Alternative	# of Open DCs	Operation cost in LAP (Z1)	Transportation cost in VRP (Z2)
1	4	$80,493.74	$74,397.86	16	13	$96,676.35	$48,711.52
2	14	$99,943.45	$46,537.74	17	14	$98,420.11	$46,942.34
3	7	$83,545.31	$59,497.13	18	9	$91,568.61	$50,742.70
4	6	$80,616.50	$69,302.92	19	7	$86,520.80	$58,509.21
5	7	$84,929.02	$58,530.48	20	7	$84,929.02	$58,743.14
6	7	$81,315.94	$64,184.07	21	13	$98,859.08	$46,907.06
7	8	$86,356.81	$55,498.99	22	10	$89,393.85	$53,304.12
8	9	$88,438.25	$54,549.90	23	5	$82,909.69	$63,657.47
9	10	$89,393.85	$52,289.53	24	9	$87,058.17	$56,529.38
10	5	$82,909.69	$62,103.26	25	8	$89,914.31	$52,583.65
11	12	$95,947.76	$46,701.68	26	10	$90,491.93	$52,466.75
12	7	$82,764.15	$63,051.12	27	7	$83,545.31	$59,979.49
13	9	$91,568.61	$50,246.43	28	13	$97,618.28	$47,885.84
14	11	$94,872.27	$50,086.08	29	8	$86,356.81	$58,523.08
15	7	$82,315.94	$65,645.07	30	9	$88,438.25	$55,496.18

6   Conclusion

In this study, we attempt to propose a two-echelon multi-objective dual sale channel supply chain network model regarding a single vender, multiple distribution centers (DCs), as well as a set of customers (physical retailers or online e-tailers). We develop a novel formulation which integrates three issues, LAP, inventory and VRP, of

SCN. This study attempts to find the location and the number of open DC, the allocation of DCs to customers, inventory replenishment and also the distribution routes for physical retailers or online e-tailers of with minimal facility and inventory operation cost and minimal transportation cost for DCSCN. NSGA-II is applied to determine a finite set of non-dominate Pareto solutions. Feasibility of the developed model was checked by presenting several small-sized random instances and solving them by proposed GA approaches. In our experiments, the proposed approach displays good behavior on the near-reality data and yields a near-optimal solution in stochastic demand environments. Several interesting phenomenon are perceived.

The model can be extended in some practical directions. Detailed sensitive analysis should be adopted to find the crucial parameters with respect to different assignments, resulting in the maximum increases/decreases on this DCSCN structure. Moreover, the proposed heuristic procedure genetic provides a variety of options and parameter settings that are worth fully examined. It is also interesting to develop more effective and elegant heuristic methods to solve the integrated model problem. For example, model can be solved by other meta-heuristic algorithms. In additions, determining the weights of the attributes in the model is important but complex. Sorting Pareto solutions is also required according to decision-makers' preferences by using multi-attribute decision making (MADM) techniques, such as Analytic Hierarch Process (AHP) or Technique for Order Preference by Similarity to Ideal Solution (TOPSIS).

References

1. Nagy, G., Salhi, S.: 'Location-routing: Issues, models and methods'. European Journal of Operational Research 177(2), 649–672 (2007)
2. Ambrosino, D., Grazia Scutellà, M.: 'Distribution network design: New problems and related models'. European Journal of Operational Research 165(3), 610–624 (2005)
3. Aksen, D., Altinkemer, K.: A location-routing problem for the conversion to the "click-and-mortar" retailing: The static case. European Journal of Operational Research 186(2), 554–575 (2008)
4. Lee, J.H., Moon, I.K., Park, J.H.: 'Multi-level supply chain network design with routing'. International Journal of Production Research 48(13), 3957–3976 (2009)
5. Alptekinoğlu, A., Tang, C.S.: A model for analyzing multi-channel distribution systems. European Journal of Operational Research 163(3), 802–824 (2005)
6. Widodo, E., Takahashi, K., Morikawa, K., Pujawan, I.N., Santosa, B.: Managing sales return in dual sales channel: its product substitution and return channel analysis. International Journal of Industrial and Systems Engineering 9(2), 121–149 (2011)
7. Du, T.C., Li, E.Y., Chou, D.: Dynamic vehicle routing for online B2C delivery. Omega 33(1), 33–45 (2005)
8. Azi, N., Gendreau, M., Potvin, J.Y.: An exact algorithm for a single-vehicle routing problem with time windows and multiple routes. European Journal of Operational Research 178(3), 755–766 (2007)
9. Guillén, G., Mele, F.D., Bagajewicz, M.J., Espuña, A., Puigjaner, L.: Multiobjective supply chain design under uncertainty. Chemical Engineering Science 60(6), 1535–1553 (2005)

10. Gaur, S., Ravindran, A.R.: A bi-criteria model for the inventory aggregation problem under risk pooling. Computers & Industrial Engineering 51(3), 482–501 (2006)
11. Liao, S.H., Hsieh, C.L., Lin, Y.S.: A multi-objective evolutionary optimization approach for an integrated location-inventory distribution network problem under vendor-managed inventory systems. Annals of Operations Research 186(1), 213–229 (2011)
12. Deb, K., Pratap, A., Agarwal, S., Meyarivan, T.: A fast and elitist multiobjective genetic algorithm: NSGA-II. IEEE Transactions on Evolutionary Computation 6(2), 182–197 (2002)

A Multimodal Approach to Exploit Similarity in Documents

Matteo Cristani and Claudio Tomazzoli

University of Verona
{matteo.cristani,claudio.tomazzoli}@univr.it

Abstract. Automated document classification process extracts information with a systematic analysis of the content of documents.

This is an active research field of growing importance due to the large amount of electronic documents produced in the world wide web and available thanks to diffused technologies including mobile ones.

Several application areas benefit from automated document classification, including document archiving, invoice processing in business environments, press releases and research engines.

Current tools classify or "tag" either text or images separately.In this paper we show how, by linking image and text-based contents together, a technology improves fundamental document management tasks like retrieving information from a database or automated documents.

We present an investigation of a model of conceptual spaces for investigation using joint information sources from the text and the images forming complex documents. We present a formal model and the computable algorithms and the dataset from which we took a subset to make experiments and relative tests and results.

1 Introduction

Nowadays the wide availability of electronic documents through the Internet or private business networks has changed the way people search for information. We deal with a huge quantity of knowledge which has to be organized and searchable to be utilized. Also for this reason in information technology research community there is an always growing interest in the field of automatic document classification. Although several innovative studies are produced every year, some topics are still to be deeply investigated. Among these, the problem of efficient classification and retrieval of documents containing both text and images has been treated in a non multidisciplinary approach.

There are several publications of efficient information retrieval from text. There are also publications about information extraction from images and even text contained in images [1], but the joint analysis of text and image information from a complex document still lacks a well documented solution. For example, if a brochure from an isolated hotel in the Dolomites describes the hotel's features and includes maps and pictures of mountainous surroundings, the categorizer

A. Moonis et al. (Eds.): IEA/AIE 2014, Part I, LNAI 8481, pp. 490–499, 2014.
© Springer International Publishing Switzerland 2014

will automatically discover the content and link the text and the images together. Then someone searching for an isolated mountain lodge within a certain price range would retrieve the brochure even if "isolated lodge in the mountains" were never mentioned in the actual text.The paper is organized as follows: Section 2 presents the model and the approach of extracting joint textual and image information; Section 3 presents the framework and the computable algorithms; Section 4 shows experiments and related results and finally in Section 5 we make some conclusions.

2   The Model

We found the model described in [2] as a valuable starting point for our model; we will use accordingly the term "mode" of a document for both text and image and we will use the "bag of word" representation for features set of both modes, but we define those contributes in a more general sense than in [2].Then we apply "noise" as suggested in [3] and we define our general model, which will be used later in the framework. Suppose we have n *documents* $\mathcal{D} = \{d_1, d_2, ..., d_n\}$ and m *tags* $\mathcal{T} = \{t_1, t_2, ..., t_m\}$. Following PLSA[1] approach, The goal is to construct a set of feature vectors $\{X_1, X_2, ..., X_n\}$ in a latent semantic space $\mathbb{R}^k$ to represent these multimedia objects having matrix $A = U\Sigma V^T$.

The Model for Multimodal Documents
We are considering both textual and visual contributions to the meaning of a document. Suppose we are given a matrix Q of content links, where $Q_{i,j}$ can represent the similarity measurement between the ith document and the jth document. Recalling the works in latest literature we have that documents can be described as *multimodal* when made of both text and visual content, each defined as "mode"; a repository that contains a set of multimodal documents is then $D = \{d_1, d_2, ..., d_n\}$. We will use these "dual" representations to compute the trans-media similarity measures as defined in [2] : $sim_{glob}(d_i, d_j) = Q_{i,j}$ using a linear combination where $\lambda_1, \lambda_2, \lambda_3, \lambda_4$ are appropriate weighting factors:

$$\lambda_1 sim_{TT}(d_i, d_j) + \lambda_2 sim_{TV}(d_i, d_j) + \lambda_3 sim_{VT}(d_i, d_j) + \lambda_4 sim_{VV}(d_i, d_j)$$

so we have that the elements in our matrix Q of similarity of multimodal content of documents can be

$$Q_{i,j} = \lambda_1 sim_{TT}(d_i, d_j) + \lambda_2 sim_{TV}(d_i, d_j) + \lambda_3 sim_{VT}(d_i, d_j) + \lambda_4 sim_{VV}(d_i, d_j) \tag{1}$$

We assume that the documents with stronger links ought to be closer to each other in the latent semantic space.

---

[1] Probabilistic Latent Semantic Analisys

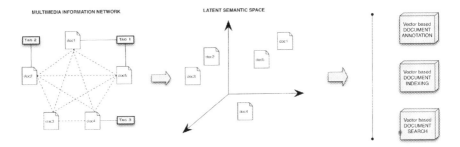

Fig. 1. Model for Multimodal Documents

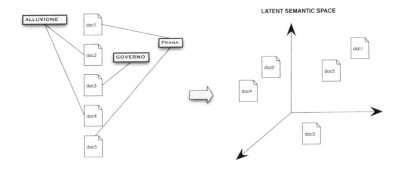

Fig. 2. documents with stronger links will be closer

Based on this assumption, we introduce the quantity Ω to measure the smoothness of documents in the underlying latent space, as in [3].

$$\Omega(X) = \frac{1}{2} \sum_{i,j=1}^{n} Q_{i,j}(X_i - X_j)(X_i - X_j)^T$$

where, $\|M\|_2^2$ is the l_2 norm of matrix M, and X_i and X_j are the ith and jth row of X. It is easy to see that by minimizing the above regularization term, a pair of documents with larger $Q_{i,j}$ will have closer feature vectors X_i and X_j in the latent space. Given D as the diagonal matrix with its elements as the sum of each row of Q and $L = D - Q$, with some matrix operations we obtain $\Omega(X) = trace(X^T L X)$ using the factorization $X = U\Sigma^k$, defining H as $H = U\Sigma_k V^T = XV^T$ and knowing that $VV^T = I$ we have

$$\Omega(X) = trace(H^T L H) \tag{2}$$

2.1   The Noisy Model

Due to the fact that we consider both textual and visual contribution to the meaning of a document, we have to consider the existence of noise in process so a noise term ε exist on the matrix Q such that

$$Q = H + \varepsilon$$

where H is the matrix which denotes the noise-free tag links, after the noise ε has been removed. The goal is to make a correctly representative H of "minimal rank". Let H_i, $1 \leq i \leq n$ denote the row vectors of H, which is the associated noise-free tag vector for the ith document.

Each tag vector represents the occurrence of the corresponding tag in the document corpus. Tag vectors of synonyms should be the same (or within a positive multiplier of one another), such as the tag vector for the synonym terms "person" and "human". Moreover, many tags probably do not independently occur in the corpus since they are semantically correlated. For example, the tag 'animal' often correlates with its subclasses such as "cat" and "tiger". This indicates from the viewpoint of linear algebra, that the tag vector of "animal" could be located in a latent subspace spanned by those of its subclasses. Since the rank of matrix H is the maximum number of independent row vectors, it follows from the above dependency among tags, that H ought to have a low rank structure. The topic vectors that represent occurrences of the associated topics in the document corpus span a latent semantic space, which contains most of tag vectors. Therefore, the rank of H should be no more than the maximum number of independent topic vectors in the latent space. Hence we can impose a low-rank prior to estimate the noise-free H from the observed noisy Q.

The *nuclear norm* of a matrix M is computed as the sum of all the singular values of the matrix. Let $\|M\|_*$ denote the nuclear norm of matrix M, then

$$\|M\|_* = \sum_i \sigma_i(M)$$

where $\sigma_i(M)$ are singular values in M. The *nuclear norm* can be used to solve the optimization problem of determining the lowest rank approximation.

So our problem can be described as finding

$$min\|Q - H\|_F + \gamma\|H\|_* \tag{3}$$

where $\|M\|_F$ is the squared summation of all elements in a matrix M (the Frobenius norm) and γ is a balancing parameter.

This problem has a unique analytical solution: recalling from above that $Q = U\Sigma V^T$ we define an operation $(x)_+ = max(0, x)$ and the matrix Σ_+^γ which is the diagonal matrix from Σ where the singular values are defines as $(\sigma - \frac{\gamma}{2})_+$. with values σ from matrix Σ.

$$\Sigma_+^\gamma = diag((\sigma - \frac{\gamma}{2})_+)$$

The solution to the problem can be described as

$$min\|Q - H\|_F + \gamma\|H\|_* = H_\gamma = U\Sigma_+^\gamma V^T \tag{4}$$

The difference with normal *Latent Semantic Indexing* is that it directly selects the largest k singular values of A where this Formulation subtracts something $(\frac{\gamma}{2})$ from each singular value and thresholds them by 0. Suppose the resulting noise free matrix H is of rank k, then the Support Vector Machine of H has form as $H = U\Sigma_k V^T$ where Σ_k is a $k \times k$ diagonal matrix. Similar with *Latent Semantic Indexing*, the row vectors of $X = U\Sigma_k$ can be used as the latent vector representations of documents in latent space. It is also worth noting that minimizing the rank of H gives a smaller k so that the obtained latent vector space can have lower dimensionality, and then the storage and computation in this space could be more efficient.

2.2   The Global Model for Multimodal Documents

Considering both contribution to the model we can make use of both Equation 2 and 3 so our problem can be completely described as finding

$$min\|Q - H\|_F + \gamma\|H\|_* + \lambda trace(H^T L H) \tag{5}$$

Here λ is another balancing parameter. In contrast to Formulation (3), Formulation (5) does not have an closed-form solution. Fortunately, this problem can be solved by the *Proximal Gradient method* known from literature which uses a sequence of quadratic approximations of the objective function in order to derive the optimal solution.

3   The Framework

3.1   The Matrix Q of Similarity for Multimodal Content

We are considering both textual and visual contributions to the meaning of a document.

We define matrix Q_t of content links, where $Q_t(i,j)$ can represent the similarity measurement between the text of the ith document and the the text of the jth document.

We define matrix Q_v of content links, where $Q_v(i,j)$ can represent the similarity measurement between the image of the ith document and the the image of the jth document.

Following PLSA approach as above specified we have $Q_t \cong U_t\Sigma_t^k V_t^T$ for the textual mode and $Q_v \cong U_v\Sigma_v^k V_v^T$ for the visual mode.

We have $S_T = U_t\Sigma_t^k$; similarly, the visual, dual representation of the textual part is: $S_V = U_v\Sigma_v^k$ We will denote the textual part of d_j by $S_T(d_j)$ and its visual part $S_V(d_j)$ which are the jth rows of matrix S_T and S_V. Recalling that in *Basic Model for Multimodal Documents* in equation 1 we had

$$Q_{i,j} = \lambda_1 sim_{TT}(d_i, d_j) + \lambda_2 sim_{TV}(d_i, d_j) + \lambda_3 sim_{VT}(d_i, d_j) + \lambda_4 sim_{VV}(d_i, d_j)$$

And *assuming that the both text and image part of a document shall define the same meaning for the document in the meaning space* we will use these partial latent semantic representations to define the single components of the equation above

$$sim_{TV}(d_i, d_j) = \|S_T(d_i) - S_V(d_j)\|_F \tag{6}$$

$$sim_{VT}(d_i, d_j) = \|S_V(d_i) - S_T(d_j)\|_F \tag{7}$$

$$sim_{TT}(d_i, d_j) = \|S_T(d_i) - S_T(d_j)\|_F \tag{8}$$

$$sim_{VV}(d_i, d_j) = \|S_V(d_i) - S_V(d_j)\|_F \tag{9}$$

This model benefits from two major aspects: it is simple to understand and it is simple to implement, both because it involves only measure of distance in a vector space.

The main assumption is that there is *one* meaning space so that features in text and features in images all refers to a set of concepts or meanings which are the same but are expressed with words and with images.

When these meanings are expressed with words the dimensionality of the feature space is different than the dimensionality of the feature space coming from the images, but using a dimensionality reduction algorithm we can reduce these different dimensions to be the same, so that we can than compute a distance.

3.2 Algorithms

First algorithm is to be used to evaluate the vector representation of text features, which are extracted from the words in text.

Algorithm 1. S_t: finds the vectorial representation of the features on texts

---

Input: A finite set $\mathcal{T} = \{t_1, t_2, \ldots, t_n\}$ of texts
Output: The Matrix S_t and the integer m, size of the textual vocabulary TW
1 $Q_t \leftarrow \emptyset$; $TW \leftarrow \emptyset$;
2 **for** $i \leftarrow 1$ **to** n **do**
3   **for** $j \leftarrow 1$ **to** $countword(t_i)$ **do**
4          $TW \leftarrow TW \cup \{w_j\}$;

5 **for** $i \leftarrow 1$ **to** n **do**
6      $Q_t(i) \leftarrow \frac{1}{countword(t_i)} \{numword_{1,i}, \ldots numword_{m,i}\}$;
7 Compute U_t, Σ_t^k, V_t as $Q_t \cong U_t \Sigma_t^k V_t^T$;
8 $S_t \leftarrow U_t \Sigma_t^k$;
9 **return** S_t, m ;

Second algorithm is to be used to evaluate the vector representation of visual features, using a bag of visual word model generated computing SIFT on images.

Algorithm 2. S_v: finds the vectorial representation of the features on images

---

Input: A finite set $V = \{v_1, v_2, \ldots, v_n\}$ of images and the integer m, size of the visual vocabulary

Output: The Matrix S_v

1   $Q_v \leftarrow \emptyset$; $FV \leftarrow \emptyset$;

2   **for** $i \leftarrow 1$ **to** n **do**

3 | $FV(i) \leftarrow$ extract SIFT features from v_i ;

4   Compute Q_v as a Bag of Visual Words Model using FV as a feature vector and GMM with m Visual Terms;

5   Compute U_v, Σ_v^k, V_v as $Q_v \cong U_v \Sigma_v^k V_v^T$;

6   $S_v \leftarrow U_v \Sigma_v^k$;

7   **return** S_v;

Third algorithm deals with the problem of combining the two sources of information a single matrix.

Last algorithm is the classical algorithm for *Proximal Gradient Approximation* applied to our matrix problem.

Algorithm 3. Q: finds the Matrix Q of similarity of multimodal content

---

Input: A finite set $D = \{d_1, d_2, \ldots, d_n\}$ of documents and numeric weight factors $\lambda_1, \lambda_2, \lambda_3, \lambda_4$

Output: The Matrix Q as in equation $\boxed{??}$

1   $V \leftarrow \emptyset$; $T \leftarrow \emptyset$; $S_v \leftarrow \emptyset$; $S_t \leftarrow \emptyset$;

2   **for** $i \leftarrow 1$ **to** n **do**

3 | $T(i) \leftarrow$ extract the text in d_i;

4 | $V(i) \leftarrow$ extract the image in d_i;

5   Compute S_t and m using algorithm 1 with input T;

6   Compute S_v using algorithm 2 with input V and m;

7   Compute $sim_{TV}(d_i, d_j)$ using equation 6;

8   Compute $sim_{VT}(d_i, d_j)$ using equation 7;

9   Compute $sim_{TT}(d_i, d_j)$ using equation 8;

10   Compute $sim_{VV}(d_i, d_j)$ using equation 9;

11   **for** $j \leftarrow 1$ **to** n **do**

12 | **for** $i \leftarrow 1$ **to** n **do**

13 | | $Q(i, j) \leftarrow$
            $\lambda_1 sim_{TT}(d_i, d_j) + \lambda_2 sim_{TV}(d_i, d_j) + \lambda_3 sim_{VT}(d_i, d_j) + \lambda_4 sim_{VV}(d_i, d_j)$

14   **return** Q;

Algorithm 4. Proximal Gradient for minimizing equation 5

Input: The Matrix Q of similarity of multimodal content and numeric weight
factors λ, γ, ϵ, $maxIter$

Output: The Matrix H denoised matrix of Q

1 $H_0 \leftarrow 0$; $\tau \leftarrow 1$

2 Compute $\sigma_{max}(I + \lambda L^T)$ as the largest singular value of $(I + \lambda L^T)$

3 $\alpha \leftarrow 2\sigma_{max}(I + \lambda L^T)$

4 **repeat**

5 $\quad K(H_{\tau-1}) \leftarrow \|Q - H_{\tau-1}\|_F + \lambda trace(H_{\tau-1}^T L H_{\tau-1})$

6 $\quad G_\tau \leftarrow H_{\tau-1} - \frac{1}{\alpha}\nabla K(H_{\tau-1}) = H_{\tau-1} - \frac{2}{\alpha}(H_{\tau-1} - Q + \lambda L^T H_{\tau-1})$

7 $\quad$ Compute $diag(\sigma)$ as singular value decomposition $G_\tau = U diag(\sigma)V^T$ and
$\quad$ remember $(x)_+ = max(0, x)$

8 $\quad H_\tau \leftarrow U diag((\sigma - \frac{\gamma}{\alpha})_+)V^T$

9 $\quad \tau \leftarrow \tau + 1$

10 **until**

11 Convergence $(rank(H_\tau) - rank(H_{\tau-1}) \leq \epsilon)$ or maximum iteration number
$(maxIter)$ achieved

12 **return** H_τ

4   Experimental Results

The test dataset is made of almost 10 years of daily issues of a local newspaper; actually under a non disclosure agreement. It is made of almost 10 years of daily issues of a local newspaper, each having 64 pages with an average of 4 articles per page.

This sums to around 800.000 documents most but not all of which have both text and image contributes. We decided to concentrate for the first experiments on a subset of articles of which we knew the tag "author" as well the others such as "topic" because in the dataset this tag was the one with less occurrences, so that we had a subset of maximum size with all tags.

In local newspapers articles are organized in a way so that page groups and hi level topics can be considered synonymous; for instance "economy" is found at pages 6,7 and "news story" at pages 9,10.

This organization is almost stable throughout the time span considered, almost 10 years of daily editions.

Authorship Attribution

We decided to consider "signed" articles and try to apply our algorithm to the problem of *authorship attribution* having around 40.000 documents and 300 Authors.

We pre-elaborated the articles so that they can be made object to a Matlab set of codes.After a few unsuccessful attempts we decided to pre-define the final dimension of the SVD dimensionality reduction algorithm.

As the resources consumed by the extraction of features from images are much greater than the ones used for text elaboration, we decided that the leading factor should be the dimension of the image features matrix.

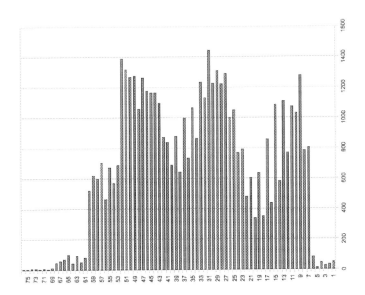

Fig. 3. *Signed* Article Distribution per page

Given that, we applied the code of algorithms 1, 2, 3, 4 obtaining results as in schema below

Information Source	Accuracy
Text	0,18
Images	0,24
Text and Images	0,32

Classification

We then decided to try our algorithm in guessing the number of the page the article was in, roughly indicating the topic.

Also in this case we found that a better result can be achieved, in fact *results are better than chance (0.0014)* and *accuracy* and *precision* are close to the ones of text only, but *recall* is higher than the one for text only, while compared to images only *accuracy* is higher and *precision* is close but *recall* is lower.

Information Source	Accuracy	F1
Text	0.0907	0.0165
Images	0.0768	0.0725
Text and Images	0.0881	0.0428

5   Conclusions

The first part of this work was dedicated to point out the overview of the research and the problems and choices we got through during the path of this study. Then we focused on the model we would use to determine different contribution to classification of the text and image information of a document; we've given the details of the definition of a meaning space using Probabilistic Latent Semantics for multimodal documents including consideration and modeling of the possible noise that shall be considered in this process and how to deal with it. Then we focused on the definitions of the distances in the meaning space and we've given the definition of computable algorithm for our model. Part of this work was then dedicated to the description of the test dataset and its possible use and organization to be useful in our research. We used it in experiments to validate the model and we presented these experiments and related results.

References

1. Ye, Q., Huang, Q., Gao, W., Zhao, D.: Fast and robust text detection in images and video frames. Image and Vision Computing 23, 565–576 (2005)
2. Ah-Pine, J., Bressan, M., Clinchant, S., Csurka, G., Hoppenot, Y., Renders, J.-M.: Crossing textual and visual content in different application scenarios. Multimedia Tools and Applications 42, 31–56 (2009)
3. Qi, C., Aggarwal, G., Tian, Q., Ji, H., Huang, T.: Exploring context and content links in social media: A latent space method. IEEE Transactions Pattern Analysis and Machine Intelligence (August 2011)
4. Kesorn, S., Poslad, K.: An enhanced bag of visual word vector space model to represent visual content in athletics images. IEEE Transactions on Multimedia (October 2011)
5. Denoyer, L., Gallinari, P.: Bayesian network model for semi-structured document classification. Information Processing and Management 40, 807–827 (2004)
6. Bouguila, N., ElGuebaly, W.: Discrete data clustering using finite mixture models. Pattern Recognition 42, 33–42 (2009)
7. Mikhailov, D.V., Emelyanov, G.M.: Semantic clustering and affinity measure of subject-oriented language texts. Pattern Recognition and Image Analysis 20, 376–385 (2010)
8. Yang, L., Geng, Y., Cai, B., Hanjalic, A.: Object retrieval using visual query context. IEEE Transactions on Multimedia (July 2011)
9. Qin, J., Yung, N.H.C.: Scene categorization via contextual visual words. Pattern Recognition 43, 1874–1888 (2010)
10. Park, G., Baek, Y., Lee, H.-K.: Web image retrieval using majority-based ranking approach. Multimedia Tools and Applications 31, 195–219 (2006)
11. Chan, W., Coghill, G.: Text analysis using local energy. Pattern Recognition 34, 2523–2532 (2001)
12. Aronovich, L., Spiegler, I.: Cm-tree: A dynamic clustered index for similarity search in metric databases. Data & Knowledge Engineering 63, 919–946 (2007)
13. Sable, C.L., Hatzivassiloglou, V.: Text-based approaches for non-topical image categorization. International Journal on Digital Libraries 3, 261–275 (2000)

Deal with Linguistic Multi-Criteria Decision Making Problem Based on Sequential Multi-person Game Methodology

Chen-Tung Chen[1], Ping Feng Pai[2], and Wei Zhan Hung[2]

[1] Department of Information Management, National United University, Miao-Li, Taiwan
ctchen@nuu.edu.tw
[2] Department of Information Management,
National Chi Nan University, Nan-Tou, Taiwan
paipf@ncnu.edu.tw, steady_2006@hotmail.com

Abstract. multiple criteria decision making (MCDM) problem handle the ranking order of alternatives with respect to various criteria in real environment. Uncertainties and vagueness should be considered because the qualitative criteria and the subjective judgment of decision-makers usually exist in the decision making process. It is reasonable for decision-makers to use linguistic variables to express their opinions. Game theory approach has been used as an efficient framework in copying with decision making problems. It can be applied extensively to solve the complex and interrelated practical decision problems. In fact, decision making approaches based on game theory have become an important research direction in decision science. The aim of this study is to develop an effective methodology for solving the game problem with linguistic variables by multiple decision makers. Based on the linguistic variable, the decision makers can easily express their opinions with respect to each criterion for each alternative. By using the backward induction method, we can find the rational solution of a game in accordance with the combination of strategies of players effectively. And then, a new decision making method, linguistic sequential multi-person multi-criteria game (LSMPMCG) model will be proposed for dealing with the fuzzy game problem in this study.

Keywords: Multi-criteria evaluation, Group decision making, Linguistic Variable, Sequential Game.

1 Introduction

There is an old saying that no one can be as an island. It means that everyone must active with others for obtaining his/her acquirement. In the real environment, everyone's execution performance usually depends on his/her behavior and competitors' activities. How to be interactive with competitors for approaching his/her goal is an important topic. In decision science field, game theory provides a general model for handing with the interactive optimization problems. Game theory is a special

A. Moonis et al. (Eds.): IEA/AIE 2014, Part I, LNAI 8481, pp. 500–510, 2014.
© Springer International Publishing Switzerland 2014

tool which can be used to analyze the interaction result of each player and have been broadly applied in business, financial, politics, education and sports etc [1-2].

Players, players' strategies and the performances (payoffs) about players' strategies are three basic components in a game. The main way to describe the interactive of players includes strategy form and extensive form. In strategy form, each play executes his/her strategy simultaneously and the payoff is decided based on the strategy combination of each player [3]. In extensive form, each play executes his/her strategy sequentially and the payoff is the final result when each player is already moved based on his/her decision.

Due to game theory is only considering one dimension, it is not suitable for satisfying the needs of the players in real environment. In this decade, the research trend is to integrate multi-criteria decision making (MCDM) with game theory to make a decision in real situations. Some literatures try to combine multi-criteria decision method (MCDM) with game theory for coping with reality problems.

Campos [4] proposed two-person zero sum fuzzy matrix game and applied fuzzy linear programming to calculate the mixed strategy probability of two players. Sakawa and Nishizaki [5] applied the max-min concept to integrate fuzzy goal and fuzzy payoffs in two-person zero sum game. Song and Kandel [3] applied fuzzy set to formulate players' goal and their competitors' strategy probability. The players' mixed strategy probability is calculated based on considering their goals and competitors' strategies probability simultaneously. The drawback of Song and Kandel's model is that their model is difficult to compute for multi-person games. Angelou and Economides [6] integrated analytic hierarchy process, game theory and real options to analyze information and communication technology (ICT) business alternatives. Reneke [7] used vector function to evaluate long term investment alternatives for predicting oil prices and environmental degradation under the conditions of risk and uncertainty. Li and Barough et al. [8] used traditional game approach to deal with two type of project construction conflict problem (prisoners' dilemma and chicken game). Li and Hong [9] developed an effective methodology for handling constrained payoffs matrix with triangular fuzzy numbers.

There are so many literatures for applying game theory to make decision. However, a few of them can integrate MCDM with game theory to execute group decision in fuzzy environment for handling with the multi-person multi-criteria game in an efficient approach. For instance, many players can influence the market in the beginning of a new industry, it naturally formulate as a multi-person game. On the other hand, the players usually become fewer after some heavily competitions. If only two players survive finally, it naturally formulate as a two-person game. We need a model to set up as multi-person game or two-person game flexibly.

In real environment, player must compete with other players. A good player not only must consider his/her strategy for approaching his/her goals but also need to forecast his/her competitors' behaviors. Due to the preference of player and his/her competitors are usually more than one and are often not the same, a good decision model must consider each player's preference flexibly.

In fact, uncertainty and fuzziness often exist in the real competitive environment because it is not easy to collect complete information and decision time is limited for making decision. A good model must provide a mechanism such as linguistic value or fuzzy number for expert to flexibly express his/her opinion.

To my best knowledge, many multi-criteria game methods can satisfy one or two decision condition. However, none of model can approach all requirements to make decision. Therefore, the goal of this paper is to develop a new decision making method, named "linguistic sequential multi-person multi-criteria game" (LSMPMCG) model, for copying with the multi-criteria decision making problems.

According to the linguistic variable, the decision makers can easily express their opinions with respect to each criterion for each alternative (strategy combination). By using the backward induction method, we can find the optimal solution of a game matrix effectively. The main advantages of LSMPMCG include effective execution process, feasible in forecasting different strategy combinations and flexible in setting criteria to make a decision.

2   Fuzzy Set And Linguistic Variable

Fuzzy set theory is first introduced by Zadeh in 1965 [10]. Fuzzy set theory is a very feasible method to handle the imprecise and uncertain information in a real world [11]. Especially, it is more suitable for subjective judgment and qualitative assessment in the evaluation processes of decision making than other classical evaluation methods applying crisp values [12]. A positive triangular fuzzy number (PTFN) $\tilde{T}$ can be defined as $\tilde{T} = (l, m, u)$, where $\tilde{T}_1 = (L_1, M_1, U_1)$ and $l > 0$. The membership function $\mu_{\tilde{T}}(x)$ of positive triangular fuzzy number (PTFN) $\tilde{T}$ is defined as [13]

$$\mu_{\tilde{T}}(x) = \begin{cases} \dfrac{x-l}{m-l}, & l < x < m \\ \dfrac{u-x}{u-m}, & m < x < u \\ 0, & otherwise \end{cases} \tag{1}$$

A linguistic variable is a variable whose values are expressed in linguistic terms. In other words, values of linguistic variables are not numbers but words or sentences in a nature or artificial language [14]. For example, "weight" is a linguistic variable whose values are very low, low, medium, high, very high, etc. These linguistic values can also be represented by fuzzy numbers. There are two advantages for using triangular fuzzy number to express linguistic variable [15]. First, it is a rational and simple method to use triangular fuzzy number to express experts' opinions. Second, it is easy to do fuzzy arithmetic when using triangular fuzzy number to express the linguistic variable.

Table 1. Linguistic Variables

Performance		Weight		
Linguistic variables	Abbr.	Linguistic variables	Abbr.	Membership function
Extremely Poor	EP	Extremely Low	EL	(0.000,0.000,0.125)
Very Poor	VP	Very Low	VL	(0.000,0.125,0.250)
Poor	P	Low	L	(0.125,0.250,0.375)
Medium Poor	MP	Medium Low	ML	(0.250,0.375,0.500)
Fair	F	Fair	F	(0.375,0.500,0.625)
Medium Good	MG	Medium High	MH	(0.500,0.625,0.750)
Good	G	High	H	(0.625,0.750,0.875)
Very Good	VG	Very High	VH	(0.750,0.875,1.000)
Extremely Good	EG	Extremely High	EH	(0.875,1.000,1.000)

Let $\tilde{T}_1 = (L_1, M_1, U_1)$ and $\tilde{T}_2 = (L_2, M_2, U_2)$ be two PTFNs. The additive operation of PTFNs can be calculated as [13].

$$\tilde{T}_1 \oplus \tilde{T}_2 = (L_1 + L_2, M_1 + M_2, U_1 + U_2) \tag{2}$$

Lee and Li [16] presented the generalized mean value method. It is very easy to compare fuzzy numbers with this method. Let $\tilde{T} = (l, m, u)$ be a TFN whose defuzzied value can be easily computed as [16-17]

$$G(\tilde{T}) = (l + m + u)/3 \tag{3}$$

$$S(\tilde{T}) = (1/18) * (l^2 + m^2 + u^2 - lm - lu - mu) \tag{4}$$

where $G(\tilde{T})$ is the generalized mean value and $S(\tilde{T})$ is the deviation of fuzzy number $\tilde{T}$, respectively.

Let $\tilde{T}_1 = (L_1, M_1, U_1)$ and $\tilde{T}_2 = (L_2, M_2, U_2)$ be two PTFNs.

(1) If $G(\tilde{T}_1) > G(\tilde{T}_2)$, then $\tilde{T}_1 > \tilde{T}_2$.

(2) If $G(\tilde{T}_1) = G(\tilde{T}_2)$ and $S(\tilde{T}_1) < S(\tilde{T}_2)$, then $\tilde{T}_1 > \tilde{T}_2$.

(3) If $G(\tilde{T}_1) = G(\tilde{T}_2)$ and $S(\tilde{T}_1) = S(\tilde{T}_2)$, then $\tilde{T}_1 \approx \tilde{T}_2$.

3   Linguistic Sequential Multi-Person Multi-Criteria Game Model

Generally speaking, the contents of linguistic sequential multi-person multi-criteria game (LSMPMCG) model include:

(1) A set of players are called $P = \{P^1, P^2, ..., P^v\}$, where v represents the number of player.

(2) A set of strategies of each player are called $S = \{S^1, S^2, \ldots S^v\}$, where $S^1 = \{s_1^1, s_2^1, \ldots, s_{m_1}^1\}$, $S^2 = \{s_1^2, s_2^2, \ldots, s_{m_2}^2\}, \ldots, S_v = \{s_1^v, s_2^v, \ldots, s_{m_v}^v\}$. The $m_1, m_2, \ldots, m_v$ are the number of strategies for player 1, player 2,..., player v. The s_π^δ represents the π th strategy of player δ.

(3) A set of strategy combinations $SC = \left[s_i^1, s_j^2, \ldots, s_k^v\right]_{m_1 * m_2 * \ldots * m_v}$.

   The strategy combination means a combination of each player's strategy.

(4) A set of criteria with respect to each strategy combination of each player $C = \{C^1, C^2, \ldots, C^v\}$ where $C^1 = \{c_1^1, c_2^1, \ldots, c_{n_1}^1\}$, $C^2 = \{c_1^2, c_2^2, \ldots, c_{n_2}^2\}, \ldots, C_v = \{c_1^v, c_2^v, \ldots, c_{n_v}^v\}$. The $n_1, n_2 \ldots, n_v$ are the number of the criteria for player 1, player 2,..., player v. The c_π^δ represents the π th strategy criteria of player δ.

(5) A set of each player' strategy combination evaluation matrix $E = \{E^1, E^2, \ldots, E^v\}$, where $E^i = \left[\tilde{x}^i_{\left(s_i^1, s_j^2, \ldots, s_k^v\right), c_\pi^i}\right]_{(m_1 * m_2 * \ldots * m_v) * n_i}$, i=1,2,...,v and $\tilde{x}^i_{\left(s_i^1, s_j^2, \ldots, s_k^v\right), c_\pi^i}$ represents the strategy performance with respect to criterion c_π^i of player i under strategy combination $\left(s_i^1, s_j^2, \ldots, s_k^v\right)$.

(6) A set of decision maker $D = \{D_1, D_2, \ldots, D_r\}$, where r represents the number of decision maker.

   In beginning, decision makers must express their opinions about the strategy performance of each strategy combination with respect to each player's criterion based on each player's standpoint.

   The $\tilde{x}^d_{\left(s_i^1, s_j^2, \ldots s_\theta^\delta, \ldots s_k^v\right), c_\pi^\delta}$ is a linguistic variable which can be formulated as PTFN=(a,b,c) where a= $^L x^d_{\left(s_i^1, s_j^2, \ldots s_\theta^\delta, \ldots s_k^v\right), c_\pi^\delta}$, b= $^M x^d_{\left(s_i^1, s_j^2, \ldots s_\theta^\delta, \ldots s_k^v\right), c_\pi^\delta}$ and c= $^U x^d_{\left(s_i^1, s_j^2, \ldots s_\theta^\delta, \ldots s_k^v\right), c_\pi^\delta}$ (Refer to Table 1) and it represents the opinion of decision maker d about the performance of strategy s_θ^δ for player δ with respect to criterion c_π^δ under the strategy combination $\left(s_i^1, s_j^2, \ldots s_\theta^\delta, \ldots s_k^v\right)$. The decision makers' opinions about the performance of strategy s_θ^δ for player δ with respect to criterion c_π^δ under the strategy combination $\left(s_i^1, s_j^2, \ldots s_\theta^\delta, \ldots s_k^v\right)$ can be integrated as

$$\tilde{x}_{\left(s_i^1,s_j^2,...,s_\theta^\delta...s_k^v\right),c_\pi^\delta}=\frac{1}{r}*\begin{pmatrix}\tilde{x}_{\left(s_i^1,s_j^2,...,s_\theta^\delta...s_k^v\right),c_\pi^\delta}^1\oplus\tilde{x}_{\left(s_i^1,s_j^2,...,s_\theta^\delta...s_k^v\right),c_\pi^\delta}^2\\\oplus...\oplus\tilde{x}_{\left(s_i^1,s_j^2,...,s_\theta^\delta...s_k^v\right),c_\pi^\delta}^r\end{pmatrix} \tag{5}$$

where $\tilde{w}_{c_\pi^\delta}^d$ is a linguistic variable which can be formulated as PTFN=(a,b,c) where $a={}^L w_{c_\pi^\delta}^d$,$b={}^M w_{c_\pi^\delta}^d$ and $c={}^U w_{c_\pi^\delta}^d$ (Refer to Table 1) and it represents the opinion of decision maker d about the weight of criterion c_π^δ .

The decision makers' opinions about the weight of criterion c_π^δ can integrate as

$$\tilde{w}_{c_\pi^\delta}=\frac{1}{r}*\left(\tilde{w}_{c_\pi^\delta}^1\oplus\tilde{w}_{c_\pi^\delta}^2\oplus...\oplus\tilde{w}_{c_\pi^\delta}^r\right) \tag{6}$$

The weighted performance of strategy s_θ^δ for player δ with respect to criterion c_π^δ under strategy combination $\left(s_i^1,s_j^2,...s_\theta^\delta...s_k^v\right)$ can be calculated as

$$\tilde{p}_{\left(s_i^1,s_j^2,...s_\theta^\delta...s_k^v\right),c_\pi^\delta}=\tilde{x}_{\left(s_i^1,s_j^2,...s_\theta^\delta...s_k^v\right),c_\pi^\delta}*\tilde{w}_{c_\pi^\delta}\quad\pi=1,2,...,n_\delta \tag{7}$$

The aggregated performance of strategy s_θ^δ for player δ under strategy combination $\left(s_i^1,s_j^2,...s_\theta^\delta...s_k^v\right)$ can be calculated as

$$\tilde{\Psi}_{\left(s_i^1,s_j^2,...s_\theta^\delta...s_k^v\right)}=\sum_{i=1}^{n_\delta}\tilde{p}_{\left(s_i^1,s_j^2,...s_\theta^\delta...s_k^v\right),c_i^\delta} \tag{8}$$

In the sequential game, the players decide their strategies sequentially. Suppose that the player 1 decides the strategy first. And then, play i (i=1,2,...,v) decides the strategy in sequential.

Backward induction is a logistic method to predict the behaviors of competitors and decide the strategy under the different strategy combinations. So, the final player decides the strategy under the different strategy combinations. And then, the player who decides the strategy before the player v (player v-1) must decide the strategy under different strategy combination according to the decision information of player v and so on. Finally, every player can realize that which is his/her reasonable strategy based on the reasonable strategies of competitors.

4   Numerical Example

Suppose that there are three enterprises P^1 , P^2 and P^3 who can produce high technology Tablet PC. Enterprise P^1 wants to decide the new strategy for developing a new Tablet PC. Enterprise P^1 possesses two development strategies and also knows that competitor P^2 has two strategies and competitor P^3 has two strategies for developing a new Tablet PC. Enterprise P^1 considers four criteria for making decision. Enterprise P^1 also knows that competitor P^2 considers four criteria for making decision and competitor P^3 considers three criteria for making decision (Refer to Table 2). Three strategy experts are invited to analyze the strategy of enterprise P^1.

Table 2. The Strategy and criteria of each enterprise

	Strategy	Criteria
P^1	S_1^1 : Develop New High Level Tablet PC	C_1^1 : mobile margin profit, C_2^1 : market share rate
	S_2^1 : Develop New Middle Level Tablet PC	C_3^1 :research cost, C_4^1 :research time
P^2	S_1^2 : Develop New High Level Mobile Phone	C_1^2 :market share rate, C_2^2 : Enterprise's total cost
	S_2^2 : Develop New Low Level Mobile Phone	C_3^2 : Enterprise's net profit, C_4^2 : research time
P^3	S_1^3 : Develop New Middle Level Mobile Phone	C_1^3 :Enterprise's net profit, C_2^3 : mobile margin profit
	S_2^3 : Develop New Low Level Mobile Phone	C_3^3 :research cost

The execution steps of linguistic sequential multi-person multi-criteria game (LSMPMCG) model can be illustrated as follows.

Step1. Each expert uses the linguistic variables to evaluate the strategy performance of each enterprise with respect to each criterion under different strategy combinations as Tables 3, 4 and 5.

Step2. Transform the linguistic evaluations of each enterprise with respect to each criterion under different strategy combinations into PTFN. And then, we aggregate the experts' opinions about the performance of each enterprise according to equation (2) and equation (5).

Step4. Each expert uses the linguistic variables to evaluate the weight of each criterion as Tables 6.

Step5. Transform the linguistic evaluation of the weight of each criterion into PTFN. And then, we aggregate the experts' opinions about the weight of each criterion according to equation (2) and equation (6).

Step7. The weighted performance of each enterprise with respect to each criterion under special kind of strategy combination can be calculated in accordance with equation (7).

Step8. The aggregated strategy performance of each enterprise under special kind of strategy combination can be calculated according to equation (8).

Step9. The aggregated strategy performance of each enterprise under special kind of strategy combination can be arranged as the multi-person multi-criteria non-cooperation game as Table 7.

According to backward induction method, enterprise p^3 must decide its strategy first under different strategy combination. According to the defuzzied values at Table 8, If enterprise p^1 chooses s_1^1 and enterprise p^2 chooses s_1^2 then enterprise p^3 will choose s_1^3 because 0.987>0.942. The strategy decision of enterprise p^3 can be shown as Table 8.

Enterprise p^2 can make decision based on rational decision of enterprise p^3 (Refer to Table 9). When enterprise p^1 chooses s_1^1 and enterprise p^2 chooses s_1^2, enterprise p^2 can predict enterprise p^3 will choose s_1^3. If enterprise p^1 chooses s_1^1 and enterprise p^2 chooses s_1^2 then enterprise p^2 can predict p^3 will choose s_1^3. Under this situation, p^2 will choose s_1^2 because the defuzzied value of performance is 1.509>1.317. The strategy decision of p^2 can be shown as Table 9.

Enterprise p^1 can make decision based on the rational decisions of enterprise p^2 and p^3 (Refer to Table 10). When enterprise p^1 chooses s_1^1, enterprise p^1 can predict enterprise p^2 will choose s_1^2 and enterprise p^3 will choose s_1^3. When enterprise p^1 chooses s_2^1, enterprise p^1 can predict enterprise p^2 will choose s_2^2 and enterprise p^3 will choose s_1^3. The performance (1.4977) of strategy combination (s_1^1, s_1^2, s_1^3) is larger than the performance (1.3571) of strategy combination (s_2^1, s_2^2, s_1^3), so enterprise p^1 chooses s_1^1. Finally, the rational equilibrium solution is that enterprise p^1 chooses s_1^1, enterprise p^2 choose s_1^2 and enterprise p^3 choose s_1^3.

Table 3. The Linguistic Variable Format of strategy performance of enterprise p^1 under different strategy combination

Strategy combination	c_1^1			c_2^1			c_3^1			c_4^1		
	D_1	D_2	D_3	D_1	D_2	D_3	D_1	D_2	D_3	D_1	D_2	D_3
s_1^1, s_1^2, s_1^3	EP	G	EG	EG	G	MG	P	VG	G	G	P	VG
s_2^1, s_1^2, s_1^3	MG	VP	VG	MP	G	EP	EG	G	MG	EP	G	G
s_1^1, s_2^2, s_1^3	P	F	MG	F	VP	P	G	F	F	MP	EG	EP
s_2^1, s_2^2, s_1^3	EG	G	P	F	EG	VG	VG	P	F	G	P	VP
s_1^1, s_1^2, s_2^3	P	MG	EP	MP	MG	P	MG	EP	VG	P	MG	P
s_2^1, s_1^2, s_2^3	MG	P	P	VG	VP	G	VG	P	EG	P	VP	G
s_1^1, s_2^2, s_2^3	EG	VG	EP	G	P	F	VP	F	P	MP	MG	G
s_2^1, s_2^2, s_2^3	VP	G	EG	P	MG	P	EP	G	MG	VG	EG	VG

Table 4. The Linguistic Variable Format of strategy performance of enterprise p^2 under different strategy combination

Strategy combination	c_1^2			c_2^2			c_3^2			c_4^2		
	D_1	D_2	D_3	D_1	D_2	D_3	D_1	D_2	D_3	D_1	D_2	D_3
s_1^1,s_1^2,s_1^3	VG	P	F	G	EG	VP	VG	MG	F	MG	EG	EG
s_2^1,s_1^2,s_1^3	MP	MG	G	VP	G	P	MG	F	EP	VG	VG	P
s_1^1,s_2^2,s_1^3	MG	F	MG	F	EP	VG	VG	MP	EG	MP	G	MG
s_2^1,s_2^2,s_1^3	P	EP	G	MP	MP	VG	VG	EG	P	MP	P	VG
s_1^1,s_1^2,s_2^3	MG	EG	P	VP	P	P	MP	VP	EG	G	F	VP
s_2^1,s_1^2,s_2^3	MG	EP	P	G	F	EP	VP	VG	MG	EG	F	VG
s_1^1,s_2^2,s_2^3	EG	G	VP	MP	MG	G	VG	VG	VG	VP	P	VP
s_2^1,s_2^2,s_2^3	VG	VP	VG	EP	G	VG	EG	G	P	VG	P	VG

Table 5. The Linguistic Variable Format of strategy performance of enterprise p^3 under different strategy combination

Strategy combination	c_1^3			c_2^3			c_3^3		
	D_1	D_2	D_3	D_1	D_2	D_3	D_1	D_2	D_3
s_1^1,s_1^2,s_1^3	MG	F	EP	G	EG	G	EP	F	G
s_2^1,s_1^2,s_1^3	MG	F	MP	VG	P	MG	F	EG	MP
s_1^1,s_2^2,s_1^3	P	G	MP	F	EG	VP	EG	MP	MG
s_2^1,s_2^2,s_1^3	MG	EP	VG	G	VG	G	G	MG	MP
s_1^1,s_1^2,s_2^3	EP	EG	MP	G	F	EG	EP	MG	F
s_2^1,s_1^2,s_2^3	P	F	MG	MG	VP	G	VG	P	MP
s_1^1,s_2^2,s_2^3	EG	G	VP	VG	G	EP	F	VG	G
s_2^1,s_2^2,s_2^3	F	F	G	G	G	F	G	EP	F

Table 6. The Linguistic format of the weight of each criterion

enterprise p^1	c_1^1	c_2^1	c_3^1	c_4^1	enterprise p^3	c_1^3	c_2^3	c_3^3
D_1	EH	MH	L	L	D_1	ML	MH	VH
D_2	VL	ML	MH	MH	D_2	MH	F	H
D_3	ML	VH	EH	MH	D_3	ML	L	F
enterprise p^2	c_1^2	c_2^2	c_3^2	c_4^2				
D_1	F	ML	EL	MH				
D_2	F	EH	H	VH				
D_3	F	H	VH	F				

Table 7. The aggregative performance in multi person multi criterion noncooperation game of each enterprise

P^1	P^3	enterprise P^2	
		s_1^2	s_2^2
s_1^1	s_1^3	((0.958,1.490,2.045), (0.958,1.516,2.054), (0.630,0.969,1.363))	((0.578,1.005,1.493), (0.844,1.339,1.884), (0.604,0.984,1.340))
	s_2^3	((0.526,0.906,1.405), (0.542,0.984,1.441), (0.599,0.932,1.295))	((0.646,1.099,1.620), (0.807,1.307,1.835), (0.708,1.094,1.503))
s_2^1	s_1^3	((0.792,1.250,1.774), (0.646,1.094,1.637), (0.620,1.005,1.398))	((0.839,1.354,1.878), (0.734,1.198,1.719), (0.719,1.109,1.554))
	s_2^3	((0.708,1.182,1.696), (0.703,1.141,1.668), (0.500,0.854,1.252))	((0.797,1.292,1.826), (0.891,1.401,1.964), (0.609,0.969,1.391))

Note: ($\bullet,\bullet,\bullet$) represents the linguistic performances of (P^1, P^2, P^3).

Table 8. The strategy decision of enterprise P^3

P^1's Strategy	P^2's Strategy	P^3's Strategy	P^3's G(T)	P^3's Decision	P^1's Strategy	P^2's Strategy	P^3's Strategy	P^3's G(T)	P^3's Decision
s_1^1	s_1^2	s_1^3	0.987	s_1^3	s_1^1	s_1^2	s_1^3	1.008	s_1^3
s_1^1	s_1^2	s_2^3	0.942		s_2^1	s_1^2	s_2^3	0.869	
s_1^1	s_2^2	s_1^3	0.976	s_2^3	s_2^1	s_2^2	s_1^3	1.127	s_1^3
s_1^1	s_2^2	s_2^3	1.102		s_2^1	s_2^2	s_2^3	0.990	

Table 9. The strategy decision of enterprise P^2

P^1's Strategy	P^2's Strategy	P^3's Strategy	P^2's G(T)	P^2's Decision	P^1's Strategy	P^2's Strategy	P^3's Strategy	P^2's G(T)	P^2's Decision
s_1^1	s_1^2	s_1^3	1.509	s_1^2	s_2^1	s_1^2	s_1^3	1.126	s_2^2
s_1^1	s_2^2	s_2^3	1.317		s_2^1	s_2^2	s_1^3	1.217	

Table 10. The strategy decision of enterprise P^1

Enterprise P^1's Strategy	Enterprise P^2's Strategy	Enterprise P^3's Strategy	Enterprise P^1's G(T)	Enterprise P^1's Decision
s_1^1	s_1^2	s_1^3	1.4977	s_1^1
s_2^1	s_2^2	s_1^3	1.3571	

5 Conclusions and Future Research

In this study, we integrate traditional game approach with multi-criteria decision method for dealing with management decision problem. In proposed method, experts can use linguistic variables to express their opinions. The linguistic variables are suitable to express the opinions in this uncertain and vague reality environment. Backward induction method is used effectively to find the rational solution of a game in accordance with the combination of strategies of each player.

In the future, there are two way to modify and extend the proposed method. First, this method is the sequential game. Maybe the simultaneous game can be modified by proposed method. Second, the multiple decision information can be applied to extend the proposed method because the enterprise can only know part of competitors' information in real environment.

Acknowledgement. This work is financial supported partially by the National Science Council of Taiwan. The grant numbers are "NSC 101-2410-H-239-004-MY2" and "NSC101-2410-H-260-005-MY2".

References

1. Bellotti, F., Berta, R., De Gloria, A., Lavagnino, E., Dagnino, F., Ott, M., Romero, M., Usart, M., Mayer, I.S.: Designing a Course for Stimulating Entrepreneurship in Higher Education through Serious Games. Comput. Sci. 15, 174–186 (2012)
2. Berga, D., Bergantiños, G., Massó, J., Neme, A.: An undominated Nash equilibrium for voting by committees with exit. Math. Soc. Sci. 54, 152–175 (2007)
3. Song, O., Kandel, A.: A Fuzzy Approach to Strategic Games. IEEE. T. Fuzzy. Syst. 7, 634–642 (1999)
4. Campos, L.: Fuzzy linear programming models to solve fuzzy matrix games. Fuzzy. Set. Syst. 32, 275–289 (1989)
5. Sakawa, M., Nishizaki, I.: Max-min solution for fuzzy multi objective matrix game. Fuzzy. Set. Syst. 67, 53–69 (1994)
6. Angelou, G.N., Economides, A.A.: A multi-criteria game theory and real-options model for irreversible ICT investment decisions. Telecommun. Policy 33, 686–705 (2009)
7. Reneke, J.A.: A game theory formulation of decision making under conditions of uncertainty and risk. Nonl. An. 71, 1239–1246 (2009)
8. Barough, A.S., Shoubi, M.V., Skardi, M.J.E.: Application of Game Theory Approach in Solving the Construction Project Conflicts. Procedia. Soc. Behav. Sci. 58, 1586–1593 (2012)
9. Li, D.F., Hong, F.X.: Solving constrained matrix games with payoffs of triangular fuzzy numbers Comput. Math. Appl., 432–446 (2012)
10. Zadeh, L.A.: Fuzzy sets. Inform. Control. 8, 338–353 (1965)
11. Xu, Z.: Group decision making based on multiple types of linguistic preference relations. Inform. Sciences. 178, 452–467 (2008)
12. Wang, W.P.: A fuzzy linguistic computing approach to supplier evaluation. Appl. Math. Model. 34, 3130–3141 (2010)
13. Chen, C.T.: Extensions of the TOPSIS for group decision-making under fuzzy environment. Fuzzy. Set. Syst. 14, 1–9 (2000)
14. Herrera-Viedma, E., Peis, E.: Evaluating the informative quality of documents in SGML format from judgments by means of fuzzy linguistic techniques based on computing with words. Inform. Process. Manag. 39, 233–249 (2003)
15. Kaufmann, A., Gupta, M.M.: Introduction to Fuzzy Arithmetic: Theory and Applications. International Thomson Computer Press, London (1991)
16. Lee, E.S., Li, R.J.: Comparison of fuzzy numbers based on the probability measure of fuzzy events. Comput. Math. Appl. 15, 887–896 (1988)
17. Bevilacqua, M., Ciarapica, F.E., Giacchetta, G.: A fuzzy-QFD approach to supplier selection. J. Purchas. Supply. Manag., 14–27 (2006)

An Evolutionary-Based Optimization for a Multi-Objective Blood Banking Supply Chain Model

Chia-Lin Hsieh

Department of Statistics and Actuarial Science, Aletheia University, No. 26, Chenli Street, Danshuei District, Taipei 251, Taiwan, ROC

Abstract. The study is focused both on the Location-Allocation Problem and the inventory control problem for a blood banking supply chain. We consider a two-echelon supply chain in which each regional blood center (RBC) sends blood to different CBCs and then delivers it to different allocated hospital blood banks (HBBs). According to the perishable characteristic of blood product, we design a two-staged approach including two models. In strategic stage, we propose model 1 to obtain the location-allocation decisions by determining (a) how many community blood centers (CBCs) should be in an area and (b) where they should be located and (c) which services should be assigned to which CBCs. In tactic stage, we implement model 2 to acquire the inventory control decisions of the optimal blood replenishment quantity and the optimal length of blood order cycle for each CBC. In additions, two objectives are used to construct model 1 so as to make the total supply chain cost the smallest and responsiveness level the biggest, not just a single objective. To solve this multiple objectives programming problem, we use a non-dominated Sorting Genetic Algorithm II (NSGA-II) to search for the Pareto set. MATLAB was implemented to solve our established models. Some computational results for the models using the actual data from all Regional Blood Organizations in Taiwan are derived.

Keywords: Facility location problem, blood banking supply chain, perishable inventory control, multiple objective evolutionary algorithm.

1   Introduction

Supply chains for time-sensitive products and, in particular, for perishable products, pose specific and unique challenges. By definition, a perishable product has a limited lifetime during which it can be used, after which it should be discarded [1]. Blood banks are an important part of health service system.' Thus, the blood banking management from a supply chain network perspective merits a fresh and updated approach. However, what makes this problem interesting and/or difficult from a research perspective? First, as [2] claimed, the main difficulty in the blood services stems from the fact that blood is a specific perishable, scarce and valuable product, with a lifetime of specific days after which it has to be discarded. Second, as [3] mentioned, the demand and the supply of blood in both their amounts and frequency are stochastic. Therefore, blood banking services need coordination. Third, as [4] states, 'the entire blood supply chain can be examined as an essentially whole system

A. Moonis et al. (Eds.): IEA/AIE 2014, Part I, LNAI 8481, pp. 511–520, 2014.

and not just a subsystem of some larger system as occurs in most other supply chains.' Such system should include an outline of the blood banking supply chain and a discussion of several strategic and tactical issues faced by this chain.

In recent years, there has been much discussion on the issue of regionalization of blood banking system, in the hope of decreasing shortages, outdates and operating costs without sacrificing blood quality. Regionalization of blood banking system is a process by which blood banks within a geographical region move toward the coordination of their activities. It is achieved by dividing a geographical area into several regions and establishing a community blood centers (CBC) in each region. The activities of all CBCs in each region are then to be coordinated. Supply generation from donors is done mainly by each regional blood center (RBC) and then the collected blood is sent and stored to CBCs that they are assigned to. Therefore, each hospital blood bank (HBB) is to obtain its primary blood supply from the CBC in its region. The HBBs in a region receive their periodic requirements from their CBCs. Our research focuses on the location-allocation and inventory aspects of regionalization of a blood banking system. The problem is to decide how many CBCs should be setup in a region, where should the CBCs be located, how to allocate HBBs to CBCs and what is the optimal inventory level to maintain in each CBC. This problem will be called the Blood Location-Inventory Problem (BLIP). In this study, we propose a bi-objective coverage-type model which takes into account two criteria (efficiency and responsiveness) for BLIP. Efficiency aims to minimize the system cost; responsiveness, on the other hand, is designed to response quickly to customer needs. Our model is not only cost-efficient but also to fulfill the more blood demands of HBBs within its coverage level to maintain availability of blood. Since the proposed blood supply chain problem is NP-hard from the perspective of optimization, a well-known NSGA-II algorithm [5] is incorporated for optimization. This paper pursues to compensate the lack of attention on the development of mathematical models and solution approach for strategic and tactical decisions in healthcare supply chain.

This paper is organized as follows. Section 2 discusses relevant literature review. Section 3 details the model formulation. Section 4 proposes a two-stage approach based on a multiple objective evolutionary algorithm for BLIP. Section 5 illustrates the computational results using the data abstracted from regional blood organizations [6] in Taiwan, and section 6 provides conclusion along with research directions.

2   Literatures Reviews

Since blood belongs to a perishable product, literature dealing with blood supply chain is relatively limited and poses specific and unique challenges. In the literature, the research on blood management mainly focuses on the efficiency and effectiveness of blood inventory management. These studies focused either on perishable goods in general or on hospital blood banking inventory problems. However, these models were rather inflexible with numerous assumptions, but made great contributions to understand the perishable inventory theory. Few studies in the application of OR/SM have appeared in the past decades to handle blood banking systems ([7] [8], [9]). Recently, several studies have applied derivations of integer optimization models such as facility location, set covering and location-allocation model, to address the optimization of supply chains design of blood or other perishable products. [4] focus on supply chain management of blood banks. Their study included an outline of the blood banking supply chain and a discussion of several tactical and operational issues faced by this chain. [10] formulate three problems using integer programming to address the location-allocation aspects of regionalization of blood services. [11] use a nonlinear goal programming model that is the integration of gravity model of continuous location models and set covering model of discrete location approaches. [12] consider a regionalized blood banking system consisting of collection sites, testing and processing facilities, storage facilities, distribution. Authors develop a generalized network optimization model. From the survey, some innovative research aspects that are noteworthy have been incorporated in our research work.

3   Problem Statement and Formulation

3.1   Problem Description and Assumptions

The blood banking supply chain consists of a regional blood center (RBC) or coordinating authority, the region's community blood centers (CBCs) or individual hospital blood banks (HBBs). Recently, hospitals found it to be more cost efficient and time effective to seek a dependable source for blood (e.g. CBC). Fig. 1 shows a hierarchical system of an organizational structure for a regional blood banking service. At the highest level, RBCs can perform all functions related to blood transfusion, which are mainly collecting, testing and processing blood products. CBCs, on the other hand, perform all main functions of storage and distribution of blood and each of them is assigned to a RBC. The problem is to decide how many CBCs should be setup in a region, where should the CBCs be located, how to allocate HBBs to CBCs and what is the optimal inventory level to maintain in each CBC. Since these problems belong to strategic and tactical level decision respectively, we decompose the entire model into two sub-problems. The first sub-problem is called a location-allocation problem but two objectives including total supply chain cost and responsiveness service level are considered. The second sub-problem is an optimal inventory control problem which applies an optimal replenishment policy based on [13] for perishable blood product. The solution procedure for BLIP is illustrated in Fig. 2.

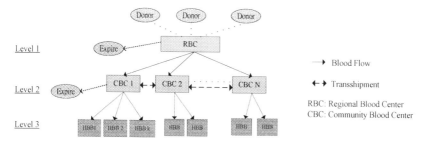

Fig. 1. A Hierarchical Organizational Structure for a Blood Banking System

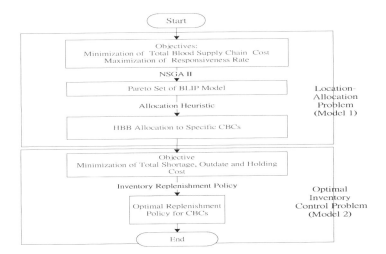

Fig. 2. The Solution Procedure for BLIP

3.2   Mathematical Models

Basic assumptions are used when modeling our problem. The potential locations of CBCs are given. The number of RBCs is exogenously determined by the decision makers. The whole blood is supplied and distributed from RBCs and stored at CBCs. The CBC makes regular shipments of whole blood to HBBs on a daily/weekly basis. The functions of blood procurement, processing, cross-matching, recycling etc. are not the issues of our concerns. Demands are aggregated to a finite number of CBCs. After CBCs receive all the requests from HBBs, the orders are filled by drawing from the inventory in a CBC. Demand for whole blood at a certain site is assumed to be shaped by the hospitals therein and it is known to be completely uncertain. It is assumed that there are sufficient capacities to supply the total whole blood from RBCs.

3.2.1   Model 1: The Location-Allocation Problem

The mathematical notation and formulation are as follows.

Indices. i is an index set for RBCs ($i \in I$). j is an index set of potential CBCs ($j \in J$). k is an index set for HBBs ($k \in K$).

Decision Variables. $\mathbf{y}_{ij}$ is the number of units of products shipped from RBC i to CBC j. $\mathbf{z}_{jk}$ is a binary variable to decide if CBC j serves HBB k. $\mathbf{x}_i$ is a binary variable to see if RBC i is chosen or not. $\mathbf{v}_j$ is a binary variable if CBC j is opened or not.

Model Parameters. S_j is he maximal amount of total blood capacity provided by *CBC* j. f_i is the fixed annual operating cost for *RBC* i and m_j is the facility operating cost of locating at DC j. D_k is the mean annual demand for whole blood at HBB k. $dist(i, j)$ and $dist(j, k)$ are the distances between RBC i and CBC j and between CBC j and HBB k, respectively. $\tau_j \stackrel{\Delta}{=} \{ k \in K \mid dist(j,k) \leq D_{max} \}$ is the set of HBBs that could be served by CBC j within the maximal covering distance D_{max}. That is, HBBs within this distance to an open CBC are considered well satisfied. c_{ij} is the unit cost of delivering blood products from the RBC i to CBC j and t_{jk} is the unit cost of shipping blood from CBC j to HBB k. In additions, there are several cost components as follows: (1) the total annual operating cost of choosing RBCs, (2) the total annual operating cost of locating CBCs, (3) the average inventory holding cost at CBCs, the annual cost of maintaining and ordering products via CBCs, (iv) the transportation cost (TC$_1$) of shipping whole blood from RBCs to CBCs, and (v) the transportation cost (TC$_2$) of shipping whole blood from CBCs to HBBs.

$$
\text{Min } Z_1 = \underbrace{\sum_{j \in J} f_j \cdot \mathbf{x}_i}_{(i)} + \underbrace{\sum_{j \in J} m_j \cdot \mathbf{v}_j}_{(ii)} + \underbrace{\sum_{i \in I} \sum_{j \in J} \frac{p_j}{2} \left(\sum_{k \in K} D_k \cdot \mathbf{z}_{jk} \right) \cdot \mathbf{y}_{ij}}_{(iii)}
$$
$$
+ \underbrace{\sum_{i \in I} \sum_{j \in J} c_{ij} \cdot dist\,(i, j) \cdot \mathbf{y}_{ij}}_{(iv)} + \underbrace{\sum_{j \in J} \sum_{k \in K} t_{jk} \cdot dist\,(j, k) \cdot D_k \cdot \mathbf{z}_{jk}}_{(v)}
\tag{1a}
$$
$$
\text{Max } Z_2 = \sum_{j \in J} \sum_{k \in \tau_j} D_k \cdot \mathbf{z}_{jk} \Big/ \sum_{i \in I} \sum_{k \in K} D_k \cdot \mathbf{z}_{jk}
\tag{1b}
$$

s.t

$$
\sum_{k \in K} \mathbf{z}_{jk} \leq M \cdot \mathbf{v}_j, \forall j, \text{ M is large enough}
\tag{2}
$$
$$
\sum_{j \in J} \mathbf{y}_{ij} \cdot \left(\sum_{k \in K} d_k \cdot \mathbf{z}_{jk} \right) \leq S_j, \forall j
\tag{3}
$$
$$
\sum_{j \in J} \mathbf{z}_{jk} \leq 1, \forall k
\tag{4}
$$
$$
\mathbf{x}_i, \mathbf{v}_j, \mathbf{z}_{jk}, \forall i, j, k \text{ binary variables;}
\tag{5}
$$
$$
\mathbf{y}_{ij} \geq 0, \forall i, j
\tag{6}
$$

Therefore, the objective function Z_1 in (1a) is to minimize the sum of facility operating, holding inventory and transportation costs. Z_2 in (1b) is the objectives referred to *responsiveness level* (RL) which is used to measure the percentage of fulfilled demand volume within an exogenously distance coverage. Constraints (2) allow the allocation to occur only if there exists a chosen CBC at the associated points, where M is a large enough number. Constraints (3) are the capacity restrictions imposing that any CBC must not exceed its annual transaction capacity; that is, any other extra HBB must not assign any opened CBC. These constraints ensure the fact that for the blood product that flows through the CBC, some is held as safety stock and the rest of it is used to satisfy HBB demands.   Constraints (4) stipulate that each HBB must be allocated no more than one of the selected CBCs. Constraints (5) are binary constraints and constraints (6) are non-negativity.

3.2.2   Model 2: The Optimal Inventory Control Problem

In a CBC or RBC, the management of inventories of whole blood and components involves a complex and interrelated set of decisions. It has been recognized that apparent benefits can be obtained HBBs by pooling resources using a CBC. It permits a

HBB to channel its energies and efforts toward the resolution of patient-related trans-fusion problem, and a HBB has the opportunity to pool widely fluctuating, largely unpredicted demands with those of other HBBs in the system [4]. [13] established an inventory model for non-instantaneous deteriorating items and obtained an optimal replenishment policy. The model of [13] goes like this: I_{max} units of item arrive at the inventory system at the beginning of each cycle. During $[0,t_d]$, the inventory level decreases only owing to stock-dependent demand rate. The inventory level is drop-ping to zero due to demand and deterioration during $[t_d,t_1]$. Then shortage interval keeps to the end of the current order cycle at T. The whole process is repeated. The total relevant inventory cost per cycle consists of the items: (i) ordering cost (A), (ii) inventory holding cost (C_1), (c) the deterioration cost (C_2), (iv) the shortage cost due to backlog (C_3) and (v) the opportunity cost due to lost sales (C_4). Therefore, the total inventory cost per unit time is given by TVC(t_1, T)= (A + C_1 + C_2 + C_3 + C_4)/T.

The following notation is similar to [13]. α and β is positive constants relating to the demand rate. δ is the backlogging parameter. θ is a constant fraction ($0 < \theta < 1$) of on-hand deteriorating inventory. Since the blood product belongs to the case of the model with instantaneous deterioration and without shortages ($t_d=0$, $\delta \to \infty$), the total inventory cost can be obtained by TVC(T) in Eq. (7) and T* is the optimal replenish-ment period, respectively. Furthermore, if we substitute T=T* into Eq. (7), we have the optimal minimal relevant cost TVC*(T*) in Eq. (8) and the optimal ordering quantity Q* in Eq. (9). Finally, the results of Eq. (8) and Eq. (9) are applied to obtain in the **inventory** decisions of our inventory control problem.

$$TVC(T) = \frac{\alpha}{T}\left\{\frac{A}{\alpha} + \frac{C_1 + \theta C_2}{(\theta + \beta)^2}[e^{(\theta + \beta)T} - (\theta + \beta)T - 1)\right\} \tag{7}$$

$$TVC^*(T^*) = \alpha M\left(e^{(\theta + \beta)T^*} - 1\right), \text{ where } M = \frac{C_1 + \theta C_2}{\theta + \beta} \tag{8}$$

$$Q^* = \frac{\alpha}{\theta + \beta}\left[e^{(\theta + \beta)T^*} - 1\right] \tag{9}$$

4   Solution Methodologies

The objectives in multi-objective optimization problem (MOP) often conflict with each other. Improvement of one objective may lead to deterioration of another. Thus, any single solution that optimizes all objectives simultaneously does not exist. In-stead, the best trade-off solutions called the Pareto optimal solutions are critical to a decision maker. NSGA-II [5] is one of the most efficient multi-objective evolutionary algorithms for MOPs. First of all, for each solution in the population, one determines how many solutions dominate it and the set of solutions to which it dominates. Then, it ranks all solutions to form non-dominated fronts according to a *non-dominated sorting* process, hence, classifying the chromosomes into several fronts. To allow for diversification, it estimates the solution density surrounding a particular solution in the population by computing a *crowding distance* operator. During selection, a crowded-comparison operator considering both the non-domination rank of an individual and its crowding distance is used to select the offspring, without losing good solutions (*elitism* strategy). NSGA-II procedure is illustrated in Fig. 3.

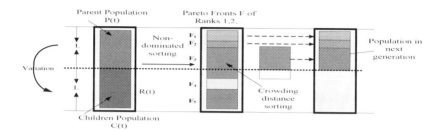

Fig. 3. NSGA-II solution scheme

Based on NSGA-II, a hybrid evolutionary algorithm is proposed. It starts by generating a random population $P(1)$ of size L. For each chromosome in $P(1)$, the algorithm evaluates its costs using the encoded solution expressions. Then, it applies *non-dominated sorting* on $P(1)$ and assigns to each chromosome a front to which it belongs. Next, the algorithm applies binary tournament selection (to form the crossover pool), crossover, and mutation operators to generate the children population $C(1)$ of size L. The Elitism strategy is then performed. First, a combined population of size $2L$ $R(t)=P(t)\cup C(t)$ is formed. Thereafter, $R(t)$ is sorted according to a partial order ($\geq_n$) which is used to decide which chromosomes are fitter. Suppose that $Z_1(.)$ indicates cost, $Z_2(.)$ indicates responsiveness level. We say that $p \geq_n q$ if $Z_1(p)\leq Z_1(q)$, and $Z_2(p)\geq Z_2(q)$. The new parent population $P(t+1)$ is formed by adding solutions from the first front till the size exceeds L. Next, it applies fitness evaluation, selection, crossover and mutation operators to generate the children population $C(1)$ of size L. Once initialized, the algorithm repeats for T generations. The chromosome representation is represented in two parts. Each part has the same length $m=|J|$ (where J is the number of CBCs) with total length of $2m$. The solution in the first part of chromosome is encoded in binary variables (v_j) where the j-th position indicates if CBC j is open (value of 1) or not. However, the second part of chromosome is encoded in integer variables where the value in it stands for the corresponding RBC assigned to it.

5   Computational Results

In this section, we present some of the results for BLIP using the actual data from all RBOs in Taiwan and discuss the numerical results obtained from above models using MATLAB. For the hybrid GA implementation, we used the following input parameters: population size = 100; maximum number of generations = 200; cloning = 20%; crossover rate = 80%; mutation rate varies from 5% to 10% as the number of generations increases. In model 1, there are 15 potential CBCs to be chosen to make our allocation plan. After aggregation, there are 58 HBBs waiting to be allocated according to the real data. The maximal covering distance D_{max} was set to be 30 km. Euclidean distance is used for measuring distribution distances. Assignment is affected by RBC and CBC's capacity limitation. The Pareto front using the NSGA-II is evaluated to find out the 'optimal' solution. Fig. 4 illustrates a good evolution approach for generating the Pareto front after 200 generations in our problem. It is revealed that the population curve converges shortly after 200 generations; they are nearly overlapped among themselves. Afterwards, no significant improvement is incurred.

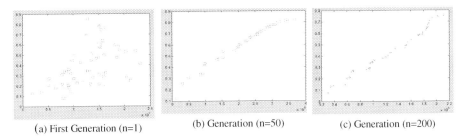

(a) First Generation (n=1)   (b) Generation (n=50)   (c) Generation (n=200)

Fig. 4. Evolutionary approach for generating Pareto front

To generate the user-defined 'optimal' solution, we define a reference point which is a vector formed by the single-objective optimal solutions. It is the best possible solution among the Pareto front that a multi-objective problem may have. Given a reference point, the problem can be solved by locating the alternatives which have the minimum distance d to the reference point. The decision maker requires determining weights w_t by prior knowledge of objectives. Then, all alternatives are ranked based on the value of d in descending orders and the alternative with the minimal value of d is the 'optimal' solution. We consider 3 scenarios by changing w_t parameters at a time as follows: (1) *equal-weight* (S_1) with $w_1=w_2=0.5$; (2) *cost-concerned* (S_2) with $w_1=0.8$, $w_2=0.2$; (3) *responsive-level* (S_3) with $w_1=0.2$, $w_2=0.8$. Fig. 5 displays the geographical locations of RBCs, CBCs and their corresponding HBBs. Fig. 5(a) illustrates the optimal assignment of alternative 10 for scenario S_1 and Fig. 5(b) shows the optimal assignment of alternative 9 for scenario S_3. Table 1 summarizes computational results of model 1 for all scenarios.

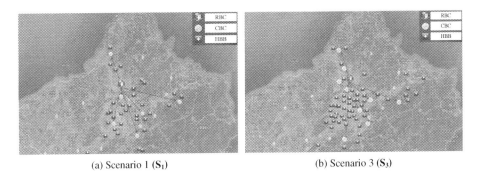

(a) Scenario 1 (S_1)   (b) Scenario 3 (S_3)

Fig. 5. Graphical display of the 'optimal' solution

In model 2, most of the cost parameters, obtained from a Regional Blood Organization, are used for the purchase and inventory holding costs (A= \$468 and C_1 = \$53.5). The deterioration cost (C_2= \$800). Since the annual blood demand values α_j are generated in Matlab using demand distributions and mean values provided from a Regional Blood Organization and similar to [14]. The demand is assumed to be randomly distributed around the mean with gamma distribution. Furthermore, θ=1/35 (0< θ<1) is selected to be the value of the on-hand deteriorating rate for the real data show that a whole blood should be discarded after 35 days. Table 2 summarizes the computational results of model 2 under the scenario S_1. It provides the optimal order quantity Q_j^* and the optimal length of order cycle T_j^* at CBC j. The value in the last column of Table 2 indicates the demand forecasting error percentage between the planned and the actual demands for each opened CBC. It is observed that there is only a minor percentage of blood shortage in CBC 10. The rest of opened CBCs under our replenishment policy could provide sufficient and appropriate quantities of blood so as to minimize the total relevant inventory cost (TVC*).

Table 1. Summary of Computational Results in Model 1

Scenarios	Objectives		Optimal Solutions				
	TC (Z_1) (million)	RL (Z_2)	% of OC	% of HIC	% of (TC1 +TC2)	Alternative	# of open CDC
S_1	1,523.4	56.09%	36.53%	38.19%	25.24%	10	6
S_2	660.2	35.87%	38.65%	39.39%	25.24%	66	2
S_3	1892.5	78.84%	42.12%	36.36%	21.46%	13	9

Table 2. Summary of Computational Results in Model 2

Scenarios	Optimal Solutions						
	Alternative	# of open CDC	Open CBC	TVC*	T* (days)	Q* (units)	Actual Demand (D)	% (error)
S_1	10	6	10	27,470	12.31	360.16	10,751	-0.67%
			12	27,638	12.35	362.38	10,701	0.05%
			21	21,342	16.00	279.99	6,380	0.11%
			25	24,751	13.80	324.70	8582	0.07%
			35	20,712	16.49	271.63	6,009	0.06%
			48	21,571	15.84	283.01	6,518	0.05%

6   Conclusions

This research presented an integrated location-inventory model for a blood banking supply chain system which incorporates both measures of cost and health care service quality. We develop a novel formulation which integrates two issues, LAP and inventory control of blood banking SCN. Health care service quality is measured by the fraction of all blood demands in HBBs that are within a specified distance of the CBCs to which they are assigned. This study attempts to find the location and the number of open CBCs, the allocation of CBCs to HBBs and optimal inventory replenishment policy. NSGA-II is applied to determine a finite set of non-dominate Pareto solutions. Through tests of the model on realistic data sets, we showed that it is important to find the right trade-off between cost and health care service quality. In this case, our proposed NSGAII-based evolutionary algorithm can generate very good alternatives to solve these problems for generating high-quality solutions quickly.

The model can be extended in some practical directions. Detailed sensitive analysis should be adopted to find the crucial parameters with respect to different assignments, resulting in the maximum increases/decreases on BLIP. It is also interesting to develop more effective and elegant heuristic methods (e.g. clustering or meta-heuristic algorithms) to solve the allocation problem. In additions, determining the weights of the objectives in the model is important but complex. It is probably required to sort Pareto solutions according to decision-makers' preferences by using multi-attribute decision making (MADM) techniques, such as Analytic Hierarch Process (AHP) or Technique for Order Preference by Similarity to Ideal Solution (TOPSIS).

References

1. Federgruen, A., Prastacos, G., Zipkin, P.H.: An allocation and distribution model for perishable products. Oper. Res. 34, 75–82 (1986)
2. Nahmias, S.: Perishable Inventory Theory: A Review. Operations Research 30, 680–708 (1982)
3. Beliën, J., Forcé, H.: Supply chain management of blood products: A literature review. European Journal of Operational Research 217, 1–16 (2012)
4. Pierskalla, W.: Supply Chain Management of Blood Banks Operations Research and Health Care. In: Brandeau, M.L., Sainfort, F., Pierskalla, W.P. (eds.), vol. 70, pp. 103–145. Springer, US (2005)
5. Deb, K., Pratap, A., Agarwal, S., Meyarivan, T.: A fast and elitist multiobjective genetic algorithm: NSGA-II. IEEE Transactions on Evolutionary Computation 6, 182–197 (2002)
6. Cordner, G.W., Scarborough, K.E.: 11 - Information in the Police Organization. In: Police Administration, 7th edn., pp. 309–344. Anderson Publishing, Ltd., Boston (2010)
7. Karaesmen, I., Scheller–Wolf, A., Deniz, B.: Managing Perishable and Aging Inventories: Review and Future Research Directions. In: Kempf, K.G., Keskinocak, P., Uzsoy, R. (eds.) Planning Production and Inventories in the Extended Enterprise, vol. 151, pp. 393–436. Springer, US (2011)
8. Ghandforoush, P., Sen, T.K.: A DSS to manage platelet production supply chain for regional blood centers. Decision Support Systems 50, 32–42 (2010)
9. Rytilä, J.S., Spens, K.M.: Using simulation to increase efficiency in blood supply chains. Management Research News 29, 801–819 (2006)
10. Şahin, G., Süral, H., Meral, S.: Locational analysis for regionalization of Turkish Red Crescent blood services. Computers & Operations Research 34, 692–704 (2007)
11. Cetin, E., Sarul, L.S.: A blood bank location model: A multiobjective approach. European Journal of Pure and Applied Mathematics 2, 112–124 (2009)
12. Nagurney, A., Masoumi, A., Yu, M.: Supply chain network operations management of a blood banking system with cost and risk minimization. Computational Management Science 9, 205–231 (2012)
13. Wu, K.S., Ouyang, L.Y., Yang, C.-T.: An optimal replenishment policy for non-instantaneous deteriorating items with stock-dependent demand and partial backlogging. International Journal of Production Economics 101, 369–384 (2006)
14. Zhou, D., Leung, L.C., Pierskalla, W.P.: Inventory management of platelets in hospitals: optimal inventory policy for perishable products with regular and optional expedited replenishments. Manufacturing & Service Operations Management 13, 420–438 (2011)

Author Index